# GENERAL
# ZOOLOGY
## LABORATORY GUIDE

**CHARLES F. LYTLE**
*North Carolina State University*

# GENERAL
# ZOOLOGY
## LABORATORY GUIDE

*thirteenth edition*

Boston   Burr Ridge, IL   Dubuque, IA   Madison, WI   New York   San Francisco   St. Louis
Bangkok   Bogotá   Caracas   Lisbon   London   Madrid
Mexico City   Milan   New Delhi   Seoul   Singapore   Sydney   Taipei   Toronto

# McGraw-Hill Higher Education

*A Division of The **McGraw-Hill** Companies*

GENERAL ZOOLOGY LABORATORY GUIDE, THIRTEENTH EDITION

 This book is printed on recycled, acid-free paper containing 10% postconsumer waste.

2 3 4 5 6 7 8 9 0 QPD/QPD 9 0 9 8 7 6 5 4 3 2 1 0

ISBN 0–07–012220–2

Vice president and editorial director: *Kevin T. Kane*
Publisher: *Michael D. Lange*
Senior sponsoring editor: *Margaret J. Kemp*
Developmental editor: *Donna Nemmers*
Marketing managers: *Michelle Watnick/Heather K. Wagner*
Project manager: *Sheila M. Frank*
Senior production supervisor: *Sandra Hahn*
Coordinator of freelance design: *Michelle D. Whitaker*
Visuals coordinator: *Jodi Banowetz*
Compositor: *GTS Graphics, Inc.*
Typeface: *10/12 Times/Roman*
Printer: *Quebecor Printing Book Group/Dubuque, IA*

Freelance cover/interior designer: *Jamie O'Neal*
Cover image: © *Diane R. Nelson*

The credits section for this book begins on page 347 and is considered an extension of the copyright page.

Some of the laboratory experiments included in this text may be hazardous if materials are handled improperly or if procedures are conducted incorrectly. Safety precautions are necessary when you are working with chemicals, glass test tubes, hot water baths, sharp instruments, and the like, or for any procedures that generally require caution. Your school may have set regulations regarding safety procedures that your instructor will explain to you. Should you have any problems with materials or procedures, please ask your instructor for help.

www.mhhe.com

# Dedication

This thirteenth edition is dedicated to the memory of my
late, beloved wife, Carol Cottingham Lytle, who assisted
and supported me for 44 years and through seven
previous editions of this book.

**Carol Cottingham Lytle**
**August 28, 1932–December 7, 1998**

# BRIEF CONTENTS

Contents    vii
Preface    xiii
Laboratory Safety    xvii
Comparative Safety of Preservatives    xix
Handling and Care of Animals in the Laboratory    xxi

1  Microscopy    1

2  Animal Cells and Tissues    15

3  Mitosis and Meiosis    35

4  Development    49

5  Protozoa    73

6  Porifera    95

7  Cnidaria    105

8  Introduction to Animal
   Morphology    121

9  Platyhelminthes    127

10  Pseudocoelomate Animals    145

11  Mollusca    159

12  Annelida    175

13  Arthropoda and
    Onychophora    189

14  Echinodermata    223

15  Chordata    233

16  Shark Anatomy    241

17  Perch Anatomy    259

18  Frog Anatomy    269

19  Fetal Pig Anatomy    295

20  Rat Anatomy    327

# CONTENTS

*Preface   xiii*
*Laboratory Safety   xvii*
*Comparative Safety of Preservatives   xix*
*Handling and Care of Animals in the Laboratory   xxi*

## 1  Microscopy   1

**Objectives   1**
**The Compound Microscope   1**
   Parts of the Microscope   2
   Magnification   3
   Resolving Power   4
   Illumination   4
   Focusing   5
   Procedure for Use of the Compound
      Microscope   6
   Returning the Microscope After Use   6
   Special Precautions   6
   Exercises Using the Compound Microscope   7
   Estimating Magnification   7
   Measuring Microscopic Objects   7
   Measurements Using an Ocular Micrometer   8
**The Stereoscopic Microscope   8**
   Exercises Using the Stereomicroscope   9
**Other Types of Microscopes   9**
   Phase Contrast Microscopy   9
   Interference Microscopy   9
   Electron Microscopy   10
*Key Terms   11*
*Internet Resources   12*
*Critical Thinking Questions   12*
*Suggested Readings   12*

## 2  Animal Cells and Tissues   15

**Objectives   15**
**The Cell Theory   15**
**Basic Cell Structure   16**
**Animal Tissues   16**
   Epithelial Tissue   17
   Connective Tissue   21
   Cartilage   25

   Bone   26
   Muscular Tissue   26
   Nervous Tissue   28
*Key Terms   32*
*Internet Resources   32*
*Critical Thinking Questions   32*
*Suggested Readings   32*

## 3  Mitosis and Meiosis   35

**Objectives   35**
**Introduction   35**
**Mitosis   36**
   The Mitotic Apparatus   36
   Stages of Mitosis   38
   Timing in the Cell Cycle   38
**Meiosis   40**
   Gametogenesis   42
   Principal Stages of Meiosis   44
   Ways to Study Meiosis   44
**Comparison of Mitosis and Meiosis   45**
*Key Terms   46*
*Internet Resources   47*
*Critical Thinking Questions   47*
*Suggested Readings   47*

## 4  Development   49

**Objectives   49**
**Introduction   49**
**Component Processes of Development   50**
   Growth   50
   Determination   50
   Differentiation   50
   Morphogenesis   50
**Gametes   50**
**Embryonic Cleavage   50**
   Influence of Yolk   51
   Patterns of Cleavage   51
   Determinate and Indeterminate Development   52
**Starfish Embryology   52**
   Summary of Early Starfish Development   54

Frog Development  54
**Chick Development  59**
    Extraembryonic Membranes  60
    Whole Mount of 24-Hour Chick Embryo  60
    Whole Mount of 48-Hour Chick Embryo  62
    Whole Mount of 72-Hour Chick Embryo  65
    Later Stages of Chick Development  67
    Living Chick Embryos (Optional Exercise)  68
*Key Terms  70*
*Internet Resources  70*
*Critical Thinking Questions  71*
*Suggested Readings  71*

## 5 Protozoa  73

**Objectives  73**
**Introduction  73**
**Classification  74**
    Phylum Sarcomastigophora  74
    Phylum Ciliophora (Ciliata)  74
    Phylum Apicomplexa  74
**An Amoeba: *Amoeba proteus*  74**
    Phylum Sarcomastigophora  74
**A Solitary Flagellate: *Euglena*  78**
    Phylum Sarcomastigophora  78
**A Holozoic Flagellate: *Peranema*  78**
**A Colonial (?) Flagellate: *Volvox*  79**
    Phylum Sarcomastigophora  79
    Other Mastigophora  81
**A Ciliate: *Paramecium caudatum*  85**
    Phylum Ciliophora (Ciliata)  85
    Other Ciliates  88
**Phylum Apicomplexa  89**
    The Malaria Parasite: *Plasmodium*  90
**Evolution of Multicellular Animals  92**
*Key Terms  93*
*Internet Resources  93*
*Critical Thinking Questions  94*
*Suggested Readings  94*

## 6 Porifera  95

**Objectives  95**
**Introduction  95**
**Morphology  96**
**Phylogeny  96**
**Skeleton  97**
**Classification  97**
    Class Demospongiae  98
    Class Calcarea (Calcispongiae)  98
    Class Hexactinellida (Hyalospongiae)  98

**Body Organization  99**
    Asconoid Sponge  99
    Syconoid Sponge  99
    Leucon-Type Sponge  101
**Freshwater Sponges  101**
**Regeneration and Reconstitution
    (Optional Exercise)  103**
    Procedure  103
*Key Terms  103*
*Internet Resources  104*
*Critical Thinking Questions  104*
*Suggested Readings  104*

## 7 Cnidaria  105

**Objectives  105**
**Introduction  105**
**Classification  106**
    Class Hydrozoa (Hydroids and
        Siphonophores)  106
    Class Scyphozoa (True Jellyfish)  106
    Class Cubozoa (Sea Wasps or
        Box Jellyfish)  106
    Class Anthozoa (Sea Anemones
        and Corals)  106
**A Hydrozoan Polyp: *Hydra*  106**
    Class Hydrozoa  106
    General Appearance and Morphology  107
    Behavior  108
    Cnidocytes and Nematocysts  108
    Histological Structure  109
    Cellular Structure  109
    Nervous System  109
    Feeding Behavior  110
    Reproduction  110
    Regeneration (Optional Exercise)  110
**A Hydromedusa: *Gonionemus*  110**
    Class Hydrozoa  110
**A Colonial Hydrozoan Polyp: *Obelia*  112**
    Class Hydrozoa  112
    Alternation of Generations  112
**A Colonial Hydrozoan: *Physalia*  113**
    Class Hydrozoa  113
**A Scyphozoan Medusa: *Aurelia*  115**
    Class Scyphozoa  115
    Reproduction and Life Cycle  115
**An Anthozoan Polyp: *Metridium*  116**
    Class Anthozoa  116
    Reproduction and Life Cycle  117
**Corals  117**
    Class Anthozoa  117

*Key Terms*   118
*Internet Resources*   119
*Critical Thinking Questions*   119
*Suggested Readings*   119

## 8  Introduction to Animal Morphology   121

**Objectives**   121
**Introduction**   121
   Organization of the Animal Body   122
   Body Symmetry   122
   Grade of Tissue Construction   122
   Body Cavity   122
   Segmentation   123
   Cephalization   123
**Hints for Dissection**   123
*Key Terms*   125
*Internet Resources*   126
*Critical Thinking Questions*   126

## 9  Platyhelminthes   127

**Objectives**   127
**Introduction**   127
**Classification**   128
   Class Turbellaria (Free-living Flatworms)   128
   Class Trematoda (Flukes)   128
   Class Monogenea (Flukes)   128
   Class Cestoda (Tapeworms)   128
**Free-living Flatworms: Class Turbellaria**   129
   A Planarian: *Dugesia*   129
**The Flukes: Class Trematoda**   133
   The Human Liver Fluke: *Clonorchis (Opisthorchis) sinensis*   133
   Cercaria Larvae   135
   The Sheep Liver Fluke: *Fasciola hepatica*   135
**The Tapeworms: Class Cestoda**   138
   The Dog Tapeworm: *Dipylidium caninum*   138
   The Dog and Cat Tapeworm: *Taenia pisiformis*   141
**Adaptations for Parasitism**   141
*Key Terms*   143
*Internet Resources*   143
*Critical Thinking Questions*   144
*Suggested Readings*   144

## 10  Pseudocoelomate Animals   145

**Objectives**   145
**Introduction**   145

**Classification**   147
   Phylum Nematoda (Nemathelminthes or Roundworms)   147
   Phylum Rotifera (Rotifers)   147
   Phylum Gastrotricha (Gastrotrichs)   147
   Phylum Kinorhyncha (Kinorhynchs)   147
   Phylum Priapulida (Priapulids)   147
   Phylum Loricifera (Loriciferans)   147
   Phylum Nematomorpha (Gordiacea or Horsehair Worms)   147
   Phylum Acanthocephala (Spiny-headed Worms)   147
**Phylum Nematoda**   147
   A Parasitic Roundworm: *Ascaris lumbricoides*   147
   A Parasitic Roundworm: *Trichinella spiralis*   150
   Hookworms: *Ancylostoma duodenale* and *Necator americanus*   151
   Pinworm: *Enterobius vermicularis*   152
**Free-living Nematodes**   152
   *Caenorhabditis:* An Important Research Animal   152
   The Vinegar Eel: *Anguillula aceti*   153
   Collecting Free-living Nematodes   154
   Some Other Important Nematodes   155
**Phylum Rotifera**   155
   A Rotifer: *Philodina*   155
   Collecting Rotifers   156
   Some Other Common Rotifers   157
*Key Terms*   157
*Internet Resources*   157
*Critical Thinking Questions*   157
*Suggested Readings*   157

## 11  Mollusca   159

**Objectives**   159
**Introduction**   159
**Classification**   160
   Class Monoplacophora   161
   Class Polyplacophora (Amphineura)   161
   Class Aplacophora   161
   Class Scaphopoda   161
   Class Gastropoda   161
   Class Bivalvia (Pelecypoda)   161
   Class Cephalopoda   161
**A Freshwater Mussel**   162
   Class Bivalvia (Pelecypoda)   162
   Feeding, Digestion, and Respiration   162
   Muscles   165

Circulation   165
Excretion, Osmoregulation,
    and Reproduction   166
Nervous Coordination   166
*Helix:* the Garden Snail (Demonstration)   167
Class Gastropoda   167
A Squid: *Loligo*   168
Class Cephalopoda   168
External Anatomy   168
Internal Anatomy (Demonstration)   170
Reproduction   170
Octopus (Demonstration)   171
Class Cephalopoda   171
*Key Terms   172*
*Internet Resources   172*
*Critical Thinking Questions   172*
*Suggested Readings   173*

## 12  Annelida   175

Objectives   175
Introduction   175
Classification   176
Class Polychaeta (Polychaete Worms)   176
Class Oligochaeta (Bristleworms)   176
Class Hirudinea (Leeches)   176
A Marine Annelid: *Nereis virens*   176
Class Polychaeta   176
Other Polychaetes   179
The Earthworm   180
Class Oligochaeta   180
External Anatomy   180
Internal Anatomy   181
Cross Sections   184
Leeches   184
Class Hirudinea   184
External Anatomy   185
Internal Anatomy (Demonstration)   186
*Key Terms   186*
*Internet Resources   186*
*Critical Thinking Questions   187*
*Suggested Readings   187*

## 13  Arthropoda and
    Onychophora   189

Objectives   189
Introduction   189
Phylum Arthropoda   190
Major Features   190
Exoskeleton   190

Classification   190
Subphylum Trilobita (Trilobitomorpha)   190
Subphylum Chelicerata   190
Subphylum Crustacea   191
Subphylum Uniramia   191
Subphylum Chelicerata   192
The Horseshoe Crab: *Limulus*   192
A Spider: *Argiope*   193
Subphylum Crustacea, Class Branchiopoda   195
A Water Flea: *Daphnia*   195
Subphylum Crustacea, Class Malacostraca   201
The Crayfish: *Procambarus*   201
Subphylum Uniramia, Class Insecta   208
A Cockroach: *Periplaneta americana*   208
A Grasshopper: *Romalea microptera*   212
Insect Metamorphosis   217
Phylum Onychophora   217
External Anatomy   219
Internal Anatomy   219
*Key Terms   220*
*Internet Resources   220*
*Critical Thinking Questions   220*
*Suggested Readings   221*

## 14  Echinodermata   223

Objectives   223
Introduction   223
Classification   224
Subphylum Echinozoa   224
Subphylum Crinozoa   224
Subphylum Asterozoa   224
The Common Starfish: *Asterias*   225
Class Asteroidea   225
A Sea Urchin   228
Class Echinoidea   228
Some Common Echinoids   229
A Sea Cucumber   229
Class Holothuroidea   229
Some Common Holothuroidea   229
*Key Terms   230*
*Internet Resources   230*
*Critical Thinking Questions   230*
*Suggested Readings   230*

## 15  Chordata   233

Objectives   233
Introduction   233
Classification   233
Group Acrania (Protochordata)   233

Group Craniata   234
**Subphylum Urochordata, Tunicates or
Sea Squirts   234**
The Tunicate Larva   234
The Adult Tunicate   234
**A Lancelet: *Branchiostoma*   236**
Subphylum Cephalochordata   236
Cross Sections   238
*Key Terms   239*
*Internet Resources   239*
*Critical Thinking Questions   239*
*Suggested Readings   240*

**16  Shark Anatomy   241**

**Objectives   241**
**The Dogfish Shark: *Squalus acanthias*   241**
External Anatomy   242
Internal Anatomy   243
*Key Terms   257*
*Internet Resources   258*
*Critical Thinking Questions   258*
*Suggested Readings   258*

**17  Perch Anatomy   259**

**Objectives   259**
**The Yellow Perch: *Perca flavescens*   259**
External Anatomy   260
Internal Anatomy   261
*Key Terms   268*
*Internet Resources   268*
*Critical Thinking Questions   268*
*Suggested Readings   268*

**18  Frog Anatomy   269**

**Objectives   269**
***Rana pipiens* or *Rana catesbeiana*   269**
External Anatomy and Behavior   270
Internal Anatomy   277
*Key Terms   293*
*Internet Resources   293*

*Critical Thinking Questions   293*
*Suggested Readings   294*

**19  Fetal Pig Anatomy   295**

**Objectives   295**
**The Fetal Pig: *Sus scrofa*   295**
External Anatomy   297
Skeletal System   297
Muscular System   298
General Internal Anatomy   298
Neck Region   305
The Coelom and Its Divisions   305
The Urogenital System   309
Circulatory System   314
Nervous System   320
*Key Terms   325*
*Internet Resources   325*
*Critical Thinking Questions   325*
*Suggested Readings   325*

**20  Rat Anatomy   327**

**Objectives   327**
**Introduction   327**
**Classification   328**
External Anatomy   328
Skeletal System   328
Muscular System (Optional)   329
Internal Anatomy   330
Digestive System   330
Respiratory System   333
Circulatory System   333
Urogenital System   339
Nervous System   340
*Key Terms   344*
*Internet Resources   344*
*Critical Thinking Questions   344*
*Suggested Readings   344*

*Credits   347*
*Index   349*

# PREFACE

A solid foundation in basic zoology is essential for students who are preparing for careers in biology, zoology, genetics, physiology, medicine, veterinary medicine, agriculture, environmental science, conservation, and many other fields. Other students also benefit from the study of animals because animals are an important part of nature, which surrounds us, and also because animals have contributed to human life and welfare in innumerable ways since the dawn of civilization. Animals affect each of us every day whether or not we are aware of it. Some knowledge of zoology is therefore essential to every educated person.

Meaningful laboratory experiences are a vital part of learning zoology, and this book is designed to facilitate laboratory study of selected animals. In the laboratory, students learn the importance of careful observation, of following specific instructions, and of seeing relationships of structure and function. By carrying out well-designed scientific experiments in the laboratory, students learn to **do** science instead of merely listening to someone **talk** about science.

In writing this book, I have tried to remember my own days as a student and the questions I had while studying various kinds of animals for the first time. I have attempted to provide descriptions, illustrations, and appropriate guidance for meaningful laboratory study of animals. Although I agree with the spirit of Louis Agassiz' admonition, "Study nature, not books," I believe that students can benefit most in their study of nature when aided by appropriate instructions and illustrations.

I can remember some of my own frustration in zoology labs when I attempted to follow some vague verbal description of anatomical structures with no illustrations or other visual aids to help me locate important structures or to give me some appropriate orientation. I have written the exercises in this book with the intention of reducing such frustration and with the intention of making student experiences in the zoology laboratory interesting, rewarding, and meaningful.

I continue to emphasize the study of living and anesthetized animals whenever feasible because students should learn that zoology is the study of animal life rather than the study of dead animals. Live animals give students the opportunity to observe and experiment with behavior and to do simple physiological experiments as well as to see the natural color and texture of body parts. Preserved specimens serve well for many anatomical studies, but students should always have the opportunity to observe and to work with living animals whenever possible. Few of us would choose a stuffed or embalmed dog or cat for a pet if given the option.

This edition of the manual continues the tradition of more than 50 years of excellence in providing students with a comprehensive introduction to zoology and to the major animal phyla. This book is written to aid students and teachers in many colleges and universities operating with different schedules, resources, and preferences, so I have intentionally included more material than can reasonably be covered in the time available in a two-semester general zoology course. We expect instructors to select those parts of the guide and those animals they deem most appropriate for their own classes. With judicious selection of chapters and of animal types, this book can also be used for one-semester and one-quarter zoology classes.

## Changes in This Edition

We have focused on making clarifications and corrections in this edition of the book and have tried to remove some old material to avoid adding to the book's length. Too often books seem to become longer and longer in successive editions, rather than better and better. With this thought in mind, we have revised a number of illustrations to improve the quality of animal dissections, to better illustrate important concepts, and to make labeling more precise.

We have also added three new sections to the end of most chapters: (1) a list of *Internet Resources,* (2) a list of *Suggested Readings,* and (3) *Critical Thinking Questions.* The Internet has become a valuable source of scientific information, and there are many valuable Internet sites with information about Zoology. Several sites containing pertinent zoological information are described at the end of each chapter. The Internet links for these descriptions are found on the McGraw-Hill Zoology web site at http://www.mhhe.com/zoology. Similarly, there are numerous books with topical information on the animals discussed within the lab exercises, and a few such books and articles are listed at the end of most chapters to aid students in further study. The Critical Thinking Questions were added at the recommendation of several teachers and reviewers to help students gain perspective from their laboratory studies.

The biohazard logo ⚠ points out any laboratory exercises where extra caution should be used due to the handling of potentially harmful materials.

Another change to this edition is that all questions within text have been set in a different typeface. We have done this to serve as a pedagogical aid to help students identify these learning points, and to enhance their learning process. We hope that students and teachers alike will find these additions useful.

## Basic Features of This Manual

In this edition we have continued the basic organization and pedagogical features of the previous edition. Important pedagogical features of the book include **boldface headings** within each chapter to indicate the major divisions of each exercise.

We also use **boldface** in the text to identify important terms (ideas, structures, processes) that students should remember and understand. The most important of these boldface terms are included in the list of **key terms** at the end of the chapter in which they are first introduced.

Several chapters provide space for students to add their own drawings of particular animals or structures to aid them in learning and remembering things observed during their laboratory study. The book also provides several blank tables and pages of graph paper for students to record and plot data from their laboratory observations and experiments.

Each chapter begins with a list of specific **objectives** that identifies important principles, concepts, and facts that students should learn as a result of their laboratory study. I have found that a specific list of laboratory objectives helps students focus their attention on the important material in each lab. I also suggest that instructors modify and add to these lists of objectives as appropriate for their own classes. Such lists of objectives can be most helpful in ensuring that students understand what they will be tested on and that the tests actually focus on students' understanding of the important principles, concepts, and processes.

Most chapters in this book start with a brief **introduction** with pertinent background material to help orient students for the exercises to follow. A **materials list** is provided showing the specimens and other materials needed for each exercise. Most chapters have one or more lists of suggested **demonstrations,** which are suggested to supplement the main studies of each exercise.

Within each chapter, student-directed questions have been placed in a unique type style to trigger students to stop and think about the animal structure or procedure being discussed.

At the end of each chapter is a list of **key terms** introduced in that chapter, as well as Internet Resources and Suggested Readings for further study. Each chapter also ends with a list of Critical Thinking Questions to help students review the important concepts and processes of the chapter. Most chapters also have blank space provided for students to add their own notes and sketches. If students use these pages to record their observations, they will have a consolidated record of their laboratory work bound in a single place instead of a scattered bunch of papers and drawings likely to be lost.

I have tried to make this laboratory manual a convenient, user-friendly companion for laboratory study. I hope every student has as much fun and satisfaction in zoology lab as I have had.

## Anatomy Films

From my many years of teaching zoology laboratories, I have learned that it is very helpful to give students an overview of the anatomy of an animal to be studied and/or dissected before they undertake the anatomical study on an actual specimen themselves. It's a lot like football players viewing game films before facing a major rival football team. They might do all right without knowing what kinds of plays the opposition typically runs and who their key players are, but they are not likely to win the championship without some good scouting information.

An excellent way to prepare for a serious anatomical study of an animal is to view a good film or video of the anatomy of that animal prior to beginning work with an actual specimen. Such preparation for the lab study gives students a better perspective and orientation and greatly increases their confidence. It also aids in their identification of anatomical structures, facilitates their recognition of relationships among various organs, and assists them in relating structure and function. Good films or videos also help students review their laboratory work in preparation for a test and in comparing the anatomy of different animals.

I have collaborated with the staff of Carolina Biological Supply Company in the development of a series of videos specifically designed to aid in the study of nine of the more complex animals included in this manual. These videos illustrate the anatomy and dissection of these nine animals and parallel the descriptions of those animals in this book.

Each video illustrates the anatomy of the animal in detail, discusses the functions of various organs and systems, and demonstrates good dissection techniques. Each video is divided into sections according to organ system so that each system can be located easily and viewed separately if desired. Several of the longer videos are too long to be productively viewed in a single session.

The videos are available from Carolina Biological Supply Company, 2700 York Road, Burlington, North Carolina, 27215. The videos and the corresponding chapters in this manual containing the exercises for the study of these animals are listed in the following table.

| Chapter | Video |
|---------|-------|
| 11 Mollusca | The Anatomy of the Freshwater Mussel (49-2365V) |
| 12 Annelida | The Anatomy of the Earthworm (49-2372V) |
| 13 Arthropoda | The Anatomy of the Crayfish (49-2403V) |
| | The Anatomy of the Grasshopper (49-2404V) |
| 14 Echinodermata | The Anatomy of the Starfish (49-2369V) |
| 16 Shark Anatomy | The Anatomy of the Shark (49-2655V) |
| 17 Perch Anatomy | The Anatomy of the Perch (49-2662V) |
| 18 Frog Anatomy | The Anatomy of the Frog (49-2704V) |
| 19 Fetal Pig Anatomy | The Anatomy of the Fetal Pig (49-3075V) |

## Acknowledgments

We are grateful to several persons for their assistance with this edition. Carol Majors of Publications Unlimited and an NC State graduate who has done many of the drawings for previous editions, provided two new drawings for this edition.

Toni Onks also provided editorial assistance with the manuscript, and Peggy Holliday of Safety and Science Education Consultants, Inc., provided a helpful review of the section on laboratory safety. Also, we appreciate the careful editing and attention that Matthew Douglas, Science Editor, gave to this thirteenth edition.

The following reviewers of the twelfth edition provided excellent suggestions and feedback for this new edition:

Barbara Abraham, Hampton University
Iona Baldridge, Lubbock Christian College
Joseph W. Camp, Jr., Purdue University North Central
Suzette Chopin, Texas A & M University
Sarah Cooper, Beaver College
Tom Dale, Kirtland Community College
Holly Downing, University of Wisconsin—Whitewater
Dennis Englin, The Masters College
W.E. Hamilton, Penn State University
Christine Holler-Dinsmore, Fort Peck Community College
Ken Hoover, Jacksonville University
Susan Keys, Springfield College
Roger Lloyd, Florida Community College
Vicky McMillan, Colgate University
Fred H. Schindler, Indian Hills Community College
Theresa Wysolmerski, The College of St. Rose

# LABORATORY SAFETY

A zoology laboratory is a place for serious scientific work and study. Students and teachers must recognize that a number of potential safety hazards are present in all science laboratories. In a zoology laboratory, the principal safety hazards are electrical circuits, potentially dangerous chemicals, hot liquids and heat sources, broken glass, live animals, and sometimes infectious agents (pathogenic bacteria, viruses, and parasites). Achieving safety in the laboratory, as in other places, requires paying attention to potential hazards and observing appropriate safety measures. Nothing we do is without some degree of risk. For example, people sometimes fall out of bed and injure themselves; many people also drown in swimming pools each year. But science laboratories can provide a safe environment when both students and teachers are aware of potential hazards and follow appropriate safety procedures.

The following list of safety rules is offered as a good start toward safe practices in the lab. This is not a complete list of safety rules, and it is certainly not a substitute for proper safety awareness for the particular lab in which the student is engaged. It is essential that students understand the need and importance of safety. All actions in a laboratory have consequences. Be protected by paying attention, listening to your instructors, and knowing about the materials and procedures necessary to perform each laboratory investigation. You should also be mindful of the activities of other students around you. Frequently it is an accident caused by another person that endangers someone in a laboratory.

## Some Basic Rules of Safety for the Laboratory

1. Use common sense.
2. Avoid horseplay in the laboratory.
3. Never eat, drink, or smoke in the laboratory.
4. Always wash your hands for at least 15 seconds and rinse them well after handling chemicals or live or preserved animals.
5. Always wear close-toed shoes in the laboratory. Sandals or open-toed shoes are not appropriate.
6. Be familiar with the location, operation, and proper use of fire extinguishers, eyewash fountains, safety showers, and other safety equipment in the laboratory.
7. Know the location of emergency exits and the evacuation routes to be used in case of an emergency.

8. Always be cautious when using electric hot plates and gas burners. You can get a serious burn by touching a hot surface or by spilling a hot liquid.
9. Use protective mittens or tongs to handle hot objects.
10. Be cautious when transferring liquids because aerosols may be formed, which can be dangerous to your eyes and lungs.
11. Wear safety goggles or other appropriate protective eye gear when performing or observing experiments or demonstrations.
12. Be familiar with the properties of, and hazards associated with, all chemicals used in the laboratory exercises. When you are in doubt about the hazards associated with any chemical, consult the Material Safety Data Sheets (MSDS) provided by the manufacturers. The appropriate MSDS for all potentially hazardous chemicals should be kept in the laboratory. Your laboratory instructor should provide you with appropriate warnings for the materials to be used, but you may also ask to see the MSDS for any chemical to be used if you feel you need further information. Laboratory instructors should teach all students to read the MSDS.
13. Beware of electrical equipment with frayed or bare wires or with faulty switches or plugs. Report such damaged items to your instructor.
14. Always work in a well-ventilated area when studying preserved specimens.
15. Make sure that all specimens you dissect are properly secured in a dissecting pan or appropriate surface. A specimen not properly secured might slip and lead to an injury from a scalpel or other sharp dissecting instrument.
16. Keep scalpel blades sharp to avoid slipping and possible injury.
17. All broken glass should be placed in a sharps container or one designated as a glass receptacle.
18. Any contact with human blood should be reported promptly to your instructor to limit your exposure to possible infection.
19. Clean all laboratory tables and other work surfaces after each use.
20. PRACTICE SAFETY AWARENESS, and remember that you are responsible for the safety of yourself and your coworkers.

# Safety Precautions When Using Preserved Animals

The chemicals used to preserve animals and parts of animals can be toxic, flammable, and/or dangerous if used improperly or under improper conditions. Ethanol, isopropanol, formaldehyde, phenol, and ethylene glycol are commonly used preservatives. Combinations of these and other solvents are contained in embalming fluids used to preserve larger animals.

It is very important for students and instructors working with preserved specimens to understand the proper precautions and conditions for safe usage of such materials. All instructors are responsible for implementing proper safety procedures when students will be using potentially hazardous chemicals and for communicating appropriate information about these materials to their students in accordance with applicable federal, state, and local regulations. In recent years these regulations have greatly increased in complexity as a result of increased public concern about environmental health and safety.

The following information, supplied through the courtesy of the Carolina Biological Supply Company, provides some excellent safety guidelines to follow when handling and dissecting preserved animal specimens. Other suppliers use similar chemicals for their preserved animal specimens. You should carefully study the safety information supplied with any preserved specimens before you begin to handle or dissect them.

To achieve the necessary level of safety in the laboratory, each instructor should be familiar with all chemicals present and the necessary precautions to be taken in using them.

Carolina provides specimens preserved in alcohol, *Carosafe*™ (contains ethylene glycol), and formalin solutions. Information is provided in the catalog regarding which particular preservative is used with a certain type of specimen. Note that specimens are never provided in a formalin preservative unless this is specifically requested by the customer. Note also that the specimens that are preserved with embalming fluids and are never treated with *Carosafe*™ are provided with a specific Material Safety Data Sheet (MSDS) prepared for that specific embalming fluid. Regardless of the preservative that is used, we recommend you follow these safety tips whenever working with preserved specimens:

1. Wear appropriate protective eyewear at all times.
2. Wear appropriate protective equipment such as gloves and lab coats.
3. Work only in a well-ventilated area.
4. Prohibit eating, drinking, and smoking in the work area.
5. In the event of contact with chemicals or specimens, wash skin with soap and water and flush eyes for 15 minutes with running water.
6. If overexposure to any chemical occurs, seek medical attention immediately.
7. Be careful with sharp objects such as pins, scalpels, and the spines and teeth of specimens.

Formalin preserved or embalmed specimens should always be used in a well-ventilated area to prevent irritation to eyes, skin, or respiratory tract. The use of goggles lessens eye irritation from formaldehyde vapors. If direct contact to eyes or skin occurs, wash thoroughly with water.

Isopropanol is very flammable, so avoid all sparks, open flames, and excessive heat.

Although it is unlikely to be ingested, ethylene glycol can be toxic if taken orally. Due to the low vapor pressure of ethylene glycol, it is very unlikely that any vapors would ever be encountered, but vapors may be a problem if the liquid is heated to excessive temperatures. We know of no reason that this should occur under normal conditions of use.

When working with preserved materials, be careful with sharp objects such as pins, scalpels, and the spines and teeth of specimens. When using a scalpel, we recommend cutting away from oneself and ensuring that fingers are kept out of the cutting path at all times.

Carolina preserved specimens are available in *Carosafe*™, an ethylene glycol–based shipping and holding fluid. *Carosafe*™ is not a fixative; it is a preservative designed to prevent mold and tissue deterioration after the tissue has been properly fixed with formalin. *Carosafe*™ is an effective substitute for the standard formalin preservative and acts to hold the unpleasant odor of formaldehyde to an absolute minimum. Additionally, Carolina preserved animals may be ordered "damp-packed." Our tradename for this improved method of packaging is Caropak. Preserved animals shipped in Caropaks have been processed with *Carosafe*™, and are as "odorless" as effective fixation and preservation techniques allow.

The following table contains further safety and health information regarding the three most common chemicals used by Carolina in the preservation process. This information is given in the form of a columnar table that contains the information required by OSHA to be present on a Material Safety Data Sheet (MSDS) under the Hazard Communication Standard (29 CFR 1910.1200).

## Comparative Safety of Preservatives

| | Formaldehyde | Isopropanol | Carosafe™ (Ethylene Glycol) |
|---|---|---|---|
| **Physical Data** | | | |
| Hazardous Components (OSHA—1994) | Methanol (TWA 200 ppm) Formaldehyde (TWA 0.75 ppm) | Isopropanol (TA 400 ppm) | Ethylene Glycol (TWA = 50 ppm Ceiling concentration) |
| Flash Point | 184° Fahrenheit (Combustible) | 53° Fahrenheit (Flammable) | 241° Fahrenheit |
| Lower Explosion Limits LEL | 7% | 2% | 3.2% |
| Fire Extinguishing Media | Alcohol Foam, Water Fog, Carbon Dioxide, Dry Chemical | Alcohol Foam, Carbon Dioxide, Dry Chemical | Water Fog, Carbon Dioxide, Dry Chemical |
| Unusual Fire or Explosion | Vapor heavier than air, may travel along ground to distant ignition source and flash back. | No unusual fire hazards noted. Closed containers exposed to fire may explode. | None |
| Threshold Limit Value (TLV) ACGIH | 200 ppm (TWA) Methanol 0.3 ppm Ceiling Formaldehyde | 400 ppm (TWA) | 50 ppm Ceiling |
| **Effects of Overexposure** | | | |
| Eyes | Vapor causes severe irritation, redness, tearing, blurred vision. Liquid may cause severe or permanent damage. | Direct contact may cause irritation. | Direct contact may cause irritation. |
| Skin (Contact) | Irritation, dermatitis, strong sensitizer. | Mild irritation possible. | Mild irritation possible. |
| Inhalation | Irritation of respiratory tract, dyspnea, headache, bronchitis, pulmonary edema, gastroenteritis. | Irritation of respiratory tract, headache, and at high concentrations, narcosis. | Reported irritant effects at extremely high (10,000 mg/cubic meter) concentrations of vapor. |
| Ingestion | May be fatal or cause blindness if ingested. LD50 (oral-rat) = 500 mg/kg (RTECS, 1986) | May cause nausea, vomiting, headaches, dizziness, gastrointestinal irritation. LD50 (oral-rat) = 5045 mg/kg (RTECS, 1986) | May be harmful or fatal if ingested. Ethylene glycol has been reported as causing liver and kidney damage when ingested. LD50 (oral-rat) = 4700 mg/kg (RTECS, 1986) |
| Chronic Effects | Listed by the National Toxicology Program (NTP) as reasonably anticipated to cause cancer in humans. Also listed by IARC and OSHA as possible human carcinogen. | Not listed as causing cancer by NTP, IARC, or OSHA. No other chronic effects noted. | Not listed as causing cancer by NTP, IARC, or OSHA. No other chronic effects noted. |
| Target Organs | If inhaled, eyes, nasal passages, throat. | None | Liver and kidneys (if ingested) |
| **First Aid Measures** | If inhaled, remove to fresh air. If not breathing, give artificial respiration. If ingested, if conscious, immediately induce vomiting. If eye or skin contact, immediately flush with flooding amounts of water for at least 15 minutes. Seek medical attention for all instances of overexposure to this chemical. | If inhaled, remove to fresh air. If not breathing, give artificial respiration. If ingested, if conscious, immediately induce vomiting. If eye or skin contact, immediately flush with flooding amounts of water for at least 15 minutes. Seek medical attention for all instances of overexposure to this chemical. | If inhaled, remove to fresh air. If not breathing, give artificial respiration. If ingested, if conscious, immediately induce vomiting. If eye or skin contact, immediately flush with flooding amounts of water for at least 15 minutes. Seek medical attention for all instances of overexposure to this chemical. |
| **Spill Control Measures** | If a spill occurs, cleanup personnel should wear full protective clothing and NIOSH-approved self-contained breathing apparatus. Eliminate sources of ignition. Keep non-essential personnel away. Absorb spilled material on vermiculite or other suitable absorbent. Containerize for disposal. | Eliminate sources of ignition. Cleanup personnel should wear proper protective clothing and equipment to avoid contact with liquid. Respiratory protection may be required. Absorb material on activated carbon or other suitable absorbent. Containerize for disposal. Flush area of spill with water. | Cleanup personnel should wear proper protective clothing and equipment to avoid contact with liquid. Absorb material on vermiculite or other suitable absorbent material. Containerize for disposal. Flush area of spill with water. |
| **Disposal** | Dispose in accordance with all applicable local, state, and federal regulations. Contact local or state waste agencies if disposal questions arise. | Dispose in accordance with all applicable local, state, and federal regulations. Contact local or state waste agencies if disposal questions arise. | Dispose in accordance with all applicable local, state, and federal regulations. Contact local or state waste agencies if disposal questions arise. |
| **Personal Protection** | Wear gloves, lab coat, splash goggles, and any other appropriate equipment suggested by the laboratory supervisor. | Wear gloves, lab coat, splash goggles, and any other appropriate equipment suggested by the laboratory supervisor. | Wear gloves, lab coat, splash goggles, and any other appropriate equipment suggested by the laboratory supervisor. |
| **Storage Information** | Store tightly closed in a location suitable for general chemical storage. | Store in a location suitable for flammable liquid storage. | Suitable for storage in a general chemical storage area. |

TWA—Time Weighted Average; ACGIH—American Conference of Governmental Industrial Hygienists; IARC—International Agency for Research on Cancer; OSHA—Occupational Safety and Health Administration; PEL—Permissible Exposure Limit; NIOSH—National Institute for Occupational Safety and Health; RTECS—Registry of Toxic Effects of Chemical Substances. LD50—Lethal Dose for 50% of a population.
Source: Carolina Biological Supply Company, 2700 York Road, Burlington, North Carolina, 27215, 910-584-0381.

# HANDLING AND CARE OF ANIMALS IN THE LABORATORY

The study of anatomy and physiology of animals is fundamental to the training of zoology students. Many students find that working with living and preserved animals is one of the most interesting and beneficial aspects of their education. Prospective employers in business and industry, and admissions committees of graduate programs, as well as medical, dental, and veterinary schools, have frequently emphasized the importance of such practical experience.

Research with laboratory animals has led to important scientific advances in physiology, genetics, behavior, nutrition, ecology, and other fields. Advances in human medicine that are direct results of experimentation involving animals include immunization against polio, diphtheria, measles, and other diseases; insulin production and therapy; blood transfusions; chemotherapy; electrocardiography, open-heart surgery, and artificial heart valves; organ transplantation; and kidney dialysis.

Major advances in veterinary medicine resulting from experimentation with animals include the development of vaccines for rabies, distemper, swine cholera, and brucellosis; medication for dog heartworms; artificial insemination, in vitro fertilization, and embryo transfer technology; methods for preserving endangered species; and surgical techniques for hip replacement. These veterinary advances have saved thousands of animal and human lives and have contributed greatly to the human food supply and to the quality of life of farm and companion animals.

Studies of animals from textbooks, photographs, charts, models, and computer simulations are good supplements, but they are not adequate substitutes for actual laboratory experience with living and preserved animals. Zoology students need to learn and practice proper methods to observe, handle, care for, experiment with, and dissect laboratory animals. Consider the dilemma of a neurosurgeon who has never observed, handled, or dissected an actual brain, but who is about to do his or her first operation on a member of your family with a brain tumor.

The handling and treatment of vertebrate animals is regulated by federal law under the Animal Welfare Act of 1966, amended subsequently in 1970, 1976, 1985, and 1990. Additional regulations governing the use and care of laboratory animals have been developed by the National Institutes of Health. Many individual states also have laws governing animal use. Invertebrate animals are generally not covered under these laws, but such animals should also be treated with care and respect as living creatures. Rare and endangered species are protected by special laws and may not be collected or used in laboratory studies except under special permits. All teachers and researchers must be familiar with these federal and state regulations and be responsible for using good judgment and for following appropriate procedures for handling and experimenting with all animals.

As a responsible citizen and a student of zoology, you should also handle living and preserved animals with care and respect. When working with both vertebrate and invertebrate animals, you should always take adequate precautions to avoid causing unnecessary stress or discomfort to the animals due to your handling or experimenting. Any animals kept in the laboratory must have a clean and appropriate environment, including adequate ventilation, food, water, and regular care. Be sure to follow the specific federal guidelines established for the care of animals kept in the laboratory for the duration of an experiment. At the end of the experiment, the animals must either be disposed of in an approved humane manner or returned to a permanent animal care facility as directed by your instructor.

Some people oppose the use of animals in the laboratory either for training or research because they believe it is unethical for humans to use animals in any way that might be harmful or detrimental to the animals for the benefit of humans or other animals. Appropriate usage of animals has been one of the most active controversies in the United States and elsewhere during the past several years.

Such opponents of animal use seek to reduce or eliminate the use of animals in teaching and research based on their convictions. They often cite alternatives to the use of animals in research and testing, such as computer simulations, models, films or videos, tissue culture, and in vitro chemical tests, as effective substitutes. While many scientists agree that alternatives to the use of animals are effective in some cases, no adequate alternatives are available in many other cases. Most scientists agree that the rational use of animals for teaching and research continues to be essential for the progress of human health and welfare. This position has been endorsed by several prestigious scientific bodies, including the American Society of Zoology, the American Association for the Advancement of Science, the Society of Sigma Xi, the National Science Teachers Association, the National Association of Biology Teachers, and several state academies of science.

The continuing controversy over the use of animals for teaching and research, as well as the escalating costs of obtaining and caring for laboratory animals, has already resulted in substantial reductions in the number of animals used for study and in research and improvements in the care and handling of animals in the laboratory. Concerns over the use of animals have also led to numerous govern-mental regulations on the use and handling of animals in the laboratory. Therefore, in addition to learning about the animals themselves, zoology students must also learn the rules and methods for the proper care and handling of the animals.

# GENERAL
# ZOOLOGY
## LABORATORY GUIDE

# 1

## Microscopy

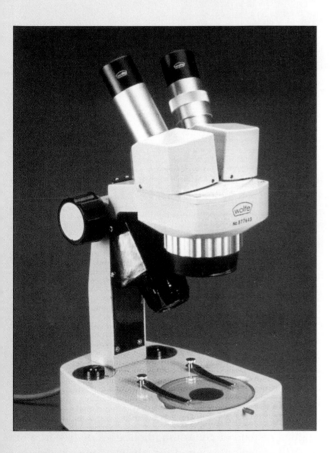

After completing the laboratory work in this chapter, you should be able to perform the following tasks:

1. Identify the main parts of a compound microscope and explain their function.
2. Define and explain focus, working distance, resolving power, and magnification.
3. Describe the proper use and care of both compound and stereoscopic microscopes.
4. Explain the difference between compound and stereoscopic microscopes and give examples of appropriate uses of each type.
5. Use both compound and stereoscopic light microscopes in the correct way.
6. Explain the operating principles of a phase contrast microscope and an interference microscope and give examples of their use.
7. Estimate magnification of a compound microscope and use the microscope to measure a microscopic object.
8. Describe the two main types of electron microscopes and give examples of their use.
9. Explain the importance of microscopes in biological studies.

## The Compound Microscope

The **compound microscope** is one of the most important and useful tools of the zoologist. It is used to study cells and cell parts, the organization of tissues, the structure of bone, and the structure of developing embryos, among many other important applications. Since many of the exercises in this course will require the use of the compound microscope, it is important to review some aspects of its construction, use, and care.

A modern compound microscope is illustrated in figure 1.1. Since there are numerous makes and models of compound microscopes in use, the microscope assigned for your use may differ slightly from the one illustrated. The operating principles and procedures, however, will be similar to those outlined later. Your laboratory instructor will point out any important differences between your microscope and the one illustrated.

A microscope is an expensive precision instrument and must be handled with care. Always carry your microscope **with both hands.** Grasp the arm of the microscope firmly with one hand and support the base with the other hand.

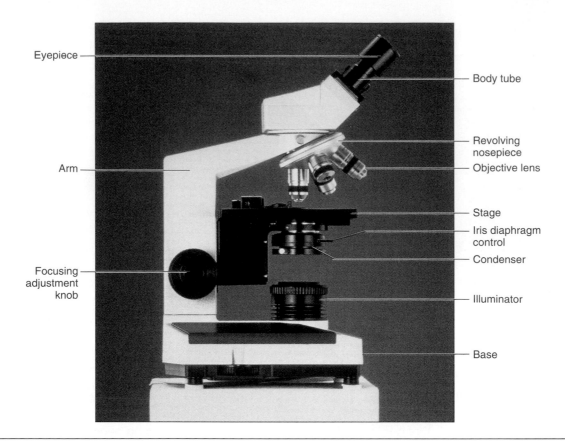

**FIGURE 1.1** Compound microscope.
Courtesy of the Olympus Corporation.

Place the microscope carefully on your table with the arm facing you. **Do not "clunk" it on the table.**

## Materials List

**Living specimens**
    *Artemia* (brine shrimp) larvae
**Prepared microscope slides**
    Letter "e"
    Frog blood
**Audiovisual materials**
    Wall charts showing parts of compound and
        stereoscopic microscopes

## Parts of the Microscope

Identify the principal parts and controls of the microscope with the aid of figure 1.1. At the top of the microscope is the **eyepiece,** or ocular lens, which is inserted in an inclined **body tube.** Microscopes with one eyepiece and body tube are monocular microscopes; those equipped with two eyepieces and body tubes for simultaneous viewing with both eyes are binocular microscopes. Below the eyepiece is the **arm** attached to the **base.** Also attached to the arm is the movable **stage,** which holds a microscope slide or other object for viewing. **Stage (slide) clips** aid in holding the slide in position on the stage. Above the stage is the **revolving nosepiece** with two or more **objective lenses.** Within the

stage is another lens system, the in-stage **condenser,** which serves to concentrate light rays from the built-in **illuminator.** On the base near the illuminator is the **light switch.** Instead of a built-in illuminator, some microscopes have a **substage mirror** to reflect light from an auxiliary light source.

Beneath the condenser, locate the **disc aperture diaphragm.** Rotating the disc diaphragm increases or decreases the amount of light on the specimen. More expensive microscopes often have an **iris diaphragm** with movable elements instead of a disc diaphragm. Raising and lowering the condenser also regulates the illumination of the specimen, although in most of your work in this course you will obtain satisfactory results by adjusting the condenser to the position that gives maximum illumination (usually near its uppermost position) and then making any further needed reductions in illumination with the disc diaphragm. This simplified method of controlling light does not produce the precise illumination required for advanced microscopy, but it produces results satisfactory for most routine purposes.

Accurate observation of a specimen requires positioning the objective lens at a specific distance from the specimen; this distance is determined by the specific construction of each objective lens and is called the **working distance** of the lens. When the objective lens is located at the proper working distance, the specimen will be in **focus.** The working distance of the lens varies **inversely** with the magnification of the objective; low-power objectives have longer working distances, and high-power objectives have shorter working distances.

Focusing the lens system is accomplished by mechanically changing the distance between the specimen and the objective lens. Coarse and fine **adjustment knobs** for this purpose are provided on the side of the arm near the base. On some modern microscopes both coarse and fine adjustments are controlled by a single knob with dual function, but most models have separate controls. *How many focusing control knobs are there on your microscope?*

When you observe a specimen through a compound microscope, several images are formed by the optical system of the microscope. Three of these images are of special importance—the **real image**, the **virtual image**, and the **retinal image** (figure 1.2).

The real image is formed at a specific distance above the objective lens. What you actually see when you look through the objective lens is the virtual image, which appears both larger and farther away than the specimen on the stage. The retinal image is formed by the rays of light striking the retina of your eye. The virtual image can be helpful to you in estimating the size of an object viewed through a microscope as explained later.

## Magnification

The principal purpose of a microscope is to magnify the image of an object. The **magnification** of an object is determined by the construction of the ocular and objective lenses of the microscope, and the total magnification is the product of the separate magnification of these two lenses.

**Example:**

10× ocular × 10× objective = 100× total magnification

Student microscopes used in introductory biology and zoology courses are commonly equipped with 10× ocular lenses, and both 10× and 43× objective lenses (some objective lenses are 40× or 44×). These two lenses are mounted on a revolving nosepiece and are called the **low-power** and **high-power** objectives, respectively. Often, a 3.5× objective lens is also used on student microscopes; this is called a **scanning lens.** It is useful for viewing relatively large objects or for preliminary location of a specimen on the microscope slide. Occasionally a 90× or 100× **oil immersion lens** may be present for viewing very small objects like bacteria. Special instructions and precautions are needed for the use of oil immersion lenses.

Magnification is changed in microscopes with multiple objectives by rotating the revolving nosepiece until the desired objective is in position below the body tube of the microscope. A newer type of compound microscope (zoom lens type) uses a system of movable lenses to produce a variable magnification rather than a series of fixed magnifications as in a microscope equipped with a rotating nosepiece.

Observe the 10× objective on your microscope and note that it is also marked 16 mm (or some similar value). This is the **working distance** of that lens; thus, this lens will be in focus at approximately 16 mm above the surface of your specimen. *What is the working distance of your high-power objective?*

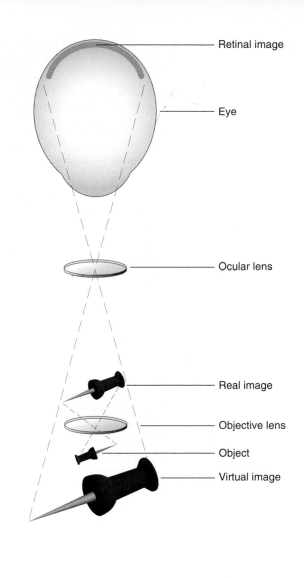

Retinal image

Eye

Ocular lens

Real image

Objective lens

Object

Virtual image

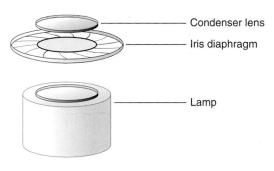

Condenser lens

Iris diaphragm

Lamp

**FIGURE 1.2** Comparison of virtual image, real image, and retinal image.

All the other parts of the microscope are accessory to the main purpose of magnification by the lenses. They consist mainly of mechanical devices to hold the specimen, to regulate the light necessary for clear vision, and to facilitate focusing.

### *Measuring the Size of an Object*

The size of an object viewed in a compound microscope can be estimated in several ways. The simplest method involves the use of a small square of graph paper or a small plastic ruler of the type found in most dissecting kits to measure the size of the virtual image. The length of the virtual image is then divided by the theoretical magnification of the ocular lens times the magnification of the objective lens.

$$\text{size of object} = \frac{\text{length of virtual image}}{\text{theoretical magnification} \times \text{objective lens magnification}}$$

This procedure gives only a rough approximation because the actual magnification of the ocular and objective lenses is slightly different from the stated theoretical magnification, and your estimate of the size of the virtual image is not precise. Instructions for estimating the size of an object with this method are given on page 7.

Microscopes are used to view small objects; therefore, you should be sure that you are familiar with the metric units listed in table 1.1 that are used to describe the dimensions of microscopic structures. Objects viewed with a stereoscopic microscope usually range from a few millimeters to a few centimeters in size. Objects viewed with a compound light microscope generally range in size from a few micrometers to a few hundred micrometers. Objects viewed with a transmission electron microscope usually range from several angstroms to several nanometers in size.

## Resolving Power

While magnification is the increase in size of the object's image, the ability of a lens or a microscope to reveal the fine detail of a specimen is called its **resolving power.** In microscopy, an increase in the apparent size of a specimen is not always accompanied by an increase in the clarity of detail within the specimen. Beyond certain limits, further magnification simply makes the apparent image of a specimen become progressively fuzzy or indistinct as it becomes larger.

The resolving power of a light microscope depends largely upon the design and quality of its objective lenses. It is actually the resolving power rather than magnification that

determines the useful magnification of a compound microscope. Furthermore, it is the wavelength of visible light that limits resolving power of a microscope. This is the reason that transmission electron microscopes are capable of much greater resolution than light microscopes. Electrons have much shorter wavelengths than the light rays of visible light.

**Resolving power is defined as the shortest distance between two points that can be visually distinguished as two separate points.** Additional magnification without increased resolution produces a larger but less distinct image. Imagine a photograph in a newspaper when the image is enlarged (magnified) to appear larger but in which the number of dots making up the photographic image remains the same. Each time the photo is enlarged, the dots are spaced farther apart, making the face in the photo appear increasingly fuzzy and indistinct (figure 1.3). This is an example of magnification without increased resolution.

## Illumination

Proper illumination of the specimen is an extremely important matter, since improper lighting of the specimen can produce poor images, inaccurate observations, and/or unnecessary eyestrain.

Most modern student compound microscopes are equipped with in-base illuminators. If you have such a microscope, you can adjust the amount of light reaching the objective lens by rotating the disc (or iris) diaphragm. Experiment with the diaphragm control, and observe the changes in amount of light as the diameter of the diaphragm opening changes. You will find that the best illumination for viewing specimens is obtained with the smallest diaphragm opening that lights the entire microscopic field.

Sometimes illumination for a compound microscope is provided by a separate lamp and a substage mirror. If you are assigned a microscope of this type, examine the two surfaces of the mirror—one surface is flat, and one surface is concave. The flat surface of the mirror is the one used most often in normal laboratory situations. When the mirror is properly adjusted, light is reflected by the mirror and passes through the condenser, the specimen, and the lenses to your eye. The concave (curved) surface of the mirror is used less frequently and only in certain situations where there is a need to concentrate the light rays. Adjust your mirror to reflect the maximum light on the specimen, and reduce the illumination as necessary by closing the iris diaphragm.

Illumination is also controlled by the substage condenser. For routine work in an introductory laboratory, the condenser should not be used to adjust illumination, but should be set at or near its uppermost position and the light intensity adjusted by means of the diaphragm.

Careful adjustment of the light beam enables you to obtain the best images with your microscope. To obtain maximum resolution and the best results, scientists use a special type of light adjustment with parallel light rays illuminating the specimen. This technique, called **Koehler illumination,** requires a lamp equipped with a field diaphragm that allows you to center and control the diameter of the light beam.

### TABLE 1.1

#### Metric Units Used in Microscopy

| | | |
|---|---|---|
| 1 angstrom (Å) | = | 0.1 nanometer (nm) |
| 10 angstroms | = | 1.0 nanometer (formerly called millimicron [mµ]) |
| 1,000 nanometers | = | 1.0 micrometer (µm) (formerly called micron [µ]) |
| 1,000 micrometers | = | 1.0 millimeter (mm) |
| 10 millimeters | = | 1.0 centimeter (cm) |

## Procedure for Koehler Illumination

1. Place a microscope slide on the stage and bring a specimen into focus using the low-power objective lens.
2. Close the field diaphragm on the light source until you can see its edges through the microscope.
3. Focus the condenser by moving it up and down until the outline of the field diaphragm is in focus.
4. Center the field diaphragm until it covers the full field of view.
5. Remove the eyepiece and observe the image coming from the objective lens. (Better results can be obtained with a centering telescope or phase telescope to replace the objective lens if one of these accessories is available.) Observe the illuminated circular field coming from the objective lens.
6. Slowly close the condenser diaphragm until you see its outline appear at the edge of the circular field. Adjust the condenser diaphragm to cover approximately the outer two-thirds of the image coming from the objective lens.
7. This setting of the field and condenser apertures produces parallel light rays to illuminate the specimen and usually gives the best balance between maximum resolution and contrast of an image.

## Focusing

To obtain a clear image of a specimen, you must carefully adjust the distance between the lenses and the specimen. This adjustment of the distance between the lenses and the specimen is called **focusing.** When a clear image of the specimen can be seen through the ocular lens, the specimen is said to be **in focus.** At this adjustment of the lens system, the specimen is located precisely at the working distance of the objective lens.

◆ Locate once again the coarse adjustment knob on your microscope. Rotate the low-power lens in place and turn the control slightly in one direction. Observe that movement of the knob increases or decreases the distance between the stage and the objective lens.

If you have a microscope with separate coarse and fine adjustment knobs, carefully try each control to see the difference in their actions. You will find that it takes several turns of the fine adjustment control to equal the effect of a single turn of the coarse adjustment control.

While testing the action of the focusing controls, watch the objective lens and the specimen stage. *Which part moves, and which part remains stationary?*

Sometimes you will find that the fine adjustment will not turn any farther and you cannot reach focus. When this occurs, you have reached the end of the range of the fine adjustment in that direction. To rectify the situation, simply turn the fine adjustment five full turns away from its stop, and refocus the image as best you can with the coarse adjustment. Now you can use the fine adjustment to improve the focus.

(a) Magnification 1×

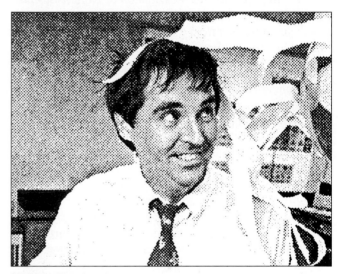

(b) Magnification 2×

(c) Magnification 4×

**FIGURE 1.3** Magnification without increased resolution.
Photograph courtesy of the Raleigh News and Observer, Raleigh, NC.

## Procedure for Use of the Compound Microscope

Good microscopy requires the adoption of work habits that facilitate observation, minimize fatigue and eyestrain, and protect the equipment from damage. Follow these steps each time you use the compound microscope.

1. Place the microscope directly in front of you on the laboratory table. Remember to carry the microscope by using both hands after removing it from the storage cabinet. Clean the objective and ocular lenses by wiping gently with a clean sheet of lens paper; never use anything but lens paper for cleaning the lenses.

2. Rotate the revolving nosepiece until the low-power (10×) objective lens clicks in place directly over the center of the condenser. While viewing the objective lens from one side of the microscope tube, **carefully** adjust the distance between the lens and the specimen stage, using the coarse adjustment control, to approximately 12 mm.

3. Select a microscope slide provided by your laboratory instructor and examine it to locate the position of the specimen on the slide. Then place the slide on the stage, with the specimen centered over the condenser lens. Make certain that the coverslip and specimen are on top of the slide.

4. Open the disc (or iris) diaphragm fully and turn on your microscope lamp. If your microscope is equipped with a substage mirror and an auxiliary lamp, adjust the positions of the lamp and the mirror until you obtain an evenly lighted, circular microscope field. Reduce the light on the specimen as necessary for clear observation by adjusting the diaphragm.

5. While observing the objective lens and the microscope slide from the side again, carefully readjust the distance between them to about 3 mm. **Do not look through the ocular while performing this step.**

6. Now look through the ocular and slowly **increase the distance between the objective lens and the specimen** until the specimen comes into focus. This may involve either raising the objective lens or lowering the specimen stage, depending upon the design of your particular microscope. **Never** focus in the opposite direction—by decreasing the distance between the objective lens and the specimen—**while looking through the ocular.** Further, center the specimen in your field of view as necessary, and bring the specimen into sharper focus with the fine focusing control.

7. Rotate the revolving nosepiece until the high-power objective lens (43×) clicks into position directly over the condenser lens. The objective lenses on most modern microscopes are factory installed and adjusted so only minor changes in focusing and centering are necessary when magnification is changed. Such lenses are **parfocal**—that is, their planes of focus and the center of their field of view are identical, or nearly so, although their working distances are different. You should find it necessary to make only a slight adjustment with the fine focusing control in order to achieve good focus after changing from the 10× objective to the 43× objective. **Use only the fine focus control** while looking through the ocular with the high-power objective in position. You will also find it necessary to open the iris diaphragm slightly when you switch from low to high power. *Why?*

## Returning the Microscope after Use

1. Rotate the revolving nosepiece to place the **low-power objective** (or scanning lens if your microscope has one) into position over the condenser lens.

2. Remove the microscope slide from the specimen stage and return it to its proper box or tray. If you have been using wet mounts, clean the specimen stage with a **clean cloth** or **cleaning tissues** as provided by your laboratory instructor.

3. Clean the objective and ocular lenses of the microscope with **lens paper.**

4. If your microscope has an in-base illuminator, turn down the power, disconnect the power cord, and carefully wind the cord as directed by your instructor.

5. Return the microscope to its storage cabinet. Remember to use both hands in carrying the microscope. Check once again to make certain that the low-power objective lens (or scanning lens) is in position **and that you have not left a slide on the stage.**

## Special Precautions

1. **Never** focus down (raise the stage or lower the objective, depending on the type of your microscope) while looking through the microscope.

2. **Always** locate the specimen under low power before switching to high power.

3. **Never** turn your microscope upside down or lay it on its side. The ocular might fall out and could be damaged.
4. **Always** keep the microscope clean and dry. **Use only lens paper to clean the lenses.**
5. **Never** turn the microscope lamp on or off before turning the power down (if you have an adjustable lamp).
6. **Always** wrap the power cord carefully around the base of the microscope or the cord hanger before putting away your microscope.
7. **Never** use the coarse focusing control when the high-power lens is in position. Focus only with your fine focusing control when using high power.
8. **Always** try to relax and keep both eyes open when using the microscope. This helps to prevent undue eyestrain. With a little practice, you can learn to concentrate on the specimen and to disregard the image received by the other eye if you are using a monocular microscope.

## Exercises Using the Compound Microscope

1. Obtain a microscope slide with the letter "e" on it. Place the slide on the stage of your compound microscope and observe under low power. *What is the orientation of the letter "e" as you view it through the compound microscope? Why does it look different from a letter "e" printed on this page?*

    Move the microscope slide slightly to the left while you look through the eyepiece. *Which way does the letter "e" seem to move?*

    Move the slide slightly away from you. *Which direction does the letter "e" move this time?* Change the objective lenses to high power by rotating the nosepiece or by changing the zoom adjustment if you have a zoom-type microscope. Observe the letter "e" under high power. *How much of the letter "e" can you see under high power?* Note that the letter seems larger and that your field of view is smaller at high power than at low power.

2. Obtain a slide of frog blood and study it first under low power and later under high power. *What is the shape of the cells? What structures can you observe in the cell?* Draw a frog blood cell on the first page of the Notes and Sketches section at the end of this chapter. Label the cell parts you recognize. You will learn more about cells and their parts in the next chapter.

## Estimating Magnification

You can estimate the magnification of your microscope by using a small square of millimeter graph paper mounted on a clean microscope slide under a coverslip.

1. Carefully move the slide to center one millimeter square in the microscope field with two sides of the square parallel to the edge of the microscope stage.

2. Place a small plastic metric ruler on the right edge of your microscope stage so that you can see the ruler with one eye while you view the slide with the other eye.

3. Keep both eyes open and focus your eyes on both images with the square of graph paper in one eye and the scale of the ruler in the other eye. It will seem strange at first to look at two different images simultaneously, but if you relax and continue to observe both objects, they will soon become superimposed in your vision.

4. When you see the two superimposed images, measure the apparent size of the superimposed millimeter square as it appears on the ruler. Since the lines surrounding the square appear relatively thick, be sure that you measure from the upper edge of one line to the upper edge of the next line to include one thickness of the line in your measurement of the square.

5. Record your measurement of the millimeter square in the Notes and Sketches section at the end of this chapter. Also in this space sketch the superimposed images that you see using both eyes.

6. Calculate the magnification under low power by dividing the apparent size of the millimeter square by its actual known size (1 mm).

**Example:**

$$\frac{103 \text{ mm (apparent size)}}{1 \text{ mm (known size)}} = 103\times$$

## Measuring Microscopic Objects

You can use a similar procedure to estimate the size of an object seen under your compound microscope. If you observe a cell under the microscope that appears to be 25 mm in diameter, simply divide by the calculated magnification to determine the actual size of the cell.

**Example:**

$$\frac{25 \text{ mm (observed size)}}{103 \text{ (magnification)}} = 0.24 \text{ mm}$$

Calculation of the size of an object viewed under high power is a bit more complicated, but follows the same principle. First, place the ruler on the stage and measure the apparent diameter of the microscope field at high power. Next, place the slide with millimeter square graph paper on it on the stage and orient the slide again so the sides of the square are parallel to the ruler. Switch to high power and move the millimeter square so that one corner is at one edge of the field. Locate some distinctive landmark on the baseline of the square, such as a bulge in the line or a large fiber, at the opposite edge of the field. Carefully move the slide to place your landmark directly across the field and find a new landmark on the opposite side of the field again. Continue this procedure until you determine how many microscope fields at high power it takes to traverse the one millimeter square.

Multiply the apparent field diameter as measured with the ruler by the number of diameters required to traverse a one millimeter square to obtain the magnification of your microscope at high power.

**Example:**

apparent diameter of field × number of fields to traverse
1 mm = magnification

or

175 mm × 2.4 field diameters = 420×

In a like fashion, you can estimate the actual size of an object by dividing its apparent size (measure the virtual image with a ruler) by the magnification under high power.

**Example:**

$$\frac{\text{apparent size}}{\text{magnification}} = \frac{160 \text{ mm}}{420\times} = 0.38 \text{ mm}$$

The values obtained by these procedures are only approximate but provide size and magnifications useful for most purposes.

## Measurements Using an Ocular Micrometer

More precise measurement of microscopic objects and estimation of magnification require additional equipment. One common method requires an **ocular micrometer,** a graduated micrometer disc that is placed in one eyepiece of the microscope. An ocular micrometer contains a millimeter scale marked off into several divisions and must be calibrated with a stage micrometer. A **stage micrometer** is a glass slide with a precision millimeter scale usually divided into 100 parts. Superimposing the micrometer disc in the eyepiece over the stage micrometer allows precise calibration of the divisions in the eyepiece and permits accurate measurements of microscopic objects.

## The Stereoscopic Microscope

The **stereoscopic microscope,** also frequently called the dissecting microscope or stereomicroscope, is another common and extremely useful laboratory instrument (figure 1.4). Stereoscopic microscopes are useful for viewing objects such as small insects, frog eggs, or large protozoa at low magnifications.

The image seen through a stereoscopic microscope is not inverted, as is the image seen through a compound microscope. Also, the internal configuration of lenses and prisms provides dual light paths and thus produces a stereoscopic or three-dimensional image. The effective magnification of such a lens system is more limited than is that of compound microscopes. Most stereomicroscopes provide magnifications in the range of 5–50×, although useful magnification of up to 100–200× can be obtained in the best quality research stereomicroscopes.

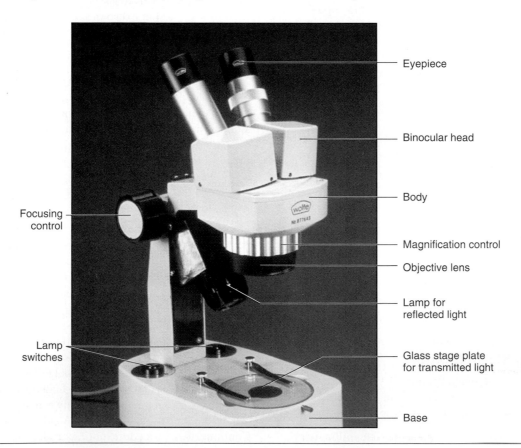

**FIGURE 1.4** Stereoscopic microscope.

Microscope photograph by H. F. Holloway. Courtesy of Wolfe Sales Corporation and Carolina Biological Supply Company, Burlington, NC.

Two designs of stereomicroscopes commonly used in teaching, as well as in research laboratories, employ different mechanical and optical systems to achieve changes of magnification. One of these types of stereomicroscopes employs a fixed-position ocular, objective paired lenses, and an internal rotating drum on which several paired prisms are mounted. Two large knobs located on opposite sides of the microscope head serve to rotate the internal drum and thus to change magnification. Focusing is accomplished by raising or lowering the objective, as in a compound microscope. The **focusing controls** are located on the sides of the microscope arm.

The other type of stereomicroscope uses a zoom-type lens system and provides a continuously variable magnification by the rotation of a cylinder located above the objective lens (figure 1.4). Focusing is accomplished in this type of stereomicroscope by means of two lateral focusing control knobs as in the previously discussed types of stereomicroscopes.

Stereomicroscopes may be equipped with either an opaque or a transparent glass stage plate or disc. When equipped with a transparent stage plate and a substage beneath the regular stage, a stereomicroscope can be used to view objects in transmitted as well as in reflected light. This is often a very useful feature for biological studies.

## Exercises Using the Stereomicroscope

1. Remove your stereomicroscope from its storage cabinet and determine which type has been provided for your use in this course. *Are there other types of stereomicroscopes present in the laboratory?* Compare them with the type you have.
2. Place your stereoscopic microscope directly in front of you on the table and focus a concentrated light on the center of the stage. Select a microscope slide with a relatively large specimen, such as a fluke, a tapeworm, an insect wing, or other suitable specimen, and examine it through the microscope.
3. Experiment with several such slides as provided by your laboratory instructor until you become familiar with the use of your stereomicroscope. Try various light adjustments on the specimen, and try both transmitted and reflected light on specimens that are transparent or translucent if your microscope is provided with a substage and a substage mirror. If you do not have a substage, you may be able to achieve a similar effect by illuminating the specimen at right angles to your plane of viewing. Place your lamp close to the tabletop and focus the light beam from the side as sharply as possible on your specimen. *Can you observe any structural details not discernible when the light comes from above or at a higher angle?*
4. Place a few newly hatched larvae of the brine shrimp *Artemia,* or of a similar small living animal, in a watch glass and examine them in both direct and transmitted light. Make a simple outline drawing on the first page of the Notes and Sketches section showing the major structures that you see.

5. When you have completed your study with the stereoscopic microscope, wipe the stage clean of any spills using a clean cloth or cleaning tissue, and return the microscope to its storage cabinet.

## Other Types of Microscopes

Compound and stereoscopic microscopes are only two of the several different types of microscopes that have been developed by scientists in their continuing efforts to observe small objects more closely and in greater detail. Three other kinds of microscopes of particular importance in zoology are the phase contrast microscope, the interference microscope, and the electron microscope. In their design, these microscopes employ physical principles that differ significantly from those employed in ordinary light microscopes.

### Phase Contrast Microscopy

Phase contrast microscopy permits the direct examination of transparent, unstained materials, including many kinds of living cells and tissues. Basically, this type of microscopy depends upon a special method of illumination that increases the contrast in an unstained specimen resulting from slight variations in the thickness and refractive index of its parts. The light passing through the specimen is manipulated in the lens system in such a way that these minor physical differences within the specimen are transformed into varying degrees of brightness and darkness. Thus, a living cell viewed through a **phase contrast microscope** has the appearance of being stained, although no chemicals that might alter its structure or kill it have been added to the cell. Figure 1.5 illustrates the type of image obtained in this kind of microscopy.

### Interference Microscopy

**Interference microscopes** are another type of instrument that use two beams of polarized light in a complex optical system to visualize structures in living, unstained cells. They also permit quantitative measurements of dry mass per unit area of microscopic objects. A special kind of interference microscope, the Nomarski interference microscope, has been especially useful in the study of cells and tissues.

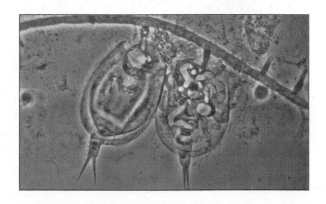

**FIGURE 1.5** Phase contrast photograph of two rotifers being digested by a carnivorous fungus.
Courtesy of Carolina Biological Supply Company, Burlington, NC.

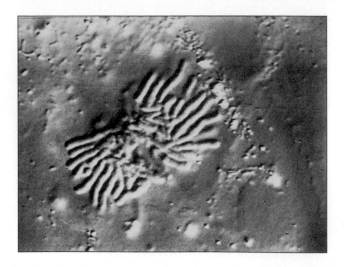

**FIGURE 1.6** Chromosomes in a newt lung cell.
Nomarski differential interference micrograph courtesy of Southern Micro Instruments.

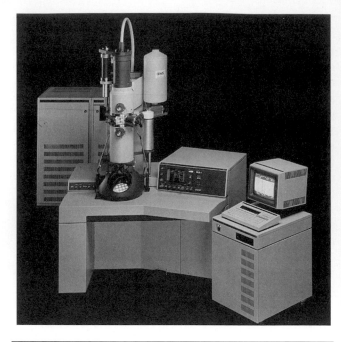

**FIGURE 1.7** Transmission electron microscope equipped with a special X-ray analyzer on the right.
Photograph courtesy of Philips Electronic Instruments, Inc.

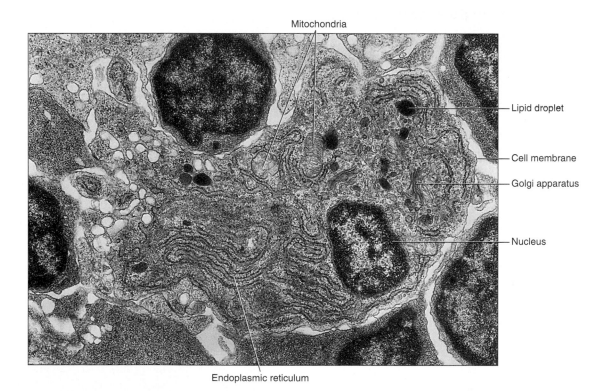

Mitochondria

Lipid droplet

Cell membrane

Golgi apparatus

Nucleus

Endoplasmic reticulum

**FIGURE 1.8** Transmission electron micrograph of a plasma cell.
Photograph by Kenneth E. Muse.

The **Nomarski interference microscope** can provide excellent images of unstained materials with a three-dimensional effect (figure 1.6). This type of microscope uses polarized light, which is split into two light beams of slightly different wavelength. Interactions of these two light beams emphasizes the hills and pits within the specimen to yield the three-dimensional effect. Interference microscopes are found mainly in research laboratories since they are expensive due to the complex system of prisms and lenses necessary to produce these optical effects.

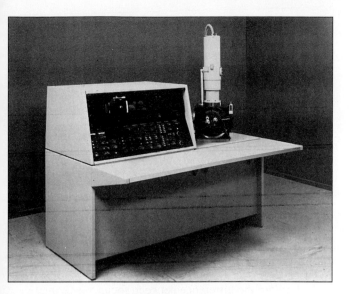

**FIGURE 1.9** Scanning electron microscope.
Photograph courtesy of Philips Electronic Instruments, Inc.

(a)

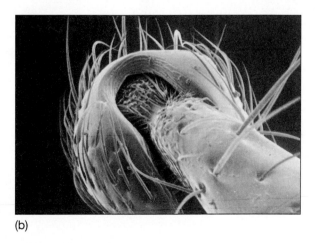

(b)

**FIGURE 1.10** (a) Scanning electron micrograph of an ant. Magnification 300×. (b) Enlarged view of the foreleg joint from the circled area in (a). Magnification 600×.
Photographs by Kenneth E. Muse.

# Electron Microscopy

Electron microscopy has become a very valuable research technique in zoology in the last few decades. In electron microscopy, a specimen is irradiated with a concentrated **beam of electrons** rather than with rays of visible light, as in a light microscope. The two main types of electron microscopes commonly used by zoological researchers are transmission electron microscopes and scanning electron microscopes.

**Transmission electron microscopes** (figure 1.7) are capable of much greater resolving power and, therefore, much higher magnifications than are ordinary light microscopes. Direct magnifications up to 200,000 diameters are commonly achieved with electron microscopes compared to a maximum of about 2,000 diameters with a compound light microscope. Photographic enlargements yield final magnifications greater than 1,000,000 diameters in many electron micrographs. Figure 1.8 is an example of a photograph taken with a transmission electron microscope.

The **scanning electron microscope** (figure 1.9) is an even newer research tool and has had many important applications in biological research during the past 25 years. The scanning electron microscope differs in principle from both light and transmission electron microscopes. This kind of electron microscope has proved to be especially useful in providing three-dimensional images of small objects, information about chemical composition, electrical properties, and structural details of the surface of specimens. The scanning electron microscope has a great depth of field (seven to ten times that of a light microscope at comparable magnifications), making possible photographs of excellent three-dimensional quality (figure 1.10).

## Key Terms

**Compound microscope**   a type of light microscope with two separate lens systems, an eyepiece and an objective lens, which together serve to magnify the image of an object. Provides magnification to about 1,000–2,000 diameters.

**Interference microscope**   a type of light microscope that uses beams of polarized light to allow visualization of low-contrast specimens such as living cells and also the determination of mass or dry weight of specimens. Nomarski differential interference microscopy employs slight differences in the wavelengths of two beams of polarized light to enhance the contrast of unstained specimens and to give an image that looks three-dimensional.

**Magnification**   the ratio of the apparent size to the actual size of an object when viewed through a microscope.

**Ocular micrometer**   a graduated disc placed in the eyepiece of a microscope that can be used to measure the size of microscopic objects.

**Parfocal lenses**   objective lenses constructed and mounted so that their focal planes are approximately the same; a specimen remains in focus and centered in the field of view when you change from one parfocal lens to another.

**Phase contrast microscope** a special type of light microscope that permits the observation of thin, unstained materials. Special lenses and illumination techniques increase the contrast in an unstained specimen due to variations in the refractive indices of its parts.

**Resolving power** the ability of a microscope to reveal fine detail in a specimen. More precisely, resolving power is defined as the shortest distance between two points that allows them to be distinguished as separate points.

**Scanning electron microscope** a type of electron microscope that provides three-dimensional images of very small objects. A concentrated beam of electrons is focused and moved along the surface of a specimen and induces the emission of secondary electrons from the specimen. These secondary electrons produce a magnified image of the specimen on a cathode-ray tube.

**Stereoscopic microscope** a type of light microscope with two separate optical paths that provide a magnified three-dimensional image. Provides useful magnifications to about 100–200 diameters.

**Transmission electron microscope** a type of electron microscope used to view specially prepared thin specimens at magnifications to about 200,000 diameters. A concentrated beam of electrons passes through the specimen and produces a pattern of light and dark areas on a phosphorescent screen because of the differential passage of electrons through portions of the specimen. Darker areas represent areas of greater electron density within the specimen.

**Working distance** the distance from the front of the objective lens of a compound microscope to the top of the specimen.

# Internet Resources

Visit the zoology website at http://www.mhhe.com/zoology to find live Internet links for each of the references listed below.

1. Website of Florida State University Molecular Expressions of Microscopy. Primer with basic information on microscopes and microscopy.

2. Web page of the Microscopy Society of America. Current information about the society and information on many kinds of microscopy.

3. Microscopy Primer Web Resources. Hundreds of links on the web.

# Critical Thinking Questions

1. List three examples of biological materials or specimens that would best be studied with each of the following types of microscope: (a) compound microscope, (b) stereoscopic microscope, (c) scanning electron microscope, and (d) transmission electron microscope. Explain why you chose the specific type of microscope for each example.

2. Explain the difference between focus, resolving power, and magnification and explain why each is important in microscopy.

3. Describe two methods for measuring objects viewed with a compound microscope and give examples for the use of each method in a biological study.

# Suggested Readings

Hunter, Elaine, P. Maloney, M. Bendagan, and M. Silver. 1993. *Practical Electron Microscopy: A Beginner's Illustrated Guide,* 2d ed. Cambridge: Cambridge University Press.

Slayter, Elizabeth M., and H.S. Slayter. 1992. *Light and Electron Microscopy.* Cambridge: Cambridge University Press.

Taylor, D. L., M. Nederlof, F. Lanni, and A.S. Waggoner. 1992. New vision of light microscopy. *American Scientist* 80(4).

# NOTES AND SKETCHES

# NOTES AND SKETCHES

# 2

## Animal Cells and Tissues

Simple squamous

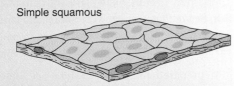

Stratified squamous

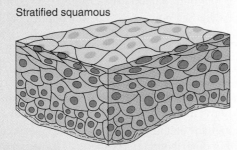

Cuboidal

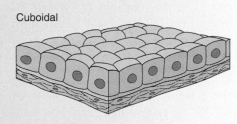

Simple columnar

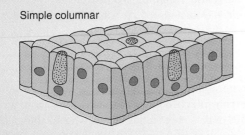

### OBJECTIVES

After completing the laboratory work in this chapter, you should be able to perform the following tasks:

1. Describe the cell theory and explain its importance in zoology.
2. Describe the principal organelles of a typical animal cell, as represented by a starfish egg, that are visible in a light microscope.
3. List six organelles typically seen in a transmission electron micrograph of a generalized (typical) animal cell that are not generally visible with a compound microscope.
4. Distinguish between a cell, a tissue, an organ, and an organ system.
5. List four main types of animal tissues and give examples of each. Identify typical examples of each in microscope slides.
6. Describe the histological structure of a compact bone and explain the role of a Haversian canal, lacuna, canaliculi, and lamellae.
7. Distinguish between smooth muscle, cardiac muscle, and striated muscle from microscopic slides or photographs.
8. Describe the structure of a vertebrate neuron and give the function of each main part.
9. Describe the composition of human blood and give the functions of erythrocytes, leucocytes, and platelets.
10. Identify erythrocytes, leucocytes, and blood platelets in microscopic preparations.

## The Cell Theory

Cells are the fundamental structural and functional units of virtually all living organisms. The body of an animal typically is made up of many different kinds of cells that are organized into tissues, organs, and organ systems that carry out certain essential functions. Thus, a knowledge of cell structure and function is essential for the proper understanding of reproduction, growth, heredity, and all other normal and abnormal animal functions.

The importance of cells is summarized in a statement called the cell theory, which is one of the most important unifying concepts in biology. This theory was developed in the nineteenth century, when naturalists were experimenting with their latest technology—the compound microscope.

The cell theory is generally attributed to two German scientists, botanist Matthias Schleiden and zoologist Theodor Schwann, who published their ideas in 1838 and 1839; however, other scientists also have contributed to the modern version of the theory.

The main points of the cell theory can be summarized as follows:

1. All organisms are composed of cells.
2. All cells come from other cells.
3. All vital functions of an organism occur within cells.
4. Cells contain the hereditary information necessary for regulating cell functions and for transmitting information to the next generation of cells.

## Materials List

Prepared microscope slides
    Starfish eggs or sea urchin eggs
    Cuboidal epithelium (rabbit kidney)
    Columnar epithelium (rabbit kidney or amphibian intestine)
    Stratified epithelium (human skin)
    Loose connective tissue
    Mammalian hyaline cartilage, cross section
    Human bone, ground, cross section
    Smooth muscle, teased
    Striated muscle
    Neurons (smear from spinal cord of cow)
    Blood film (human), Wright stain
    Ciliated epithelium (demonstration)
    Adipose tissue (demonstration)
    Cardiac muscle (demonstration)
    Amphibian blood (demonstration)
Fresh cartilage (from frog sternum, ends of long bone, etc.)
Cross sections of long bones from pig, cow, or other mammal

## Basic Cell Structure

Our knowledge of cell structure has increased dramatically in the last few decades because of the availability of electron microscopes with much higher magnification and resolution than light microscopes, as described in Chapter 1. Figure 2.1 illustrates the appearance of a relatively simple cell seen in

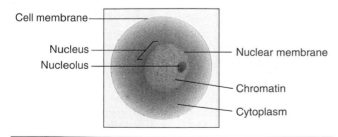

**FIGURE 2.1**   Starfish egg.
Courtesy of Carolina Biological Supply Company, Burlington, NC.

an ordinary compound light microscope of the type usually available in general zoology laboratories. Figure 2.2 is a diagram illustrating the parts of an animal cell seen in a transmission electron micrograph. Note the much greater structural detail in the electron micrograph. We know that animal cells are composed of many kinds of **organelles,** the distinctive parts of cells that carry out specific functions. Several types of cellular organelles can be visualized only in electron micrographs because they are too small to be seen with a compound microscope.

Since electron microscopy is a very complex process, in this exercise we will concentrate on those aspects of cell and tissue structure observable with a compound microscope and compare our findings with information available from more sophisticated techniques and instruments.

Good illustrations of generalized (unspecialized) animal cells are provided by the unfertilized eggs of many animals. For our introductory study of cells, we will use the unfertilized eggs of the starfish. Sea urchin eggs are very similar to starfish eggs and serve equally well. Prepared microscope slides containing many stained starfish eggs will be provided for your study.

◆  Obtain a prepared microscope slide and, under low power on your compound microscope, observe the numerous starfish eggs. Select a well-stained cell similar to that illustrated in figure 2.1 and center the cell in your field of view. Rotate the high-power objective into position, regulate the light as needed, and readjust the focus. Observe the two well-differentiated parts of the **cell,** the central **nucleus** and the surrounding **cytoplasm.** Externally, the cytoplasm is bounded by a thin **cell membrane.** Although the cell membrane of animal cells is seen only as a thin outer boundary of the cell, it plays a very important role in the functions of the cell. The special properties of the cell membrane control the passage of materials into and out of the cell. Animal cells lack the thickened cellulose cell walls outside the cell membrane that are usually found in plant cells.

◆  Also on your slide, identify the spherical nucleus, bounded by the nuclear membrane and containing numerous darkly staining masses of chromatin material. Within the nucleus of the starfish egg, find the darkly staining nucleolus. In table 2.1, list the cell organelles that you are able to identify with your compound microscope.

Several additional cell organelles can be seen in micrographs of cells obtained with a **transmission electron microscope** as well as many more structural details of those organelles visible with a light microscope. The drawing in figure 2.2 illustrates an animal cell as seen in a transmission electron micrograph, and table 2.2 lists several of these organelles and their functions.

## Animal Tissues

**Tissues** are groups of cells with a common embryonic origin that work together to perform a certain function or functions. Tissues, in turn, are organized into **organs,** which consist of

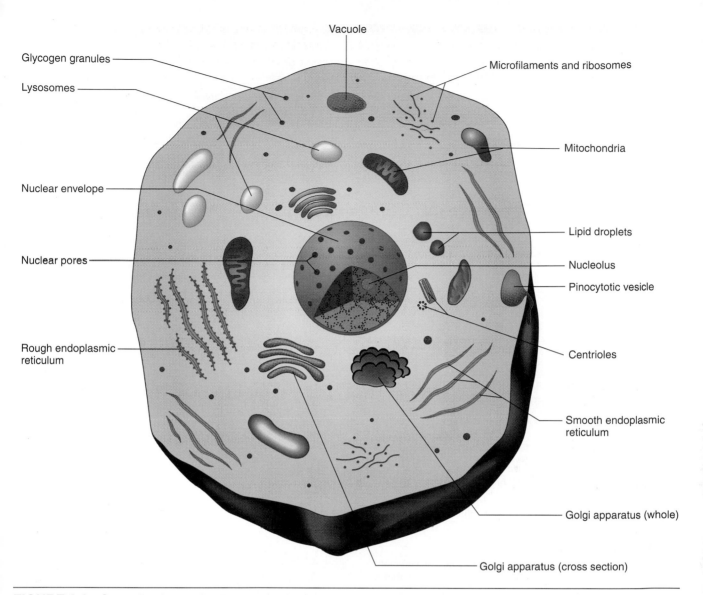

**FIGURE 2.2**  Generalized animal cell, as seen with the aid of a transmission electron microscope.

one or more kinds of tissues grouped into a structural and functional unit. Organs with related functions work together as **organ systems** in most types of animals. An **organism** consists of several integrated organ systems.

Animal tissues are usually divided into four main types: (1) **epithelial tissue,** (2) **connective tissue,** (3) **muscular tissue,** and (4) **nervous tissue.** Each of these tissue types shares certain common functions, and each type is represented by two or more subtypes found in various tissues and in various kinds of animals. The study of tissues, including their structure and function, is called **histology.**

## Epithelial Tissue

The principal function of epithelial tissues is to cover and protect surfaces. In addition to covering the outside of the body like the outer layers of your skin, other kinds of epithelial tissues line internal cavities and ducts, form glands, and aid in the transport of materials through, from, and into ducts and canals. Various kinds of epithelia are characterized mainly by the shape and arrangement of their cells. **Simple epithelia** consist of single layers of cells, and **stratified epithelia** contain several layers of cells.

| TABLE 2.1 |
|---|
| **Cellular Organelles Observed with Light Microscope** |
| 1. _____ |
| 2. _____ |
| 3. _____ |
| 4. _____ |
| 5. _____ |
| 6. _____ |

# TABLE 2.2

## Principal Cell Organelles Seen in Transmission Electron Micrographs

| | |
|---|---|
| Plasma membrane | The bilayer lipoprotein membrane that forms the outer boundary of a cell, regulates the passage of materials into and out of the cell, and allows the cell to interact with its environment. |
| Endoplasmic reticulum | A system of membrane-bound compartments in eukaryotic cells that are involved in the synthesis and transport of materials through the cell. Rough endoplasmic reticulum has many ribosomes bound to its membranes, and smooth endoplasmic reticulum has no ribosomes associated with its membranes. |
| Golgi apparatus | A complex stack of flattened membranous sacs and vesicles in eukaryotic cells that serves to store, modify, and sort secretory products received from the endoplasmic reticulum. |
| Lysosomes | Specialized products of the Golgi apparatus in eukaryotic cells consisting of membrane-bound vesicles containing hydrolytic enzymes that can digest foreign materials or aid in the breakdown of old cell organelles. |
| Mitochondria | Ovoid or cylindrical organelles of eukaryotic cells with a double membrane surrounding an inner matrix. Serve as the principal site for ATP synthesis. |
| Peroxisomes | Membrane-bound organelles in eukaryotic cells that contain enzymes that catalyze the transfer of hydrogen and break down hydrogen peroxide. |
| Cytoskeleton | System of minute tubules and fibrils that provide structural support for the cell, aid in movements of other organelles, and function in cell movement. |
| Centrioles | A pair of cylindrical structures found in animal cells composed of nine triplet microtubules surrounded by an amorphous area called the centrosome with which the centrioles appear to organize microtubule assembly. |
| Ribosomes | Particles made up of RNA and protein that serve as the site of protein synthesis in the cytoplasm; may either be bound to the membrane of rough endoplasmic reticulum or free. |
| Nuclear envelope | Double lipoprotein membrane that encloses the nucleus during interphase but disappears during mitosis. |
| Nucleus | Large organelle surrounded by the two-layered nuclear envelope. Stores, replicates, and transfers information stored in DNA. |
| Nucleolus | Cluster of ribosomes in the interphase nucleus. |

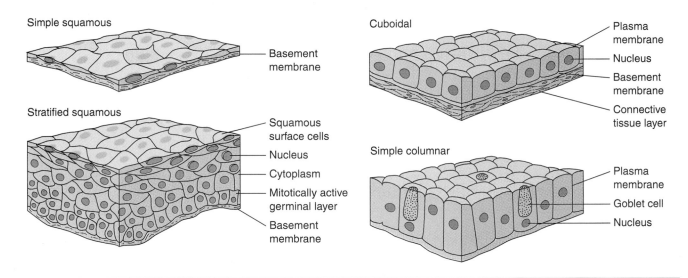

**FIGURE 2.3** Types of epithelium.
Courtesy of Carolina Biological Supply Company, Burlington, NC.

## Simple Epithelia

Three common types of simple epithelium are **squamous epithelium, cuboidal epithelium,** and **columnar epithelium** (figure 2.3). Squamous epithelium consists of a single layer of thin, flattened cells. Simple squamous epithelium lines such cavities as the air sacs of the lungs, kidney tubules, the coelom, capillary walls, and the inner lining of blood vessels where the exchange of materials by the absorption and the diffusion of gases is especially important to the underlying cells. For example, amphibians have a thin, moist skin covered with squamous epithelium, which permits easy passage of oxygen and carbon dioxide (figure 2.4).

◆ Obtain a prepared slide of squamous epithelium cells and draw a few adjacent amphibian epithelial cells in the space provided in figure 2.5 showing the structures you can observe under the light microscope.

**Cuboidal epithelium** is another type of epithelial tissue found lining several kinds of ducts such as kidney tubules, salivary glands, and the secretory follicles of thyroid glands.

These cells are sturdier than squamous epithelial cells and can withstand the frequent abrasions from the passage of materials through the ducts.

◆ Examine a slide of cuboidal epithelium (figure 2.6) from the kidney tubules of a rabbit or other appropriate tissue. The height and width of the single layer of cuboidal cells are about equal. Observe the prominent

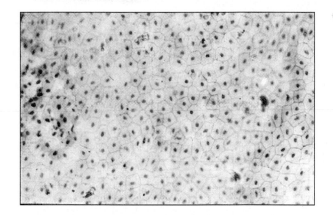

**FIGURE 2.4** Squamous epithelium from frog skin.
Courtesy of Carolina Biological Supply Company, Burlington, NC.

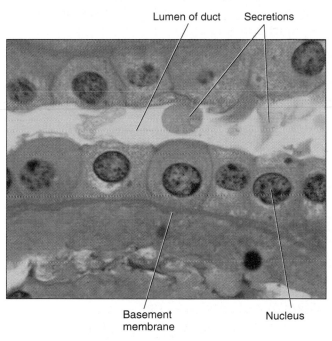

**FIGURE 2.6** Cuboidal epithelium.
Courtesy of Carolina Biological Supply Company, Burlington, NC.

**FIGURE 2.5** Student drawing of amphibian epithelial cells.

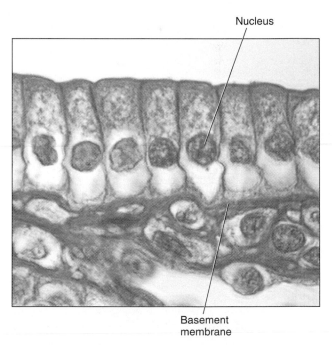

**FIGURE 2.7** Columnar epithelium.
Courtesy of Carolina Biological Supply Company, Burlington, NC.

nuclei and the darkly staining **basement membrane** found at the base of the cells. Electron microscopic studies of the basement membrane have revealed it to be a complex structure consisting of a network of very fine collagen filaments and other complex protein molecules with carbohydrate side chains.

**Columnar epithelium** (figure 2.7) consists of a layer of tall, closely packed cells that line most of the digestive tract of many vertebrate animals and are also found lining the excretory ducts in many glands.

◆ Obtain a slide with columnar epithelium from the intestine of a frog or other animal and identify the nuclei, the darkly staining outer brush border, and the basement membrane. The brush border consists of many tiny microvilli, tiny fingerlike extensions of the cell that extend into the lumen or opening of the intestine. *What do you think the function of the microvilli might be?*

Columnar epithelia from the intestine and from the trachea often contain goblet cells, which release their secretions directly into the lumen (figure 2.8). Some columnar epithelia, such as those lining the oral cavity of the frog and the trachea of mammals, also bear numerous cilia. *What function might these cilia have?*

### Stratified Epithelia

Stratified epithelia are made up of several layers of cells stacked on top of each other. They are found on surfaces in which wear and abrasion occur. The most common type is stratified squamous epithelium like that found lining the mouth, esophagus, and vagina. The cells lining your mouth cavity are a good example of this tissue type. Figure 2.9 is a phase contrast micrograph of a single living human squamous epithelium cheek cell. This cell was removed from the outer layer of the stratified epithelium lining the mouth.

Some types of stratified epithelia are hardened by the secretion of **keratin,** a tough protein. Human skin is a good example of stratified epithelium in which the outer layers of cells have been keratinized (figures 2.10 and 2.11). This type of epithelium consists of several layers of squamous epithelium that become progressively flattened and hardened as they rise to the surface and are constantly worn and sloughed off.

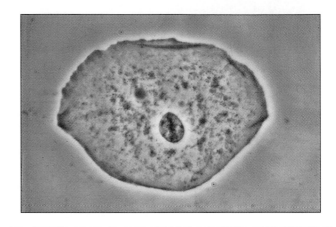

**FIGURE 2.9**   Phase contrast photograph of a living cell from human cheek epithelium.
Courtesy of Carolina Biological Supply Company, Burlington, NC.

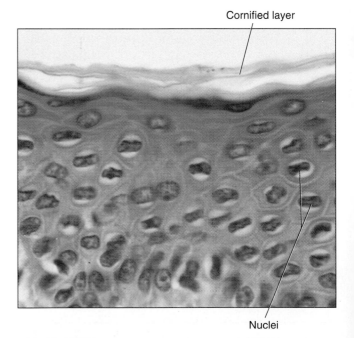

**FIGURE 2.10**   Stratified squamous epithelium, human skin, cross section.
Courtesy of Carolina Biological Supply Company, Burlington, NC.

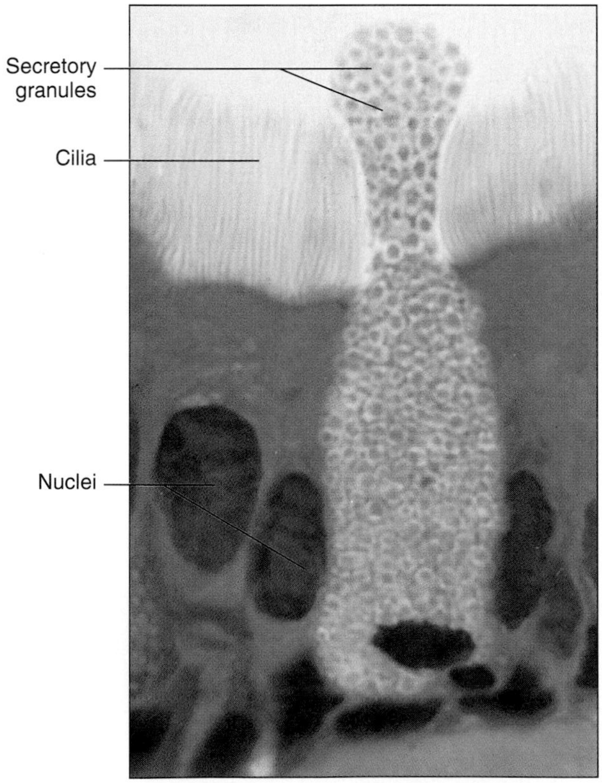

**FIGURE 2.8**   Goblet cell in columnar epithelium of monkey trachea.
Courtesy of Carolina Biological Supply Company, Burlington, NC.

Other types of stratified epithelia found on other surfaces are made up of columnar or cuboidal cells.

◆ Obtain a slide of human skin and observe the complex structure of the multilayered integument, or outer covering, of the human body.

## Connective Tissue

All connective tissues exhibit relatively large amounts of nonliving, intercellular substance produced by the living cells. This intercellular substance may be liquid, semisolid, or solid. You will study four examples of connective tissue in this exercise: **blood, loose connective tissue, cartilage,** and **bone.**

Blood and lymph are rather different connective tissues because of their important role in transporting materials to and from cells, but they do connect parts of the body in a very real physiological sense. The other types of connective tissues provide physical support and protection for various parts of the animal body.

### Blood and Lymph

The blood of living vertebrates is a red liquid that is constantly in motion as it circulates through a closed system of tubes—the blood vessels. Lymph is formed by intercellular tissue fluid that bathes the cells of the body and is collected in a closed system of lymphatic ducts. The lymphatic system returns the fluid to the veins near the heart.

Blood comprises about 7 percent of the human body weight. It consists of a straw-colored liquid, the **plasma,** and several types of **blood cells** suspended within it. In permanent, stained preparations, the plasma is not seen. Many of the blood cells, in particular the red cells, may be distorted by the reagents used in the preparation of the slides.

### Human Blood

Suspended in the plasma of human blood are the **red** and **white cells** and structures called **blood platelets.** Several types of human blood cells are shown in figure 2.12.

◆ Study the types of human blood cells in the prepared slides provided for you.

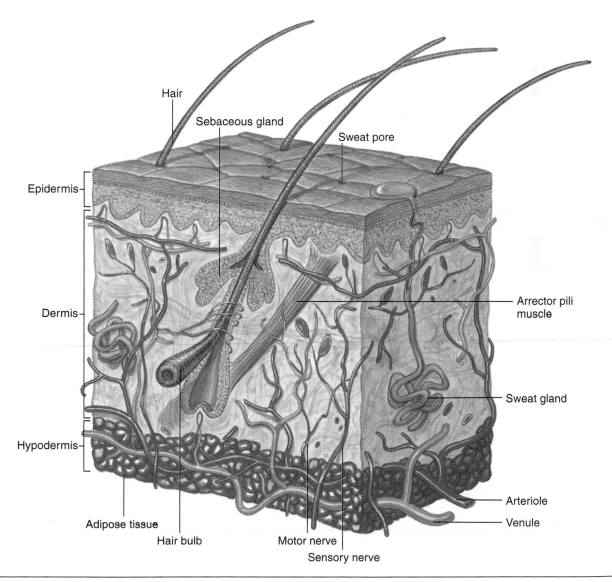

Hair
Sebaceous gland
Sweat pore
Epidermis
Dermis
Arrector pili muscle
Hypodermis
Sweat gland
Adipose tissue
Hair bulb
Motor nerve
Sensory nerve
Arteriole
Venule

**FIGURE 2.11** Drawing of human skin, cross section.

Animal Cells and Tissues  **21**

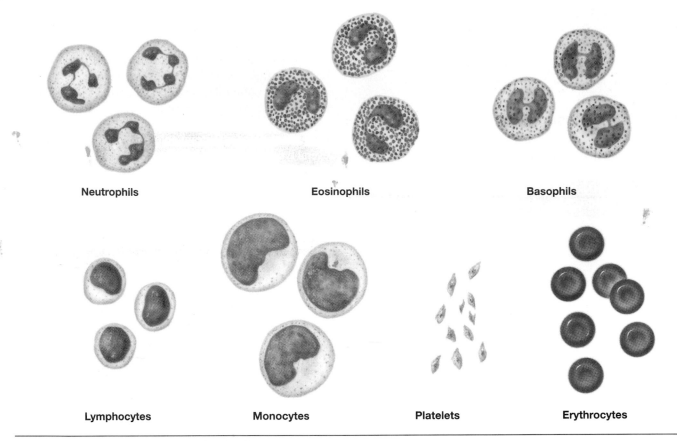

| Neutrophils | Eosinophils | Basophils |
| Lymphocytes | Monocytes | Platelets | Erythrocytes |

**FIGURE 2.12** Types of human blood cells.

**FIGURE 2.13** Human erythrocytes.
Scanning electron micrograph by Kenneth E. Muse.

1. **Erythrocytes.** Observe that human red blood cells, or erythrocytes, are small, circular, biconcave (concave on two sides), and **lack nuclei.** Their chief function is to carry oxygen to the cells of the body. They contain large amounts of the iron-containing protein **hemoglobin,** which combines with oxygen in the lungs and releases it in the body tissues. The biconcave shape of human red blood cells is clearly shown in figure 2.13, which is a photograph taken with a scanning electron microscope. Figure 2.14 shows cross sections of two red blood cells within a capillary of a mouse in a transmission electron micrograph.

◆ Note the outer membrane and the uniform dark, granular cytoplasm of the erythrocytes. Observe the lack of nuclear material in both cells. The absence of a nucleus in mature red blood cells is characteristic of all mammals.

2. **Leucocytes.** There are five main types of leukocytes or white blood cells: **lymphocytes, monocytes, neutrophils, basophils,** and **eosinophils.** Each type serves to combat infections in its own way. Several types of leucocytes are illustrated in figure 2.15.

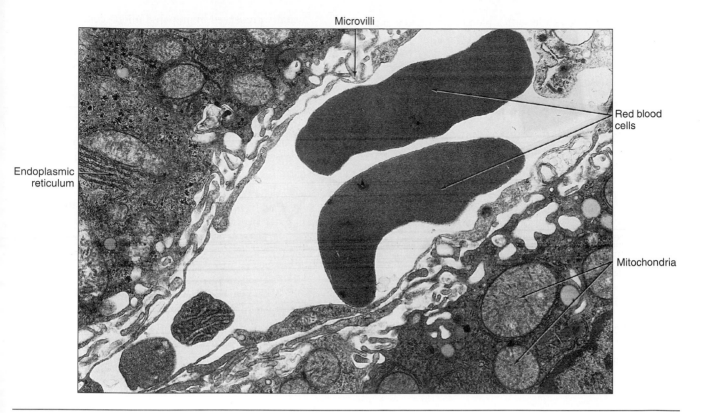

**FIGURE 2.14**   Erythrocytes in capillary of mouse liver. Magnification 13,750×.
Electron micrograph by Kenneth E. Muse.

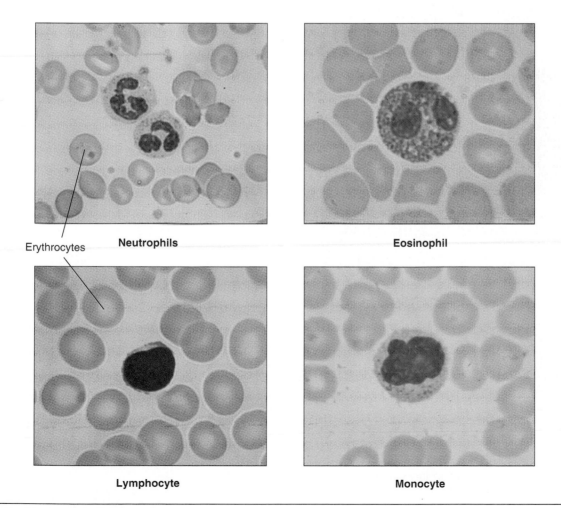

**FIGURE 2.15**   Leucocytes, four types.
Courtesy of Carolina Biological Supply Company, Burlington, NC.

- Obtain a prepared slide with a blood smear and identify as many types of leucocytes as you can find.

  A cross section of a leucocyte of a mouse is shown in figure 2.16.

- Observe its large nucleus and the complex structure of its cytoplasm. Compare this complex structure of the cytoplasm with the relatively simple appearance of the erythrocytes in figure 2.15. *Why do you think that mammalian leucocytes would have a more complex structure than mammalian erythrocytes? How is their structure related to their function?*

  Figure 2.17 is a scanning electron micrograph of a **macrophage** from the human lung. Note the many irregular extensions from the surface of the cell. Macrophages are large phagocytic cells formed from monocytes, which engulf and break down bacteria and cellular debris.

3. **Blood platelets.** Platelets are small, nonnucleated bits of cytoplasm that bud off from large cells (megakaryocytes) in the bone marrow. They are about one-third the size of erythrocytes and are not usually preserved in prepared microscope slides. Platelets function in **blood clotting** by temporarily plugging punctures that may occur in blood vessels and also by releasing substances that trigger later chemical reactions necessary for clotting.

## Loose Connective Tissue

Loose connective tissue (figure 2.18) consists of scattered cells surrounded by a clear, jellylike **ground substance** and two types of fibers: thin **elastic fibers** and thicker bundles of **nonelastic** (collagenous) **fibers.** Loose connective tissue is often called areolar tissue. It is found in tendons attaching muscles to bones, and other tough tissues connecting the various parts of the body in humans and other vertebrate animals. Nerves, blood vessels, and cells also lie embedded in loose connective tissue.

- Obtain a slide of loose connective tissue and observe the cells, the elastic and nonelastic fibers, and the apparently open spaces where the ground substance has been dissolved during the preparation of the slide. Identify the thin, wavy, elastic fibers and the thicker nonelastic fibers. Other types of connective tissue include **adipose** (fat) **tissue, cartilage,** and **bone.**

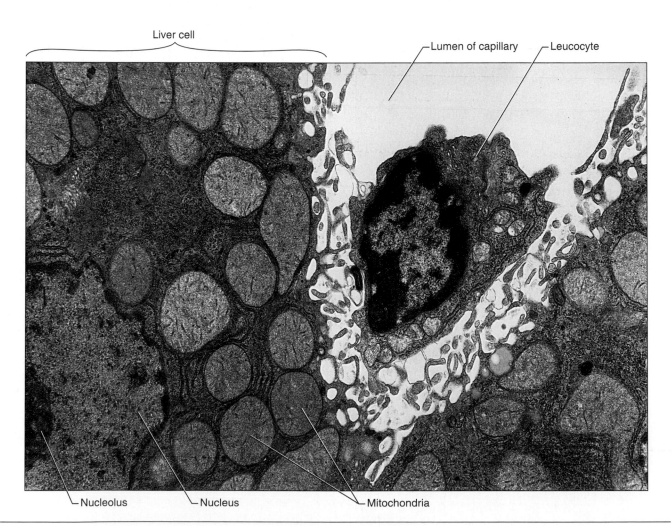

**FIGURE 2.16** Leucocyte in capillary of a mouse. Magnification 21,000×.

Electron micrograph by Kenneth E. Muse.

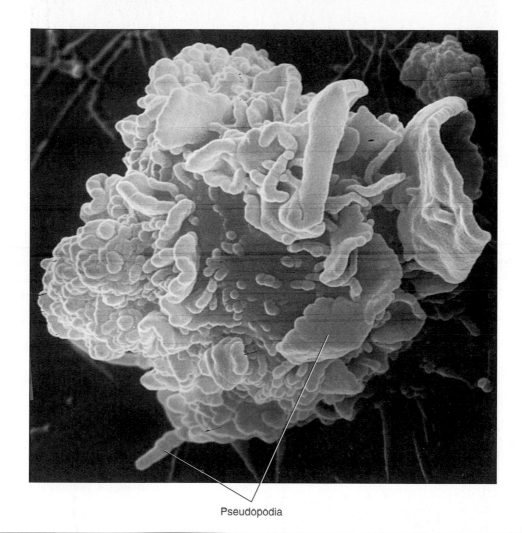

Pseudopodia

**FIGURE 2.17**  Macrophage from human lung. Magnification 20,000×.
Scanning electron micrograph by Kenneth E. Muse.

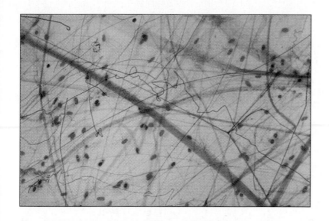

**FIGURE 2.18**  Loose connective tissue.
Courtesy of Carolina Biological Supply Company, Burlington, NC.

◆ Draw examples of elastic and nonelastic fibers in figure 2.19 and label each type.

### Cartilage

Cartilage (figure 2.20) consists of a firm but elastic **matrix** secreted by numerous cartilage cells embedded within the matrix. **Hyaline cartilage** is found at the ends of long bones.

Cartilage cells are found in open spaces or **lacunae** (singular: lacuna) scattered within the matrix (chondrin), which they secrete. Most cartilage cells are isolated in a lacuna separated by a relatively large amount of intercellular material. Occasionally, however, you may find two recently divided cells within a single lacuna. Following their division, the two new cartilage cells begin to secrete more chondrin, and each new cell creates its own new lacuna. This is how cartilage tissue grows and how it may be repaired after injury. Athletes often suffer torn cartilages in their knees. If the injury is severe, it may require surgery to remove the damaged cartilage. Small tears in the cartilage, however, may be partially repaired by limited regrowth and the formation of scar tissue by the cartilage.

◆ Study a microscope slide of hyaline cartilage and observe the scattered cells, lacunae, and chondrin matrix. *Can you find two recently divided cartilage cells in your microscope slide? How would you identify them as recently divided cells?*

◆ Also study the demonstration of fresh cartilage from the end of a bone or the sternum. Feel its tough rubbery nature. *How is this physical property related to the function of cartilage?*

heart and the lungs are also protected by the bony framework of the rib cage embedded in the thoracic wall.

The intercellular matrix of bone also plays an important physiological role in the storage of calcium, which can be withdrawn from the bone and returned to circulation in soluble form when the calcium level in the blood is lowered. This is one reason why it is important for growing children and older persons to maintain an adequate intake of dietary calcium to avoid osteoporosis, the weakening of bone from excessive loss of calcium.

The bone marrow found in the center of long bones contains tissue that plays an important role in the formation of blood elements and in immunity.

◆ Examine a thin section of **compact bone** (figure 2.21) prepared from a long bone, such as the humerus or femur of a human or other mammal, and identify the following structures. (1) the central **Haversian canal** through which passes many small blood vessels and nerves; (2) the **concentric layers** of bone (lamellae); (3) the **lacunae,** or spaces that house the bone cells or osteocytes; and (4) the numerous fine **canaliculi,** which serve to interconnect the lacunae and the Haversian canals.

## Muscular Tissue

Muscular tissue is specialized for **contraction** and therefore has the capacity to perform mechanical work. Three different types of muscular tissue are distinguished: **smooth muscle, skeletal muscle,** and **cardiac** (heart) **muscle.**

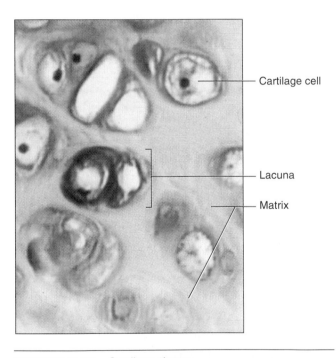

**FIGURE 2.19**  Student drawing of elastic and nonelastic fibers.

**FIGURE 2.20**  Cartilage tissue.
Courtesy of Carolina Biological Supply Company, Burlington, NC.

Labels: Cartilage cell, Lacuna, Matrix

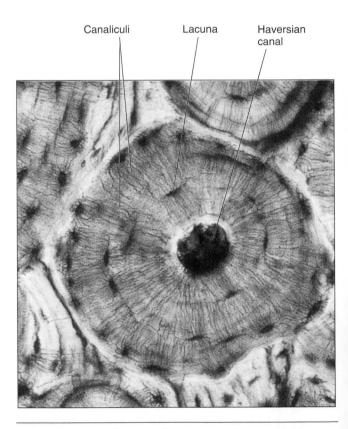

Labels: Canaliculi, Lacuna, Haversian canal

**FIGURE 2.21**  Compact bone, ground, cross section.
Courtesy of Carolina Biological Supply Company, Burlington, NC.

### Bone

**Bone** plays an important role in the mechanical support of the body and in protecting vital parts from injury. The skull and the vertebral column of humans, for example, play dual roles in the support and protection of the brain and the spinal cord as well as support for the main part of the body. The

## Smooth Muscle

Smooth muscle consists of elongated, spindle-shaped cells with a single, central nucleus (figure 2.22). This type of muscle is sometimes referred to as nonstriated muscle because it lacks the cross striations seen in both striated and cardiac muscle.

Smooth muscle forms the simplest type of muscle tissue and is generally found in parts of the body not under voluntary control where rapid movement or contraction is not essential, such as in the walls of the digestive tract, in the walls of blood vessels, and in the walls of the urinary bladder and the uterus.

◆ Examine a slide of smooth muscle that has been teased apart to show the individual spindle-shaped cells or a section through some smooth muscle that shows the individual cells cut longitudinally. Locate the tapered smooth muscle cells and note the location of the nucleus. Identify the cell membrane and the fine longitudinal threads, the **myofibrils,** in the cytoplasm. Observe also the demonstration slides of smooth muscle from other types of material.

Draw several smooth muscle cells in figure 2.23 and label the **nucleus, cell membrane,** and **myofibrils.**

## Skeletal Muscle

The large muscles attached to various parts of the skeleton are composed of skeletal, or voluntary, muscle (figures 2.24 and 2.25). This type of muscular tissue is made up of long cylindrical fibers containing many nuclei. *What is unusual about the location of the nuclei in skeletal muscle as compared to most other kinds of cells?* The multinucleate (syncytial) condition in skeletal muscle arises during embryonic development of the muscle as a result of the fusion of many mononucleate cells.

◆ Note the conspicuous cross striations in these fibers and the outer limiting membrane, called the **sarcolemma.** Observe also the fine longitudinal **myofibrils** running lengthwise through the skeletal muscle fibers.

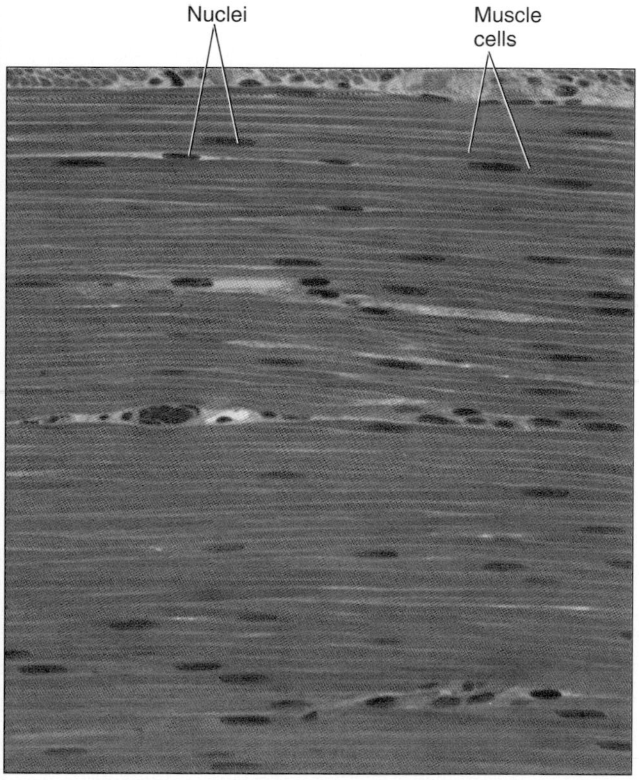

**FIGURE 2.23**   Student drawing of smooth muscle cells.

Nuclei     Muscle cells

**FIGURE 2.22**   Smooth muscle, mammal.
Courtesy of Carolina Biological Supply Company, Burlington, NC.

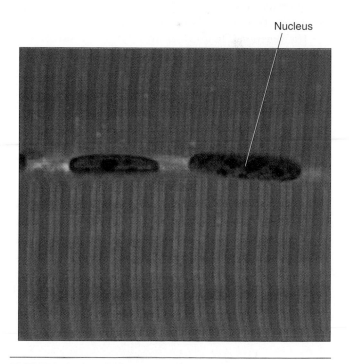

Nucleus

**FIGURE 2.24**   Striated muscle, mammal.
Courtesy of Carolina Biological Supply Company, Burlington, NC.

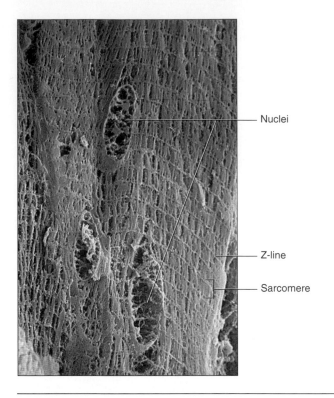

FIGURE 2.25 Skeletal muscle. Scanning electron micrograph of freeze fracture section of skeletal muscle. Note muscle fibrils, cross striations, and ovoid spaces indicating location of nuclei near surface of muscle fibers. Magnification 3,700×.
Photograph by Louis de Vos.

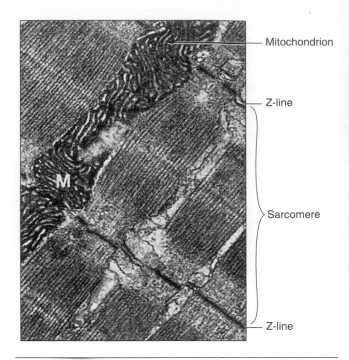

FIGURE 2.26 Striated muscle fibrils.
Electron micrograph by Kenneth E. Muse.

The individual myofibrils of striated muscle also have a very distinctive banded or striated structure. Because of their small size, individual myofibrils can be studied only with an electron microscope. Figure 2.26 is an electron micrograph showing parts of four adjacent myofibrils.

The functional unit of the myofibril is the **sarcomere,** which extends between two adjacent Z-lines. During muscle contraction, the sarcomeres shorten (the distance between Z-lines decreases). Investigations have shown that the contraction of muscle is due to interactions between two contractile proteins, **actin** and **myosin.** These two proteins make up a substantial portion of each myofibril.

The large **mitochondria** provide energy in the form of ATP (adenosine triphosphate) to power the contraction.

### Cardiac Muscle

A third type of muscle tissue, **cardiac muscle,** is found in the walls of the heart of vertebrate animals. Cardiac muscle (figures 2.27 and 2.28) consists of striated muscle fibers, which branch and reunite (anastomose) with other fibers to form a continuous network of muscle fibers. Cardiac muscle contains cross striations like striated muscle, but the fibers of cardiac muscle are divided into cell-like units by many **intercalated discs,** which partially divide the fibers. This type of muscle, like smooth muscle, is involuntary, and throughout the life of the organism, cardiac muscle contracts and relaxes rhythmically and automatically. In fact, the heart of some of the lower vertebrates, such as a frog or a turtle, can be removed from the body and placed in a physiological salt solution where it will continue to beat for many hours or even days.

◆ Study the demonstration slides of human and other vertebrate cardiac muscle and observe: (1) the branching and anastomosing network of muscle fibers, (2) the cross striations, (3) the scattered nuclei, (4) the outer sarcolemma, and (5) the intercalated discs.

## Nervous Tissue

The nervous system of vertebrate animals consists of the brain, spinal cord, and nerves. Nervous tissue consists of highly specialized cells that carry impulses from one part of the body to another. A nerve cell, together with its branches or processes, which in some cases are many centimeters long, is called a **neuron.** A neuron consists of a cell body containing the nucleus and two or more elongated nerve processes (figure 2.29). These processes are **axons** and **dendrites.** Impulses in vertebrate neurons can travel only in one direction. Dendrites carry impulses **toward** the cell body and axons carry impulses **away** from the cell body.

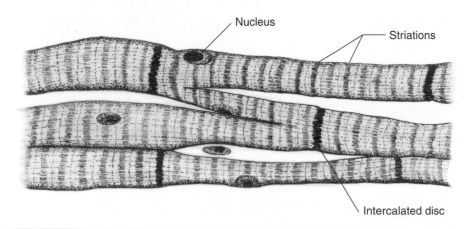

**FIGURE 2.27** Cardiac muscle.

Nerve fibers are covered by one or more sheaths. All peripheral nerves are covered by a thin neurolemma, and most peripheral nerves are also covered by a myelin sheath, which may be several layers thick (figure 2.30). Myelinated nerves are capable of rapid transfer of impulses because of their special electric properties.

**Sensory neurons** carry impulses from sensory receptors to the spinal cord or other parts of the central nervous system. **Motor neurons** carry impulses away from the spinal cord or brain and end at muscle cells or other effectors. Motor nerves that end at muscle cells have specialized endings called **motor end plates,** which aid in transmission of stimuli to the muscle cells (figure 2.31).

◆ Examine also a slide with a stained smear preparation of the gray matter from the spinal cord of a cow. The details of the individual neurons are more readily seen in such a preparation than in sections. Observe the large neurons (easily seen under low power). Compare with figure 2.32. Select a typical cell and note the **cell body, nucleus, nucleolus,** numerous **dendrites,** and the longer **axon.** Observe the numerous fine fibrils, the **neurofibrils,** within the cytoplasm of the cell body and extending into the processes.

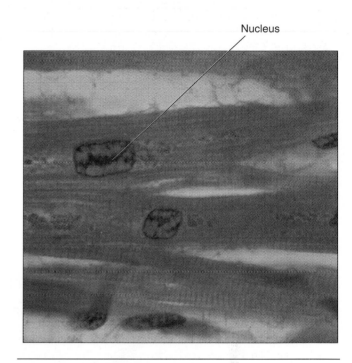

**FIGURE 2.28** Cardiac muscle, mammal.
Courtesy of Carolina Biological Supply Company, Burlington, NC.

 ## Demonstrations

1. Ciliated epithelium (microscope slide)
2. Adipose (fat) tissue (microscope slide)
3. Fresh cartilage (from the sternum of a frog, ends of a long bone, etc.) to illustrate gross structure, toughness, and flexibility
4. Cross sections (1–2 inches thick) of long bone to illustrate gross structure of bone
5. Cardiac muscle (microscope slide)
6. Amphibian blood (microscope slide)

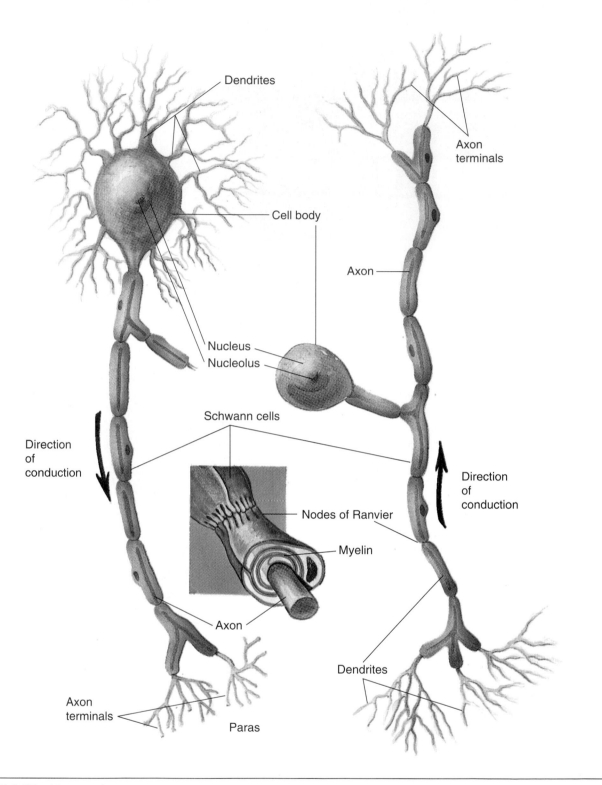

**FIGURE 2.29** Motor and sensory neurons.

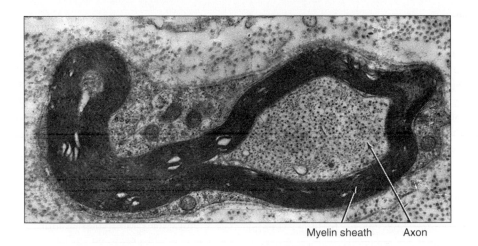

Myelin sheath    Axon

**FIGURE 2.30**   Neuron in ganglion of cat.
Electron micrograph by Kenneth E. Muse.

Axon

Muscle fibers

**FIGURE 2.31**   Motor end plate, teased, snake.
Courtesy of Carolina Biological Supply Company, Burlington, NC.

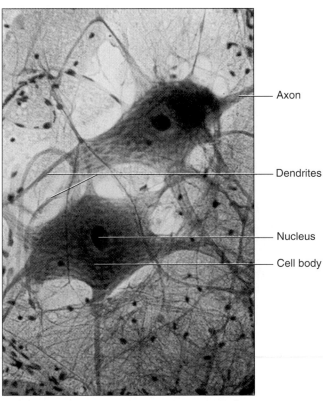

Axon

Dendrites

Nucleus

Cell body

**FIGURE 2.32**   Neurons from spinal cord smear from a cow.
Courtesy of Carolina Biological Supply Company, Burlington, NC.

## Key Terms

**Bone** hardened, mineralized type of connective tissue that provides support and protection in vertebrate animals.

**Cardiac muscle** type of muscle found in the heart wall of vertebrate animals. Consists of a branching and anastomosing network of striated, multinucleate muscle fibers.

**Cartilage** a firm and elastic type of connective tissue that provides support in vertebrate animals.

**Cell membrane** outer limiting membrane of an animal cell; composed of phospholipids, proteins, cholesterol, and carbohydrates.

**Connective tissue** type of animal tissue that binds, supports, and protects other body parts. Includes several kinds of tissue, such as loose connective (areolar) tissue, adipose (fat) tissue, cartilage, and bone.

**Epithelial tissue** type of animal tissue that lines the outer surface, inner cavities, and ducts of the body. May be simple or stratified (multilayered) and may be hardened with keratin.

**Erythrocytes** red blood cells whose chief function is the transport of oxygen; contain large amounts of hemoglobin. May be nucleated (as in frogs and salamanders) or without a nucleus in the mature stage (as in humans).

**Leucocytes** white blood cells that are usually colorless in life and that perform many essential functions, including engulfment of foreign particles, production of antibodies, and wound healing. Several distinct types of leucocytes are usually present in the blood of an animal.

**Motor end plate** specialized ending of a motor neuron on a muscle cell; serves to transmit stimulus from nerve to muscle.

**Muscular tissue** type of tissue specialized for contraction; contains fibrils constructed of contractile proteins. Three types are distinguished—smooth muscle, skeletal muscle, and cardiac muscle.

**Nervous tissue** type of tissue specialized for the conduction of electrical impulses; important in the coordination of body activities. The basic functional unit is the neuron, or nerve cell.

**Neuron** the basic, functional unit of the nervous system of animals. The neuron is a single nerve cell, which consists of a cell body with a nucleus and two or more long extensions, or processes (axons and dendrites).

**Nucleus** the central organelle of an animal cell, which contains the genetic material (chromosomes) and controls the metabolism of the cell.

**Plasma** the liquid portion of the blood in which the formed elements (blood cells and platelets) are suspended.

**Platelet** a small, nonnucleated body in the blood of humans and other mammals. Formed by megakaryocytes in the bone marrow and serves mainly to plug leaks in blood vessels and to release chemical substances that initiate clotting.

**Skeletal muscle** type of muscle tissue found attached to parts of the skeleton. Consists of long, cylindrical fibers with a cross-banded or striated appearance in microscopic preparations; each fiber contains many nuclei. Skeletal muscle is also known as striated or voluntary muscle since many skeletal muscles are under voluntary control.

**Smooth muscle** type of muscle tissue found associated with internal organs in higher animals. Consists of spindle-shaped cells with a single central nucleus. Also known as involuntary muscle.

## Internet Resources

Visit the zoology website at http://www.mhhe.com/zoology to find live Internet links for each of the references listed below.

1. Prokaryotes, Eukaryotes, & Viruses Tutorial. A discussion of the 6 kingdoms, and the basic functions of the eukaryotic cell organelles.
2. Jay Doc Histo Web. The University of Kansas (the Blue Jay's) Histology site. You can chose from various systems to view photomicrographs and electron micrographs of histological sections.
3. Mitosis and Meristems. Really terrific site with cartoon-like diagrams describing cellular parts, and mitosis.
4. Introduction to Cell Biology. Nearly all textual, but much good information on cellular structure and function.
5. MIT site showing and describing cellular parts. Quite detailed with many links.

## Critical Thinking Questions

1. Compare the structure and function of the three types of muscle tissue. Describe where each type would be found in the human body.

2. Several kinds of tissues play a protective role in the body of a vertebrate animal such as a dog, cat, or human. List four kinds of tissue that serve a protective role and explain how each kind serves to protect.

3. List four different kinds of cells found in human blood, briefly describe how they can be identified in a microscopic preparation, and give their principal function.

## Suggested Readings

Fawcett, D.W. 1981. *The Cell*. Philadelphia: WB Saunders. Excellent text/atlas of electron micrographs of animal tissues.

Fawcett, D.W., and E. Raviola. 1994. *Bloom and Fawcett's Textbook of Histology,* 12th ed. Philadelphia: Lippincott Raven.

Gray, P. 1958. *Handbook of Basic Microtechnique*. New York: McGraw-Hill.

# NOTES AND SKETCHES

# NOTES AND SKETCHES

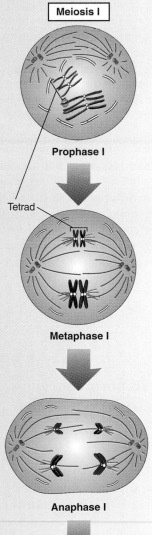

**Meiosis I**

Prophase I

Tetrad

Metaphase I

Anaphase I

Telophase I

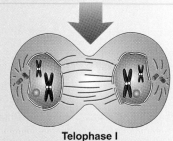

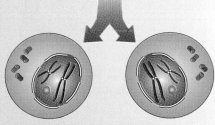

**Daughter cells: Interkinesis**

## OBJECTIVES

After completing the laboratory work in this chapter, you should be able to perform the following tasks:

1. Explain the basic differences between mitosis and meiosis.
2. Describe the cell cycle and explain the principal events of its four main stages.
3. Briefly describe the structure and function of chromosomes, centromeres, spindle fibers, aster rays, centrioles, and centrosomes.
4. Identify the principal mitotic stages of an animal cell in microscope slides or photographs.
5. Explain the chief events that occur in prophase, metaphase, anaphase, and telophase of mitosis.
6. Define *random sample* and explain its significance in biological research.
7. Describe a method for estimating the relative duration of the various stages of mitosis in a population of animal cells.
8. Explain the terms *haploid* and *diploid* and how they relate to the process of sexual reproduction in animals.
9. Distinguish between a chromosome and a chromatid.
10. List the principal stages of meiosis and identify each stage in microscopic preparations or illustrations of animal cells.

## Introduction

All animals depend on cell division for their growth and repair processes. Each cell has a precise set of genetic information built into its chromosomes. This information is essential for the proper functioning of each cell and for prescribing the characteristics of the next generation of cells. The division of the **nucleus** of the cell and the precise distribution of the duplicated chromosomes between the two new cells in this type of cell division is called **mitosis.**

Meiosis is a specialized type of cell division that usually occurs during the formation of the gametes or sex cells of multicellular animals. During meiosis, the normal diploid (2n) chromosome number of the somatic (body) cells is reduced by half to the typical (n) chromosome number of the gametes. Meiosis is extremely important for the survival and evolution of animals because it provides for recombinations of genes during each generation. Thus, variations occur among the offspring of each generation, and natural selection can operate to select the better adapted individuals.

# Mitosis

The division of nuclei by mitosis is exhibited by the somatic cells in most plants and animals. During its life span, a cell passes through a regular sequence of physiological events called the **cell cycle** (figure 3.1). This sequence includes several distinct stages, each characterized by certain metabolic activities of the cell. In actively dividing cells, the cycle may last only a few hours; in other cells, the cycle may last for days or weeks. At the completion of the cell cycle, a new generation of cells is produced by division of the parent cell.

The cell cycle is made up of four phases: $G_1$, S, $G_2$, and M. Together $G_1$, S, and $G_2$ constitute **interphase.** The **$G_1$ (gap) phase** is a period of active protein synthesis and the formation of new cell organelles, like mitochondria, Golgi bodies, ribosomes, and endoplasmic reticulum. It is a phase of rapid cell growth. The **S (synthesis) phase** is the time when DNA and other molecules making up the chromosomes, such as histones, are synthesized. Duplication of the DNA strands and of the chromosomes also occurs during S phase. In the **$G_2$ (gap) phase,** the cell synthesizes the pro-

teins, actin and tubulin, plus the enzymes and other materials necessary for the mitotic spindle. **Mitosis (M phase)** follows the $G_2$ phase and involves the division of chromosomes and the formation of two new nuclei. Mitosis consists of four distinct stages, but actually takes up only about 5–10 percent of the complete cell cycle in most cells.

The genetic material in the nucleus is duplicated in the **nondividing cell prior to the initiation of mitosis.** Thus the amount of DNA is doubled through active synthesis of DNA at this time. Therefore, the chromosomes are **already doubled** when they first become visible during prophase, the first stage of mitosis. The term *mitosis* refers specifically to the process of nuclear division, the orderly distribution of the chromosomes between two daughter nuclei, starting with prophase and continuing through metaphase, anaphase, and telophase. Technically, therefore, mitosis occurs *after* DNA synthesis is completed and chromosome duplication has been completed.

Nuclear division of a cell is usually coupled with division of the cytoplasm (**cytokinesis**). The fact that these two processes do not always occur together, along with evidence from numerous experiments that have demonstrated that various chemical and physical treatments of dividing cells have different effects on mitosis and cytokinesis, clearly shows that different **chemical** and **physical** processes are involved in nuclear division (mitosis) and in cytoplasmic division (cytokinesis).

An important point to remember is that cell division is a **dynamic** series of events during which the cell undergoes dramatic and often rapid physiological and morphological changes. The so-called stages of mitosis merely represent a few morphologically identifiable points in this continuum.

## The Mitotic Apparatus

During mitosis, a new structure, called the **mitotic apparatus,** is formed within the dividing cell (figure 3.2). The mitotic apparatus plays an essential role in mitosis and provides the mechanism by which the duplicated chromosomes of the dividing cell are distributed between the daughter cells. A knowledge of the structure of the mitotic apparatus and its principal parts will help you to understand the process of

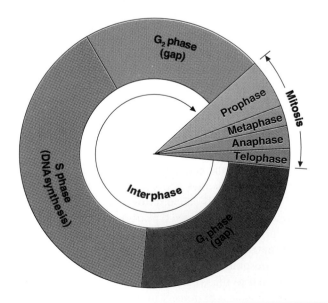

**FIGURE 3.1** The cell cycle.

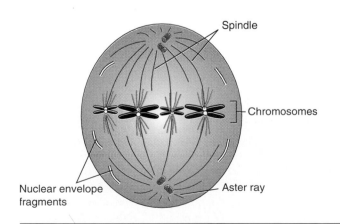

**FIGURE 3.2** Mitotic apparatus.

mitosis and to learn how genetic continuity is maintained throughout many generations of cells. The mitotic apparatus is of vital importance to you; without it, a corn plant would not grow, a cut finger would not heal, a broken leg would not mend, and you would be unable to have children.

There are six main components of the mitotic apparatus: the asters, the chromosomes, the centromeres, the centrioles, the centrosomes, and the spindle fibers. Most dividing animal cells exhibit all six of these components. Most plant cells and certain invertebrate animal cells, however, lack **centrioles** and **aster rays.** Five of these six parts of the mitotic apparatus appear to play important roles in the process of mitosis, but the function of one of these structures, the asters, is still uncertain. The functions of these various components will be considered later in this exercise.

Several important events take place in a living cell during the early stages of mitosis. These events include:

1. The breakdown of the **nuclear membrane** and the mingling of the nuclear contents with the cytoplasm of the cell.
2. The condensation of the chromosomal material within the nucleus to form discrete, visible chromosomes.
3. The separation of the centrioles and their migration to opposite sides of the cell.
4. The formation of spindle fibers and aster rays.

The spindle fibers and aster rays are formed by the coalescence or condensation of relatively small protein molecules already existing within the cytoplasm of the cell. Thus, the formation of the components of the mitotic apparatus represents a recombination of molecules preexisting in the parent cell rather than the synthesis of new protein molecules. There is relatively little synthesis of new molecules during the process of mitosis; instead, most of the new molecules formed within the cell are synthesized during the interphase.

### Chromosomes

**Chromosomes** are more or less elongated structures present throughout the life cycle of a cell. Their structure, however, appears quite different when observed at various times during the life cycle of a cell. During mitosis, the chromosomes can be seen as short, rodlike structures that are formed by condensation or contraction of very fine, threadlike structures present in the nucleus prior to mitosis (figure 3.3). Thus, the appearance of chromosomes, like the appearance of the other components of the mitotic apparatus during the early portions of mitosis, results mainly from the reorganization of preexisting materials rather than from the actual synthesis of new materials.

Figure 3.4 is a photograph of the metaphase chromosomes in a male human cell with the typical 46 chromosomes. Observe the different sizes and shapes of the chromosomes. Each species of animal (and plant) has a characteristic number of chromosomes. In humans, this number is 46; in the fruit fly *Drosophila,* it is eight; in the dog, it is 78; and in the cat, it is 60.

The unique set of chromosomes of an individual characterized by the number, size, and shape of the chromosomes is called a **karyotype.** In human genetics the use of karyotypes is very helpful in identifying hereditary genetic conditions.

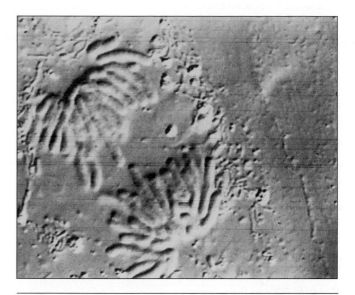

**FIGURE 3.3** Anaphase chromosomes in a newt lung cell. Nomarski differential interference photomicrograph.
Courtesy of Southern Micro Instruments.

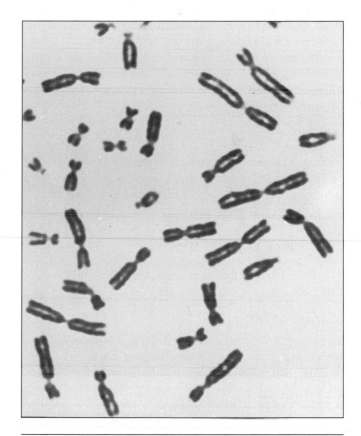

**FIGURE 3.4** Chromosomes from a male human lymphocyte in metaphase.
Photograph by Wendell Mackenzie.

The chromosomes contain the hereditary units, or genes, which specify each of the characteristics of a cell and determine all of its capabilities. Thus, the chromosomes are responsible for the control of cell metabolism and enable cells to differentiate and to organize into specialized tissues, organs, and organ systems.

### Centromeres

A **centromere** is a specific segment of a chromosome that serves to attach the chromosome to a spindle fiber by the formation of a protein complex called the kinetochore. During the latter portion of mitosis, the chromosome pairs move toward opposite **poles** of the **mitotic spindle.** Research studies indicate that the chromosomes appear to be both pulled and pushed toward the poles by separate actions of the microtubular proteins of the spindle. Other studies have demonstrated that damaged chromosomes in cells injured by X rays or by chemical agents sometimes separate into two or more fragments. Those chromosomal fragments that lack centromeres do not become attached to the chromosome and do not move toward the poles. Such observations clearly demonstrate the important functional role played by the centromeres in mitosis.

### Spindle Fibers and Asters

Special research techniques have been devised that allow scientists to isolate and to study the mitotic apparatus from living cells. Biochemical studies on such preparations of isolated mitotic apparatuses have demonstrated that the **spindle fibers** and **asters** are microtubules constructed mainly of two common contractile proteins, actin and tubulin. These proteins are involved in the movements of all eukaryotic cells. Some research evidence indicates that the asters may influence the location of the cleavage plane in cytokinesis.

### Centrioles and Centrosomes

**Centrioles** are tiny structures usually found in pairs near the nucleus in nondividing animal cells with a well-formed nucleus. The centrioles migrate toward opposite sides of the cell during the early stages of mitosis and seem to form centers or foci for the spindle fibers and aster rays at each end of the mitotic apparatus. Surrounding each centriole is an amorphous area with no visible structures called the **centrosome,** which serves as the organizer for the formation of microtubules into the mitotic spindle. Most plant cells have centrosomes that serve the same purpose, but plant cells typically lack centrioles.

## Stages of Mitosis

An important point to remember is that mitosis involves a dynamic series of events during which the cell is undergoing dramatic, and often rapid, physiological and morphological changes. When studying prepared microscope slides of cells in mitosis, you are merely seeing cells that have been killed and stained at specific points in this continuous process. For convenience in describing the process, mitosis is usually divided into four main stages: **prophase, metaphase, anaphase,** and **telophase** (figure 3.5). The stage between successive mitoses is called **interphase.**

◆ Many rapidly growing tissues provide good material for the study of mitotic cell division. Obtain a microscope slide prepared from some appropriate tissue such as the whitefish blastula, early embryos in the uterus of the roundworm *Ascaris,* or the skin of an amphibian tadpole. Study first one or more cells in interphase. Observe the structure of the nucleus and the arrangement of the chromatin material. *Can you identify definite chromosomes? How many* **nucleoli** *do you find? Is the number of nucleoli the same in all interphase cells?*

Observe other cells on the slide and select cells in each of the main stages of mitosis for further study. During your study, try to follow the actual sequence of stages as described and try to visualize the changes that occur during the transition from one "stage" to the next. Draw on figure 3.6 each of the following stages of mitosis.

1. **Early prophase stage** during which the **chromatin** material shortens to form long, coiled, threadlike chromosomes.
2. **Middle prophase stage** with relatively thick chromosomes.
3. **Late prophase stage** in which the chromosomes are further shortened and thickened. Under high magnification, late prophase chromosomes can be seen to consist of two separate strands, the chromatids, joined by a single centromere.
4. **Metaphase stage** showing the chromosomes arranged in a disclike pattern on the **equatorial plane** and attached to fibers of the mitotic spindle.
5. **Anaphase stage** during which the centromeres divide and the two chromatids of each chromosome move apart to opposite poles. The anaphase stage is relatively brief.
6. **Early telophase stage** showing the full set of chromosomes at each end of the elongated cell and the beginning of a **cleavage furrow** around the middle of the cell.
7. **Middle telophase stage** in which the individual chromosomes start to uncoil and lengthen, and begin to appear less distinct.
8. **Late telophase stage** during which the nuclei in the two daughter cells reorganize. The chromosomes disappear, nuclear membranes and the nucleoli reappear, and the separation of the daughter cells is completed.

## Timing in the Cell Cycle

The different stages in the cell cycle are not of equal duration; some stages are relatively long, while others are very brief. The actual duration of mitosis and the relative duration of individual stages of mitosis vary widely from one cell type to another. Cells of a particular type also vary in their rates of division depending upon numerous physiological and environmental conditions. Nevertheless, you can obtain an estimate of the relative duration of the various stages in a population of cells and learn something about the kinetics of cell division by employing some relatively simple experimental

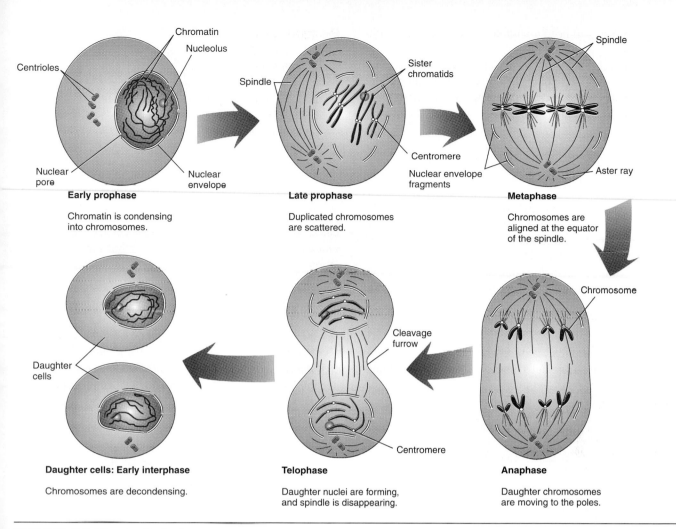

FIGURE 3.5 Stages of mitosis.

FIGURE 3.6 Student drawings of animal mitosis.

techniques. You will need a whitefish blastula slide for this experiment.

Consider the whitefish embryo from which your mitosis slide was prepared as a **population of dividing cells** and see what can be determined about the kinetics of cell division, the relative duration of various stages of mitosis, and the relationship between mitosis and interphase in the life cycle of a cell. The whitefish embryo represents a population of relatively homogeneous cells; during the earliest stages of development, the embryo consists of few cells, which divide more or less synchronously. As development continues and the number of cells in the embryo increases as a result of cell division, the degree of synchrony decreases, and division becomes progressively more randomized.

Your whitefish embryo slide represents a slice through a fish embryo containing several thousand cells. It therefore represents a sample taken from a larger population of cells. Still, the number of cells on your slide is too great for you to count readily in the limited time available in the lab. Select a still smaller sample of cells from the embryo to yield some numbers that you can use to estimate some characteristics of the population of cells that made up the original whitefish embryo.

◆ Take your whitefish embryo slide and select a random sample of 50 cells. *How can you be sure that you obtain a random sample of cells?* Record the number of cells in each of the four mitotic stages and those in interphase and carefully record the results of your count in the Notes and Sketches section at the end of this chapter.

## RANDOM SAMPLE

A **random sample** can be defined most simply as a sample from a population in which **every member** of the population has an **equal chance** to be included. Thus, a sample in which the individuals selected are determined by use of a table of random numbers (or the random number generator in a computer) would be a random sample. A sample in which the individuals selected are determined by taking every tenth individual would not necessarily be a random sample. *Why?*

**Note:** Count only cells in which the nucleus appears in the section and in which enough nuclear material (chromosomes, mitotic spindle, etc.) can be seen to allow accurate identification of the stage.

Since the cells in your slide were all killed (fixed) at approximately the same time, they represent a sample of the fish embryo cells stopped in action at a particular point in time. **Thus, the frequency of cells in the various stages is proportional to the relative duration of the stages.**

### Procedure

1. Select a random sample of 50 cells from your whitefish embryo slide.
2. Identify the stage of each cell in your random sample (prophase, metaphase, anaphase, telophase, or interphase).
3. Record the stage of each cell in your sample in the Notes and Sketches section at the end of this chapter.
4. Construct a bar graph on figure 3.7 to illustrate the results of your count showing the number of cells in each stage.

*In which stage did you find the most cells? The fewest cells? Which stage, therefore, would you conclude is the longest in duration? The shortest? What are the relative lengths of the other stages as estimated from your sample?*

◆ Compare the data from your slide with those obtained by other students from their slides. *How do their counts compare with yours? How much variation does there appear to be in the counts on different slides? What are some of the sources of variation that may account for the differences in your counts? Are your conclusions regarding the relative lengths of the various stages the same as those of other students in the class, or do they differ? Why?*

Now, combine the data from your count with those of the other students in your lab section to obtain estimates of the characteristics of another population of cells. The population represented by these pooled data consists of the cells in several (20–24) different whitefish embryos, and the 50 cells counted by each student represent a series of subsamples taken from that larger population of cells.

From these pooled data, construct another bar graph on figure 3.8 to show the number of cells in each stage. Estimate the relative lengths of the various stages in the new cell population from these pooled data and compare the results with those from your count of a single slide. *How do your results compare with and how do your conclusions differ from those based on data from a single whitefish blastula slide? Explain.*

## Meiosis

Meiosis is a special kind of nuclear division that ensures the constancy of chromosome number in the cells of succeeding generations of organisms. Sexually reproducing animals form male and female **gametes** at some point in their life history. **Fertilization** normally occurs at a later time in the life cycle and involves the fusion of male and female gamete nuclei. Thus, in order to maintain a constant number of chromosomes in successive generations (and to avoid doubling the chromosome number each time), some mechanism is necessary to provide a reduction (halving) of chromosome number between successive fertilizations. The process that results in the reduction in chromosome number is called **meiosis.**

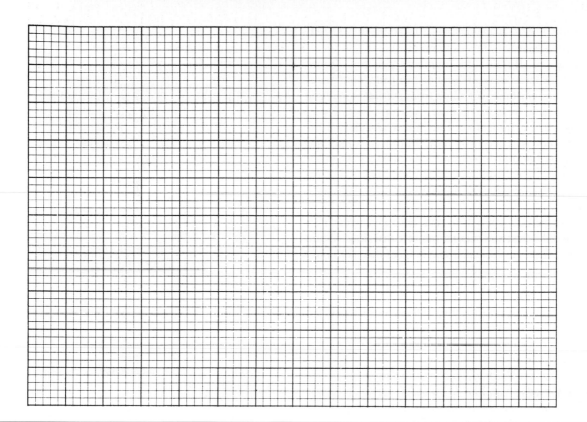

**FIGURE 3.7** Graph of the distribution of cells in various stages of division. Data from your slide.

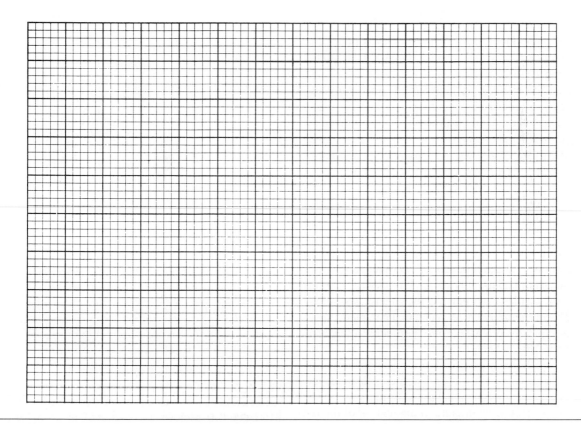

**FIGURE 3.8** Graph of the distribution of cells in various stages of division. Pooled data from the whole class.

The somatic cells of every species of animal have a definite and characteristic number of chromosomes. This is referred to as the **diploid** (2n) number of chromosomes because the chromosomes are arranged in pairs. One member of each chromosome pair came from the father and one chromosome came from the mother.

During the formation of gametes in animals, the number of chromosomes is reduced by half, and the resulting gametes have the **haploid** (n) chromosome number. The subsequent fusion of the two haploid gametes (egg and sperm) during fertilization results in a return to the diploid chromosome number.

Meiosis generally consists of two successive nuclear divisions called the **first** and **second meiotic divisions** (figure 3.9). Meiosis differs in two important respects from ordinary mitosis.

1. The final number of chromosomes in a gamete resulting from meiosis is **only half** that of the parent cell, and each gamete receives only **one chromatid** from each homologous pair of chromosomes that was present in the original parent cell.
2. During the reduction in number, the chromosomes are **assorted at random** so that each gamete receives a chromatid from one or the other member of each homologous pair. This random assortment of genetic material during meiosis plays a very important role in heredity.

**Homologous chromosomes** are the paired chromosomes found in diploid cells that are very similar in size and shape, but differ both in origin (one comes from the father and one from the mother) and in genetic composition. (The father and mother usually contribute different sets of alleles to the offspring.)

Meiosis, like mitosis, is a dynamic process during which the cells are undergoing continuous change. Nonetheless, a good understanding of the process can be achieved by describing it as a sequence of two nuclear divisions, each with four distinct stages and with an intervening interkinesis stage between the first and second meiotic divisions.

## Gametogenesis

Meiosis occurs during the formation of male and female gametes in animals. This process takes place in the **gonads,** the testes and ovaries. The formation of male gametes is called **spermatogenesis,** and the products of spermatogenesis are **spermatozoa,** a term commonly shortened to sperm. Certain cells in the testis, called early **germ cells,** undergo spermatogenesis and form sperm. Each germ cell that completes spermatogenesis typically produces four spermatozoa, each of which is genetically distinct.

In the ovary, the early germ cells undergo **oogenesis** and form eggs, or **ova.** Three of the four daughter cells produced in oogenesis become nonfunctional **polar bodies** and do not form ova. These three cells migrate to the margin of the remaining ovum and make no genetic contribution to any resulting offspring. Thus the process of oogenesis normally

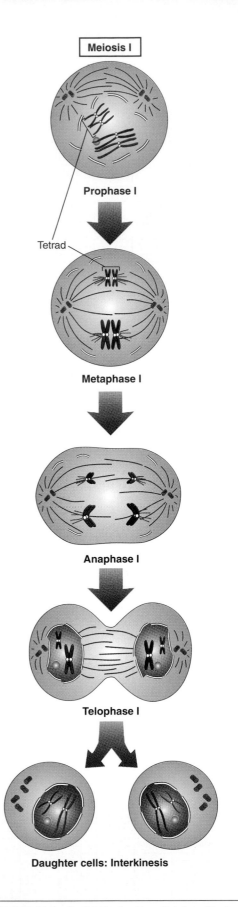

(a)

**FIGURE 3.9** Meiosis. (*a*) Reduction division. Red chromosomes are from one parent; blue chromosomes are from the other parent. (*b*) Nonreduction division.

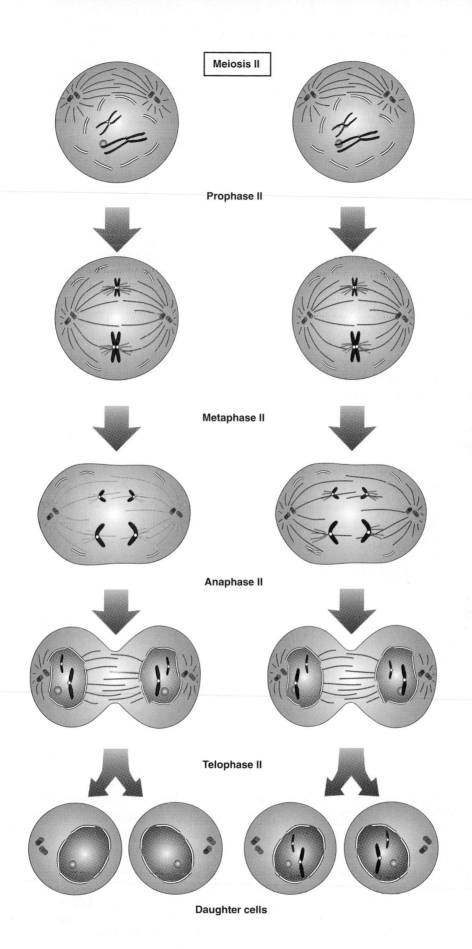

Meiosis II

Prophase II

Metaphase II

Anaphase II

Telophase II

Daughter cells

(b)

produces only one viable ovum capable of being fertilized and three nonfunctional polar bodies from each germ cell that matures. Interestingly, however, in very rare cases through some poorly understood process, one of the polar bodies may also form into a viable ovum and be fertilized. There are known cases of human twins who developed in this way.

## Principal Stages of Meiosis

Animal meiosis is usually studied in biology and genetics laboratories using microscope slides prepared from the testes of grasshoppers or the roundworm *Ascaris* and with microscope slides from the ovaries of *Ascaris*. The description of the principal stages of meiosis that follow apply to materials from any of these sources.

### First Meiotic Division (Reduction Division)

*Prophase I.* Many of the events in the first meiotic prophase are similar to prophase in mitosis. The chromosomes become visible as the very thin chromatin strands coil and condense. The nucleoli and the nuclear membrane disappear, and the mitotic spindle appears.

The key difference between the first meiotic prophase and prophase in mitosis is that, in meiosis, each pair of homologous chromosomes comes closely together in an intimate pairing process called **synapsis.** By the end of the first meiotic prophase, each homologous chromosome pair is seen as two double-stranded chromosomes closely held together. This structure, formed as a result of synapsis and consisting of two chromosomes with a total of four chromatin strands, is called a **tetrad.** Tetrads are found only during prophase I of meiosis. Note the two chromosomes at the end of prophase I as illustrated in figure 3.9. Each chromosome consists of two **distinct strands,** or chromatids, attached to a **single centromere.**

*Metaphase I.* At the onset of metaphase I, the synapsed chromosome pairs move together as a unit to the equatorial plane. Each chromosome pair is attached to a single spindle fiber by its **two adjacent centromeres.**

*Anaphase I.* The important difference between this stage and the corresponding stage in ordinary mitosis is that there is **no division of centromeres** in anaphase I of meiosis. The centromeres of the homologous chromosomes simply move apart, and the two double-stranded chromosomes of each pair migrate toward opposite poles. Thus, **half** of the chromosomes move to one pole, and **half** of the chromosomes move to the opposite pole. Note the two chromosomes migrating toward each pole in anaphase I, as shown in figure 3.9.

*Telophase I.* The chromosomes reach the poles of the mitotic spindle in each of the daughter cells, the spindle disappears, and new nuclear membranes appear around the reforming nuclei in the daughter cells. The chromosomes begin to elongate, gradually fade from view, and nucleoli reappear within the nuclei.

### Interkinesis

*Interkinesis.* Between the two successive divisions in meiosis is a brief stage called **interkinesis.** This stage is generally similar to an interphase between mitotic divisions, but there is **no duplication of genetic material** (no DNA synthesis) in interkinesis.

### Second Meiotic Division (Nonreduction Division)

*Prophase II.* The second meiotic division is essentially like an ordinary mitotic division. There is no **synapsis** in prophase II; the double-stranded chromosomes reappear and move independently toward the equatorial plane.

*Metaphase II.* Each double-stranded chromosome attaches **separately** to a spindle fiber.

*Anaphase II.* The **centromeres divide** at the end of metaphase II, and during anaphase II the newly separated, single-stranded chromosomes move toward opposite poles.

*Telophase II.* The new chromosomes in the nuclei that appear in the daughter cells are **single-stranded** and contain only **half** the number of chromosomes as in prophase I.

## Synapsis and Crossing Over

Two very important processes occur during prophase I of meiosis. These processes are synapsis and crossing over. **Synapsis** is the close pairing of homologous chromosomes to form tetrads. As a result of this close pairing, adjacent chromatids often become entangled. The entanglement of chromatids can be observed in microscopic preparations and are called **chiasmata.**

When the homologous chromosomes begin to migrate to opposite poles of the mitotic spindle at the end of metaphase I, chromatids often exchange segments because of these entanglements. Such exchange of segments between chromatids is called **crossing over.** Crossing over is an important source of new genetic combinations in the resulting sperm and ova and in the offspring that may result from subsequent fertilization.

## Ways to Study Meiosis

The actual observation of meiosis is often difficult for an inexperienced observer because of the small size of the chromosomes in most kinds of cells. Also it is difficult to make good microscopic preparations showing cells in clearly recognizable stages of meiosis. The chromosomes in most cells are small, they are often numerous, and it may require study of several slides to find good examples of meiotic stages. For these reasons, meiosis is often studied in introductory biology and zoology courses using a series of carefully selected demonstration microscope slides.

### Microscope Slides

◆ Study the demonstration materials on meiosis provided in the laboratory and **draw** selected stages of meiosis in the space provided in figure 3.10 as directed by your laboratory instructor.

**FIGURE 3.10** Student drawings of selected stages of meiosis.

## Chromosome Simulation Kits

◆ Chromosome simulation kits are another good way to study meiosis. Several types of kits are available, but the most common type consists of a string of beads with a magnet in the middle that represents the centromere of a chromosome. Different colors of beads can be used to help students distinguish different chromatids, to simulate the movements of chromatids and chromosomes during process of meiosis, and to see the effects of crossing over. Ask your instructor if chromosome simulation kits are available.

## Films and Videos

◆ One of the best ways to gain an understanding of the process of meiosis is to study a film or video showing time-lapse sequences of meiosis. Several good films and videos are available that show the dynamic nature of meiosis, and they can be studied both in the laboratory and in libraries or learning centers where available. Ask your instructor what films and videos may be available.

## Internet Sites

◆ Several Internet sites are available to aid students in studying meiosis. If you have access to the Internet, try some of the Internet topics listed at the end of this chapter for additional study aids. Some of them include short film clips of meiosis, good-quality microscopic images of cells in meiosis, and/or interesting simulations to help you understand the process of meiosis.

## Comparison of Mitosis and Meiosis

An excellent way to review and to check your knowledge and understanding of mitosis and meiosis is to consider the similarities and differences between these two processes. A common question appearing on biology examinations asks students to compare and contrast mitosis and meiosis. When asked to compare and contrast, you should always discuss **both** the similarities and differences between the two processes or structures cited in the question. Table 3.1 provides a summary of similarities and differences between mitosis and meiosis.

### TABLE 3.1
#### Comparison of Mitosis and Meiosis

| Description | Mitosis | Meiosis |
|---|---|---|
| Type of cells involved | somatic | reproductive |
| Number of nuclear divisions | 1 | 2 |
| Number of cells produced | 2 | 4 |
| Chromosome number | | |
|     Before | diploid (2n) | diploid (2n) |
|     After | haploid (n) | haploid (n) |
| Synapsis occurs (close pairing of homologs) | no | yes |
| Crossing over occurs | no | yes |
| Genetically identical cells produced | yes | no |

## Key Terms

**Aster** includes all of the aster rays surrounding one pole of the mitotic apparatus in an animal cell. May serve to influence the location of the cleavage plane during cytokinesis. Absent in plant cells.

**Aster ray** one of the fibrils, or rays, making up an aster.

**Centriole** self-replicating tubular organelles usually found in pairs adjacent to the nucleus of an interphase cell surrounded by an amorphous centrosome. Also found centered in the asters of the mitotic apparatus of most animal cells.

**Centromere** a narrow region of a chromosome that binds the two chromatids together prior to separation in meiosis and attaches the spindle fibers to the chromosome by the formation of a kinetochore during mitosis.

**Centrosome** the amorphous area surrounding a centriole in animal cells. Serves as the organizing center for the mitotic spindle.

**Chiasma (plural: chiasmata)** the crossings of chromatids observed in microscopic preparations during synapsis in prophase I of meiosis.

**Chromatid** one strand of a duplicated chromosome.

**Chromatin** genetic material in the interphase nucleus; represents the chromosomes in a long, thin, threadlike form.

**Chromosome** filamentous structure that carries the genetic material of the cell (DNA). Chromosomes can be very long and uncoiled in interphase, shorter with double strands (chromatids) in prophase of mitosis, and still shorter with a single strand in anaphase.

**Cleavage furrow** indentation of the cell membrane around the equator of an animal cell at the beginning of cytokinesis.

**Crossing over** The exchange of segments of chromatids during the separation of homologous chromosomes following metaphase I of meiosis. Such exchanges result from the entanglements of chromatids in prophase I. Crossing over is an important source of genetic diversity in sexually produced offspring.

**Cytokinesis** division of the cytoplasm of a cell.

**Diploid (or 2n)** cells containing both members of each homologous pair of chromosomes.

**Equatorial plane** by analogy with the earth, a plane that passes through the middle of the cell, equidistant from the poles and perpendicular to the line connecting the poles.

**Gametes** the specialized haploid reproductive cells produced by sexually reproducing animals; the ova and spermatozoa (eggs and sperm).

**Gametogenesis** the production of male and female gametes (eggs and sperm) in sexually reproducing animals.

**Germ cells** unspecialized cells found in the gonads of sexually reproducing animals that develop into the gametes (eggs and sperm).

**Gonads** the reproductive organs of sexually reproducing animals, the testes and ovaries, that produce gametes.

**Haploid (or n)** cells containing only one member of each homologous pair of chromosomes.

**Homologous chromosomes** chromosomes with the same size and shape, and carrying genetic material for the same characteristics. One member of each homologous pair comes from each parent.

**Interkinesis** the period intervening between the first and second meiotic divisions. No chromosomes are duplicated and no DNA is synthesized during interkinesis.

**Interphase** the stage between successive nuclear (mitotic) divisions, consisting of $G_1$, S, and $G_2$ phases. It is the stage during which the cells are metabolically active.

**Karyotype** the unique chromosome complement of an individual organism characterized by the number, size, and configuration of the chromosomes.

**Meiosis** a special type of nuclear division in which the chromosome number is reduced from 2n to n by separating the members of the homologous pairs of chromosomes.

**Mitosis** nuclear division resulting in two new nuclei with the same genetic complement as the original nucleus.

**Mitotic apparatus** a special structure formed during mitosis consisting of the spindle fibers, asters, and centrioles.

**Mitotic spindle** all of the spindle fibers collectively.

**Nonreduction division** the second meiotic division. This division follows interkinesis and resembles a mitotic division. There is no reduction in chromosome number.

**Nuclear membrane (nuclear envelope)** the double membrane surrounding the nucleus in an interphase cell.

**Nucleus** a membrane-bound organelle containing the genetic material of a cell and which controls cell metabolism.

**Oogenesis** the formation of ova or eggs. Includes meiosis and the further development into functional ova.

**Ovum (plural: ova)** the haploid female gamete or egg.

**Polar bodies** nonfunctional nuclei produced in the process of oogenesis that migrate to the periphery of the functional ovum. Three polar bodies and one functional ovum are usually formed during oogenesis.

**Poles** opposite ends of a cell where spindle fibers converge during mitosis and meiosis.

**Reduction division** the first meiotic division, during which the number of chromosomes in a cell is reduced by half.

**Spermatogenesis** the formation of the spermatozoa in sexually reproducing animals.

**Spermatozoon (plural: spermatozoa)** the haploid male gamete produced by spermatogenesis. Often shortened to "sperm."

**Spindle fiber** one of the microtubular filaments extending between the poles of the cell in mitosis and meiosis; made of contractile protein.

**Synapsis** the pairing of homologous chromosomes in prophase I of meiosis; the centromeres adhere to each other at this time.

**Tetrad** a group of four chromatids of a pair of homologous chromosomes formed by synapsis during prophase I of meiosis.

## Internet Resources

Visit the zoology website at http://www.mhhe.com/zoology to find live Internet links for each of the references listed below.

1. The Biology Project: Cell Biology. An excellent site with information on mitosis, meiosis, fertilization in Ascaris, and other related topics.
2. Mitosis and Meiosis–Internet Resources. All of the links you could imagine, including many with excellent photographs.
3. Introduction to Cell Biology. Nearly all textual, but much good information on cellular structure and function.
4. Mitosis and Meristems. Really terrific site with cartoon-like diagrams describing cellular parts, and mitosis.

## Critical Thinking Questions

1. Compare the four phases of the cell cycle. Explain how each phase prepares the cell for the next phase.

2. What would happen if you treated a cell actively undergoing mitosis with a chemical that inhibited cytokinesis but did not affect mitosis? What would be produced?

3. Why is it important to use a random sample in an experiment if we wish to compare two populations? Describe two ways you could select a random sample from a population of dogs.

4. Where in your body would you most likely find cells in active mitosis? Where would you find cells undergoing meiosis?

5. Compare and contrast the first and second divisions of meiosis. Why is it important that animals have two meiotic divisions rather than just one division as in mitosis?

6. The process of spermatogenesis usually produces four viable sperm from each parent cell, while the process of oogenesis usually produces only one viable egg. Does this mean that spermatogenesis is a more **important** source of genetic variability than oogenesis? Why or why not?

7. Why is the process of crossing over important to the survival and evolution of a sexually reproducing species?

## Suggested Readings

McIntosh, J.R., and M.P. Koonce. 1989. Mitosis. *Science* 246:622–28.

McIntosh, J.R., and K.L. McDonald. 1989. The mitotic spindle. *Scientific American* 261:48–56.

Murray, A.W., and M.W. Kirshner. 1991. What controls the cell cycle? *Scientific American* 264:56–63.

# NOTES AND SKETCHES

# NOTES AND SKETCHES

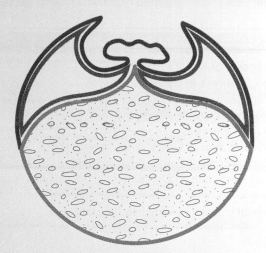

**Two-day embryo**

**Five-day embryo**

**Fourteen-day embryo**

## OBJECTIVES

After completing the laboratory work in this chapter, you should be able to perform the following tasks:

1. List and explain the component processes of development.
2. List and describe the main types of cleavage observed in animal embryos.
3. Distinguish between determinate and indeterminate development.
4. Describe the structure of a spermatozoon of a frog or other representative animal and identify its principal structures visible in a light microscope.
5. List and describe the major events in starfish or sea urchin development from fertilization to the gastrula stage and identify representative stages in microscopic preparations.
6. Identify the principal structures in the blastula and gastrula stages of a starfish or sea urchin.
7. Discuss the organization of a frog egg and tell how it differs from that of a starfish egg.
8. Describe the major events in frog development from fertilization to the tadpole stage and identify representative stages from living or preserved specimens.
9. Discuss the organization of a chick egg and its adaptations for development on land.
10. Identify the four extraembryonic membranes surrounding a chick embryo and explain the function of each.
11. Identify the principal structures seen in whole mounts of 24- and 48-hour chick embryos.

## Introduction

Animal development usually begins with the fertilization of an egg by a sperm. The nuclei of the egg and sperm fuse, and the male and female parents' genes share in determining the characteristics of the offspring. Compared to other biological processes, embryonic development is relatively slow. New cells, tissues, and organs make their appearance in the embryo over a period of hours, days, or weeks.

Embryonic development can be divided into five major phases: (1) **gametogenesis,** the formation of the haploid male and female gametes (sperm and eggs); (2) **fertilization,** activation of the egg and fusion of the sperm and egg nuclei to form the diploid zygote; (3) **cleavage,** the subdivision of the zygote into many cells by mitosis; (4) **gastrulation,** the formation of germ layers; and (5) **organogenesis,** the initiation and differentiation of specific organs.

In this exercise, we shall study examples of development from three different animals to illustrate different aspects of embryonic development and some variations in the development of different kinds of animals. The **starfish** illustrates the earliest stages of development, the **chick** serves to illustrate the later stages of development and adaptation for terrestrial life, and the **frog** is well suited for the study of development from fertilization to hatching.

# Component Processes of Development

Embryonic development consists of a complex series of processes by which a new organism arises from an egg. Following the fertilization of an egg with a sperm, the fertilized egg or **zygote** grows and increases in the complexity of its structure, function, and behavior. Different kinds of organisms exhibit many differences in the details of their development, but there are some important basic similarities in the development of all organisms.

We can identify four major component processes of development in organisms: growth, determination, differentiation, and morphogenesis.

## Growth

**Growth** is the increase in mass of the organism through the addition of new cells and/or an increase in size of existing cells.

## Determination

An unfertilized egg has the potential to form a complete embryo with many kinds of cells, tissues, and organs. As development proceeds, and the egg becomes fertilized and divides to form new cells, individual cells and tissues become progressively restricted in the structures that they are able to form. This progressive limitation of the prospective fate of a cell or tissue is called **determination.** Determination precedes the process of differentiation.

## Differentiation

The progressive increase in the complexity of organization and specialization of individual cells and tissues is called **differentiation.** A fertilized egg of a frog, for example, forms many new cells and tissues during its development. This progressive increase in the number of biochemically and morphologically specialized cells and tissues is called differentiation.

## Morphogenesis

Another important process of development is **morphogenesis.** A living organism is not merely a bag of assorted parts heaped together in a random fashion. The parts of every living organism are arranged in a specific pattern and bear definite relationships to one another. Morphogenesis includes

those **movements** of cells and tissues through which the characteristic structures (both external and internal) of an organism are formed. For example, the movements of cells and masses of cells in an embryo to form a wing or a limb are important processes of morphogenesis. Similarly, the heart of a chick and of a human embryo forms first as a simple tubular structure that twists into the shape of an S before becoming the four-chambered organ we know as the heart.

## Materials List

**Living specimens**
    Frog embryos at various stages
    Frog sperm
**Prepared microscope slides**
    Starfish embryos
    Chick embryo, 24-hour, whole mount
    Chick embryo, 48-hour, whole mount
    Frog ovary with ova, cross section (demonstration)
    Frog testis, cross section (demonstration)
    Selected slides of early frog development
      (demonstration)
**Chemicals**
    Amphibian Ringer's solution

# Gametes

**Gametes** are the mature germ cells, **eggs** and **spermatozoa.** Both living sperm and stained microscope slides should be available for your study of male gametes. Observe the rapid movements of the living sperm in a wet mount.

◆ On a stained microscope slide, identify the anterior **head,** the narrower **middle piece,** and the long posterior **tail.** The sperm of different animal species vary considerably in size and shape, particularly in the shape of the head. Draw a sperm cell in figure 4.1 and label each of these parts.

Eggs, or ova, are much larger than sperm and contain varying amounts of stored food materials for the nourishment of the developing embryo. Since the eggs of the starfish, frog, and chick differ substantially in this regard, they will be described separately as we study the development of each animal.

# Embryonic Cleavage

Shortly after fertilization, the zygote undergoes a series of rapid mitotic cell divisions. This process of dividing the zygote into many cells is called **embryonic cleavage.** The pattern of cleavage differs among various groups of animals, and these patterns have been conserved during the course of evolution. For this reason, scientists have long been interested in cleavage patterns as possible evidence of evolutionary relationships among animals. One of the factors that influences the pattern of cleavage is the amount and distribution of nutrients or **yolk** in the egg.

## Influence of Yolk

Eggs can be classified into four main types based on their yolk contents: (1) **isolecithal eggs,** as found in starfish and humans, which have relatively little yolk and in which the yolk is uniformly distributed throughout the egg; (2) **mesolecithal eggs,** as found in the frogs, toads, and salamanders, which have a moderate amount of yolk, and which have a concentration of yolk in the vegetal (lower) hemisphere; (3) **telolecithal eggs,** as found in birds and reptiles, which have a large amount of yolk, and the cleaving portion of the embryo is restricted to a small disc at one end of the egg; and (4) **centrolecithal eggs,** as found in insects, which have much yolk and in which the actively developing portion of the embryo forms a thin layer around the outside of the large central yolk mass.

Among the animals chosen for this exercise, therefore, we have examples of three egg types:

**Starfish egg**—isolecithal

**Frog egg**—mesolecithal (although some textbooks classify frog eggs as "moderately telolecithal")

**Chick egg**—telolecithal

## Patterns of Cleavage

The cleavage of an embryo can be **complete** or **incomplete,** depending on the relative amount of yolk contained in the egg. Eggs with little or moderate amounts of yolk, such as a starfish or a frog, divide completely. This is **holoblastic cleavage.**

### Spiral Cleavage

Animals with holoblastic cleavage also demonstrate differences in the directions of subsequent cleavage planes. Many flatworms, nematodes, most molluscs, and many annelids exhibit **spiral cleavage** (figure 4.2). The first two cleavage planes are vertical and produce blastomeres of equal size. The third cleavage plane is horizontal and produces four blastomeres of unequal size, four smaller micromeres nearest the **animal pole** on the upper surface, and four larger macromeres nearest the **vegetal pole** on the lower surface of the egg. Prior to the third cleavage the mitotic spindle of the blastomeres rotates 45 degrees, causing the micromeres to lie over the furrows between the macromeres. Following the third cleavage and continuing after several subsequent cleavage divisions, the mitotic spindles rotate 90 degrees in the opposite direction. This produces several layers of blastomeres with each layer lying over the furrows between the blastomeres just below.

### Radial Cleavage

Starfish, sea urchins, and frogs (also sponges, cnidarians, and cephalochordates) exhibit **radial cleavage** (figure 4.3). In this type of cleavage, the first two cleavage planes are horizontal and produce blastomeres of equal size. The third plane is horizontal and produces four smaller micromeres and four larger macromeres. The four micromeres contain less yolk and surround the animal pole, which remains uppermost; the four macromeres contain more yolk and surround the

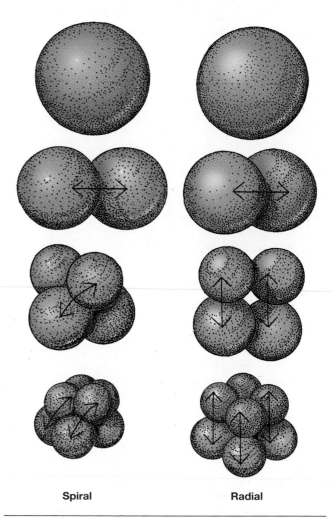

**Spiral**          **Radial**

**FIGURE 4.1** Student drawing of sperm.

**FIGURE 4.2** Spiral and radial cleavage.

vegetal pole opposite the animal pole. The blastomeres remain radially arranged around the animal-vegetal axis, and there is no rotation of the blastomeres as in spiral cleavage.

### Other Types of Cleavage

Mammals (including humans) also have holoblastic cleavage, but their cleavage pattern is **rotational** (figure 4.3). After the second cleavage, one pair of blastomeres comes to lie at right angles to the other. Squid, octopi, and other cephalopod molluscs have a **bilateral cleavage** (figure 4.3) pattern in which the first cleavage plane divides the embryo into right and left halves and establishes a plane of bilateral symmetry.

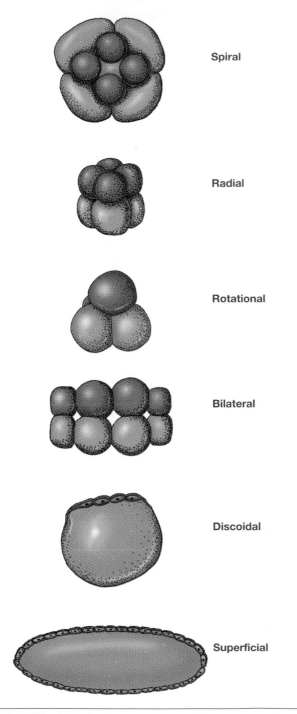

**FIGURE 4.3** Cleavage patterns in different types of eggs.

In other animals with abundant yolk in the eggs, only part of the embryo is divided, and much of the yolky part of the embryo remains undivided. One example is the telolecithal eggs of birds, which have **discoidal cleavage** (figure 4.3) with cleavage restricted to a small disc of cells at one end of the embryo. Another example is found among the arthropods in which yolk is concentrated in the center of the egg and cleavage is restricted to the outer surface. This is known as **superficial cleavage** (figure 4.3).

## Determinate and Indeterminate Development

Determination of the prospective fate of embryonic cells occurs at different times in various types of animals. In certain animals determination occurs as early as the first cleavage division. Experimental separation of these first two blastomeres leads to the formation of incomplete embryos. The separated blastomeres only develop into the structures that they would normally form in an intact embryo. They do not compensate for the missing parts. Each blastomere develops only into those structures it would form *had the other blastomere been in place*. This is an example of **determinate development** (also called mosaic development) in which the prospective fate of the embryonic cells is fixed during early cleavage. Determinate development is typical of protostomes including flatworms, nematodes, annelids, and other animals often with spiral cleavage.

In other animals, determination occurs much later in development. Experimental separation of blastomeres from early cleavage stages leads to complete but smaller embryos. After separation, the blastomeres still have the ability to compensate for the missing parts and form a complete but miniature embryo. This is an example of **indeterminate development** (also called regulative development) and is characteristic of the deuterostomes, including the echinoderms and chordates.

## Starfish Embryology

Starfish embryos are often used for introductory studies of development because they illustrate clearly the basic pattern of early development of multicellular animals (figure 4.4). Sea urchin embryos are also frequently used, and their development is very similar to that of starfish in these early stages. Sea urchins also have the added advantage that living eggs and embryos are relatively easy to obtain for laboratory study. With living sea urchin embryos, it is possible to observe most of the early developmental events described later with an ordinary compound microscope.

◆ Obtain a whole-mount microscope slide with stained starfish embryos at various stages of development. Examine the slide first under the low power of your microscope and select good representatives of the main stages described in the following section.

Color transparencies are excellent aids for the study of early embryology. They can be viewed easily in a small slide

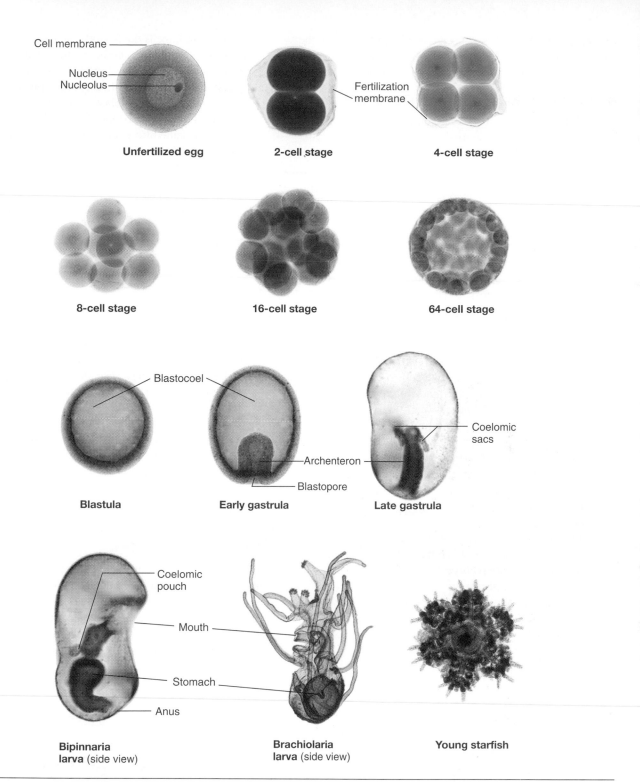

**FIGURE 4.4** Starfish development.
Courtesy of Carolina Biological Supply Company, Burlington, NC.

viewer or with the aid of a projector. Be sure to study the projection slides and any other demonstration materials available to assist in your understanding of starfish development.

## Summary of Early Starfish Development

1. **Unfertilized egg.** Observe the large **nucleus** and the darkly staining **nucleolus.** A large food reserve, which provides energy for early development of the starfish embryo, is present as yolk in the cytoplasm. Surrounding the egg, or ovum, is a **cell membrane.**

    Prior to fertilization, the nuclear membrane breaks down, and the contents of the nucleus mingle with the cytoplasm. After penetration of the cell membrane by a spermatozoan, the haploid sperm nucleus migrates through the cytoplasm and fuses with the haploid egg nucleus. Also following fertilization, a new membrane is formed around the egg. This new membrane rises slightly above the surface of the egg and can often be observed in microscopic preparations. It is called the fertilization membrane.

2. **Two-cell stage.** Shortly after fertilization, the fertilized egg divides into two cells by **mitosis.** This is the first of a number of rapid cell divisions that take place during the next several hours. This series of cell divisions is called **embryonic cleavage.**

3. **Four-cell stage.** Observe the fertilization membrane enclosing the four **blastomeres** (embryonic cells).

4. **Eight-cell stage.** *What is the orientation of the cleavage plane that produced this stage relative to the last cleavage plane?*

5. **Sixteen-cell stage.**

6. **Thirty-two-cell stage.** Note the rapidly diminishing size of the blastomeres. *What has happened to the overall size of the embryo?* It is called the fertilization membrane.

7. **Morula.** Solid mass of cells formed after many cleavages. No central cavity.

8. **Blastula.** Several more cleavage divisions occur, leading to the formation of a hollow ball of ciliated cells. This stage is the blastula, and after escaping from its enveloping membrane, the ciliated blastula swims about freely. The central cavity is the **blastocoel.**

9. **Early gastrula** (figure 4.5a). A few hours later, the cells at one end of the blastula begin to push (or be pulled) into the blastocoel. This infolding of the embryo is called **invagination** and results in the formation of a two-layered structure. These two layers are **primary germ layers,** and similar layers are formed by virtually all multicellular animals. (Sponges are one of the few peculiar exceptions to this rule.) The layers are an outer **ectoderm** layer, which forms the skin and the nervous system of the starfish, and an inner **endoderm** layer, which primarily forms the lining of the digestive tract of the adult starfish (and a few accessory organs of the digestive tract).

10. **Midgastrula** (figure 4.5b and c). Invagination continues, and a hollow tube of endoderm extends into the blastocoel. The opening in the center of the endodermal tube is the **archenteron** (primitive gut) of the starfish embryo. The external opening of the archenteron is the **blastopore** (future anus of the starfish).

11. **Late gastrula.** The endodermal tube later expands near its inner end and forms two lateral pouches. These lateral pouches become part of the internal body cavity or **coelom** of the starfish. The walls of these lateral pouches become differentiated and form the third primary germ layer, the **mesoderm.**

## Frog Development

Ripe, unfertilized eggs of the grass frog, *Rana pipiens,* are relatively large (averaging about 1.75 mm in diameter) and have a moderate amount of yolk (mesolecithal type of egg). Remember that the frog is a semiaquatic animal; its eggs must be deposited in water, and development of the embryos and tadpole larvae can take place only in an aquatic medium.

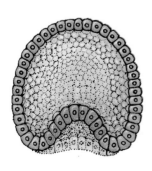

(a) **Early gastrula**
(longitudinal section)

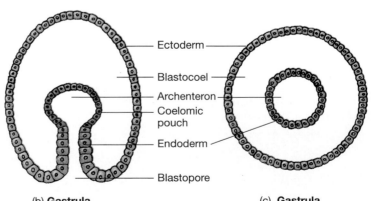

Ectoderm
Blastocoel
Archenteron
Coelomic pouch
Endoderm
Blastopore

(b) **Gastrula**
(longitudinal section)

(c) **Gastrula**
(cross section)

**FIGURE 4.5** Starfish gastrula.

◆ Obtain a few unfertilized frog eggs in water and study them under your stereomicroscope (figure 4.6). Observe that the eggs are darkly pigmented. *How is the pigment distributed on the surface of the eggs? Is the distribution of pigment uniform or nonuniform?*

The center of the dark portion of the surface of the egg is called the **animal pole.** The center of the lighter surface (opposite the animal pole) is called the **vegetal pole.** The line connecting these two poles is the **animal-vegetal axis.** Shortly after fertilization the egg rotates so that the animal pole is uppermost. *How does this differ from the orientation of the animal-vegetal axis of the unfertilized eggs in your sample?*

Closely applied to the surface of the unfertilized egg is a transparent **vitelline membrane,** which lifts from the surface of the egg following fertilization and thereafter is known as the **fertilization membrane.** Surrounding the vitelline membrane are three **jelly coats** made up largely of albumen, a protein secreted by the oviduct during the passage of the egg from the ovary (figure 4.6). The jelly coats serve to protect the egg from injury and infection.

Fertilization is external in the frog. During mating, the male frog grasps the female with his forelimbs, and as the female discharges her ripe eggs into the water, the male releases sperm over them. This mating posture is called **amplexus** (figure 4.7).

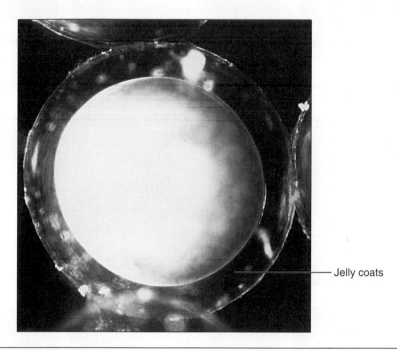

Jelly coats

**FIGURE 4.6** Unfertilized frog egg showing jelly coats.
Courtesy of Carolina Biological Supply Company, Burlington, NC.

**FIGURE 4.7** Male and female frogs in mating position.
Courtesy of Carolina Biological Supply Company, Burlington, NC.

Shortly after fertilization, the vitelline membrane rises to form the fertilization membrane, and later a **gray crescent** area appears on one side of the egg along the margin of the heavily pigmented zone. The gray crescent is a more lightly pigmented zone of the egg cortex and appears on the side opposite the point of sperm entrance. The center of the gray crescent area marks the future **posterior end** of the embryo; thus, the future **anterior-posterior axis** of the frog embryo is determined at the time of sperm entry into the egg.

Development is fairly rapid following fertilization (figures 4.8 and 4.9), and within a few hours at normal temperatures, the fertilized egg becomes divided into many cells, a hollow

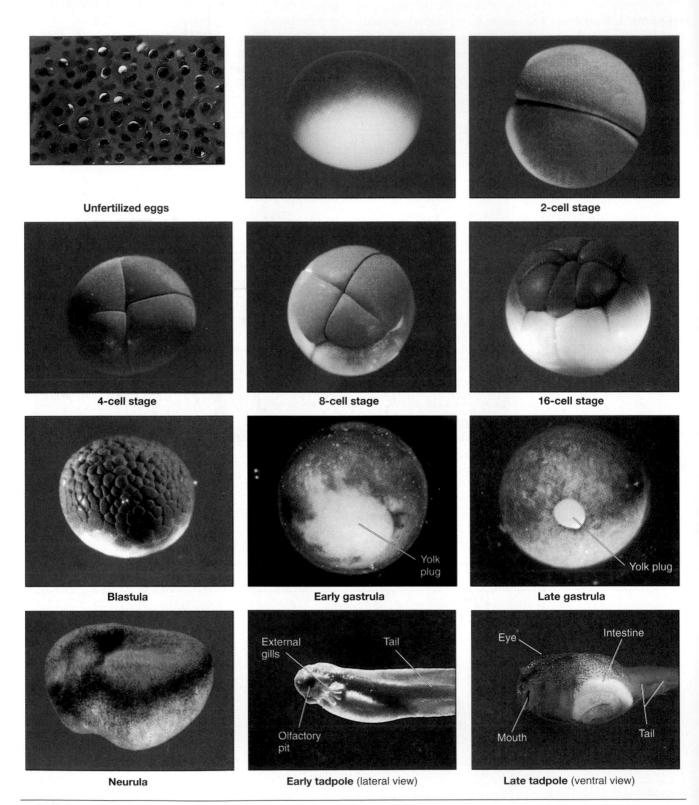

**FIGURE 4.8** Frog development.
Courtesy of Carolina Biological Supply Company, Burlington, NC.

blastula is formed, and gastrulation is completed. Cleavage is **holoblastic** (all of the egg cytoplasm becomes divided) and **unequal** (the cells nearest the animal pole are smallest in size and those nearest the vegetal pole are largest in size). *How is this related to the distribution of yolk in the frog egg?*

A few hours after gastrulation, the first visible elements of the nervous system appear, first with the **neural plate** followed by the elevation of the **neural folds** on each side of the neural groove to form the **neural tube** within which the dorsal nerve cord begins to differentiate.

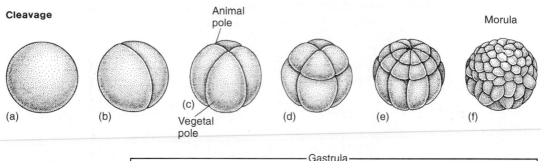

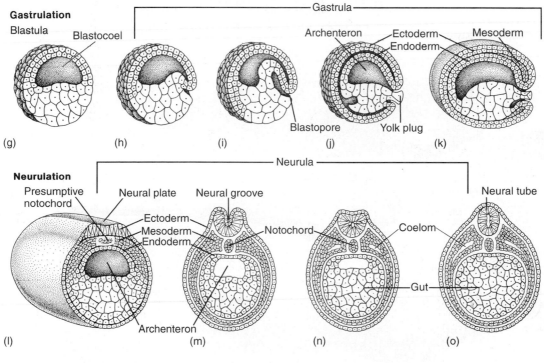

**Cleavage**

Animal pole

Morula

(a)  (b)  (c)  (d)  (e)  (f)

Vegetal pole

**Gastrulation**

Gastrula

Blastula  Blastocoel  Archenteron  Ectoderm  Mesoderm
Endoderm

Blastopore  Yolk plug

(g)  (h)  (i)  (j)  (k)

**Neurulation**

Neurula

Presumptive notochord  Neural plate  Neural groove  Neural tube

Ectoderm  Notochord  Coelom
Mesoderm
Endoderm

Gut

Archenteron

(l)  (m)  (n)  (o)

(a)

**FIGURE 4.9** Frog development.

*(continued on next page)*

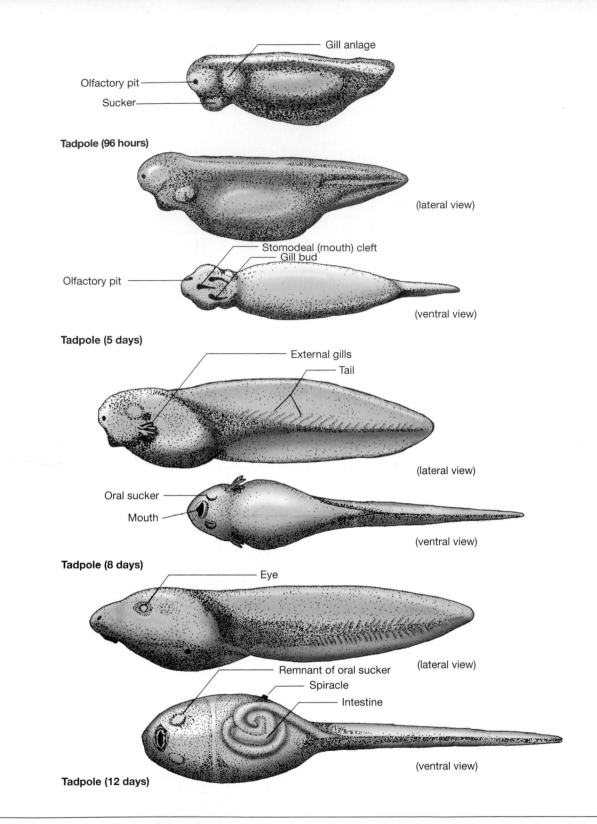

Gill anlage

Olfactory pit

Sucker

**Tadpole (96 hours)**

(lateral view)

Stomodeal (mouth) cleft
Gill bud

Olfactory pit

(ventral view)

**Tadpole (5 days)**

External gills

Tail

(lateral view)

Oral sucker

Mouth

(ventral view)

**Tadpole (8 days)**

Eye

(lateral view)

Remnant of oral sucker

Spiracle

Intestine

(ventral view)

**Tadpole (12 days)**

(b)

**FIGURE 4.9** *(continued)* Frog development.

Following the neural tube stage, the posterior portion of the embryo elongates to begin formation of the embryonic tail. This is the **tail bud stage,** which occurs at about 84 hours postfertilization at 18°C (figure 4.10).

Table 4.1 illustrates the schedule of development in *Rana pipiens* at 18°C. Other common species of frogs, toads, and salamanders follow a generally similar pattern of development but differ in the times required to reach the various stages. Several representative stages of development in *Rana pipiens* are shown in figure 4.8.

◆ Study the frog embryos provided in your laboratory and compare the various stages with figures 4.8 and 4.9. Living embryos are best for study if they are available, although plastic-embedded and preserved embryos can also be used. Supplement your study of the living or preserved whole embryos by observing the microscopic demonstrations provided to illustrate the internal structure of the embryos at different stages. When you have completed your study, you should be able to describe and explain the major events in the development of a frog from an unfertilized egg to an adult.

## Demonstrations

1. Microscope slide showing developing eggs in the uterus
2. Microscope slide showing frog testis with developing sperm
3. Wet mount of living frog sperm in 10% amphibian Ringer's solution
4. Microscopic slides with cross sections and/or sagittal sections of selected early developmental stages of the frog through hatching
5. Procedures for inducing ovulation in the frog by injection of pituitary extract and artificial insemination of frog eggs in vitro

# Chick Development

The eggs and embryos of birds exhibit several major differences from those of amphibians. Birds have eggs adapted for development on land. This type of egg is called an **amniotic**

## TABLE 4.1

### Schedule of Development in *Rana pipiens* at 18°C

| Stage | Hours |
|---|---|
| Fertilization | 0 |
| Formation of gray crescent | 1 |
| Rotation (animal pole now uppermost) | 1.5 |
| First cleavage (2 cells) | 3.5 |
| Second cleavage (4 cells) | 4.5 |
| Third cleavage (8 cells) | 5.5 |
| Blastula | 18 |
| Early gastrula (dorsal lip stage) | 26 |
| Midgastrula | 34 |
| Late gastrula (yolk plug stage) | 42 |
| Neural plate | 50 |
| Neural folds | 62 |
| Ciliary movement | 67 |
| Neural tube formation | 72 |
| Tail bud stage | 84 |
| Muscular contractions | 96 |
| Heartbeat | 5 days |
| Gill circulation, hatching (ruptures fertilization membrane) | 6 days |
| Circulation in tail fin | 8 days |
| Internal gills formed, opercular fold present | 9 days |
| Operculum closed, covers internal gills | 12 days |
| Metamorphosis into frog | 3 months |

After Shumway and Rugh.

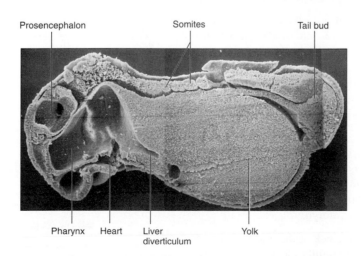

**FIGURE 4.10** Frog development. Sagittal section of the tail bud stage of a developing frog embryo.
Scanning electron micrograph of a freeze fracture preparation by Louis de Vos.

(or cleidoic) **egg** because the embryo is enclosed in special extraembryonic membranes that protect and support the development of the embryo in a liquid environment. Other important adaptations of the eggs of birds (and reptiles) that allow them to develop on land are the inclusion of a large food supply and a hard outer shell. These features along with the extraembryonic membranes ensure an optimal environment during embryonic development.

## Extraembryonic Membranes

Four **extraembryonic membranes** are formed by a developing chick embryo: yolk sac, chorion, amnion, and allantois. Each of these membranes forms in a specific way, and each performs a distinctive role in protecting the embryo (figure 4.11).

The **yolk sac** forms as a pouchlike outgrowth from the developing gut. It grows around the yolk, releases enzymes to digest the yolk, and transports the digested yolk products through its blood vessels to the developing embryo.

The **chorion** and **amnion** are sheets of living tissue that grow out of and around the embryo. These sheets ultimately join above the embryo and enclose the embryo in a double sac. Both the chorion and the amnion consist of two tissue layers, ectoderm and mesoderm. After closure of the amnionic folds, the amnion becomes filled with a watery fluid. Thus, the amnion maintains an aqueous environment to protect the growing embryo from desiccation and serves as a physical protection for it.

The **allantois,** like the yolk sac, arises as a saclike outgrowth from the ventral surface of the gut. It has quite a different function, however. In birds, the allantois collects and stores metabolic waste products (largely crystals of uric acid). The allantois also grows and fuses with the chorion to form the **chorioallantoic membrane.** The chorioallantoic membrane is highly vascularized and facilitates the exchange of gases between the embryo and the external environment.

## Whole Mount of 24-Hour Chick Embryo

◆ Study first a whole mount of a 24-hour chick embryo. Consult figures 4.12 and 4.13, and study the wholemount slide under your stereomicroscope.

> **Caution:** Whole mounts of embryonic stages are relatively thick and should never be viewed under high power on your compound microscope.

◆ Note the approximate size and shape of the embryo. Observe the transparent area immediately surrounding the embryo and an opaque outer area. The inner portion of this opaque area contains many blood vessels that bring nutrients to the developing embryo.

◆ Locate the **headfold** of the chick at the anterior end of the embryo, lying free and slightly elevated above the underlying membrane. Within the head, observe that the brain at this stage consists of a **neural tube,** which is

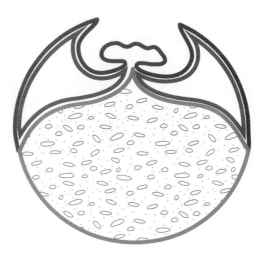

**Two-day embryo**

**Five-day embryo**

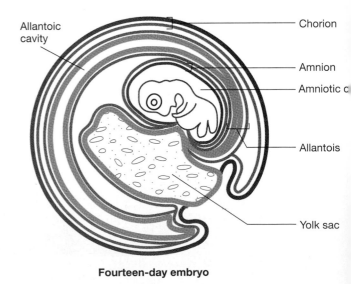

**Fourteen-day embryo**

**FIGURE 4.11** Chick development, formation of extraembryonic membranes. Ectoderm is blue, endoderm is yellow, and mesoderm is red.

continued posteriorly with the open **neural plate** consisting of a median **neural groove** and two lateral **neural folds.** Note also that the tissue of the neural folds is continuous laterally with the surface ectoderm, thus demonstrating the ectodermal origin of the nervous tissue.

Posterior to the neural plate is the **primitive streak,** the site of invagination of the surface ectoderm and the formation of the mesoderm layer. The primitive streak, thus, is functionally similar to or analogous to the blastopore of the frog and other amphibian embryos.

◆ Find the **notochord,** which lies beneath the neural tube and the neural groove. Note that anteriorly, the notochord appears as a well-defined rod, but more posteriorly it appears as a wide band of less dense tissue.

Differentiation of the notochord, as of the neural structures, proceeds from anterior to posterior. Likewise, the **somites,** blocks of developing muscle tissue, differentiate from anterior to posterior. New somites are formed by the coalescence and differentiation of mesoderm cells behind previously formed somites.

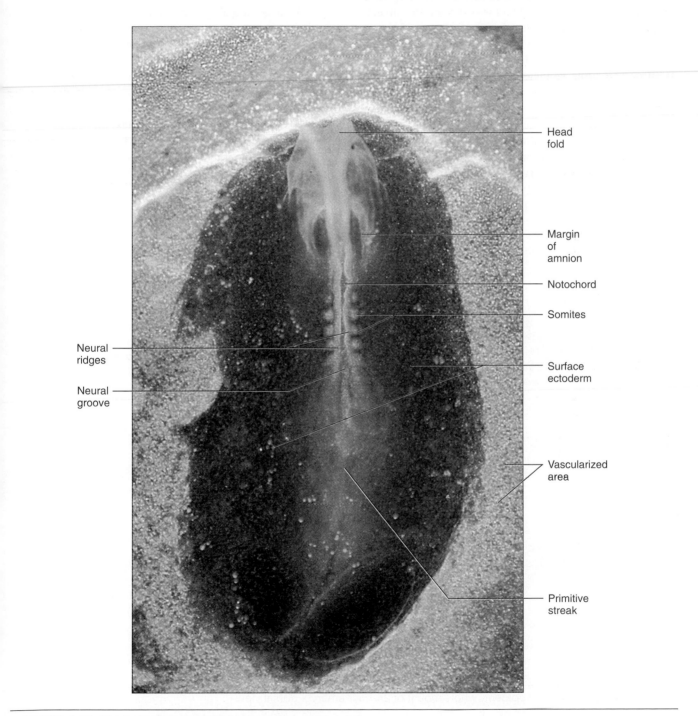

**FIGURE 4.12** Chick development, 24-hour embryo, dorsal view.
Courtesy of Carolina Biological Supply Company, Burlington, NC.

## Whole Mount of 48-Hour Chick Embryo

◆ Study a whole mount of a 48-hour chick embryo and observe some of its major features (figures 4.14 and 4.15). Note that the **head** and the **heart** are much more well-developed than at 24 hours. Observe that the **eyes** are clearly differentiated and that they exhibit a developing lens.

The brain is enlarged and is now divided into five divisions: the anterior **telencephalon** followed by the **diencephalon,** the **mesencephalon,** the **metencephalon,** and the **myelencephalon.** On the sides of the myelencephalon are a pair of **otic vesicles** that later form the ears.

The heart in a 48-hour embryo is a twisted tube with two well-established chambers: a **ventricle** and an **atrium.** The **sinus venosus** leading to the atrium is also beginning to develop. By 48 hours of development, the heart is actively pumping, and circulation of the blood within the embryo and the surrounding vitelline area is well established. The heart connects with three pairs of large arteries, which represent the first three **aortic arches.** A large **dorsal aorta** extends posteriorly and connects laterally with two large **vitelline arteries.** *Why is the circulation of blood in the vitelline area important to the development of the chick?*

The head and the anterior part of the embryo are now covered by the **amnion,** which continues to grow posteriorly. The number of **somites** has increased to 24 pairs, which lie along each side of the neural tube. At the posterior end of the embryo, the caudal fold of the amnion has begun to grow forward to enclose the posterior portion of the embryo. Much later, the anterior and posterior portions of the amnion grow together and completely envelop the embryo.

The head of a chick embryo at 48 hours is turned to its right side. Beginning at about 38 hours of incubation, a series of movements of the embryo are initiated that change the orientation of the developing embryo to the underlying yolk mass. These movements of the embryo consist of two types, flexion and torsion. Flexion involves bending of the body along its longitudinal axis and starts at the anterior end with a bending forward of the head. **Flexion** (bending) of the body continues slowly from anterior to posterior so that in later stages the longitudinal axis of the body is bent into the shape of a C.

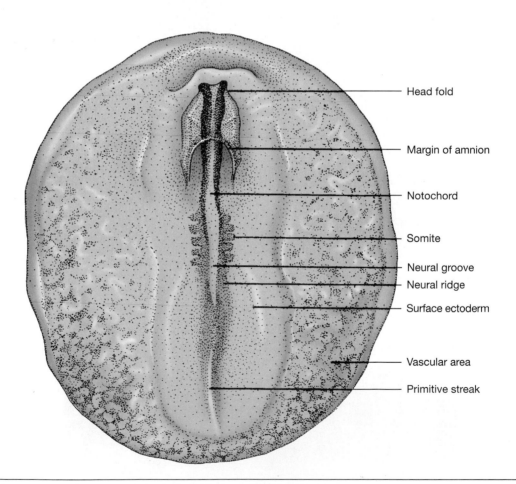

**FIGURE 4.13** Chick development, whole mount of 24-hour embryo, dorsal view.

— Head fold

— Margin of amnion

— Notochord

— Somite

— Neural groove

— Neural ridge

— Surface ectoderm

— Vascular area

— Primitive streak

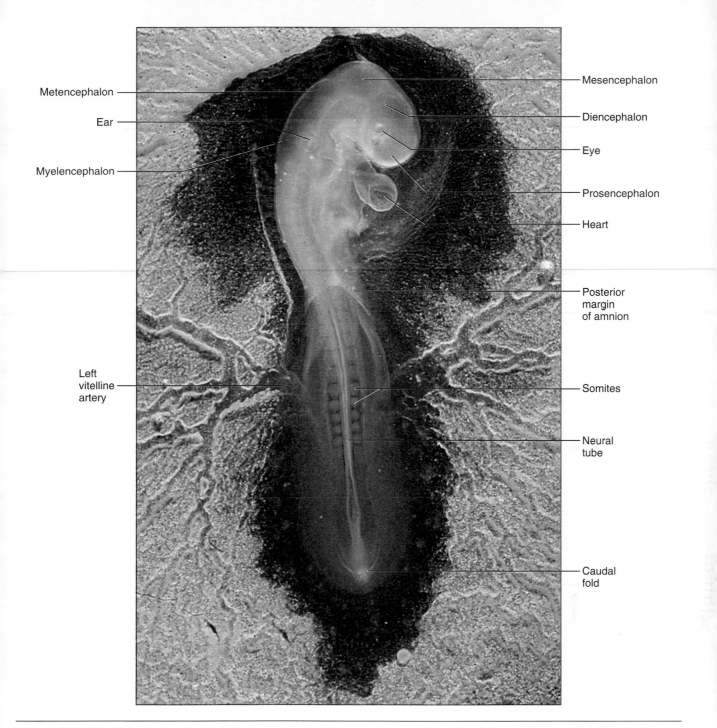

Metencephalon

Ear

Myelencephalon

Left vitelline artery

Mesencephalon

Diencephalon

Eye

Prosencephalon

Heart

Posterior margin of amnion

Somites

Neural tube

Caudal fold

**FIGURE 4.14** Chick development, 48-hour embryo.
Courtesy of Carolina Biological Supply Company, Burlington, NC.

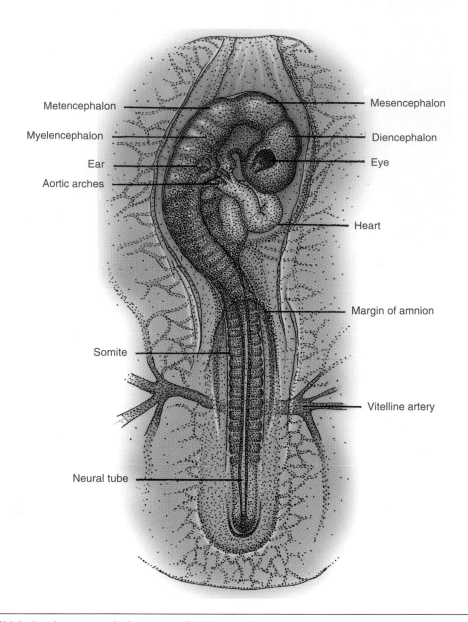

Metencephalon

Myelencephalon

Ear

Aortic arches

Mesencephalon

Diencephalon

Eye

Heart

Margin of amnion

Somite

Vitelline artery

Neural tube

**FIGURE 4.15** Chick development, whole mount of 48-hour embryo.

Remember, however, that the embryo consists of a small discoidal mass of tissue lying on top of a large mass of yolk (figure 4.16). Therefore, the flexion of the body is accompanied by **torsion,** a twisting of the longitudinal axis, which starts with the twisting of the head toward its right side (figure 4.14). Torsion continues from anterior to posterior, and by 96 hours (four days) of incubation, the entire embryo lies with its anterior surface directed to the right and with its left side adjacent to the yolk.

Flexion of the longitudinal (spinal) axis of the developing embryo, which bends the embryo into a C shape, is not unique to bird embryos, but is also characteristic of the embryos of reptiles and mammals. The well-known fetal position of human embryos in the later stages of development is the result of such flexions.

## Whole Mount of 72-Hour Chick Embryo

A 72-hour chick embryo shows further enlargement of the head, heart, and other anterior structures (figures 4.17 and 4.18). The eyes are larger and more fully developed, the **otic vesicle** (ear) is enlarged, and the beginnings of the nostrils, the **nasal pits,** are formed on the ventral surface of the telencephalon. Several **nerve ganglia** can be observed adjacent to the spinal cord, and four **gill arches** with corresponding **gill clefts** are present. In addition to the atrium, ventricle, and sinus venosus, the heart has developed a muscular **truncus arteriosus** that carries blood to the aortic arches. Later, partitions develop inside the heart to divide the single ventricle and the single atrium into the typical **four-chambered heart** of adult birds and mammals. Two large **vitelline veins** can be observed adjacent to the **vitelline arteries.** *What is the function of these veins?*

Somites now number 36, and the torsion (twisting) of the body has progressed posteriorly beyond the level of the heart. Small masses of tissue form the **limb buds** for the wings adjacent to somites 15 to 20 and for the legs adjacent to somites 27 to 32.

The fourth extraembryonic membrane, the **allantois,** appears in a 72-hour chick as a sac extending from the ventral surface of the hindgut near the tail bud. The allantois continues to grow, and later, parts of the allantois fuse with the chorion to form the **chorioallantoic membrane** that

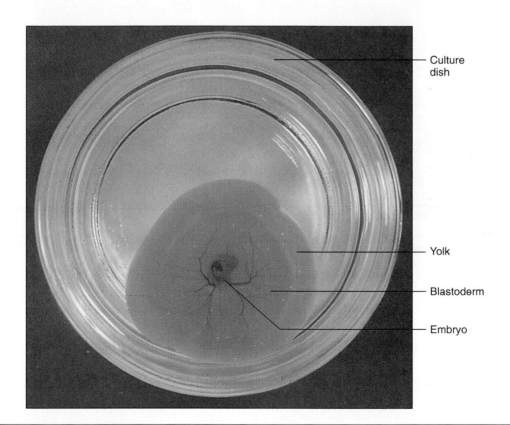

Culture dish

Yolk

Blastoderm

Embryo

**FIGURE 4.16** Chick development, living chick embryo removed from shell and placed in a culture dish with saline solution.
Courtesy of Carolina Biological Supply Company, Burlington, NC.

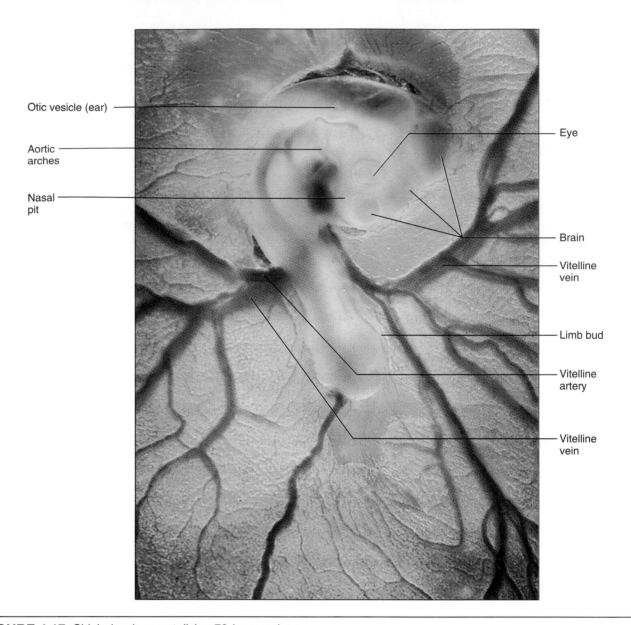

Otic vesicle (ear)

Aortic arches

Nasal pit

Eye

Brain

Vitelline vein

Limb bud

Vitelline artery

Vitelline vein

**FIGURE 4.17** Chick development, living 72-hour embryo.
Courtesy of Carolina Biological Supply Company, Burlington, NC.

lines the inside of the shell and plays an important role in gas exchange for the embryo. The allantois also serves as a reservoir for the deposit of waste products formed by the metabolism of the developing embryo.

## Later Stages of Chick Development

Most of the major organs of the body have been established by 72 hours. Later development involves continued growth of the embryo and further differentiation of various organs (figures 4.19 and 4.20). The brain continues to enlarge and to develop several accessory structures. The spinal cord becomes enclosed along its entire length, and many pairs of spinal nerves form along the spinal cord. Later, these spinal nerves grow out and connect with the developing peripheral nervous system.

The digestive system forms specialized regions, and outgrowths from certain parts of the gut form the thyroid gland, the lungs, the liver, and the pancreas. At the anterior end of the tubular gut, a new opening forms to become the mouth. At the posterior end, the gut forms an opening into a common chamber with the excretory and reproductive system, the **cloaca.** In the middle region, the yolk sac is connected with the ventral part of the gut, allowing yolk to pass into the gut for nourishment of the growing embryo. Later, the yolk is used up, and the ventral wall of the gut closes.

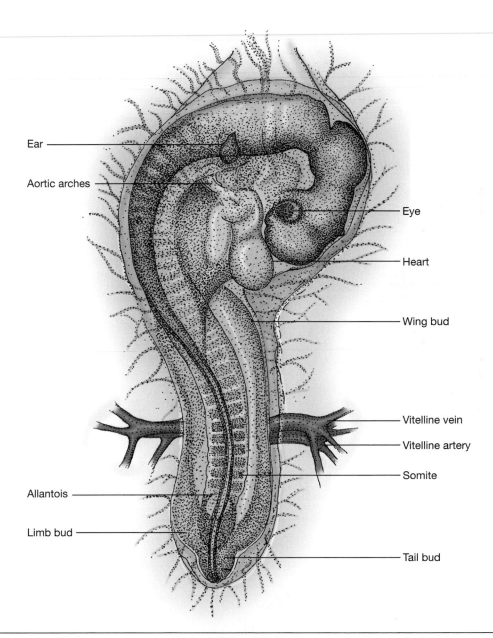

**FIGURE 4.18** Chick development, whole mount of 72-hour embryo.

## Living Chick Embryos (Optional Exercise)

Living chick embryos are easily obtained and make excellent material for laboratory study of development. During its early stages of development, the chick embryo occupies a relatively small portion of the inside of the eggshell. Most of the space is taken up by the stored food materials, the yolk and the white of the egg. The portion of the egg that will become the chicken is composed at first of a small disc of rapidly dividing cells (called the **blastoderm**), which later undergoes a process of gastrulation, and develops a head, nervous system, circulatory system, and other organ systems (figure 4.16).

Stages from two to four days (48–96 hours) are most satisfactory for study in an introductory zoology laboratory (figures 4.16 to 4.20). Obtain some fertilized eggs that have been incubated for two to four days and study the principal features of the embryo at each of these stages.

### Directions for Opening Eggs

1. Put a penciled X on the egg to mark the side that was on top in the incubator tray. This step is important because the blastoderm rotates to the top of the egg during incubation. Otherwise, you may have difficulty finding the blastoderm after you crack the egg.
2. Fill a culture dish half full of warm (37°C) chick Ringer's solution. Ringer's solution contains 9.0 g NaCl, 0.4 g KCl, 0.24 g CaCl, and 0.2 g NaHCO$_3$ in a liter of distilled water.
3. Gently crack the eggshell on the side opposite the location of the embryo (i.e., opposite the X) by striking it carefully against the edge of the finger bowl.

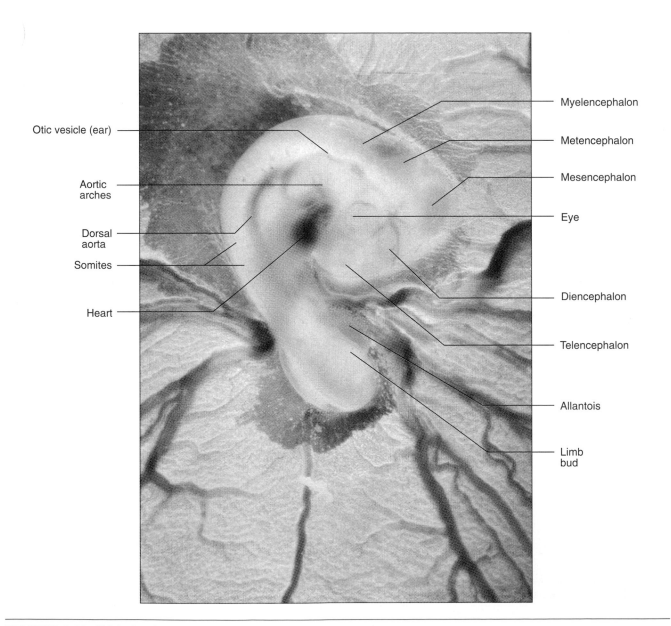

**FIGURE 4.19** Chick development, living 96-hour embryo.
Courtesy of Carolina Biological Supply Company, Burlington, NC.

4. With your thumbs placed over the X, lower the egg and your fingers into the solution and carefully pry open the cracked surface with your fingers, using your thumbs as a pivot.

5. **Caution:** If you open the shell too slowly, the sharp edges of the shell may sever the delicate membranes of the embryo as it slides out of the shell. Likewise, if you open the shell too quickly, you can also damage these membranes.

6. If there is no embryo on the surface of the yolk, try another egg. Be sure to look on the sides and bottom of the egg also, in case the egg rotated as you opened the shell. A plastic spoon is a good instrument to aid in rotating the egg.

◆ Study the embryo and identify as many structures of the embryo as possible using figures 4.16 to 4.20 as a guide.

Repeat your study with other stages of development as available. Make notes on your observations in the Notes and Sketches section at the end of the chapter. Make a list of the most obvious features of each stage that you study.

**Caution:** Be sure to dispose of all eggs, eggshells, and other waste at the end of your study as directed by your instructor. Also, remember to wash your glassware when you have finished with it.

### Demonstrations

1. Living chick embryos after 33, 56, 72, and 96 hours of incubation (in chick Ringer's solution)
2. Microscope slides with selected cross sections of 33-hour chick embryos
3. Microscope slides with whole mounts of 72- and 96-hour chick embryos

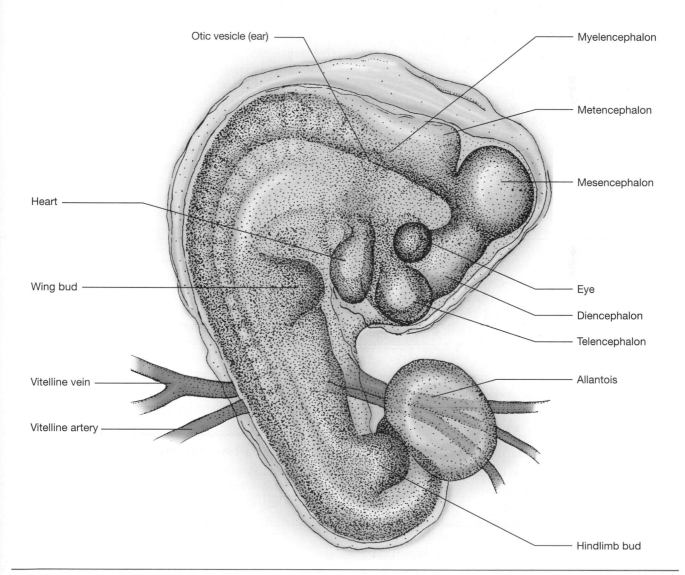

**FIGURE 4.20** Chick development, whole mount of 96-hour embryo.

## Key Terms

**Allantois**   one of the extraembryonic membranes found in the chick; forms as an outgrowth of the gut. Functions in gas exchange and for the storage of metabolic wastes.

**Amnion**   an extraembryonic membrane found in the chick; consists of layers of ectoderm and mesoderm. Encloses the developing embryo and provides a fluid environment to protect the embryo.

**Archenteron**   the primitive gut of an embryo; lined by endoderm tissue.

**Blastopore**   the opening in the gastrula through which the ectodermal cells invaginate. Often becomes the mouth or the anus of the adult, depending on the type of animal.

**Blastula**   hollow-ball stage in early embryonic development following the morula stage and preceding the gastrula stage.

**Centrolecithal egg**   type of egg with abundant, centrally located yolk, such as an insect egg.

**Chorion**   an extraembryonic membrane composed of ectoderm and mesoderm layers; protects the developing embryo.

**Cleavage**   the period of rapid cell divisions in an embryo following fertilization; leads to the formation of the morula and blastula stages.

**Determinate development**   type of development in which the fate of embryonic cells is fixed during early cleavage; also called mosaic development.

**Determination**   the progressive limitation of the developmental potential of an embryo, tissue, or cell.

**Differentiation**   formation of a specialized tissue or cell type from a simpler tissue or cell type.

**Ectoderm**   primary germ layer found on the exterior of an early embryo; forms the nervous system, integument, and certain other tissues in the adult.

**Endoderm**   primary germ layer lining the gut of an early embryo; forms the lining of the digestive tract, liver, pancreas, thyroid, and certain other organs in the adult.

**Extraembryonic membranes**   membranes that form external to the embryo of the chick and most other terrestrial vertebrate animals; serve to enclose and protect the developing embryo. Includes the chorion, amnion, allantois, and yolk sac in the chick.

**Fertilization**   the fusion of the male and female nuclei, which initiates embryonic development. The sperm cell must first penetrate the cell membrane of the egg and migrate to the egg nucleus.

**Gametes**   specialized sex cells, eggs and sperm, necessary for sexual reproduction.

**Gametogenesis**   the formation of eggs and sperm; involves meiosis and cellular differentiation of the male and female sex cells to form gametes.

**Gastrula**   developmental stage following the blastula. Invagination or inward movement of cells at this stage leads to differentiation of the three primary germ layers: endoderm, ectoderm, and mesoderm.

**Gastrulation**   formation of the gastrula.

**Growth**   increase in size or mass of an embryo or individual.

**Indeterminate development**   type of development in which the fate of embryonic cells is not fixed early in development; also called regulative development.

**Invagination**   infolding of cells through the blastopore during gastrula formation; leads to the differentiation of endoderm and mesoderm tissues.

**Isolecithal egg**   type of egg with a small amount of yolk, as in a starfish egg.

**Mesoderm**   one of the primary germ layers; formed from invaginated ectoderm cells or as outpockets of the gut. Develops into the muscles, bone, connective tissues, circulatory system, and many other structures in the adult.

**Mesolecithal egg**   type of egg with a moderate amount of yolk, as in the frog egg.

**Morphogenesis**   the molding of a structure during embryonic development by cell and tissue movements.

**Morula**   stage formed during the latter part of embryonic cleavage; consists of a solid ball of cells.

**Notochord**   cartilaginous supporting rod parallel to the dorsal nerve cord of all chordates. May be replaced by a vertebral column in later stages of development.

**Organogenesis**   the formation of a specific organ, such as the heart, during embryonic development.

**Primary germ layer**   one of the three tissue layers differentiated early in development—ectoderm, mesoderm, and endoderm.

**Radial cleavage**   pattern of cleavage in which the blastomeres are arranged radially around the central animal-vegetal axis; characteristic of deuterostomes.

**Somite**   a block of mesodermal tissue along the nerve cord in an embryo; differentiates into segmental muscles and other mesodermal tissues during development.

**Spiral cleavage**   type of cleavage in which the blastomeres rotate in a spiral fashion after the third and several subsequent cleavage divisions because of shifts in the axis of the mitotic spindle; characteristic of protostomes.

**Telolecithal egg**   type of egg with a large amount of yolk and an embryo restricted to one end (for example, a chick egg). The embryo develops on the surface of the yolk.

**Yolk**   stored food reserves in an egg and embryo; rich in lipids and proteins.

**Yolk sac**   one of the extraembryonic membranes of the chick and many other vertebrates; forms as an outgrowth of the gut and encloses the yolk.

## Internet Resources

Visit the zoology website at http://www.mhhe.com/zoology to find live Internet links for each of the references listed below.

1. Bill Wasserman's Developmental Biology Page. This site provides many links to sites on fruitfly, nematode, zebrafish, Arabidopsis, mammal, and sea urchin developmental biology.

2. Embryo Development. This site supported by the University of Pennsylvania provides images of embryos, representations

of developmental stages, and descriptions of developmental processes.

3. Amphibian Embryology. This site is a tutorial on amphibian embryology. Click on an icon to learn more about particular stages of development.

4. Developmental Biology Online. Extensive information about cleavage, gametogenesis, development of the frog and chick, and other topics.

## Critical Thinking Questions

1. Compare the contributions of the egg and sperm to the developing embryo. Are they both essential? Why? What evidence can you cite for your answer?

2. Discuss the roles of growth, determination, differentiation, and morphogenesis in development. Why is each process important for the normal development of an embryo?

3. Discuss the contributions of each of the three primary germ layers to the body of an adult animal.

4. Discuss the principal similarities and differences in the embryonic development of a frog and of a chick. Why do you think the development of these animals is so different?

## Suggested Readings

Epel, D. 1977. The program of fertilization. *Scientific American* 37(Nov.):128–138.

Gerhart, J., et al. 1986. Amphibian early development. *BioScience* 36:541–40.

Patten, B.M. 1971. *Early Embryology of the Chick,* 5th ed. New York: McGraw-Hill. The classic descriptive embryology of the chick.

Rugh, R. 1951. *The Frog: Its Reproduction and Development.* New York: McGraw-Hill. The classic description of frog embryology.

Wolpert, L. 1991. *The Triumph of the Embryo.* Oxford: Oxford University Press.

## NOTES AND SKETCHES

# NOTES AND SKETCHES

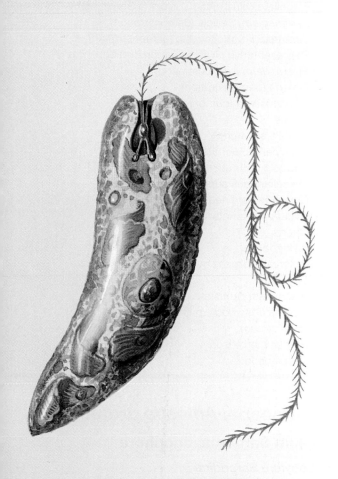

After completing the laboratory work in this chapter, you should be able to perform the following tasks:

1. Identify the major structures in a specimen of *Euglena* and tell the function of each structure.
2. Describe the structure of a *Volvox* spheroid and of an individual *Volvox* cell.
3. Describe the reproductive processes and life cycle of *Volvox*.
4. Identify the principal structures in a specimen of *Amoeba proteus.*
5. Demonstrate the techniques for preparing a wet mount and a hanging drop for microscopy and explain the uses of each.
6. Identify the main structures in a specimen of *Paramecium* and give the function of each structure.
7. Describe the functions of cilia in the feeding and locomotion of *Paramecium*.
8. Compare the locomotion of *Euglena, Amoeba,* and *Paramecium*.
9. Discuss the possible role that protozoa may have had in the evolution of animals.
10. Describe the life cycle of *Plasmodium* and identify its principal stages in microscope slides or photographs.
11. Explain the relationship of *Plasmodium* to the disease malaria.

## Introduction

Protozoa are eukaryotic unicellular organisms, generally microscopic, that live as single individuals or in simple colonies. They are sometimes called acellular since their bodies are not divided into cells. Protozoa are not animals according to modern classification, but many members of the protozoa show animal-like features. Protozoa, along with the algae and certain slime molds formerly considered fungi, are currently placed in the Kingdom Protista, and within this kingdom several major groups of protozoa are recognized as separate phyla.

For many years, protozoa were classified as animals and placed in the Phylum Protozoa. The term *protozoa,* however, no longer has taxonomic status and thus is no longer capitalized. It is still generally used as a common name for this group of interesting and important organisms. Seven phyla are included in the classification of protozoa adopted in 1980 by the Society of Protozoologists.

Protozoa exhibit great intracellular complexity with many internal organelles that are analogous to the organs and organ systems of higher animals. Regardless of their taxonomic status, however, protozoa are an important assemblage of organisms with many members that exhibit animal-like characteristics. Most scientists believe that the multicellular animals evolved from some group or groups of ancestral protozoa. For these reasons, protozoa are studied in most general zoology courses and are included in this book. More than 70,000 species of protozoa have been described, including species widely distributed in many different kinds of moist or wet habitats; in fresh, marine, and brackish waters; in sewage; in moist soil; in or on the bodies of many species of animals; and in or on some plants.

In this chapter, we shall consider representatives of four common and well-differentiated groups of protozoa: the **Mastigophora,** the **Sarcodina,** the **Ciliophora** (formerly called Ciliata), and the **Apicomplexa.**

# Classification

## Phylum Sarcomastigophora

Protozoa with locomotion by means of flagella and/or pseudopodia; with a single type of nucleus.

### Subphylum Mastigophora (Flagellata)
Locomotion by one or more flagella. Many members of this group are photosynthetic and exhibit other plantlike features. Examples: *Euglena, Volvox, Trypanosoma* (blood parasites of man and other animals), *Gonyaulax,* and *Gymnodinium* (dinoflagellates, often implicated in the red tides of coastal waters).

### Subphylum Sarcodina
Locomotion by pseudopodia. Examples: *Amoeba proteus, Arcella* and *Difflugia* (testate amoebae), *Entamoeba histolytica* (a human parasite), *Globigerina* (a foraminiferan), and *Actinosphaerium* (a heliozoan).

## Phylum Ciliophora (Ciliata)

Protozoa with cilia or ciliary organelles present in at least one stage of the life cycle; with two distinct types of nuclei (macronucleus and micronucleus). Examples: *Paramecium, Tetrahymena, Euplotes, Vorticella, Stentor, Blepharisma,* and *Trichodina.*

## Phylum Apicomplexa

Protozoa typically lacking locomotory organelles (except for gametes in some groups); with a characteristic set of anterior organelles called the apical complex (visible only with the electron microscope); microspores present at some stage in the life cycle; all species parasitic. Formerly included in the Class Sporozoa.

## Materials List

**Living specimens**
  *Amoeba proteus*
  *Euglena*
  *Volvox*
  *Paramecium caudatum*
**Prepared microscope slides**
  *Arcella* (demonstration)
  *Difflugia* (demonstration)
  *Entamoeba histolytica* (demonstration)
  *Actinosphaerium* (demonstration)
  *Globigerina* (demonstration)
  *Peranema* (demonstration)
  Symbiotic flagellates from termite or wood cockroach (demonstration)
  Dinoflagellates (demonstration)
  *Volvox,* cell walls (demonstration)
  Flagellates illustrating Volvocine Series (demonstration)
  *Paramecium,* pellicle (demonstration)
  *Paramecium,* trichocysts (demonstration)
  Representative ciliates (demonstration)
  *Plasmodium* (demonstration)
  *Eimeria* (demonstration)
  *Trypanosoma* (demonstration)
**Chemicals**
  Lugol's solution
  Methylene blue, 0.1% solution
  Protoslo (or methyl cellulose solution)
  Miscellaneous supplies
  Congo red stained yeast cells
**Audiovisual materials**
  Charts
  Amoeba
  Colonial protozoa
  *Euglena*
  Freshwater protists
  Malaria (*Plasmodium* life cycle)
  *Paramecium*
  Parasitic protozoa
  Trypanosomes

# An Amoeba: *Amoeba proteus*

## Phylum Sarcomastigophora

### Subphylum Sarcodina
*Amoeba proteus* (figure 5.1) is a protozoon found in ponds and streams. It often occurs on the undersides of plant leaves and among diatoms and desmids. The transparent amoeba constantly changes shape by extending pseudopodia, footlike extensions of the cytoplasm, which serve for locomotion and in food capture. *Amoeba proteus* feeds on bacteria, small algae, and small protozoons.

In feeding, an advancing pseudopodium flows over one or more food organisms to trap the food in a water-filled cup.

The opening of the food cup then narrows until the food is completely enclosed in a food vacuole.

◆ Prepare a wet mount to study living amoebae under your compound microscope.

***Preparing a Wet Mount and Hanging Drop.*** Obtain a clean microscope slide and add a drop of amoeba culture solution to the center of the slide. Take care to withdraw the drop of culture solution from the bottom of the culture dish or jar with a clean eyedropper or pipette. The amoebae are slightly heavier than the culture solution and are usually concentrated on the **bottom** of the vessel. Add a few bits of broken coverslip or grains of sand around the edge of the drop to protect your specimens from being crushed, then carefully cover your preparation with a coverslip. This preparation is called a **wet mount.**

Another method of studying living protozoa is to prepare a **hanging drop.** With this method you place a drop of the amoeba culture on a coverslip and invert the coverslip over the cavity of a depression slide. Both types of preparations can be observed for long periods of time if the outside edge of the coverslip is coated with petroleum jelly before it is placed in position on the microscope slide. Be sparing in your use of the petroleum jelly; avoid getting it into the culture droplet and on your microscope lens.

Study your preparation under the low power of the compound microscope to observe the general appearance of the amoeba. For best observation of a living amoeba, reduce the illumination to a minimum with the iris diaphragm since living specimens are nearly transparent and almost invisible in bright light. Search your slide carefully to locate a specimen before discarding it or asking your instructor for a new preparation.

A phase contrast microscope is an excellent tool for viewing living amoebae and other protozoa since various parts of the cell appear lighter or darker depending on differences in their refractive indices. This type of a microscope reveals many structural details in thin preparations of living cells and/or tissues without staining.

◆ Locate an actively moving amoeba and note its constantly changing shape. The long, fingerlike projections are **pseudopodia** ("false feet"). Observe the lack of permanent orientation of the body of an amoeba; any portion may temporarily be anterior, posterior, right, or left.

◆ With the aid of figure 5.1, identify and study the following structures found in the amoeba.

1. **Endoplasm**—the inner granular region that forms the bulk of the cytoplasm.
2. **Ectoplasm**—the thin layer of clear cytoplasm that surrounds the endoplasm.
3. **Cell membrane**—the outer membrane surrounding the amoeba. Sometimes also called the plasmalemma.
4. **Plasmagel**—the stiff, jellylike, granular outer layer of colloidal endoplasm in the **gel state.**

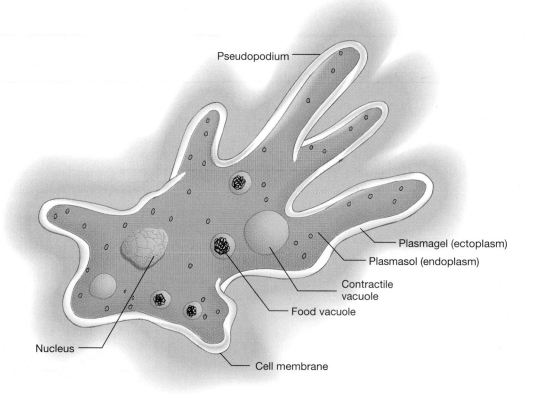

**FIGURE 5.1** *Amoeba proteus.*

5. **Plasmasol**—the central mass of colloidal endoplasm in a fluid, or **sol state.** Note the streaming movements within the plasmasol.

6. **Nucleus**—a transparent structure with no fixed position in the cell. It has the shape of a biconcave disc (concave on two sides) and often exhibits a folded or wrinkled appearance. Examine also the nucleus in a stained microscope slide of *Amoeba proteus*. Observe the darkly staining granular chromatin material within the nucleus.

7. **Contractile vacuole**—a clear vacuole found in the endoplasm that collects excess water from the surrounding cytoplasm and discharges it outside the body. Shortly after one contractile vacuole discharges its contents at the cell surface, a new contractile vacuole forms. ***Where are the new contractile vacuoles formed?*** Formerly, it was believed that the contractile vacuole also played an important role in the excretion of waste products from protein metabolism, but recent studies suggest that contractile vacuoles function primarily in maintaining water balance in the cell (osmoregulation).

8. **Food vacuoles**—vacuoles containing bits of ingested food and the digestive enzymes that act to break down these food materials into soluble materials that can be utilized by the amoeba. ***How are the food vacuoles formed? How are the undigested contents of a food vacuole disposed of after digestion has taken place?***

***Amoeboid Movement.*** An amoeba moves about by extending pseudopodia into which some of the innermost cell contents flow. Various kinds of amoebae form pseudopodia of different size and form. Pseudopodia are important in feeding, support, and locomotion. The mechanism of amoeboid movement has been studied by many scientists because of its intriguing nature and because similar movements occur in many other kinds of cells, including human leucocytes. Also, scientists believe that amoeboid movement may be closely related to the phenomenon of cytoplasmic streaming, movement of cell contents that occur in virtually all kinds of living cells.

The movement of an amoeba is accomplished by the forward flow of the relatively liquid **plasmasol** from the center of the amoeba toward and into an expanding pseudopodium. Around the periphery of the pseudopodium, the plasmasol changes into a stiff **plasmagel.** Thus, the plasmasol moves the pseudopodium forward, and the plasmagel serves to fix it in position.

Biochemical and biophysical studies have demonstrated that the mechanism of amoeboid movement is similar to that in muscle contraction. **Contractile proteins** similar to the actin and myosin found in vertebrate muscles are present in the cytoplasm of an amoeba. We now know that amoeboid movement results from folding, unfolding, polymerization, and depolymerization of these proteins.

◆ Study the locomotion of an *Amoeba* on your microscope slide and the pattern of its internal protoplasmic

movements. In an active specimen, locate and carefully follow the movement of some granules in the plasmasol at the temporary posterior end. Observe how the plasmasol of the endoplasm changes into plasmagel, which flows forward and then changes into the gel state again, just behind of the tip of the forming pseudopodium.

***Reproduction.*** The reproduction of *Amoeba proteus* occurs only through the asexual process of binary fission. The nucleus and cytoplasm of a parent cell divide to form two daughter cells approximately equal in size. Thus, each of the daughter cells is genetically identical to the parent cell, excluding the rare occurrence of a mutation in one of the daughter cells.

### Other Sarcodina

Many members of this group of protozoa are more specialized than *Amoeba*. *Pelomyxa carolinensis* is a large multinucleate amoeba often studied in zoology classes (figure 5.2). Numerous species of amoebae live in shells or tests, which they secrete, or which they form from sand grains or other materials (figure 5.3). *Difflugia* and *Arcella* are two common

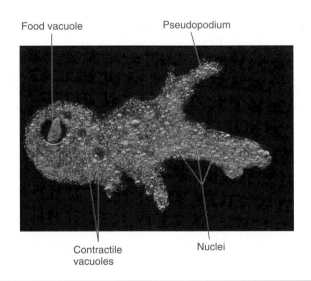

Food vacuole      Pseudopodium

Contractile vacuoles      Nuclei

**FIGURE 5.2** *Pelomyxa carolinensis.* A large multinucleate amoeba.
Courtesy of Carolina Biological Supply Company, Burlington, NC.

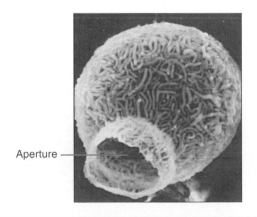

Aperture

**FIGURE 5.3** Test of a freshwater amoeba.
Scanning electron micrograph by F.W. Harrison.

testate amoebae found in freshwater ponds and streams. Other species of amoebae are parasites or symbionts in the digestive tracts of various animals. *Entamoeba histolytica,* an important intestinal parasite of humans (figure 5.4), is the cause of amoebic dysentery, a disease often spread by drinking water or by eating raw vegetables contaminated by human wastes in parts of the world with poor sanitary facilities.

Some freshwater members of the Subphylum Sarcodina have many long, thin pseudopodia supported by axial rods of microtubules. *Actinosphaerium* (figure 5.5) is a common freshwater example of a group called heliozoans because of the resemblance to the sun and its rays of sunlight.

Members of three marine classes of this subphylum, called **radiolarians,** form skeletons or tests of silicon and/or strontium compounds and exhibit many beautiful shapes (figure 5.6). The radiolarians are among the oldest known protozoa, and their tests are abundant in marine sediments in many parts of the world.

The **foraminiferans** (figure 5.7), representing another class of sarcodines, are an ancient and important group of marine sarcodines that form tests of calcium carbonate or other materials. The shells of foraminiferans accumulate on the sea bottom and contribute to the formation of chalk and limestone. England's White Cliffs of Dover are made

up largely of foraminiferan tests, as is much of the Bedford limestone found in Indiana and Illinois, and some of the limestone that was used to build the Egyptian pyramids.

The distribution of certain other species of foraminifera in rock samples is very important to petroleum geologists as indicators of ancient environmental conditions that may have been favorable for the formation of petroleum and thus provide important clues to the possible location of petroleum pools.

## Demonstrations

1. Models and charts of *Amoeba proteus*
2. Culture of *Amoeba* under a stereomicroscope
3. Microscopic slides with testate and parasitic representatives of the Subphylum Sarcodina, such as *Arcella, Difflugia, Entamoeba histolytica, Actinosphaerium* (a heliozoan), and *Globigerina* (a foraminiferan)

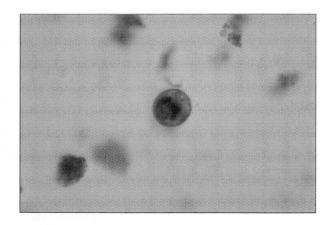

**FIGURE 5.4**   *Entamoeba histolytica.*
Courtesy of Carolina Biological Supply Company, Burlington, NC.

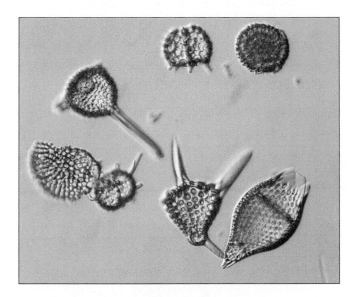

**FIGURE 5.6**   Radiolarian tests.
Courtesy of Carolina Biological Supply Company, Burlington, NC.

**FIGURE 5.5**   *Actinosphaerium.*
Courtesy of Carolina Biological Supply Company, Burlington, NC.

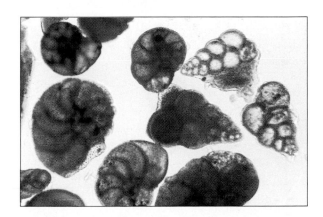

**FIGURE 5.7**   Foraminiferan tests.
Courtesy of Carolina Biological Supply Company, Burlington, NC.

# A Solitary Flagellate: *Euglena*

## Phylum Sarcomastigophora

### Subphylum Mastigophora

*Euglena* (figure 5.8) is a common green flagellate often found in the greenish surface scum of standing or slowly moving water. *Euglena* is an enigmatic organism with a curious mixture of plant and animal characteristics and, therefore, sometimes is considered to represent a borderline case between the plant and animal kingdoms. *Euglena* is smaller than *Amoeba* and *Paramecium* and, therefore, the details of its internal structure are more difficult to observe.

◆ Prepare a wet mount from a culture of living *Euglena* and observe the locomotion of an active specimen under your compound microscope.

The active swimming movements result from the beating of the long flagellum, which pushes the organism through the water. A second, shorter flagellum is present within the flagellar pocket, but does not aid in the swimming movements. At certain times *Euglena* also exhibits another type of wormlike locomotion during which waves of contraction pass along the body in a characteristic fashion. This type of locomotion is peculiar to *Euglena* and related organisms and is appropriately termed **euglenoid movement** or **metaboly.** It appears to result in part from the elasticity of the thick outer covering of the body, the pellicle.

◆ After the wet mount begins to dry out, temporarily immobilizing some of your specimens, study the anatomy of a stationary *Euglena*. You will also find it useful to supplement your observations with the study of a prepared microscope slide.

◆ Identify the following structures under high power on your compound microscope: (1) **pellicle,** the thick outer covering of the body; (2) **chloroplasts** with green chlorophyll; (3) **nucleus,** exhibiting a large central **endosome** in stained preparations; (4) **gullet;** (5) **contractile vacuole;** (6) a red **stigma,** or eye spot; (7) the long anterior **flagellum;** and (8) **paramylum grains,** a type of starch that represents stored food materials.

*Euglena* is quite sensitive to light, and changing light intensity affects its response. In weak light, *Euglena* tends to be **positively phototrophic** (attracted to the light); in strong light, *Euglena* tends to be **negatively phototrophic** (repelled by the light); and sudden changes in light intensity often stun *Euglena* so that it remains stationary.

◆ After you have completed your observations of the living specimen, add a drop of Lugol's solution (iodine and potassium iodide). This solution will kill the specimen and stain the flagellum to make it more visible.

As suggested by the presence of chloroplasts, the nutrition of *Euglena* is normally autotrophic; organic molecules (sugars) are synthesized from inorganic nutrients absorbed from the medium. Light from the sun provides the energy necessary for this process.

Biochemical tests have shown the paramylum granules to be a form of starch similar to that found in plants. Thus, both the presence of the chloroplasts and the storage of a plantlike form of starch both indicate a close relationship of *Euglena* and its relatives to the plant kingdom. Some species of *Euglena* are also able to survive, grow, and reproduce in the dark with no visible evidence of chloroplasts, chlorophyll, or stored food materials. ***How might such organisms obtain their food?***

## A Holozoic Flagellate: *Peranema*

*Peranema* (figure 5.9) is a common flagellate often studied because of its prominent anterior flagellum, which is larger than the flagellum of most species of *Euglena*. *Peranema* (25–80 microns in length) is slightly smaller than most species of *Euglena* (50–100 microns in length) and is often found in stagnant pools of fresh water along with species of *Euglena*. Like *Euglena*, *Peranema* is a solitary flagellate, but unlike *Euglena*, *Peranema* lacks chloroplasts and ingests solid foods through its cytostome.

◆ Make a wet mount of *Peranema* on a clean microscope slide adding some Protoslo or methyl cellulose solution to minimize movement of your specimens, and observe a living specimen. Observe the prominent anterior **flagellum** that extends straight out from the anterior

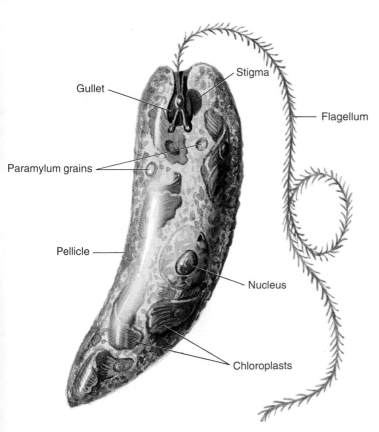

Gullet

Stigma

Flagellum

Paramylum grains

Pellicle

Nucleus

Chloroplasts

**FIGURE 5.8** *Euglena.*

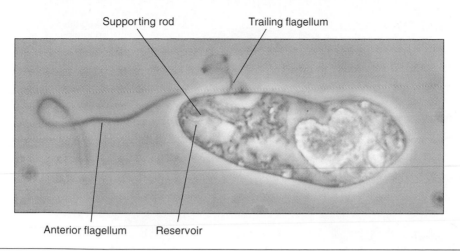

Supporting rod    Trailing flagellum

Anterior flagellum    Reservoir

**FIGURE 5.9** *Peranema.*
Courtesy of Carolina Biological Supply Company, Burlington, NC.

end. Note that most of the anterior flagellum is stiff except for a short portion near the anterior tip, which tends to be very flexible and active. *Peranema* actually has two flagella but one of them adheres closely to the pellicle and is difficult to observe in most ordinary microscopic preparations.

Observe how *Peranema* usually moves along the substrate in a gliding fashion, not clearly resulting from the action of its flagellum. The actual mechanism of its locomotion is not really understood.

Find the flask-shaped **reservoir** near the base of the two anterior flagella. Food particles pass through a tiny **cytostome** into this reservoir before being enclosed in **food vacuoles** that move into the cytoplasm. Unlike *Euglena,* the nutrition of *Peranema* is completely carnivorous or holozoic. Food consists of bacteria and small protozoa that are engulfed whole through the cytostome. Supporting the reservoir and extending into the adjacent cytoplasm are two rodlike structures. When *Peranema* feeds on bacteria and other small organisms, the cytostome can open widely, and these rods support this area while a food vacuole is formed.

Locate the single **nucleus** near the middle of the body. Sexual reproduction has never been described in this flagellate, and new individuals appear to arise only through **binary fission** from a parent organism.

# A Colonial (?) Flagellate: *Volvox*

## Phylum Sarcomastigophora

### *Subphylum Mastigophora*

*Volvox* (figure 5.10) is a common green alga that often occurs in great numbers in freshwater ponds and lakes. Although *Volvox* is photosynthetic and is sometimes considered to be an alga, it is often studied in zoology courses because it illustrates the organization of a simple colonial (or multicellular) organism well and because of its usefulness in illustrating **one popular theory for the evolution of multicellular organisms from unicellular ances-**

**tors.** It also demonstrates some basic similarities between plants and animals.

The spherical green *Volvox* is large enough to be seen with the naked eye swimming near the surface of a pond or in a laboratory culture.

◆ Obtain some living *Volvox* from a culture and prepare a wet mount. Be sure to add a few bits of broken coverslip or sand grains to protect the spherical *Volvox* bodies from being crushed by the weight of the coverslip. Observe the spheroid shape of *Volvox,* its swimming movements, and its prominent green color.

The spherical *Volvox* bodies are usually called colonies in textbooks, but recent studies have suggested that they are more similar to multicellular individuals. The coordination of cells and cellular function within the spheroid body is much greater than in most other colonial algae and protozoons; thus, recent workers use the term **spheroids** for the *Volvox* body.

The hollow *Volvox* spheroids average about 0.5 mm in diameter and have many small green cells embedded in their outer walls (figure 5.11). Each of the tiny body cells of *Volvox* contains a **nucleus,** a **contractile vacuole,** a green **chloroplast,** and two whiplike **flagella.** Some or all of the cells (depending on the species) may also have a red **stigma** (light-sensitive spot). The flagella project outward from the surface, and their beating keeps the spheroids in a constant spinning motion. Although the somatic cells of *Volvox* are very small and difficult to observe except with special microscopic preparations, the adjacent cells in some species of *Volvox* are connected by thin **cytoplasmic bridges** or strands. Other species of *Volvox,* however, lack these intercellular connections.

◆ Add a drop of 0.1% methylene blue solution to a wet mount of *Volvox* and study the slide under high power on your compound microscope. *Can you observe any cytoplasmic bridges?* Try cutting down the light using the disk diaphragm on your microscope to make the transparent bridges more visible.

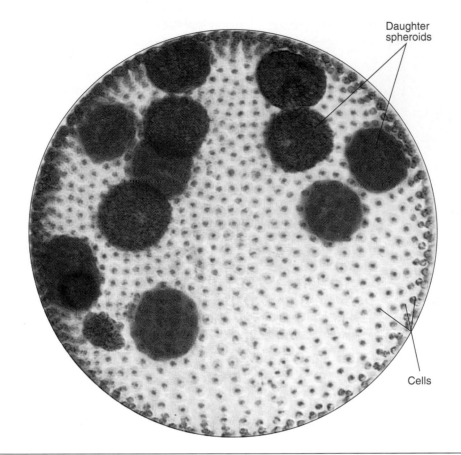

**FIGURE 5.10** *Volvox.* Asexual spheroid.
Courtesy of Carolina Biological Supply Company, Burlington, NC.

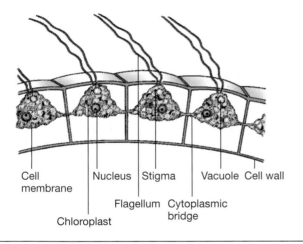

Cell membrane   Nucleus   Stigma   Vacuole   Cell wall
Flagellum   Cytoplasmic bridge
Chloroplast

**FIGURE 5.11** *Volvox* cell structures, longitudinal section of cells at surface of spheroid.

The green color of the spheroid individuals results from the presence of a chloroplast in each cell. *What can you therefore infer about the nutrition of* **Volvox**?

◆ Within the spheroid you should be able to observe one or more large reproductive cells or **gonidia.** Reproduction in *Volvox* involves both sexual and asexual processes. In asexual development, embryos are formed from the gonidia. Locate several gonidia within the in-

terior of an asexual parent individual (figure 5.12). They are smaller in size than the multicellular embryos formed from them.

◆ During asexual development, the gonidia undergo a series of cell divisions remarkably like those seen in the embryonic development of many animal species. See if you can locate several different developmental stages of gonidia in the living specimens provided.

The beginning of sexual reproduction can be recognized when certain gonidia form male and/or female spheroids. Most species of *Volvox* have separate male and female individuals (i.e., are **dioecious,** the sexes occurring in "two houses"), but some species produce both eggs and sperm in the same individual and are thus **monoecious** ("one house"). Figure 5.12 also shows both male and female individuals.

◆ The living cultures available for study in the laboratory are usually all asexual. You should study the demonstration chart to learn more about the life cycle of *Volvox.*

◆ Study the structure of *Volvox* using the living specimens provided in the laboratory. In a drop of water on a clean microscope slide, observe the swimming of *Volvox* spheroids first under low power and then add a coverslip over the water drop and observe more details of the structure of the colony under higher magnification.

Observe somatic cells, gonidia, eggs, sperm, and zygotes, using both the living specimens and the demonstration materials as necessary.

## Other Mastigophora

Other important mastigophorans include the dinoflagellates, symbiotic flagellates that inhabit the digestive tracts of termites and the wood roaches, some peculiar flagellates that may be related to sponges, and several important parasites of humans.

**Dinoflagellates** are common in both fresh and marine waters; many species form a characteristic outer covering, called a **test,** made of cellulose. Some freshwater dinoflagellates can cause an unpleasant odor or taste in drinking water, while many other species appear to be innocuous. *Ceratium* (figure 5.13*a*) is a common dinoflagellate that inhabits both fresh- and saltwater habitats. It has a characteristic shape with one long anterior horn, one to four posterior horns, and a transverse groove around the middle to its test. One transverse flagellum lies in the transverse groove and a longer flagellum trails posteriorly.

*Gonyaulax* and *Gymnodinium* are two marine dinoflagellates often associated with **red tides** of coastal waters of North America, Europe, and Africa that sometimes result in massive fish kills. A recently described species of dinoflagellate, *Pfisteria piscida,* produces a powerful toxin and has been found to be responsible for several fish kills along the Atlantic coast of the United States (figure 5.13*b*). There is now substantial evidence that blooms of *Pfisteria piscida* also constitute a serious health threat to persons fishing, boating, or swimming in or near coastal waters along the mid-Atlantic coast of the United States.

Some flagellates live in a **symbiotic association** within the digestive tracts of wood roaches and termites (figure 5.14). Experiments have shown that the termites lack the digestive enzymes necessary to digest the cellulose in the wood they eat. The flagellates ingest splinters of wood and form food vacuoles around them. Later, digested products

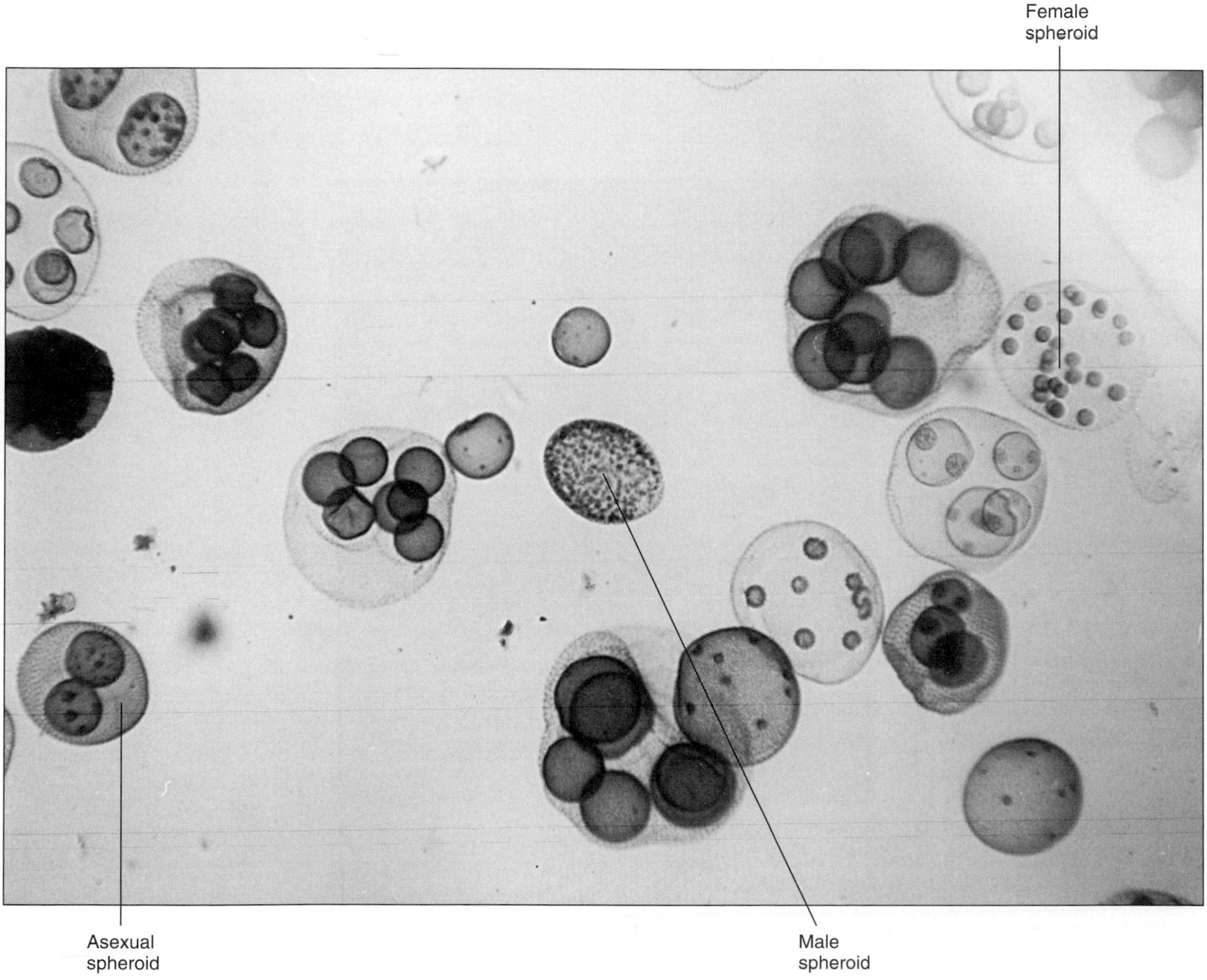

Female spheroid

Asexual spheroid

Male spheroid

**FIGURE 5.12** *Volvox,* male, female, and asexual spheroids.
Photograph by Barbara Grimes.

Protozoa   **81**

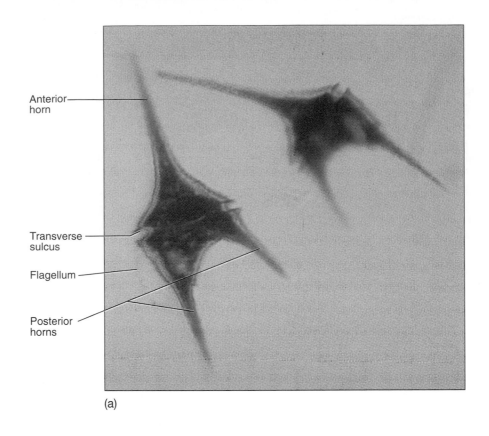

Anterior horn

Transverse sulcus

Flagellum

Posterior horns

(a)

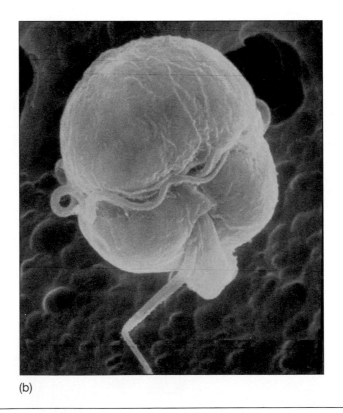

(b)

**FIGURE 5.13** (a) *Ceratium,* a dinoflagellate.
Courtesy of Carolina Biological Supply Company, Burlington, NC.
(b) *Pfisteria piscida,* a toxic dinoflagellate responsible for fish kills and a human health threat in Atlantic coastal waters.
Scanning micrograph courtesy of Dr. JoAnn Burkholder.

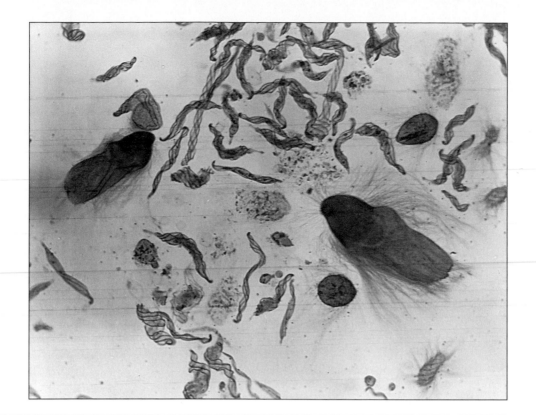

**FIGURE 5.14** Symbiotic flagellates from termite gut.
Courtesy of Carolina Biological Supply Company, Burlington, NC.

from the breakdown of the cellulose are released from the protozoa and provide nutrients for the termites. Termites from which the flagellates are experimentally removed soon die of starvation no matter how much wood they ingest. The flagellates benefit from the continuous supply of cellulose and from the suitable anaerobic environment of the host hindgut. Such a mutually beneficial symbiotic relationship is called **mutualism.**

**Symbiotic relationships** may take many forms; for example, the relationship may be beneficial to one or both partners (mutualism), beneficial to one partner and harmful to the other partner (parasitism), or beneficial to one partner and neutral to the other partner (commensalism). Numerous variations in these relationships occur in various types of symbiosis. *Why would a relationship in which neither partner benefitted be unlikely to survive?*

*Proterospongia* (figure 5.15) is a colonial flagellate with species that closely resemble the flagellated collar cells, or choanocytes, characteristic of sponges (see Chapter 6). Some biologists have suggested that the sponges may have evolved from some ancient protozoon similar to *Proterospongia.*

Several flagellates are important parasites of humans, including *Trypanosoma, Leishmania,* and *Giardia.* **African sleeping sickness** is a devastating disease affecting hundreds of thousands of people and resulting in thousands of deaths each year. This disease is caused by the blood parasite, *Trypanosoma brucei,* and is spread by the tsetse fly. **Chagas disease,** caused by *Trypanosoma cruzi,* affects

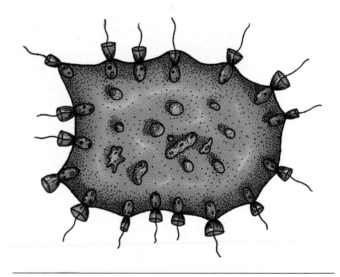

**FIGURE 5.15** *Proterospongia.*

millions of people in Central and South America. Chagas disease is transmitted by bites from the "kissing bug," a common biting insect found in these areas. *Trypanosoma* has a thin, **undulating membrane** connecting its single long, whiplike flagellum with its body (figures 5.16 and 5.17).

◆ Observe also the microscope slide of trypanosomes on demonstration. Since trypanosomes are very small, they must be viewed under high magnification with an **oil immersion lens** to see the undulating membrane and flagellum.

White blood cell

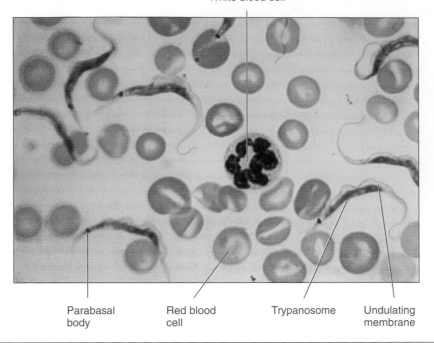

Parabasal body      Red blood cell      Trypanosome      Undulating membrane

**FIGURE 5.16** *Trypanosoma,* blood smear.
Courtesy of Carolina Biological Supply Company, Burlington, NC.

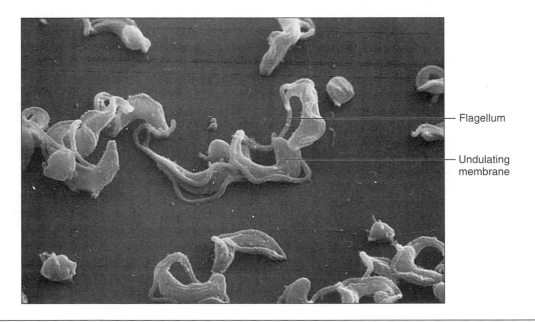

Flagellum

Undulating membrane

**FIGURE 5.17** *Trypanosoma.* Magnification 3,525×.
Scanning electron micrograph by Louis de Vos.

*Leishmania* is spread from other mammals to humans by sand flies and causes serious skin lesions and may weaken the immune system.

*Giardia lamblia* is a common inhabitant of the human digestive tract and serious infections can result in abdominal pain and diarrhea. It is transmitted by drinking water contaminated by human feces.

## Demonstrations

1. Large flagella in other Mastigophora, such as *Peranema*
2. Microscope slide of trypanosomes
3. Microscope slide of symbiotic flagellates from digestive tract of termite or wood roach
4. Microscope slide of dinoflagellates
5. Microscope slide showing cell walls in *Volvox*
6. Chart illustrating the life cycle of *Volvox*

## A Ciliate: *Paramecium caudatum*

### Phylum Ciliophora (Ciliata)

*Paramecium* (figure 5.18) is a large, common, ciliated protozoon often found in water containing bacteria and decaying organic matter. There are several species of *Paramecium* that differ in various details of structure and that range in length from about 120–300 microns. The description provided here is based upon *Paramecium caudatum,* a species frequently used for laboratory study and experimentation, but it will also apply, with minor variations (such as body size and number of micronuclei), to the study of other species of *Paramecium.*

◆ Obtain a drop of *Paramecium* culture in a clean pipette and make a wet mount on a clean microscope slide with a similar-sized drop of methyl cellulose solution (or other similar agent) to slow movement. Methyl cellulose is a viscous material and serves mechanically to slow the swimming of the fast-moving *Paramecium.* Add a coverslip and observe your preparation under low power with your compound microscope. Note the form, color, and behavior of the animals in your preparation. Observe the slipper-shaped body with an **oral groove** beginning at the anterior end and running diagonally across the anterior portion of the animal. At the posterior end of this groove is the **cytostome,** or "cell mouth," through which food particles are passed as a result of the action of the specialized oral cilia lining the oral groove.

Observe that *Paramecium* is much more complex in its structure than *Amoeba.* Select a large, immobile, or slowly moving specimen, and with the aid of figure 5.18, identify and study the following structures.

1. **Cilia**—the numerous cylindrical protoplasmic extensions that cover the surface of the *Paramecium* and that function in locomotion and in food gathering.
2. **Pellicle**—the thick outer covering of the body through which the cilia project. The pellicle has a complex structure, but its details are difficult to observe without special techniques. Figure 5.19 shows some of the surface depressions in the bilayered pellicle.
3. **Trichocysts**—tiny, rodlike structures embedded in the cortical (outer) cytoplasm beneath the pellicle. When properly stimulated, the trichocysts discharge

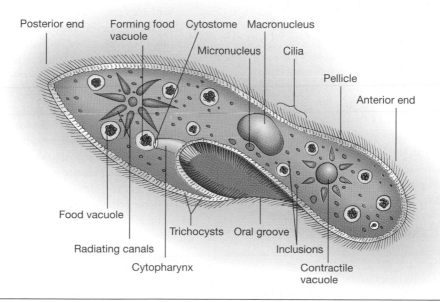

**FIGURE 5.18** *Paramecium caudatum.*

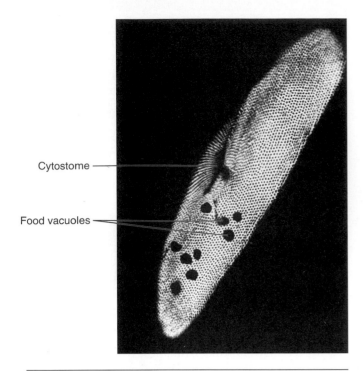

Cytostome

Food vacuoles

**FIGURE 5.19** *Paramecium*, nigrosin stain to illustrate sculpturing of bilayered pellicle.
Photograph by Barbara Grimes.

their contents and form long threads. There is some evidence that the trichocysts may serve as a defense against predators, and they also serve to anchor the animal during feeding. In other types of ciliated protozoa, trichocysts have been found to have additional functions. Observe the microscopic demonstration of discharged trichocysts.

4. **Macronucleus**—the large nucleus located near the center of the cell. Since it is transparent in a living animal, the structure of the macronucleus is best studied in a stained preparation. Experiments have demonstrated that the macronucleus controls most metabolic functions of the cell.

5. **Micronucleus**—a smaller nucleus located close to and lying partly within a depression on the oral side of the macronucleus. The micronucleus is involved primarily in the reproductive and hereditary functions of the animal. This presence of two distinct types of nuclei is called **nuclear dimorphism** and is a condition found only in the Phylum Ciliophora. *Paramecium caudatum* has only a single micronucleus, but some other species of *Paramecium* have two or more micronuclei. As with the macronucleus, the structure of the micronucleus is best studied in a prepared microscope slide.

6. **Contractile vacuoles**—two clear, slowly pulsating vesicles located near each end of the body. Each contractile vacuole is surrounded by several **radiating canals** (not often seen in ordinary student preparations), which collect water from the surrounding cytoplasm. Observe the behavior of the

contractile vacuoles. *Are they fixed in position? Do they contract alternately or simultaneously?* The function of the contractile vacuoles in *Paramecium* is similar to that in *Amoeba* (i.e., the vacuoles collect and discharge excess water from the cell). Freshwater protozoa often have contractile vacuoles; marine protozoa generally lack them. *How would you explain this difference?*

7. **Cytostome** (cell mouth)—a permanent opening near the posterior end of the oral groove through which food is passed.

8. **Cytopharynx**—a short tube extending from the cytostome posteriorly and downward into the cytoplasm where food vacuoles are formed.

9. **Food vacuoles**—vacuoles located within the cytoplasm where they are carried by the streaming movements of the cytoplasm. Undigested materials are discharged through the **cytopyge,** or anal pore, located posterior to the oral groove.

## Feeding

*Paramecium* is a filter-feeding organism that normally feeds on bacteria and yeast cells collected by a specialized food-collecting apparatus. An **oral groove** extends diagonally back along the body to a funnel-shaped **cytopharynx.** Food is swept along the oral groove by the action of specialized cilia lining the groove, is passed through the circular **cytostome** at the opening of the cytopharynx, and is passed through the cytopharynx into a newly forming **food vacuole.**

◆ Prepare a wet mount with a drop of *Paramecium* culture to study the feeding process. Add a small amount of Congo red stained yeast with the tip of a toothpick or clean dissecting needle. Try to pick up the smallest amount of yeast possible on the toothpick; too much yeast will cloud your preparation and obscure the *Paramecium.*

With this preparation, you can study the movement of the food particles, the formation of food vacuoles, and the subsequent movement of the food vacuoles within the cytoplasm. After the food vacuoles are formed, digestive enzymes are released into them, and chemical digestion of the food particles begins. Note the color change in the vacuoles as the enzymes work. The color change is due to a change of pH in the vacuoles. *Where do the digestive enzymes come from? Why don't they digest the other materials in the cell such as mitochondria and ribosomes?* The diffusible products of digestion are released into the cytoplasm, and the undigestible remains are discharged at a specific site on the surface of the animal. This site is the **cytopyge,** or cell anus.

## Cilia and Flagella

Most of the surface of *Paramecium* is covered by thin, hair-like projections called **cilia** (singular: cilium). Cilia are extensions of the cortical (outer) cytoplasm of the cell and play important roles in feeding and locomotion.

A great deal has been learned in recent years about the structure and function of cilia. These studies have revealed that cilia are closely related to the flagella (singular: flagellum) found on the surface of other kinds of protozoa. The structural differences between cilia and flagella are minor. When the projections are short and numerous, they are called cilia. When they are long and few, they are flagella. Cilia generally exhibit a relatively simple back and forth movement. The movements of flagella are often more complex and may involve a series of helical waves propagated along the flagellum.

Both cilia and flagella have a common basic structure. A cross section reveals an outer membrane enclosing a circle of **nine pairs of microtubules and two single microtubules** in the center of the cilium or flagellum. This basic pattern is found in all cilia and flagella, not only among the protozoa but also on the gills of molluscs, the ciliated epithelium lining the trachea of vertebrates, and the tail of spermatozoa.

Biochemical studies have also demonstrated that the movements of cilia and flagella involve **contractile proteins** similar to those found in striated muscle. This is another important illustration of the basic similarity of all living organisms.

### Reproduction

*Paramecium* reproduces by a simple type of asexual reproduction in which the parent divides into two equal daughter cells. This type of asexual reproduction is termed **transverse fission** and is found in many kinds of protozoa. Living specimens are occasionally seen in the process of fission, but the details of fission are best studied in a stained microscope slide (figure 5.20).

◆ Obtain a prepared slide of *Paramecium* in fission and observe the nuclei. During fission, the micronucleus first divides by **mitosis,** and the macronucleus later divides by **amitosis.** No visible chromosomes are formed in the macronucleus; the macronucleus simply constricts, and the two portions separate. Macronuclear division is followed by cytoplasmic division (cytokinesis). The process of fission may be completed rapidly, and under optimal conditions, *Paramecium* can reproduce asexually two or more times per day.

Unlike *Amoeba, Paramecium* can also reproduce sexually. The specialized type of sexual process exhibited by *Paramecium* is called **conjugation** (figure 5.21). During this process, two individuals come together, adhere by their oral surfaces, undergo a complex series of changes in both the macronuclei and the micronuclei, exchange a single pair of micronuclei (one from each cell), separate, and resume asexual reproduction. Following the exchange of micronuclei in each *Paramecium,* the newly introduced micronucleus fuses with another (nonmigrating) micronucleus. Thus, there is an exchange of hereditary material and a subsequent fusion of hereditary material from the two parents, a situation analogous to that of ordinary sexual reproduction studied earlier.

◆ Examine a prepared slide or a demonstration of *Paramecium* in conjugation.

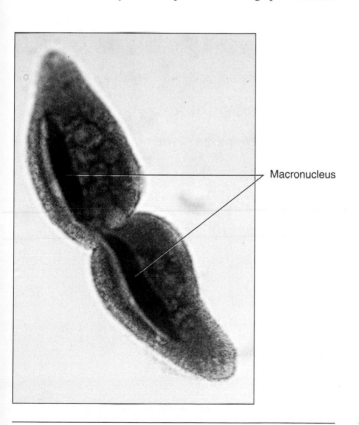

Macronucleus

**FIGURE 5.20**  Binary fission in *Paramecium.*
Courtesy of Carolina Biological Supply Company, Burlington, NC.

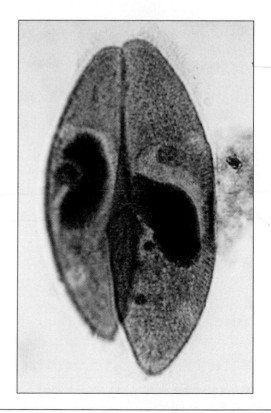

**FIGURE 5.21**  *Paramecium* in conjugation.
Courtesy of Carolina Biological Supply Company, Burlington, NC.

## Other Ciliates

The Phylum Ciliophora includes a large and diverse group of protozoa with many important and well-known species and genera. One well-known form is *Stentor* (figure 5.22), a large, trumpet-shaped ciliate common in freshwater lakes, streams, and ponds where it temporarily attaches to submerged sticks, stones, and vegetation. Some common species of *Stentor* contain a blue or green pigment. *Stentor* has been the subject of many experimental studies. Its macronucleus looks like a string of beads. Another characteristic feature is a spiral array of complex ciliary organelles (membranelles) leading from its apical end to the cytostome.

*Euplotes* (figure 5.23) is another well-known ciliate that has been used in many experimental studies. Found in both fresh and salt water, *Euplotes patella*, the most common American species, has a flattened ovoid body averaging about 90 μm by 52 μm in size. Cilia in *Euplotes* are restricted to certain regions of the body, and groups of adjacent cilia fuse together to form a row of triangular **membranelles** along the oral groove leading to the cytostome. Several **cirri,** also composed of several fused cilia, are located in specific locations on the ventral side of the body and move in a coordinated fashion, which allows the organism to move along the substrate as if walking on little legs.

*Euplotes* has a ribbonlike C-shaped macronucleus and a small spherical or ovoid micronucleus. **Conjugation** occurs in fashion similar to that described for *Paramecium;* asexual reproduction occurs by binary fission.

*Didinium* (figure 5.24) is a barrel-shaped predaceous ciliate with a voracious appetite. It feeds on other ciliates including *Paramecium*. A hungry *Didinium* can eat a *Paramecium* every two hours.

*Spirostomum* (figure 5.25) is a long, wormlike ciliate that has contractile fibrils that seem to function in a way similar to striated muscle fibrils. *Tetrahymena* (figure 5.26) is a small, ovoid ciliate that has been used in many experimental studies of biochemistry and genetics.

*Vorticella* (figure 5.27) is a sessile form with a long, contractile stalk that attaches to submerged stones, shells,

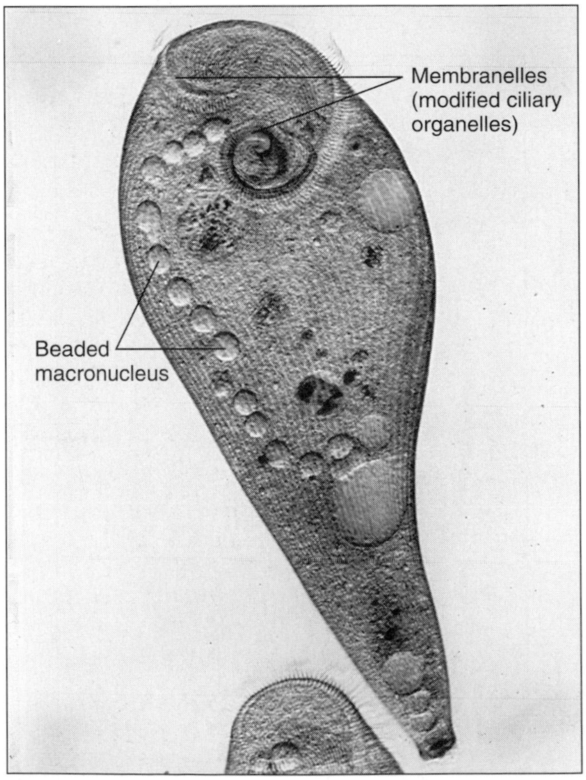

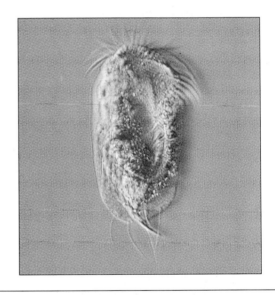

**FIGURE 5.23** *Euplotes,* living, unstained.
Courtesy of Carolina Biological Supply Company, Burlington, NC.

**FIGURE 5.22** *Stentor.*
Courtesy of Carolina Biological Supply Company, Burlington, NC.

[Figure 5.22 labels: Membranelles (modified ciliary organelles); Beaded macronucleus]

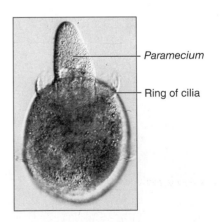

[Figure 5.24 labels: Paramecium; Ring of cilia]

**FIGURE 5.24** *Didinium,* a carnivorous ciliate ingesting a *Paramecium.*
Courtesy of Carolina Biological Supply Company, Burlington, NC.

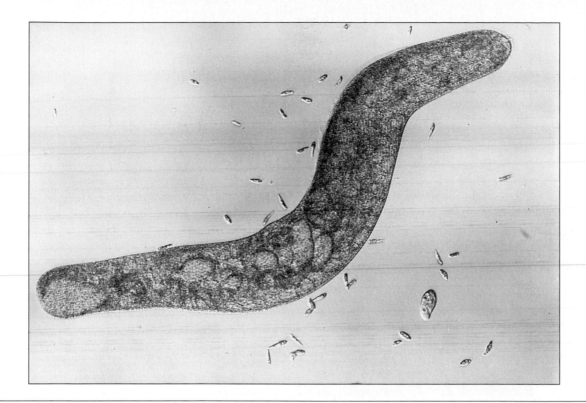

**FIGURE 5.25** *Spirostomum,* a ciliate with contractile fibers and a flexible pellicle that exhibits a peculiar wormlike locomotion.
Courtesy of Carolina Biological Supply Company, Burlington, NC.

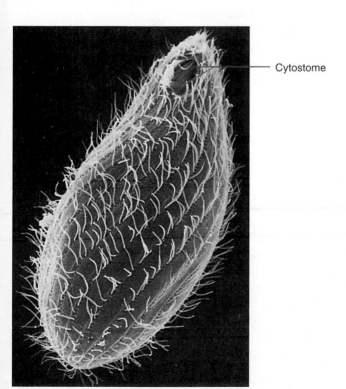

Cytostome

**FIGURE 5.26** *Tetrahymena.*
Scanning electron micrograph by Jolanta Nunnley.

plants, animals, and other objects. Other relatives of *Vorticella,* such as *Carchesium* and *Zoothamnion,* form similar stalked colonies.

*Podophyra* is a suctorian, a specialized group of sessile ciliates that have **suctorial tentacles** in their mature stages and are predators of other ciliates. Cilia are found only on juvenile stages of the suctorians.

## Demonstrations

1. Models and charts of *Paramecium*
2. Stained slide showing pellicle of *Paramecium*
3. Stained slide to show discharged trichocysts
4. Stained slides with representative members of the Class Ciliata, such as *Stentor, Euplotes, Tetrahymena, Vorticella, Didinium, Blepharisma, Trichodina,* and *Podophyra*

## Phylum Apicomplexa

Members of this group were formerly included among the Sporozoa, but recent investigations have indicated that they should be considered to be a separate phylum. All members of this phylum are parasitic on other organisms.

Many species in this group parasitize invertebrate animals such as earthworms, crabs, and oysters, but the most important species are parasites of vertebrates, including humans.

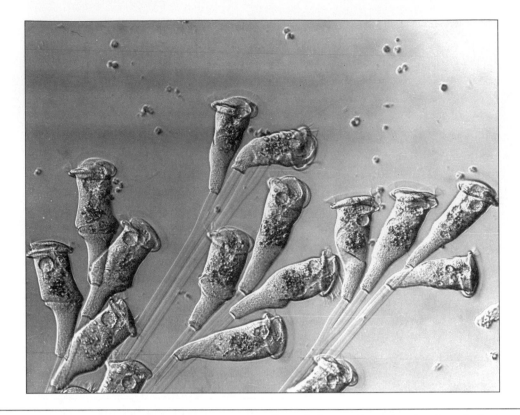

**FIGURE 5.27** *Vorticella.* A sessile ciliate with a long contractile stalk for attachment and a ring of cilia surrounding the mouth.
Courtesy of Carolina Biological Supply Company, Burlington, NC.

Among the most important sporozoans are *Plasmodium* and *Eimeria*. Several species of *Plasmodium* cause various forms of **malaria** in humans and other animals.

*Eimeria* is a genus of sporozoan parasites that causes **coccidiosis** in birds, rabbits, and other animals. Coccidiosis is a serious disease that has great economic impact on the poultry industry. *Eimeria*, like *Plasmodium*, has a complex life cycle that includes both sexual and asexual forms.

◆ Study the demonstrations illustrating the life history and importance of *Eimeria* and coccidiosis.

## The Malaria Parasite: *Plasmodium*

More than 50 species of *Plasmodium* have been described. All are parasites of vertebrate animals, including amphibians, reptiles, birds, and mammals. Four species cause human malaria, *P. vivax*, *P. ovale*, *P. malariae*, and *P. falciparum*. The life cycles of these species are basically similar.

Malaria, one of the most serious and debilitating of human diseases, has had an important role in human history from the fall of the Roman Empire to the war in Vietnam. Although modern medicine has made some progress in eliminating malaria, the disease is yet to be conquered. It is most prevalent in tropical and semitropical areas, and today it still costs millions of lives and trillions of dollars each year.

### Life Cycle of Plasmodium

The life cycle of *Plasmodium* (figure 5.28) is complex like those of other apicomplexa and includes several generations with both sexual and asexual reproduction. The life cycle can best be understood by starting with the zygote in the gut of a mosquito, one of the two hosts necessary for the completion of the life cycle.

The zygote becomes motile and passes through the lining and wall of the stomach or midgut of the mosquito and is now called an ookinete. The ookinete then rounds up and encysts on the outside of the gut wall and is called an oocyst. After several days of growth, the oocyst divides internally to form several hundred sporozoites. The sporozoites escape by rupturing the external wall of the oocyst and migrate through the hemocoel to the salivary glands of the mosquito.

When a mosquito bites a human, the sporozoites and the mosquito's salivary secretions are injected into this host. Sporozoites that find their way into the human bloodstream are eventually carried to the liver, where the sporozoites enter host cells. Inside the host liver cells, the sporozoites transform into amoeboid multinuclear schizonts and feed upon the contents of the host liver cells. The schizonts reproduce asexually to form many merozoites (the next stage in the life cycle), escape from the liver cells, and, in some cases, invade other liver cells to repeat the process.

Merozoites escape into the bloodstream and penetrate erythrocytes to initiate the erythrocytic phase of the life

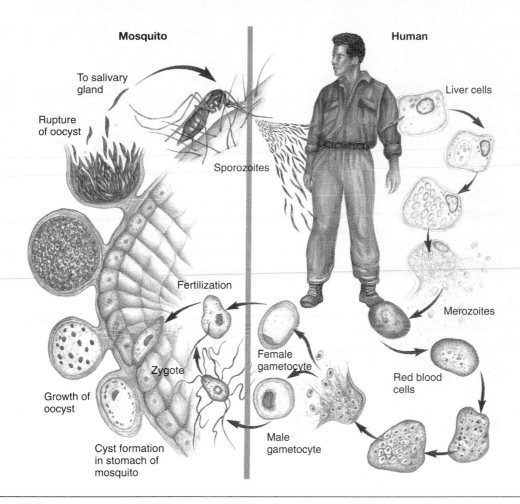

**Mosquito**

To salivary gland

Rupture of oocyst

Sporozoites

**Human**

Liver cells

Fertilization

Zygote

Growth of oocyst

Cyst formation in stomach of mosquito

Female gametocyte

Male gametocyte

Merozoites

Red blood cells

**FIGURE 5.28** Life cycle of *Plasmodium*.

cycle. Parasitic stages in the liver prior to entry into the erythrocytes make up the extraerythrocytic stage. The developing *Plasmodium* stages inside the erythrocytes exhibit a characteristic morphology, as seen in Giemsa-stained microscope preparations, and are recognized by their red nucleus and blue ring-shaped cytoplasm (figure 5.29). This characteristic morphology is very helpful in the laboratory diagnosis of malaria. The merozoites in the erythrocytes undergo further asexual reproduction within the erythrocytes. Later, the merozoites rupture the wall of the erythrocytes, escape into the blood, and enter most erythrocytes. This multiplication process can be repeated several times, so that an enormous number of parasites are produced within the host. The rupture of the erythrocytes by the merozoites also releases accumulated toxic wastes from the parasites, and these wastes cause the symptomatic chills and fever commonly associated with human malaria.

Some of the merozoites develop into sexual forms instead of repeating the asexual merogony. These sexual forms develop within the erythrocytes and become microgametocytes (male) and macrogametocytes (female) (figure 5.30). These stages are the progenitors of the male and female gametes, and represent the start of the sexual portion of the life cycle.

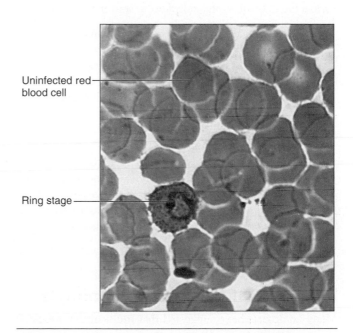

Uninfected red blood cell

Ring stage

**FIGURE 5.29** *Plasmodium*. Ring stage inside human erythrocyte.
Courtesy of Carolina Biological Supply Company, Burlington, NC.

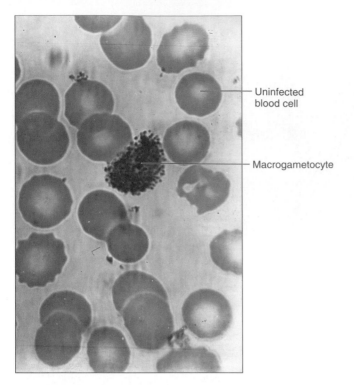

Uninfected
blood cell

Macrogametocyte

**FIGURE 5.30** *Plasmodium.* Developing
macrogametocyte in human erythrocyte.
Courtesy of Carolina Biological Supply Company, Burlington, NC.

If a mosquito bites an infected host and ingests infected
erythrocytes, the gametocytes pass into the mosquito's stomach, where they mature into microgametes and macrogametes. The microgametocytes develop flagellalike
outgrowths, which break free, become motile, and fertilize
macrogametes. The fertilized macrogametes, or zygotes, then
invade the gut wall of the mosquito to start the cycle again.

◆ Study the microscope slides with blood smears prepared
with blood from humans infected with *Plasmodium vivax*
or a similar species. Identify as many stages in the life cycle of *Plasmodium* as you can from your own slides and
the demonstration slides. Observe the changes in morphology of the parasite during its development in the human erythrocytes. With the aid of figure 5.28, try to relate
the portion of the *Plasmodium* life cycle in human erythrocytes to the other parts of the life cycle completed in the
mosquito and in the exoerythrocytic stages in the human.
***Why do you think* Plasmodium *requires two hosts to
complete its life cycle? What special adaptations for life
as a parasite can you identify in* Plasmodium?**

◆ List several of these adaptations for parasitism in table
5.1.

## Evolution of Multicellular Animals

Many zoologists have suggested that multicellular animals,
or the metazoa, evolved from some ancient protozoan ancestor. The particular protozoan stock that may have led to multicellular animals is now only a matter of speculation
because this major evolutionary step must have taken place

in the ancient seas of the Precambrian era more than 700
million years ago. No good fossil evidence has been found to
document the transition from one to many cells, so zoologists have often searched among living animals to seek possible models that might help our understanding of the early
evolution of animals.

The two most widely discussed proposals for the origin
of animals during recent years are the **colonial theory** and
the **syncytial theory.** The colonial theory is based largely on
ideas first put forth over 100 years ago by Ernst Haeckel.
Haeckel proposed that an ancient flagellated protozoon
evolved into a simple multicellular animal resembling the
planula larva of a cnidarian through a process of increasing
colonialization from a unicellular flagellated ancestor. The
green flagellate *Volvox* and several other related colonial
phytoflagellates have sometimes been used to illustrate how
colonies of increasing complexity might have arisen from
unicellular forms. Other scientists have suggested that
amoeboid forms may have divided and formed aggregates
leading to multicellular forms.

Recently some authors have focused attention on the
choanoflagellates (figure 5.15), a small group of solitary and
colonial protozoans that bear a striking similarity to the
choanocytes of sponges (see figure 6.1). This similarity has
long led many scientists to suspect that the choanoflagellates
probably were the ancestors of the sponges. Choanoflagellates have a single flagellum surrounded by a thin collar
formed by microvilli. Recent studies have revealed the presence of similar cells in other kinds of animals and, in fact,
some ultrastructural evidence on the structure of mitochondria and the flagellar apparatus of these cells has led some
biologists to speculate that choanoflagellates may have
given rise to all metazoa. See also the discussion of the origin of animals in Chapter 6.

The syncytial theory, based largely on the writings of J.
Hadzi and E. Hanson, suggests that a multinucleated ciliated
protozoan gave rise to the metazoa by development of cell
membranes that separated the many nuclei and allowed increased specialization and further evolution of metazoans
through development of a multicellular form resembling a
simple flatworm.

| TABLE 5.1 |
| --- |
| Adaptations for Parasitism Exhibited by *Plasmodium* |
| _____ |
| _____ |
| _____ |
| _____ |
| _____ |
| _____ |
| _____ |

Both the colonial theory and the syncytial theory assume that the metazoa are **monophyletic,** arising from a single ancestor. Other zoologists, however, have proposed that the metazoa may not actually be monophyletic, but that multicellularity may have evolved independently several times, making the metazoa **polyphyletic** in origin. Multicellularity does appear to have evolved independently in each of the currently recognized five kingdoms. Among the Monera, there are multicellular forms in the Cyanobacteria (blue-green algae, etc.) and in the Actinomycetes. Among the Protista, there are large multicellular forms included in the green, red, and brown algae. Most fungi and all plants and animals are multicellular; hence, the transition from unicellular to multicellular must have occurred in the evolution of living organisms not once but many times.

Interesting as these theories about the evolution of multicellularity in animals may be, none of them adequately explains the origin of the metazoa; for each of them, there are specific objections and some critical unanswered questions. The real story of the early evolution of the metazoa remains a secret of those ancient seas.

## Key Terms

**Binary fission**  method of asexual reproduction in which an organism constricts and separates into two smaller new individuals.

**Cilium** (plural: cilia)  cylindrical cytoplasmic extension from the surface of certain protozoa and of some kinds of metazoan cells. Serves in locomotion and feeding of protozoa. Similar to a flagellum, but generally shorter and more numerous. Both cilia and flagella are supported by internal microtubules arranged in a characteristic nine outer plus two central pattern.

**Coccidiosis**  disease of birds, rabbits, and other animals caused by the sporozoan *Eimeria* and related sporazoans (Phylum Apicomplexa).

**Colonial theory**  suggested explanation for the evolution of metazoa from some unicellular protozoan ancestor through the aggregation of unicellular forms into a colony.

**Conjugation**  a specialized type of mating, nuclear exchange, and nuclear reorganization characteristic of ciliates; a form of sexual reproduction.

**Contractile vacuole**  an organelle found in many freshwater protozoa that serves in osmoregulation (water balance).

**Cytostome**  the "cell mouth" found in many protozoa, including ciliates, some flagellates, and some apicomplexa. In ciliates, the cytostome is often surrounded by specialized ciliary feeding organelles.

**Dioecious**  occurrence of sex organs in different individuals; that is, there are distinct male and female individuals of a species.

**Flagella** (singular: flagellum)  cylindrical cytoplasmic extensions from the surface of certain protozoa and some metazoan cells. Function in locomotion and feeding of mastigophorans. Similar to cilia but longer and usually fewer per cell.

**Gonidia** (singular: gonidium)  specialized reproductive cells in *Volvox*.

**Macronucleus**  the large metabolic nucleus typical of ciliates. Often has a characteristic shape. Divides amitotically by pinching in two. Contains many duplicated sets of genes (polyploid).

**Malaria**  disease of humans and other animals. Caused by members of the protozoan genus *Plasmodium*, which invade the blood and other tissues of the hosts.

**Micronucleus**  the small reproductive nucleus in ciliates. Some ciliates have more than one micronucleus. Usually divides by ordinary mitosis.

**Monoecious**  occurrence of male and female reproductive organs in the same individual; hermaphroditic. Not usually capable of self-fertilization.

**Monophyletic**  having a common evolutionary origin; sharing a common ancestry.

**Nuclear dimorphism**  having two distinct types of nuclei within the same cell. Characteristic of the ciliated protozoa, Phylum Ciliophora.

**Polyphyletic**  having different evolutionary origins; no common ancestry.

**Pseudopodium** (plural: pseudopodia)  protoplasmic extension of an amoeboid cell; the "false foot" of the Sarcodina used for feeding and locomotion. Various Sarcodina have pseudopodia specialized for specific purposes. Also present in other kinds of amoeboid cells such as leucocytes or white blood cells in many kinds of animals.

**Stigma**  light-sensitive spot found in certain protozoa, such as *Euglena* and *Volvox*.

**Syncytial theory**  suggested explanation for the possible evolution of metazoa from some protozoan ancestor as a result of partitioning a multinucleated ciliate form. The resulting metazoan is thought to resemble a primitive flatworm.

## Internet Resources

Visit the zoology website at http://www.mhhe.com/zoology to find live Internet links for each of the references listed below.

1. Society of Protozoologists. This site contains links to images of protozoans, databases, culture collections, and discussion groups.

2. Protozoa. Various lectures on different protozoan groups are located at this site.

3. National Center for Infectious Diseases. This CDC site has many links to information on bacterial, viral, protozoan, and worm-related diseases (primarily affecting humans).
   Other links:
   Trypanosomiasis.
   Malaria.

4. *Giardia lamblia.* A FDA supported page with information on *Giardia lamblia,* a flagellated protozoan that causes diarrhea in humans.

5. *Entamoeba histolytica.* A FDA supported page with information on *Entamoeba histolytica,* a protozoan that

affects humans and other mammals, causing gastrointestinal distress and diarrhea.

6. Introduction to the Ciliata. UC Berkeley page on ciliates.

7. The World Health Organization Division of Control of Tropical Diseases. Provides extensive information on malaria and other tropical diseases and the protozoans that cause them. Very interesting site—lots to learn here!

## Critical Thinking Questions

1. Compare the locomotion of a mastigophoran (flagellate), a ciliophoran (ciliate), a sarcodine (amoeba), and an apicomplexan (sporozoan). Explain how the type of locomotion is important in the nutrition, ecology, and behavior of each type of protozoon.

2. Identify three important protozoan parasites of humans and discuss their importance in human health.

3. Discuss the possible role of protozoa in the evolution of multicellular animals.

## Suggested Readings

Anderson, O.R., and M. Druger. 1997. *Explore the World Using Protozoa*. Arlington, VA: National Science Teachers Association and Society of Protozoologists.

Jahn, T., E.C. Bovee, and F.F. Jahn. 1979. *How to Know the Protozoa*. Dubuque, IA: WCB/McGraw-Hill.

Lee, J.J., S.H. Hutner, and E.C. Bovee. 1985. *Illustrated Guide to the Protozoa*. Lawrence, KS: Society of Protozoologists.

# NOTES AND SKETCHES

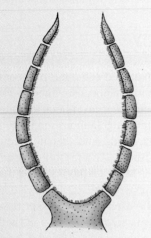

**Asconoid**

**Syconoid**

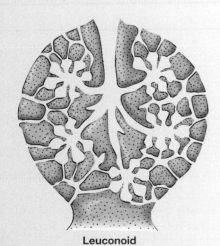

**Leuconoid**

After completing the laboratory work in this chapter, you should be able to perform the following tasks:

1. Briefly characterize the Phylum Porifera.
2. Describe the basic organization of the sponge body.
3. Explain the skeletal elements of sponges and their composition.
4. List and distinguish the three classes of sponges.
5. Differentiate between asconoid, syconoid, and leuconoid types of canal systems found in sponges.
6. Describe the structure of a choanocyte and explain the importance of choanocytes in sponges. Explain where choanocytes are found in asconoid, syconoid, and leuconoid sponges.
7. Describe the pattern of water flow in a syconoid sponge and explain the importance of water currents in sponges.
8. Explain the role of gemmules in sponges.
9. Explain the significance of experiments on the reaggregation of sponges from dissociated sponge cells.
10. Discuss the evolutionary origin of sponges and their possible phylogenetic relationships with other groups of organisms.

## Introduction

The Phylum Porifera includes the sponges, a group of sedentary animals so different from other types of animals that they were long thought to be plants. Sponges are among the most primitive of multicellular animals. They have a simple type of body organization, with a porous body permeated by a system of water canals through which water is pumped by action of special flagellated cells, the choanocytes (figure 6.1). The body of a sponge consists of a network of epithelial cells, canals, and chambers embedded in a loose connective tissue.

Most of the more than 5,000 described species of sponges are marine, but there are also about 150 species of sponges that occur in fresh waters. Formerly, certain types of marine sponges were of considerable economic importance and commonly used for bathing and cleaning purposes. Commercial sponge fishing was a profitable industry in several parts of the world, started by the Greeks in the Aegean Islands. Later

**FIGURE 6.1** Choanocyte.

**FIGURE 6.2** Bath sponge. One-fourth natural size. Class Demospongiae.

Courtesy of Carolina Biological Supply Company, Burlington, NC.

however, a hurricane in 1985 stirred up the waters off the west coast of Florida and the sponge beds began to grow again, giving new life to the sponge industry in that area. *Why do you think a hurricane might have an effect on the growth of sponges and other marine life?*

New interest in sponges has arisen in recent years with the discovery of useful pharmacological compounds that can be extracted from some species of sponges. An important anti-cancer drug has been extracted from a Caribbean sponge.

## Morphology

The external surface of the sponge body is covered by an epithelial layer of flattened cells, the **pinacocytes.** Certain of the inner canals and chambers are lined by **choanocytes.** Between the inner and outer layers lies the **mesohyl** (formerly called mesenchyme), a gelatinous matrix with several types of amoeboid cells, fibrils, and skeletal elements of **spicules** and/or **spongin fibers.** The mesohyl resembles a type of connective tissue. Most textbooks state that sponges lack true tissues. Some specialists, however, believe that the **pinacoderm** and mesohyl do represent tissues, despite the fact that these layers are not homologous with the tissue layers found in higher animals. An inner layer of flagellated choanocytes generates water currents through the internal canal systems. These water currents are essential in the life of the sponge because they carry food particles and oxygenated water into the sponge and waste products as well as gametes and/or larvae out of the sponge.

Sponges are sedentary filter feeders, able to capture tiny food particles measuring from about 0.5 to 50 μm from the seawater. The capture of food, which consists chiefly of fine, suspended organic particles and tiny planktonic organisms, is accomplished mainly by the choanocytes. Some food is passed to internal amoebocytes, which may in turn pass it on to other cells. All digestion is intracellular (within individual cells). *How efficient do you think such a system of food transfer might be?*

## Phylogeny

Sponges lack organized multicellular organs, including nerves and muscles. In one sense, the water canal system of sponges is like an externalized circulatory system, but all the life processes of sponges take place within individual cells or groups of cells. Because of their lack of differentiated organs and because of their various other morphological and developmental peculiarities, sponges have long been believed to represent an early offshoot from the main line of animal evolution and not to be closely related to any more advanced animal types.

Scientists familiar with sponge choanocytes and **choanoflagellate** protozoans many years ago suggested that sponges might have evolved from choanoflagellates or some choanoflagellate ancestor. The choanoflagellates are represented by a small group of living protozoa with flagellated collar cells that resemble the choanocytes of sponges and

sources of sponges included the Mediterranean, the West Indies, the Gulf of Mexico, and the Caribbean Sea where the warm, shallow waters and rocky bottoms were favorable for the growth of bath sponges (figure 6.2). In the United States, Florida was the center of sponge fishing, and in 1887 sponges were the most important fishery product from Florida.

Overfishing, sponge diseases, and pollution have taken their toll on sponge beds around the world. As a result, there is now relatively little commercial harvesting of natural sponges. Synthetic sponges have largely replaced this once important natural product in the marketplace. Interestingly,

include both solitary species and simple colonies. It has generally been assumed by most scientists that the metazoa evolved from some flagellated protozoan ancestor, but the specific type of ancestral protozoan continues to be a topic for speculation.

Some recent electron microscopic studies on sponge choanocytes, choanocytelike protozoans, and similar cells found in some more advanced multicellular animals, however, have led to a more far-reaching suggestion. Some scientists now suggest that **both** the sponges and other multicellular animals (metazoa) may have evolved from some type of choanoflagellate, but that the two groups emerged at different times. They point to the lack of fossil sponges in the oldest geological formations and have suggested that this lack of fossil remains indicates that the sponges evolved much later in geological time than generally believed.

◆ Review the discussion of the evolution of multicellularity at the end of Chapter 5.

## Skeleton

The skeleton of sponges is relatively complex in comparison to the general organization of the body of sponges. The skeleton is internal and consists of individual mineralized elements, called **spicules,** made up of calcium carbonate, or silicon salts, and/or a network of organic fibers composed of fibrillar collagen and/or spongin fibers.

Spicules occur in a variety of forms (figures 6.3 and 6.4) and are important in the classification of sponges. Spongin is a tough, fibrous protein chemically similar to collagen, but unique to the sponges.

## Classification

The division of the phylum into classes is based largely on the nature of the skeleton, although some recent textbooks have also included methods of reproduction in the classification of sponges. Three or four classes of sponges are recognized by different scientists. The **Demospongiae,** horny sponges; the **Calcarea,** calcareous sponges; and the **Hexactinellida,** glass sponges, are well-defined and generally recognized. A fourth class, the **Sclerospongiae,** coraline sponges, is recognized by some recent specialists. This class was established after the discovery of an unusual group of sponges about 40 years ago off the coast of Jamaica. Similar sponges have subsequently been found on coral reefs and in associated undersea caves in other places around the world. These sponges have an unusual, coral-like skeleton with a calcareous matrix containing siliceous spicules and spongin fibers. Other scientists, however, include the sclerosponges in the class Demospongiae.

This disagreement about the number of classes illustrates an **important principle of taxonomy** and the nature of higher taxonomic categories. Higher categories are established and defined on the basis of evidence considered and evaluated by scientists specializing in a particular group of animals. As new information is discovered and as new or old specialists reevaluate the evidence, higher taxonomic categories may be established, redefined, or abolished according to the best professional judgment of those specialists.

**Species** are generally defined as discrete, natural categories made up of a group of individuals sharing a common gene pool. This concept of the nature of a species makes the species less subject to changes due to the discovery of new evidence or changes of professional opinion than are taxonomic categories above the species level. In contrast to this concept of a species, the definitions of **higher categories** such as genus, family, order, class, phylum, and kingdom are more subject to change as new evidence accumulates.

**Classification, like all of science, should be viewed as a status report;** the current classification of any group of organisms represents best the current state of knowledge and opinion of specialists on that group of organisms. As new evidence is obtained, scientists are better able to understand and to describe living organisms and biological processes.

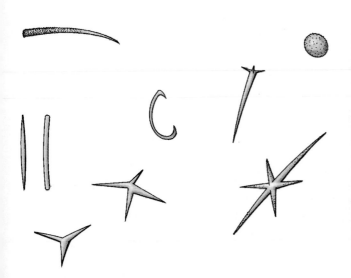

**FIGURE 6.3**    Examples of spicule shapes.

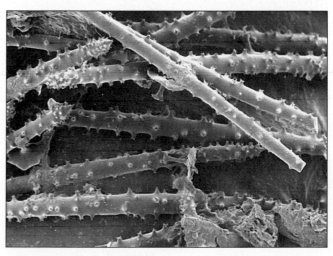

**FIGURE 6.4**    Spicules from a freshwater sponge. Magnification 7,000×.
Scanning electron micrograph by F.W. Harrison.

# Class Demospongiae

## Horny Sponges

Members of this class possess a skeleton made up of a network of spongin fibers (a structural protein secreted by certain sponge cells), siliceous spicules, both, or neither. Most members of this class are marine, but two families are found in freshwater streams, ponds, and lakes. All commercial sponges belong to this class. All members of this class have a leuconoid canal system. Examples: *Spongia, Haliciona, Microciona,* and the deep-sea sclerosponges (all marine); *Spongilla* (freshwater).

# Class Calcarea (Calcispongiae)

## Calcareous Sponges

Sponges with a skeleton consisting of many small spicules made of calcium carbonate embedded in a loose, jellylike matrix. All species are marine. All three types of canal systems (asconoid, syconoid, and leuconoid) are found among members of this class. Examples: *Scypha, Leucosolenia.*

# Class Hexactinellida (Hyalospongiae)

## Glass Sponges

Sponges with a skeleton composed of siliceous spicules, usually with six rays, as the class name implies. The spicules are often fused together into a continuous network. The tissues of glass sponges are a syncytial network of fused amoeboid cells. All glass sponges are marine, and most species are found in deep areas of the world oceans. Because of their remote habitat, little is known about their biology. Members of this class have either syconoid or leuconoid canal systems. Examples: *Euplectella* (Venus's flower basket) and *Hyalonema.*

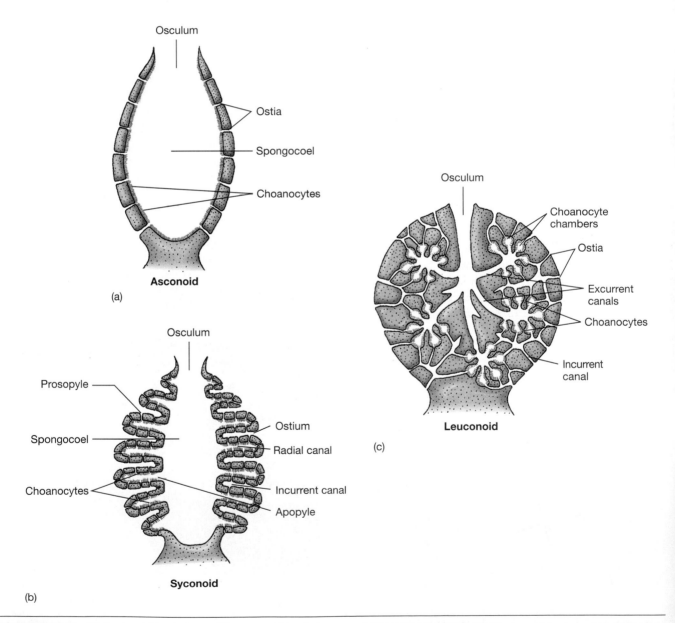

**FIGURE 6.5** (*a*) Asconoid body type, (*b*) syconoid body type, and (*c*) leuconoid body type.

**Preserved specimens**
  *Leucosolenia*
  *Scypha*
  Bath sponge
  Glass sponge (demonstration)
  Assorted marine sponge types (demonstration)
  Freshwater sponges (demonstration)
**Prepared microscope slides**
  *Leucosolenia,* whole mount
  *Scypha,* cross section
  *Scypha,* eggs and embryos (demonstration)
  Sponge spicules (demonstration)
  Spongin fibers (demonstration)
  Sponge gemmules (demonstration)

## Body Organization

Morphologically, the bodies of sponges exhibit three distinct types based on the organization of their internal canal systems. These three types of canal systems are designated as **asconoid, syconoid,** and **leuconoid** (figure 6.5). It is important to recognize that these three types of canal systems represent morphological organization and are not the basis for differentiating the three classes of extant sponges. Actually, most extant sponges are leuconoid.

### Asconoid Sponge

*Leucosolenia* (figure 6.6) is a small colonial sponge of the asconoid type. Examine a preserved specimen or a microscopic whole mount with your stereoscopic microscope and observe the following features.

1. A system of horizontal tubes that bears numerous upright branches.
2. The upright branches that represent individual sponges of the colony.
3. Buds formed on the sides of the individual sponges.
4. The terminal opening, or **osculum,** at the upper end of each sponge. Water passes out of the sponge through this opening.
5. The **spongocoel,** a large central cavity within the sponge. This cavity is lined by the specialized, flagellated collar cells or **choanocytes,** which create water currents within the sponge.
6. Water enters the sponge through many **ostia,** tiny pores that penetrate the body wall.
7. Numerous triradiate (three-rayed) **spicules** may be seen embedded in the wall.

### Syconoid Sponge

*Scypha* (figures 6.7 and 6.8), formerly also called *Sycon* and *Grantia,* is a small, slender sponge of the syconoid type.

◆ Obtain a preserved specimen and note the size and shape of the body. Examine specimens cut in half longi-

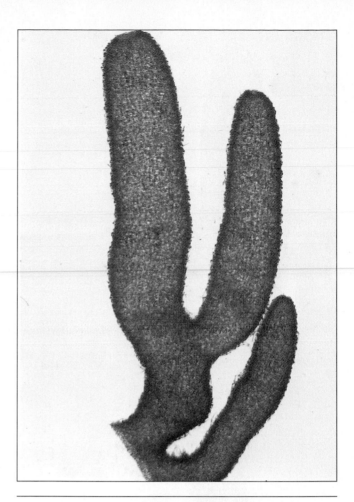

**FIGURE 6.6** *Leucosolenia.*
Courtesy of Carolina Biological Supply Company, Burlington, NC.

**FIGURE 6.7** *Scypha,* typical cluster of sponges.
Courtesy of Carolina Biological Supply Company, Burlington, NC.

tudinally and observe the spongocoel, the radial canals leading from the spongocoel into the body wall, and the short collar region formed by a cylinder of giant spicules, which leads to the osculum. On the surface, observe the numerous cortical spicules that extend outward through the pinacocyte layer, which covers the external surface of the sponge.

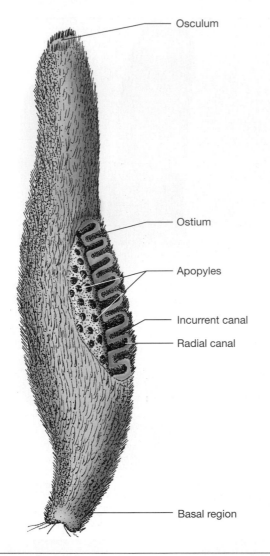

Osculum

Ostium

Apopyles

Incurrent canal

Radial canal

Basal region

**FIGURE 6.8** *Scypha.* External view with portion of body wall removed to show interior detail.

◆ Study a stained cross section of *Scypha* and identify the **spongocoel,** the **radial canals,** and the **incurrent canals,** which open to the exterior of the sponge through small incurrent pores.

The openings between the radial and excurrent canals are the **prosopyles;** the excurrent canals empty into the spongocoel through the **apopyles.** Much of the tissue within the sponge body consists of a loose mesohyl. Details of the cross section of the body wall of *Scypha* are illustrated in figure 6.9.

The outer surface of the sponge is covered by a layer of thin, flat cells, called **pinacocytes.** A similar layer of pinacocytes lines the spongocoel. The **choanocytes** (figure 6.10) are found lining the radial canals, which empty into the spongocoel through the apopyle (figure 6.11). These choanocytes are small and difficult to identify in most microscopic preparations.

You should be able to find some large undifferentiated **amoeboid cells** within the mesohyl. Some of these amoeboid cells develop into eggs. Sponges lack differentiated gonads, and eggs and embryos are often found embedded in the body wall in microscopic cross sections of mature sponges. **Sperm** develop from choanocytes and are carried by water currents into other sponges. *Scypha,* like many sponges, is **hermaphroditic,** but the eggs and sperm develop at different times.

◆ Examine your cross section and see if you can identify some eggs and embryos.

Sponges reproduce both sexually and asexually. Marine sponges may reproduce asexually by fragmentation, in which part of a sponge colony may break off and be washed away to grow into a new colony elsewhere. In sexual reproduction, the fertilized eggs of sponges develop into a ciliated larva that is released into the plankton and that may later settle on a suitable substrate and develop into a small sponge.

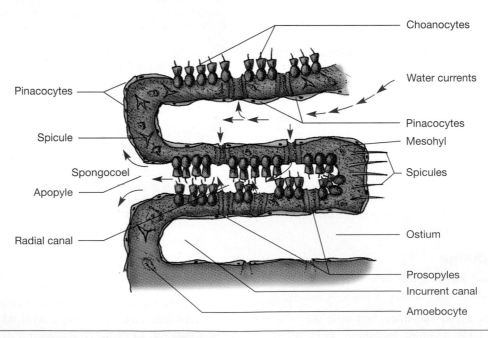

Pinacocytes

Spicule

Spongocoel

Apopyle

Radial canal

Choanocytes

Water currents

Pinacocytes

Mesohyl

Spicules

Ostium

Prosopyles

Incurrent canal

Amoebocyte

**FIGURE 6.9** Cross section through the body wall of *Scypha.*

Courtesy of Carolina Biological Supply Company, Burlington, NC.

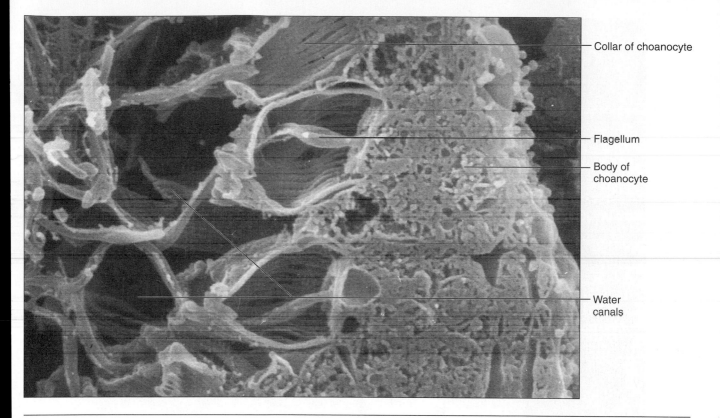

Collar of choanocyte

Flagellum

Body of choanocyte

Water canals

**FIGURE 6.10** Cross section through choanocytes of *Ephydatia,* a freshwater sponge. Magnification 21,000×.
Scanning electron micrograph of a freeze fracture section by Louis de Vos.

## Leucon-Type Sponge

Leucon-type sponges are structurally the most complex and also the most common body type among living sponges. All freshwater sponges and most marine sponges are leucon-type.

◆ Study a portion of a preserved specimen of a bath sponge and note its rubbery texture and the complex system of branching canals.

Located between the incurrent canals and the excurrent canals are numerous small, spherical **flagellated chambers.** Collar cells are found lining only these tiny flagellated chambers in leuconoid sponges.

◆ Study also a dried specimen of a bath sponge and note how its texture differs from that of the preserved specimen. Only the network of **spongin fibers** remains in the dried sponge.

◆ Observe the microscopic demonstration slide of spongin fibers.

◆ Compare the three types of canal systems in figure 6.5 and in the sponge materials provided in the laboratory. Remember, sponges are sedentary and thus dependent on water currents passing through their canal systems for their food supply and for gas exchange. Even sponges have to respire. *Which type of canal system is likely to be most efficient in gas exchange and food capture? Why? Do you think this might have some importance in the evolution of sponges?*

### Demonstrations

1. Eggs and developing embryos of *Scypha* (microscope slide)
2. Assorted spicules (microscope slide)
3. Spongin fibers (microscope slide)
4. Glass sponge and examples of other types of sponges
5. Skeletons of glass sponges
6. Preserved and dried samples of bath sponges and other types of sponges

## Freshwater Sponges

Although most sponges are marine, a few species live in freshwater streams, ponds, and lakes. Freshwater sponges (figure 6.12) are much less prominent than their ubiquitous marine relatives. The freshwater sponges grow as tufts or small irregular masses encrusting sticks, stones, or submerged plants. Most species are yellow or brown, but a few species are green because of symbiotic algae that live within the sponge.

All freshwater sponges (and a few marine sponges) form internal asexual buds called **gemmules** (figures 6.13 and 6.14). Gemmules are resistant stages, which serve to carry the sponge through the winter or aid in survival during a drought. Observe the microscope slide of gemmules on demonstration. *What important features of the gemmule aid in the survival of the species?*

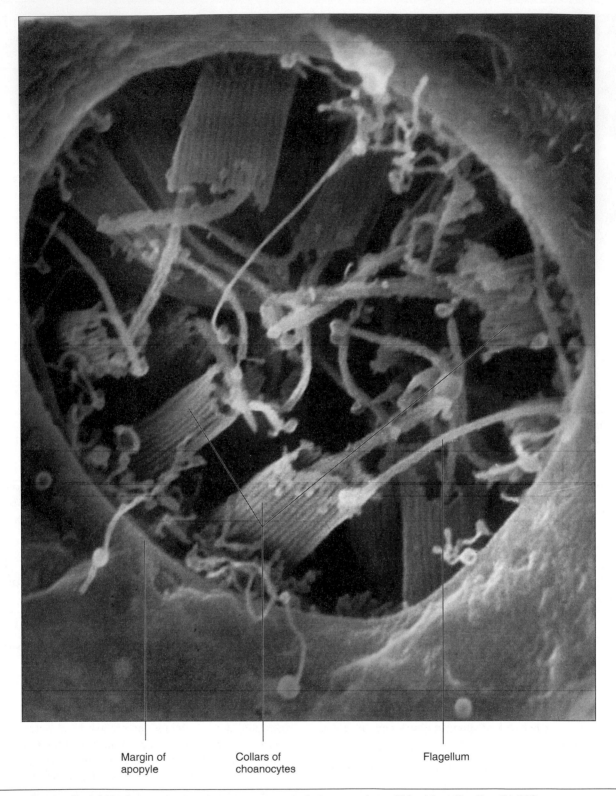

Margin of          Collars of          Flagellum
apopyle            choanocytes

**FIGURE 6.11**  Opening of apopyle showing arrangement of choanocytes within. Magnification 21,000×.
Scanning electron micrograph by Louis de Vos.

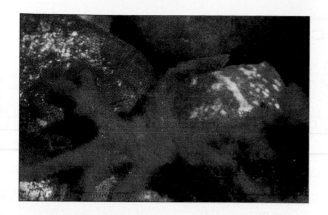

**FIGURE 6.12** Freshwater sponge, *Spongilla*.
Courtesy of Carolina Biological Supply Company, Burlington, NC.

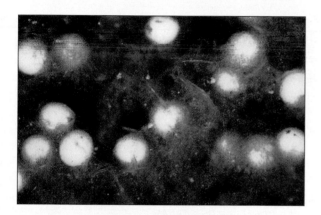

**FIGURE 6.13** Gemmules in freshwater sponge.
Courtesy of Carolina Biological Supply Company, Burlington, NC.

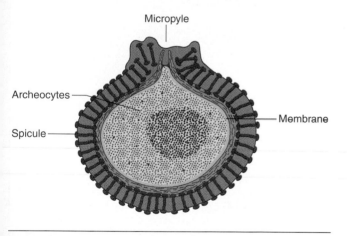

**FIGURE 6.14** Cross section of a gemmule of a freshwater sponge.

## Regeneration and Reconstitution (Optional Exercise)

Sponges are especially noted for their powers of regeneration. A small part of a sponge can regenerate a complete sponge. Many years ago, H.V. Wilson showed that pieces of a living sponge pressed through a fine cloth mesh to separate the sponge into individual cells and clusters of cells could reassemble and develop into a complete new sponge. More experiments with the separation and reassociation of sponge cells have provided important information about the nature and mechanism of cell-recognition processes and the organization of tissues during development.

If living sponges are available in your laboratory, you may be able to repeat this famous experiment. *Microciona*, a common red sponge on the Atlantic coast of the United States, was used in the original experiments. You will need a small finger bowl, a watch glass, a clean microscope slide, a pipette, some silk mesh cloth, and some seawater for this experiment.

### Procedure

1. Fill the finger bowl about two-thirds full of seawater of the proper ionic strength (depending on the source of the sponges) and place the watch glass on the bottom of the dish. Place the microscope slide on top of the watch glass.
2. In a separate bowl of seawater, prepare a suspension of cells and fragments of *Microciona* by pressing small pieces of sponge through a fine silk bolting cloth or nylon mesh of about 150–200 micrometers in diameter.
3. Pipette a small amount of this cell suspension onto the slide and allow the cells to settle.
4. Carefully lift out the slide and observe the cells. Make similar observations at intervals during the next 24–48 hours. Watch for the initiation of cellular aggregation and the thin protoplasmic extensions (filopodia) put out by the small clumps of aggregating cells. Sketch the cells and clumps of cells as a record of your observations.

### Demonstrations

1. Gemmules (microscope slide)
2. Living or preserved freshwater sponges

## Key Terms

**Asconoid**   the simplest type of sponge canal system with a central spongocoel lined with choanocytes and with many ostia opening directly into the spongocoel.

**Choanocyte**   a special type of flagellated collar cell characteristic of sponges that lines the flagellated water chambers and canals.

**Gemmule**   a dormant stage in the life cycle of freshwater sponges and a few species of marine sponges. Formed as internal asexual buds, they aid the sponges in surviving adverse environmental conditions.

**Leuconoid**   a complex type of sponge body form with an intricate system of internal water canals. Choanocytes line certain small chambers within the system.

**Mesohyl**   the loose, gelatinous matrix in a sponge body containing several types of amoeboid cells, fibrils, and

The figure labels in Figure 6.14: Micropyle, Archeocytes, Spicule, Membrane

skeletal elements (spicules and/or spongin fibers). Located between the outer pinacoderm and the inner choanocyte layer. Formerly called mesenchyme.

**Pinacocyte**    a thin, flattened type of cell that lines the outer surface of sponges and most inner surfaces not lined by choanocytes.

**Pinacoderm**    outer layer of sponge body made up of pinacocytes.

**Spicules**    individual skeletal elements that make up the skeleton of most sponges. Consist mainly of calcium carbonate or silicon salts and exhibit many different shapes.

**Spongin fibers**    a loose network of protein fibers that form part or all of the skeleton of most horny sponges (Class Demospongiae); also occur in the mesohyl of other sponges and in the walls of gemmules.

**Syconoid**    a type of sponge canal system that exhibits a central spongocoel into which numerous radial canals lined with choanocytes empty.

## Internet Resources

Visit the zoology website at http://www.mhhe.com/zoology to find live Internet links for each of the references listed below.

1. Introduction to Porifera. Information on living and fossil species of sponges, with color photographs.
2. Phylum Porifera. Information about the structure, reproduction, and ecology of sponges, with color photographs.
3. Porifera Web Page. Information on sponge research, including color photographs.

## Critical Thinking Questions

1. Many years ago, sponges were often thought to be plants rather than animals. Discuss why early naturalists might have made such an error.

2. Discuss how natural selection may have influenced the development of complex canal systems in sponges.

3. Sponges are very common and abundant in marine waters but only a few species are found in fresh water, and these freshwater species are neither common nor abundant. What factors might be responsible for this large difference in number of species and abundance of sponges between the sea and fresh water?

## Suggested Readings

Bergquist, P.R. 1978. *Sponges.* London: Hutchinson & Co.

De Vos, L., K. Rutzler, N. Boury-Esnault, C. Donadey, and J. Vacelet. 1991. *Atlas of Sponge Morphology.* Washington, D.C.: Smithsonian Institution Press.

Stevely, J.M., J.C. Thompson, and R.E. Warner. 1978. *The Biology and Utilization of Florida's Commercial Sponges.* Tallahasee Technical Paper no. 8, Florida Sea Grant Technical Report.

Tucker, J.B. 1985. Drugs from the sea spark renewed interest. *Bioscience* 35:541–45.

# NOTES AND SKETCHES

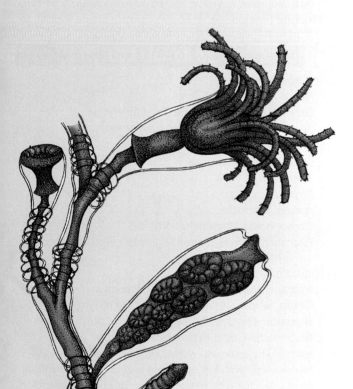

After completing the laboratory work in this chapter, you should be able to perform the following tasks:

1. List the four classes of cnidarians and briefly characterize each class.
2. Describe the general morphology of *Hydra*.
3. Identify the main morphological features of *Hydra* on a microscope slide or illustration.
4. Discuss the structure and function of nematocysts and explain their importance in *Hydra* and other cnidarians.
5. Describe the feeding reaction of *Hydra* and its control.
6. Describe the reproductive processes of *Hydra* and be able to identify mature male and female individuals.
7. Identify the chief morphological features of the medusa of *Gonionemus* and explain the function of each part.
8. Describe the life cycle of *Gonionemus*.
9. Identify the main morphological features of a mature *Obelia* colony and explain the function of each.
10. Describe the life cycle of *Obelia* and explain how it illustrates the alternation of generations commonly found in cnidarians.
11. Describe the main morphological features of the scyphozoan medusa *Aurelia* and explain the function of each.
12. Describe the life cycle of *Aurelia*.
13. Describe the structure of the sea anemone *Metridium* and explain the function of its principal parts.

## Introduction

Cnidarians (formerly called coelenterates) are the simplest animals with definite tissues. The cnidarian body has **diploblastic construction;** that is, it consists of two well-defined tissue layers and a third intervening layer of gelatinous material, the **mesoglea,** that varies in structure among the three classes of cnidarians. The outer **epidermis** layer covers the external surface of the body, and the inner **gastrodermis** layer lines a single internal body cavity. In the simplest cnidarians, Class Hydrozoa, the mesoglea is thin and gelatinous; in the Class Scyphozoa and the Class

Cubozoa, the mesoglea is fibrous and contains amoeboid cells; and in the Class Anthozoa, the mesoglea has many amoeboid cells and thus resembles a true mesenchyme layer. The cnidarians are especially noted for their prominent **radial symmetry** (figure 7.1).

The name Cnidaria is derived from the **cnidocytes,** special cells that produce the **nematocysts** or "stinging capsules" characteristic of this phylum. The alternate phylum name, Coelenterata, refers to the large **coelenteron,** or gastrovascular cavity, also characteristic of the group.

Two basic body forms are exhibited by the Cnidaria: an attached **polyp stage** and a free-swimming **medusa stage.** Many species have both a polyp stage and a medusa stage, and their life cycles involve an **alternation** of these two body forms or "generations." Two other important distinguishing characteristics of the phylum include **tentacles** around the mouth and a diffuse **nerve net,** which provides a modest degree of nervous coordination.

# Classification

The phylum is divided into four classes:

## Class Hydrozoa (Hydroids and Siphonophores)

Animals usually have both polyp and medusa stages in the life cycle, medusae with a velum, and gonads on the radial canals of the medusae. The mesoglea is thin and gelatinous; there is a well-defined epidermis and gastrodermis. Freshwater and marine species. Examples: *Obelia, Hydra, Gonionemus,* and *Physalia* (Portuguese man-of-war).

## Class Scyphozoa (True Jellyfish)

Most are large marine jellyfish with an abundant mesoglea, a reduced or absent polyp stage, and without a velum in the medusae. The mesoglea contains fibers and amoeboid cells. All species are marine. Examples: *Aurelia, Chrysaora* (sea nettle), and *Cyanea.*

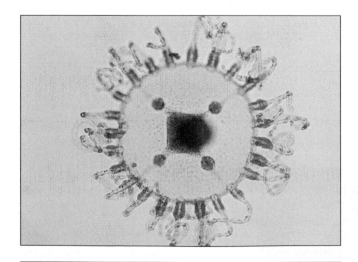

**FIGURE 7.1** Medusa of *Obelia.*
Carolina Biological Supply Company, Burlington, NC.

## Class Cubozoa (Sea Wasps or Box Jellyfish)

All species have small marine medusae with conical bells of four flattened sides and a velum or velumlike structure around the margin of the bell. Small polyps metamorphose directly into medusae. Common in all tropical seas, especially in the Indo-Pacific and Western Pacific. Their stings are dangerous to humans; the nematocysts contain some of the strongest toxins known. Examples: *Carybdea, Chironex,* and *Chiropsalmus.*

## Class Anthozoa (Sea Anemones and Corals)

These solitary or colonial animals only have a polyp stage (the medusa is absent), a pharynx or gullet is present, gastrovascular cavity partitioned by septa. Mesoglea with many amoeboid cells. All speices are marine. Examples: *Metridium* (sea anemone), *Astrangia* (coral), *Gorgonia* (sea fan), and *Renilla* (sea pansy).

## Materials List

Living specimens
 *Hydra*
 *Daphnia* or *Artemia* larvae (food for *Hydra*)
 Sea anemones in aquarium (demonstration)
Preserved specimens
 *Gonionemus,* medusae
 *Metridium*
 *Physalia*
 *Aurelia,* medusae
 Ctenophores (demonstration)
 Representative Hydrozoa, Anthozoa, and Scyphozoa (demonstrations)
Prepared microscope slides
 *Hydra,* cross section, longitudinal section
 *Obelia,* whole mount of polyp, whole mount of medusa
 Nematocysts, discharged (demonstration)
 *Gonionemus,* statocysts (demonstration)
 *Hydra,* whole mount, with buds (demonstration)
 *Hydra,* male with testes, whole mount (demonstration)
 *Hydra,* female with ovary, whole mount (demonstration)
 Green *Hydra* with symbiotic algae, whole mount (demonstration)
 Planula larva, whole mount (demonstration)
 *Aurelia* marginal sense organs, scyphistoma, ephyra larva, strobila (demonstrations)
Chemicals
 1% Acetic acid
 0.01% Methylene blue solution
 Pond water

# A Hydrozoan Polyp: *Hydra*

## Class Hydrozoa

*Hydra* (figure 7.2) typifies the polyp form of a cnidarian. *Hydra* lives in freshwater streams, lakes, and ponds. It is usually attached to submerged sticks, stones, or vegetation and

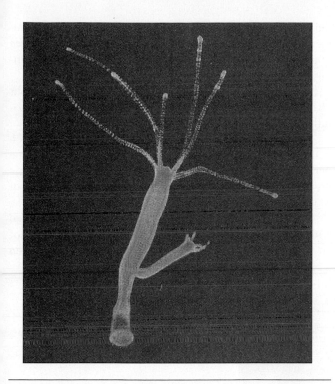

**FIGURE 7.2** Living *Hydra.*
Courtesy of Carolina Biological Supply Company, Burlington, NC.

feeds on various small aquatic animals. About fourteen species of *Hydra* are known to occur in the United States.

Although *Hydra* serves as a good example of the polyp form of a cnidarian, it differs in several important ways from most members of the Class Hydrozoa: (1) *Hydra* is solitary, whereas most hydrozoan polyps are colonial; (2) *Hydra* has no separate medusa stage, which most hydrozoan species do have; (3) the polyp of *Hydra* bears gonads, unlike most hydrozoan polyps; (4) *Hydra* is mobile, unlike most hydrozoan polyps, which are sessile; and (5) *Hydra* lives in fresh water, whereas most hydrozoans are marine.

## General Appearance and Morphology

◆ Examine a living specimen of *Hydra* in a dish of pond water. Caution: Be sure to use pond water and not tap water, since most tap water contains trace amounts of copper and other substances toxic to *Hydra*.

Locate the following structures on your specimen with the aid of figure 7.3: (1) the **basal disc** at the lower end, which serves for attachment; (2) the cylindrical **body;** (3) a circle of **tentacles** at the free end (*how many?*); (4) the **hypostome,** an elevation between the bases of the radially arranged tentacles; (5) the **mouth** in the center of the hypostome; (6) **buds,** the products of asexual reproduction, may also be present; and

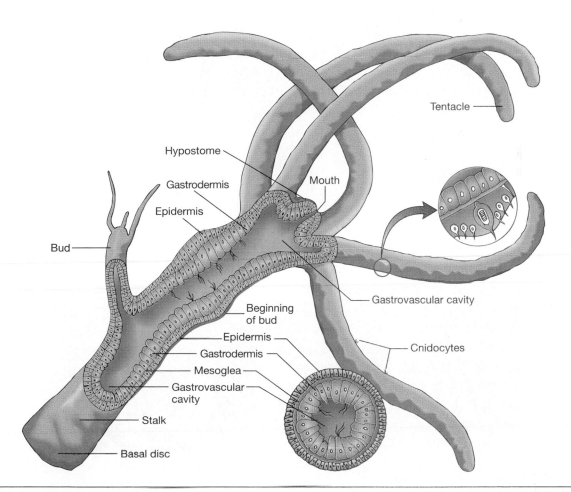

**FIGURE 7.3** *Hydra,* body construction.

Cnidaria **107**

(7) **ovaries** or **testes** are also present on the middle portion of the body in mature specimens (figure 7.9).

## Behavior

◆ *Does your specimen change in shape?* Touch one of the tentacles with the tip of your dissecting needle. *What is the reaction? What methods of locomotion are used by* **Hydra?** Observe *Hydra* feeding in an aquarium, or add a few *Daphnia* or washed brine shrimp (*Artemia*) larvae to your dish near the specimen to observe feeding.

## Cnidocytes and Nematocysts

After you have studied the basic form and behavior of your specimen, place it on a clean microscope slide in a drop of water and carefully add a coverslip. Take care not to crush the animal with the weight of the coverslip. Observe the numerous **cnidocytes,** which appear in clusters on the surface of the tentacles (figure 7.4). Each cnidocyte is a cell containing a **nematocyst,** or stinging capsule.

The cnidocytes of *Hydra* can be stained to aid in your observations by adding a drop of 0.01% methylene blue solution to the edge of the coverslip.

*Hydra* has four different kinds of nematocysts, each with a distinctive structure and function (figure 7.5). Some penetrate and paralyze, some adhere, and some entangle prey organisms. Nematocysts are important in the capture of food, in locomotion, and in attachment.

Nematocysts are complex **secretory products** formed within developing cnidocytes. Each nematocyst consists of an outer protein **capsule,** and a long, coiled **tube** often

**FIGURE 7.4** *Hydra.*
Scanning electron micrograph by Louis de Vos.

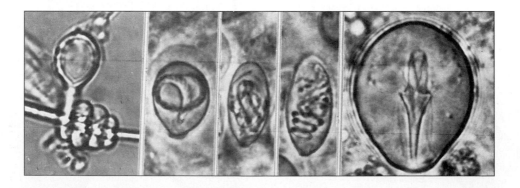

**FIGURE 7.5** *Hydra.* Nematocyst types.
Courtesy of Carolina Biological Supply Company, Burlington, NC.

armed with **spines** or **barbs.** Toxins, enzymes, and other chemicals are also contained within the capsule of certain kinds of nematocysts.

Tap on the coverslip of your wet mount of *Hydra* tentacles to induce a discharge of the nematocysts. When properly stimulated, nematocysts empty their contents with a rapid discharge as the coiled tube shoots out. In addition to tapping the coverslip to induce nematocyst discharge, you can also stimulate nematocyst discharge by adding a drop of dilute acetic acid (1% solution).

Study the discharged nematocysts under high power or with an oil immersion lens, if available. Observe the outer capsule, the long thread or tube, and the large spines or barbs at the base of the tube. *How many kinds of nematocysts can you find on your slide?*

Observe also the microscopic demonstration of discharged nematocysts.

## Histological Structure

*Hydra* and other members of the Class Hydrozoa exhibit the simplest histological structure among the Cnidaria (figure 7.3).

◆ Examine stained slides of cross and longitudinal sections of *Hydra,* and note the following: (1) **epidermis,** the outer, thinner epithelial layer of cells; (2) **gastrodermis,** the inner layer of cells; (3) **mesoglea,** a very thin, noncellular layer between the epidermis and gastrodermis; and (4) **gastrovascular cavity,** or enteron, the internal cavity lined by the gastrodermis.

## Cellular Structure

*Hydra* is significantly more complex than a sponge, not only in its general structure but also in the degree of differentiation of cellular structure and functions. Several different types of cells may be distinguished in the stained preparations. In the epidermis, under high power, try to distinguish the following cell types: (1) The large and abundant **epitheliomuscular cells,** which possess contractile processes or fibers at their base, all running lengthwise. *What specific functions do these fibers perform?* (2) The small **interstitial cells** at the bases of the epitheliomuscular cells. (3) The **cnidocytes.** (4) The mucus-secreting **gland cells** abundant on the pedal disc. Also among the epidermal cells are many small nerve cells (described in the next section). They can be seen only in specially stained slides.

In the inner gastrodermal layer, note the following cell types: (1) The abundant, large **digestive cells** with many vacuoles. These cells bear one or two flagella, and ingest food particles for intracellular digestion. They are also epitheliomuscular in character and possess contractile fibers that run transversely along their bases, thus providing a circular musculature. *What is the function of the contractile fibers in the gastrodermal cells?* Compare this function with the function of the fibers in the epidermis. (2) The **gland cells,** which secrete either mucus or digestive enzymes. Mucus-secreting cells are abundant in the hypostome region. (3) The **interstitial cells** at the bases of the gastrodermal

cells. Several other types of cells are also present but are difficult to observe except in specially stained preparations.

## Nervous System

Specialized nerve cells, or **neurons,** are located among the cells of both the epidermal and gastrodermal layers of *Hydra.* These neurons may have two or more processes that connect via **synaptic junctions** with other neurons or with various types of receptor or effector cells.

Figure 7.6 shows an isolated multipolar neuron from the body wall of *Hydra.* Special techniques are required to see cnidarian neurons with either a light microscope or an electron microscope.

The interconnecting network of neurons in *Hydra* form a **nerve net** that lacks concentrations of neurons that form ganglia or a brain. This diffuse nerve net is characteristic of *Hydra* and other cnidarians. Nervous impulses tend to spread in a radiating pattern from the point of origin or stimulation because of the structure of synapses between adjacent neurons. Many of these synapses are symmetrical and are **nonpolarized,** thus allowing impulses to flow in both

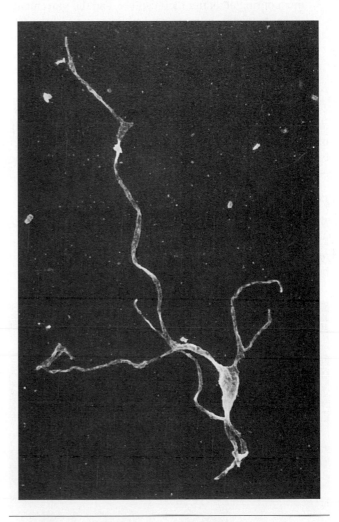

**FIGURE 7.6** Isolated multipolar neuron from body column of *Hydra* with several branches or processes.
Scanning electron micrograph by J. A. Westfall and L. G. Epp from *Tissue and Cell* 17(2):165.

directions. Some polarized synapses also occur in cnidarians. Recent studies have revealed similar diffuse nerve nets in the nervous system of some higher animals, including the digestive systems of annelid worms and humans. The rhythmic peristaltic movements of your stomach and intestine after you eat are coordinated by such a nerve net.

Neurons also have processes connecting via synapses with epithelial muscular cells, gland cells, and nematocysts. Substantial evidence suggests that nematocyst discharge is at least partly under nervous control.

## Feeding Behavior

*Hydra* is a carnivore and feeds on living crustaceans, rotifers, insect larvae, and other small animals (figure 7.7). When properly stimulated, *Hydra* exhibits a characteristic feeding response. You can observe the feeding behavior by placing a healthy, unfed *Hydra* in a clean watch glass containing about 10 ml of pond water (or *Hydra* culture solution).

◆ Add a few (6–12) washed brine shrimp (*Artemia*) larvae near the *Hydra* and record your observations. Note how the food organisms are captured; what happens to them during and after their capture; and the movements of the tentacles, hypostome, and other parts of the *Hydra*. Use the second hand on your watch to time various parts of this complex behavioral reaction of *Hydra*. Record your observations in the Notes and Sketches section at the end of this chapter.

Scientists have shown that the feeding reaction of *Hydra* is normally caused by body fluids oozing from the body of its prey, which has been pierced by nematocysts. Further experiments have shown that a feeding reaction can also be elicited by a solution of a tripeptide, reduced glutathione.

## Reproduction

*Hydra* reproduces asexually by budding and sexually by the production of eggs and sperm (figure 7.8). The *Hydra* in most laboratory samples are asexual, and you can often find budding specimens among them. In nature, *Hydra* reproduce most of the year by budding, and at certain times form gonads—testes or ovaries—which appear as thickenings of the body wall (figure 7.9). Most species of *Hydra* are **dioecious** (sexes are separate), although a few species are **monoecious** (both sexes are in one individual). Observe the demonstration materials of male and female hydra with gonads.

## Regeneration (Optional Exercise)

*Hydra* and other cnidarians have great powers of regeneration. Place a *Hydra* in a watch glass and cut across the middle of the body with a sharp scalpel or razor blade. Separate the two pieces into different watch glasses and observe their development during the next several days. Observe the formation of a new hypostome, mouth, and tentacles by the *basal half* of your *Hydra*. **What happens to the other half?** Keep a record of your observations in the Notes and Sketches section at the end of this chapter.

### Demonstrations

1. *Hydra* with testes (microscope slide)
2. *Hydra* with ovaries (microscope slide)
3. *Hydra* budding (microscope slide)
4. Green *Hydra* with symbiotic algae (microscope slide)

## A Hydromedusa: *Gonionemus*

### Class Hydrozoa

*Gonionemus* is a small marine **hydromedusa,** or hydrozoan jellyfish, common in coastal waters in many parts of the world. The medusa of *Gonionemus* is the adult or sexually mature stage and serves to illustrate the typical structure of a hydrozoan medusa (figure 7.10).

◆ Examine a preserved specimen in a watch glass partly filled with water and note its umbrellalike form. The specimens are delicate and must be handled with care.

**FIGURE 7.7** *Hydra* ingesting *Daphnia.*
Courtesy of Carolina Biological Supply Company, Burlington, NC.

**FIGURE 7.8** *Hydra* with bud and fertilized egg.
Courtesy of Carolina Biological Supply Company, Burlington, NC.

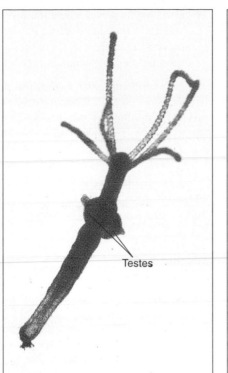
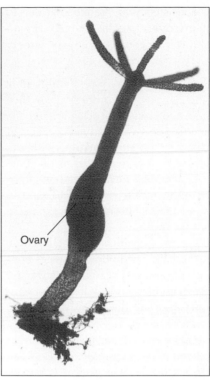

Testes

Ovary

**FIGURE 7.9** *Hydra.* Male and female with gonads.
Courtesy of Carolina Biological Supply Company, Burlington, NC.

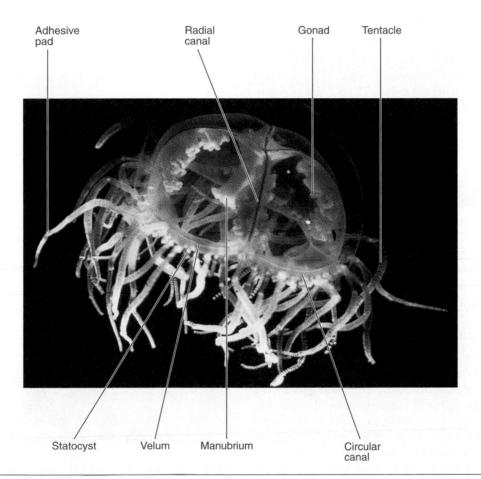

Adhesive pad    Radial canal    Gonad    Tentacle

Statocyst    Velum    Manubrium    Circular canal

**FIGURE 7.10** *Gonionemus* medusa.
Courtesy of Carolina Biological Supply Company, Burlington, NC.

Note the jellylike consistency of the medusa, due to the thick layer of **mesoglea** within the body (figure 7.10). The bulk of the medusoid body is largely mesoglea.

The medusa usually swims with its convex **exumbrellar surface** upward and the concave **subumbrellar surface** downward. Observe the numerous **tentacles** around the margin of the bell. Extending downward from the center of the subumbrellar surface is the **manubrium,** with the **mouth** at its tip surrounded by four **oral lobes.** Around the inner margin of the bell, extending inward, is a thin circular flap of tissue, the **velum,** which is believed to aid in swimming. At the base of the manubrium is an expanded portion of the manubrium, the "stomach." Find the four **radial canals** extending from the stomach to the **circular canal** at the margin of the bell. The hollow tentacles connect with the circular canal, and their cavities represent a continuation of the gastrovascular cavity. Observe the numerous **nematocyst batteries** and the **adhesive pads** on the tentacles.

At the bases of the tentacles are round, pigmented structures that serve as light-sensitive photoreceptors. Between the tentacle bases are the **statocysts,** which serve as balancing organs.

Observe the folded gonads attached to the subumbrellar surface of the radial canals. The gonads on the specimen you are studying are either ovaries or testes, since *Gonionemus,* like most cnidarians, is dioecious. Gametes produced by the gonads are released into the sea, and the fertilized eggs develop into a ciliated **planula larva.** Later, the planula larva settles and attaches to a submerged object in the sea and transforms into a microscopic **polyp.** The polyp may reproduce asexually by budding and, under certain conditions, may form tiny medusa buds that detach and grow into mature medusae, thus completing the life cycle and the alternation of generations.

## Demonstrations

1. Statocysts at margin of *Gonionemus* (wet mount or microscope slide)
2. Planula larva (microscope slide)

# A Colonial Hydrozoan Polyp: *Obelia*

## Class Hydrozoa

*Obelia* is a colonial marine cnidarian that illustrates the **complex life cycle** with alternating polyp and medusa stages found in many cnidarians (figure 7.11). Such species are members of a large group of common colonial organisms called **hydroids.** Hydroids are often found attached to shells, rocks, sea grass, and other objects in the intertidal zone but they also occur on floating algae and on the sea bottom. The polyp generation predominates in most hydroids, and the medusa is small and often short-lived.

◆ Examine a stained whole mount of the polyp or asexual stage of *Obelia* and study its organization. Like many colonial animals, *Obelia* exhibits **polymorphism**—the morphological specialization of its members.

You can distinguish two different kinds of individuals in an *Obelia* colony: feeding polyps or **hydranths** and reproductive polyps or **gonangia.** The feeding polyps bear tentacles armed with nematocysts, a mouth, a hypostome, and a delicate outer covering, the **hydrotheca** (an extension of the **perisarc**) (figure 7.12). Reproductive polyps consist of a central **blastostyle** on which **medusa buds** develop, and a thin outer covering, the **gonotheca.** At the distal end of the gonotheca is an opening, the **gonopore,** through which the newly liberated medusae escape. Note that the gonangia have no mouth or tentacles. *How do they receive their nutrition?*

The hydranths and gonangia are attached to a main stem, or **hydrocaulus,** which consists of a cylindrical tube of living tissue, the **coenosarc,** and an outer secreted covering, the **perisarc.** In cross section, the coenosarc resembles a cross section of *Hydra* with an outer epidermis, an inner gastrodermis, and a thin intervening layer of mesoglea.

## Alternation of Generations

The life cycle of *Obelia* is fundamentally similar to that of *Gonionemus* and illustrates the alternation of generations characteristic of the Phylum Cnidaria (figure 7.12).

The alternation of the asexual polyp generation and the sexual medusa generation is basically different from the alternation of generations in plants. Both the polyp and

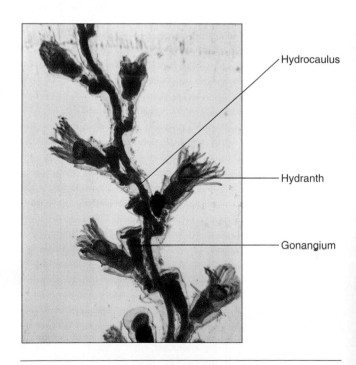

**FIGURE 7.11**  Portion of an *Obelia* colony.
Courtesy of Carolina Biological Supply Company, Burlington, NC.

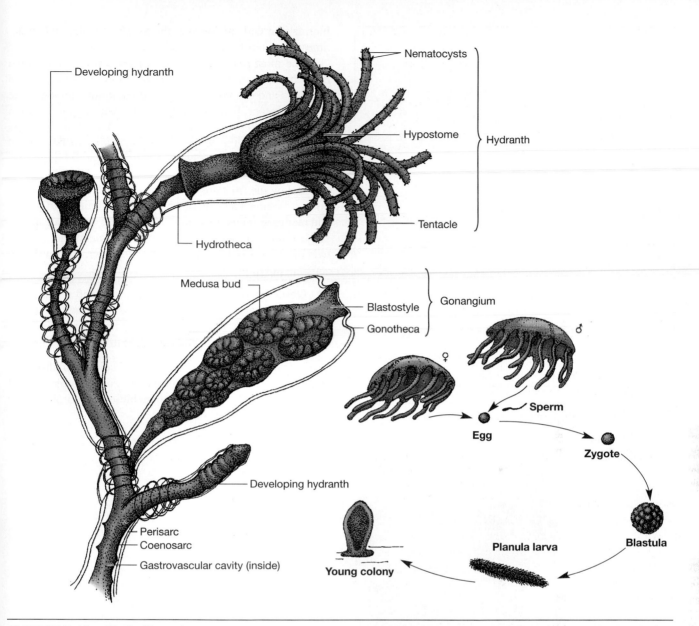

Labels on figure:
- Developing hydranth
- Nematocysts
- Hypostome
- Hydranth
- Tentacle
- Hydrotheca
- Medusa bud
- Blastostyle
- Gonangium
- Gonotheca
- ♀
- ♂
- Sperm
- Egg
- Zygote
- Developing hydranth
- Blastula
- Perisarc
- Coenosarc
- Gastrovascular cavity (inside)
- Planula larva
- Young colony

**FIGURE 7.12** *Obelia,* life cycle.

medusa generations of cnidarians are **diploid** (2n), while in plants the alternation is between a **haploid** (n) gametophyte generation and a **diploid** (2n) sporophyte generation.

In the life cycle of *Obelia,* the polyp generation produces medusa buds within its gonangia. The tiny, short-lived medusae escape to become part of oceanic plankton and produce either eggs or sperm. Fertilized eggs develop into ciliated **planula larvae,** which swim about in the sea for a time and settle to transform into a new polyp. Buds transformed by asexual reproduction of the polyp do not detach, thus forming a colony.

The sexual stage of *Obelia* is a free-swimming medusa somewhat similar to *Gonionemus,* but much smaller in size.

◆ Observe a stained whole mount of an *Obelia* medusa and observe its structure. Draw a picture of an *Obelia* medusa in figure 7.13.

## Demonstration

Preserved specimens of *Tubularia, Physalia,* and/or other hydrozoans

# A Colonial Hydrozoan: *Physalia*

## Class Hydrozoa

The Portuguese man-of-war (figure 7.14), *Physalia,* is a complex, colonial hydrozoan. A single colony may consist of as many as 1,000 individuals and may include several types of polypoid and medusoid forms. What looks like a single animal is in fact a superorganism made up of hundreds of feeding, reproductive, and defensive individuals closely joined in a colony. The familiar iridescent colonies of *Physalia* are commonly found along the beaches of

**FIGURE 7.13** Student drawing of *Obelia* medusa.

Florida, the Gulf of Mexico, the South Atlantic, and sometimes as far north as Cape Cod.

The most prominent feature of *Physalia* is a gas-filled **float** above which is a sail-like **crest.** *Physalia* is transported by winds and oceanic currents, and its normal habitat is the open sea rather than the sandy beach where it is most often seen (and sometimes felt!) by bathers. The sting from the nematocysts on the tentacles of *Physalia* can be painful when touched but is rarely dangerous, except to highly sensitive individuals.

Below the float are numerous suspended **tentacles** and other structures made up of several kinds of modified polyps and medusae. Thus, *Physalia* is an unusual cnidarian since a single colony contains both polypoid and medusoid individuals of several types closely joined together, in contrast to separate polypoid and medusoid generations.

◆ Observe a preserved or a plastic-embedded specimen of *Physalia* and identify the float, the crest, and the tentacles. See how many types of individuals you can identify among the structures attached to the float.

*Velella* and *Porpita* are two related colonial hydrozoans often seen on our Pacific coast and less often in the Gulf of Mexico and South Atlantic. Both exhibit polymorphism and habits similar to *Physalia*.

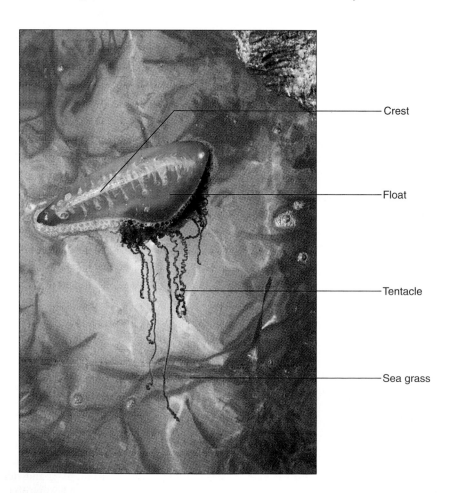

Crest

Float

Tentacle

Sea grass

**FIGURE 7.14** *Physalia.* Portuguese man-of-war. Approximately one-fourth life size.
Photograph by C. F. Lytle.

# A Scyphozoan Medusa: *Aurelia*

## Class Scyphozoa

*Aurelia* (figure 7.15) is a common, widely distributed marine medusa, or jellyfish. Large specimens may reach 30 centimeters (12 inches) in diameter. The polyp form, called a **scyphistoma,** is small, sessile, and lives attached to rocks and other submerged objects in shallow coastal waters.

◆ Study a preserved specimen of *Aurelia* to learn about the organization of a scyphozoan medusa. You will not need to dissect the specimen, since the transparent body readily shows the most important features. Handle the specimen with care.

Observe the four-part **radial symmetry** and locate the four long **oral arms** arising from the corners of the square mouth. Along the arms, find the many short **oral tentacles** that help to capture food (small planktonic animals), which are then moved toward the mouth along the **ciliated groove** on the oral side of each arm. After passing through the mouth, the food enters the gastrovascular cavity. Internally the gastrovascular cavity is divided into four **gastric pouches.** A ring of **gastric filaments** within each gastric pouch immobilizes or kills any food organisms still alive. The gastric filaments bear many nematocysts.

Four horseshoe-shaped **gonads** surround the ring of gastric filaments within the four gastric pouches. Depressions on the subumbrellar surface of the bell beneath the gonads are called **subgenital pits.** Their function is unknown.

Observe the complex branching system of **radial canals** that distribute food materials from the gastric pockets to other parts of the bell, and an outer **circular canal** around the margin of the bell. Also, around the margin of the bell, locate the eight marginal **sense organs.** These marginal organs are sensitive to touch and balance.

## Reproduction and Life Cycle

Mature *Aurelia* medusae release gametes from the gonads into the gastrovascular cavity. ***How does this differ from the hydrozoan medusae?*** The gametes exit from the mouth, and fertilization is external.

The fertilized eggs or zygotes develop into ciliated **planula larvae,** which may be retained for a time on the oral arms of the medusa and later settle to the sea bottom. There the larvae develop into small, trumpet-shaped polyps, called **scyphistomas.**

Under appropriate environmental conditions, the scyphistoma transforms into a **strobila** (figure 7.16). The strobila develops and releases (by transverse fission) a series of saucer-shaped **ephyra larvae,** which bear marginal sense organs and other medusoid features. The ephyra larvae gradually transform into adult medusae to complete the life cycle.

◆ Study the demonstration materials and draw a scyphistoma in figure 7.17.

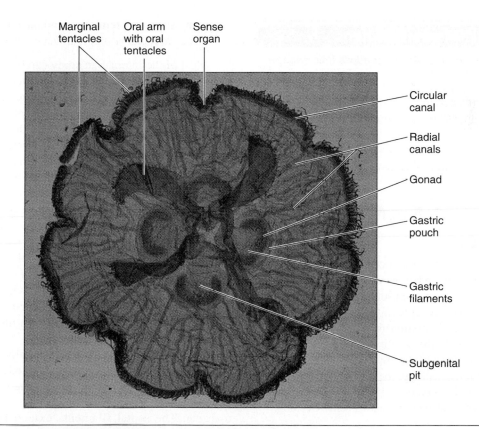

Marginal tentacles
Oral arm with oral tentacles
Sense organ
Circular canal
Radial canals
Gonad
Gastric pouch
Gastric filaments
Subgenital pit

**FIGURE 7.15** *Aurelia,* medusa, oral view.
Courtesy of Carolina Biological Supply Company, Burlington, NC.

**FIGURE 7.16** *Aurelia,* strobila.
Courtesy of Carolina Biological Supply Company, Burlington, NC.

**FIGURE 7.17** Student drawing of an *Aurelia* scyphistoma.

## Demonstrations

1. Marginal sense organ of *Aurelia* medusa (microscope slide)
2. Scyphistoma of *Aurelia* or other scyphozoan (preserved or microscope slide)
3. Planula and ephyra larvae of *Aurelia* (microscope slide)
4. Strobila stage of *Aurelia* (microscope slide)

## An Anthozoan Polyp: *Metridium*

### Class Anthozoa

Sea anemones are typically sessile cnidarians that attach to rocks, shells, pilings, and other hard substrates in the sea. Some species, however, burrow in soft sediments or are free-swimming. All nonetheless represent the polyp form of the cnidarians; the medusa generation is totally lacking in this class. The anemones and other Anthozoa represent the highest degree of specialization of the cnidarian polyp. The basic features of the anthozoan polyp are well illustrated by the common North Atlantic anemone, *Metridium* (figure 7.18). Similar or related genera also occur in the Pacific Ocean.

◆ Select a preserved specimen of *Metridium* and identify the **mouth** in the center of the **oral disc** surrounded by

many short **oral tentacles,** and the **basal disc,** which attaches to rocks or other hard substrates.

Most of the internal anatomy can be observed in specimens that have been bisected longitudinally or horizontally. It is not necessary, therefore, to further dissect the specimens. Study precut preserved specimens and locate the following structures. Find the tubular **gullet** leading internally from the mouth to the large **gastrovascular cavity.** One or more **ciliated grooves** (siphonoglyphs) should be found along the edge of the gullet. Usually the cilia within these grooves beat inward to provide an oxygenated current of water into the gastrovascular cavity. Cilia on the remainder of the gullet wall beat outward to remove wastes and foreign particles from the gastrovascular cavity. During feeding, however, the beat of the cilia along the gullet wall is reversed and aids in moving food particles into the gastrovascular cavity. Observe the feeding of living anemones in an aquarium (if available) to better understand the interaction of the tentacles and ciliary currents on the oral disc and in the gullet during the feeding process.

The gastrovascular cavity is partially divided into sections by thin vertical walls of tissue called **septa.** Some septa attach to the central gullet; others (partial or incomplete septa) bear thickened **septal filaments** on their free inner margins. The septal filaments extend below the partial septa as thin, twisted, threadlike **acontia.** The acontia bear numerous nematocysts and act to subdue living prey taken

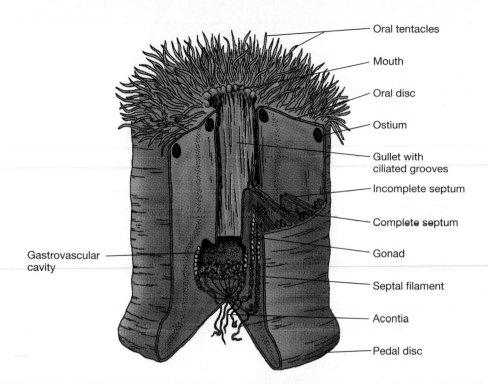

Oral tentacles

Mouth

Oral disc

Ostium

Gullet with ciliated grooves

Incomplete septum

Complete septum

Gonad

Septal filament

Acontia

Pedal disc

Gastrovascular cavity

**FIGURE 7.18** *Metridium,* partly dissected. Approximately life size.

into the gastrovascular cavity. Distal to the septal filaments on the partial septa, find the **gonads,** which appear as thickened ridges parallel to the septal filaments. On the septa near the oral disc, locate the **ostia,** round openings that allow circulation of fluid between adjacent sections of the gastrovascular cavity.

## Reproduction and Life Cycle

Reproduction in *Metridium* and most other anemones is both asexual and sexual. Some anemones reproduce asexually by splitting longitudinally (**longitudinal fission**), but the main asexual means of reproduction in *Metridium* is called **pedal laceration.** Bits of tissue from the pedal disc are split from the anemone as the animal moves along the substrate. These tissue pieces later regenerate an entire small anemone, literally in the footsteps of its parent.

Sexual reproduction occurs seasonally when gametes are released from the gonads on the partial septa into the gastrovascular cavity. The gametes are released and are fertilized in the sea. The fertilized eggs develop into free-swimming **planula larvae.** After a period spent as planktonic larvae, the planulae settle on a hard substrate and metamorphose into an anemone.

## Corals

### Class Anthozoa

Most anthozoans are corals, the largest and best known of which are the **stony** or **scleractinian corals** (Subclass Zoantharia) and the **octocorals** (Subclass Alcyonaria), which include the soft and horny corals. Corals, like all anthozoans, have only a polyp form. There is no medusa stage in the life cycle.

The polyp of a stony coral resembles a sea anemone, although the individual polyps are almost always smaller than those of anemones. The coral polyp sits in a **cup** on the surface of a calcareous exoskeleton secreted by the lower portion of the column and the basal disc. Extending inward from the wall of the cylindrical cup are several **calcareous septa,** which extend from the sides and base of the cup into folds of the basal tissue of the polyp. These tissue-covered septa partially subdivide the gastrovascular cavity inside the polyp into several chambers.

Most stony corals are colonial, and adjacent polyps are connected by lateral extensions of the body wall that cover the intervening stony skeleton of the colony. These sheets of lateral tissue also contribute to the formation of the skeleton by their secretions and serve to connect the gastrovascular cavities of adjacent polyps.

Many growth forms occur among the stony corals, and most species exhibit characteristic skeletons. A few species are solitary and occur as large individual polyps, like *Fungia* from the Pacific Ocean.

Some stony corals contain **symbiotic algae** (zooanthellae), and many interesting studies have been conducted on the physiological and biochemical interactions of these symbionts. Among the stony corals are several **reef-building species** largely responsible for the formation of coral reefs in warmer parts of the oceans, including those in the Bahama Islands, off the Florida Keys, in Belize, and of the Great Barrier Reef of Australia.

Some examples of stony corals are *Fungia,* a solitary coral; *Astrangia,* the Atlantic star coral; *Oculina,* the eyed coral; and *Diploria* and *Meandrina,* brain corals.

The octocorals exhibit strong **octamerous** (eight-part)

**FIGURE 7.19**  Student drawing of corals.

**radial symmetry** and have an **endoskeleton** consisting of separate microscopic pieces (spicules). A tough, horny organic material is also present in some species. This group is especially prominent in tropical waters and includes the sea fans, the sea whips, the sea pens, the sea pansies, the organ pipe coral, and the precious red coral (*Corallium*) used for jewelry.

◆ Study the demonstration materials illustrating several forms of coral and draw several different types in figure 7.19. Identify each type of coral that you draw.

## Demonstrations

1. Living sea anemones and/or corals in a marine aquarium
2. Representative preserved anemones
3. Assortment of preserved corals and dried coral skeletons

## Key Terms

**Alternation of generations**  the alternation of the sessile polyp and free-swimming medusa generations typical of the life cycle of the cnidarians.

**Cnidocytes**  specialized cells of cnidarians that produce and contain the stinging nematocysts.

**Coenosarc**  the living portion of the tubular connecting portions of colonial cnidarians like *Obelia*. Consists of a simple cylinder of an outer epidermal tissue layer, an inner gastrodermal tissue layer, and an intermediate mesoglea surrounding a central gastrovascular cavity.

**Dioecious**  condition of an animal with male and female sex organs borne in different individuals.

**Diploblastic construction**  the two-layered construction typical of the cnidarians. Consists of an outer epidermal tissue layer and an inner gastrodermal layer with an intervening noncellular mesoglea layer.

**Epidermis**  outer tissue layer protecting the surface of an animal from its environment.

**Gastrodermis**  inner tissue layer of animals bordering the digestive cavity.

**Gastrovascular cavity**  a central cavity of an animal that serves both for digestion and circulation, with a single mouth opening that serves both as an entrance and an exit. A type of incomplete digestive system (without an anus).

**Gonangium**  a type of reproductive individual in colonial Hydrozoa, such as *Obelia;* produces free-swimming medusae.

**Hydranth**  feeding individual in a colonial hydrozoan, as in *Obelia*.

**Medusa stage**  free-swimming stage in the life cycle of many cnidarians. Usually bears gonads and produces gametes.

**Mesoglea**  the gelatinous layer between the epidermal and gastrodermal layers of cnidarians; simple and noncellular in the Hydrozoa, containing cells and/or fibers in the Scyphozoa and Anthozoa.

**Monoecious**  condition of bearing both male and female sex organs in one individual. Usually not self-fertilizing.

**Nematocysts**  the stinging capsules produced by the cnidocytes of cnidarians. A key characteristic of the phylum.

**Nerve net**  the diffuse, interconnected network of nerve cells in the cnidarians. Lacks ganglia or other nervous centers. A very primitive type of nervous system.

**Perisarc**  the nonliving outer covering secreted by the coenosarc of colonial hydrozoans. Surrounds the interconnecting coenosarc.

**Planula larva**  a simple, ciliated, sausage-shaped larval form produced by many cnidarians. Develops from the zygote or fertilized egg.

**Polymorphism**   having two or more distinct body forms in a single species, as the several types of individual polyps found in *Obelia, Physalia,* and many other cnidarians.

**Polyp stage**   the sessile (and usually asexual) stage in the life cycle of many cnidarians.

**Radial symmetry**   a body plan in which all body parts are arranged symmetrically around a central axis.

**Scyphistoma**   the inconspicuous polyp stage in the life cycle of certain scyphozoan medusae.

## Internet Resources

Visit the zoology website at http://www.mhhe.com/zoology to find live Internet links for each of the references listed below.

1. Hawaii Coral Reef Network. Many links to topics such as reefs and mangroves to coral paleoclimatology. Links to coral reef sites worldwide.

2. Animal Diversity Web, University of Michigan. Phylum Cnidaria. Pictures of cnidarians and descriptions, links to other pages on each of the four classes of cnidarians.

3. Cnidarian WWW Server. Many aspects of cnidarian biology are described here, from morphology to molecular evolution. This site, supported by the University of California—Irvine, has many links to other sites.

4. Introduction to the Cnidaria. University of California at Berkeley Museum of Paleontology. Links to the fossil record, life history and ecology, systematics, and more on morphology. Many links to the different classes and each site typically include at least one picture of a member of the taxon.

## Critical Thinking Questions

1. Why do we say that cnidarians are the simplest animals with definite tissues? (Do you think the strict definition of "tissue" excludes other possibilities concerning cellular organization and cooperation?)

2. Speculate on the evolution of alternation of generations (polyp forms and medusa forms). Could the medusa form simply be an upside-down polyp? Or is the polyp a rightside-up medusa? If so, which form do you suppose evolved first?

3. Why do you suppose that radial symmetry did not become a big "evolutionary hit" among the animals? What drawbacks might such a symmetry have over bilateral symmetry?

4. Discuss the various nematocyst types and their functions. Do you think this variety on a common theme is an example of a basically good design being preadapted for many different possible functions?

5. Explain how the Portuguese man-of-war might be viewed as a "superorganism" (i.e., the polyps have taken on different functions to form a colonial animal.)

## Suggested Readings

Ax, P. 1989. Basic phylogenetic systemization of Metazoa. In K.B.B. Fernholm and H. Jornvall (eds.), *The Hierarchy of Life*. Amsterdam: Elsevier, pp. 453–70.

Fautin, D.G., and R.N. Mariscal. 1991. Cnidaria: Anthozoa. In F.W. Harrison and J.A. Westfall (eds.), *Microscopic Anatomy of Invertebrates, volume 2: Placozoa, Porifera, Cnidaria, and Ctenophora*. New York: Wiley-Liss.

Harbison, G.R. 1991. Ctenophora. In S.D. Cairns, et al (eds.), *Common and Scientific Names of Aquatic Invertebrates from the United States and Canada: Cnidaria and Ctenophora*. Bethesda, MD: American Fisheries Society, pp. 1–75, pl. 1–4.

Levine, J.S. 1985. *Undersea Life*. New York: Steward, Tabori & Chang.

Meglitsch, P.A. and F.R. Schramm. 1991. *Invertebrate Zoology*, 3d ed. New York: Oxford University Press.

Nielsen, C. 1995. *Animal Evolution: Interrelationships of the Living Phyla*. Oxford: Oxford University Press.

Pennak, R.W. 1978. *Fresh-Water Invertebrates of the United States*, 2d. ed. New York: John Wiley & Sons.

# NOTES AND SKETCHES

# CHAPTER 8

## Introduction to Animal Morphology

**Acoelomate**
Platyhelminthes
(flatworms)

**Pseudocoelomate**
Nematodes
(roundworms)

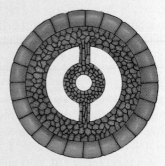

**Coelomate**
Mollusks
Annelids
Arthropods
Echinoderms
Chordates

## OBJECTIVES

After completing the laboratory work in this chapter, you should be able to perform the following tasks:

1. Explain why the study of animal morphology is important.
2. Describe the types of symmetry found among animals and give examples of each type.
3. Describe the grades of tissue construction found among animals and give examples of each type.
4. Describe animal types according to their central body cavity and give examples of each type.
5. Describe the coelom and explain why the presence of a coelom may have contributed to the evolution of higher animals.
6. Discuss how cephalization may have contributed to the evolution of animals.
7. Explain the purposes of dissection and the importance of good dissection techniques.
8. List the basic instruments needed for dissection, explain the use of each, and demonstrate their proper use and care.
9. Distinguish between common anatomical terms: anterior and posterior; dorsal and ventral; and transverse plane, frontal plane, and sagittal plane.

## Introduction

Animal morphology is the study of the form of animal bodies. Animals come in many sizes, shapes, and forms. Careful study of animals reveals many similarities and differences among them. Thoughtful analysis of these similarities and differences can reveal evolutionary relationships among animals. Such studies are fundamental to establishing a natural classification that reflects the patterns of animal evolution.

The study of morphology is almost always closely linked with studies of function. Specific organs and other structures are preserved, lost, or modified depending upon the adaptive value these organs or structures have for the survival of each species. Morphological studies, along with biochemical, cytological, paleontological, and other types of study, contribute to our understanding of animals and their evolution.

In this chapter we will discuss some fundamental principles of the organization of the bodies of animals. An understanding of these principles will help you to appreciate

**121**

how animals are constructed, to recognize the similarities and differences among the major groups of animals, and to gain some new insights into the process of animal evolution.

Since dissection is an important means for learning about the internal anatomy of animals, we will also provide some suggestions for dissection later in this chapter.

## Organization of the Animal Body

All animals are multicellular, and their eukaryotic cells are organized into tissues, organs, and organ systems with varying degrees of complexity. Some of the morphological features considered important by most zoologists include: (1) **body symmetry,** (2) **grade of tissue construction,** (3) **type of body cavity,** (4) **segmentation,** and (5) **cephalization.**

## Body Symmetry

Animals differ in the arrangement of body parts, and symmetry can best be described in relation to certain reference planes or axes. Three main types of symmetry are observed among animals: (1) **asymmetry,** (2) **radial symmetry,** and (3) **bilateral symmetry.**

### Asymmetry

Irregular arrangements of body parts. Such animals have no plane of symmetry that can serve to divide the animal into similar halves. Example: many sponges.

### Radial Symmetry

Body parts arranged around one central axis; any plane passing through the central axis divides the body into similar (mirror image) halves. Examples: most medusae (jellyfish), sea stars, sand dollars.

### Bilateral Symmetry

Body parts divided into equal (mirror image) halves by a single plane of symmetry. Examples: flatworms, fish, humans.

Most animals are bilaterally symmetrical; radial symmetry is most common among the Cnidaria and the Echinodermata. In the cnidarians, radial symmetry is primary; both the larvae and adults usually show obvious radial symmetry. In the echinoderms, however, radial symmetry is secondary; the larvae are typically bilaterally symmetrical during their development, and only after metamorphosis do adults become radially symmetrical. This change in symmetry is commonly attributed to the sedentary habits of most adult echinoderms.

## Grade of Tissue Construction

Animals differ in the organizational complexity of their cells and tissues. Some animals consist of a loosely organized colony of cells, while others have many types of cells organized in very specific ways into well-integrated tissues. The origin and development of animal tissues can be observed by careful study of embryological development. Three grades of tissue organization are found among animals: (1) **tissue grade,** (2) **diploblastic construction,** and (3) **triploblastic construction.**

### Tissue Grade Construction

Found in animals whose multicellular bodies are composed of cells organized into simple tissues but are lacking organs and organ systems. Examples: sponge, Phylum Porifera.

### Diploblastic Construction

Found in animals with two distinct tissue layers derived from the embryonic germ layers ectoderm and endoderm. Examples: *Hydra,* Phylum Cnidaria.

### Triploblastic Construction

Found in animals with tissues derived from three embryonic germ layers: **endoderm, ectoderm,** and **mesoderm.** Examples: flatworm, Phylum Platyhelminthes; earthworm, Phylum Annelida; and dog, Phylum Chordata. Most animals show triploblastic construction. Each of these germ layers develop into specific types of tissues in larval and adult organisms. Endoderm forms the lining of the gut and several internal glands derived from the gut. Ectoderm forms the outer layer of the body (integument) and all nervous tissue including the brain. Mesoderm gives rise to muscle, bone, connective tissue, and blood.

## Body Cavity

Most large animals have a central body cavity, while some small animals lack a central cavity but have their central space filled with loosely packed cells. Further study of anatomy reveals three basic groups of animals based on their central body cavity: (1) **acoelomate,** (2) **pseudocoelomate,** and (3) **coelomate** (figure 8.1).

### Acoelomate

Animals whose central space is filled with loosely packed cells (**mesenchyme**). Body fluids percolate through irregular spaces between cells and carry nutrients to the cells and assist in removing wastes from the cells. Examples: flatworm, Phylum Platyhelminthes.

### Pseudocoelomate

Animals with a central body cavity, the **pseudocoelom,** derived from the embryonic blastocoel. In the adult organism, this cavity lies between the endoderm and mesoderm tissues; there is no specialized mesodermal peritoneum lining around the pseudocoelom. Examples: roundworm, Phylum Nematoda; rotifer, Phylum Rotifera.

### Coelomate

Animals with a central body cavity that develops within mesoderm tissue. A specialized mesodermal peritoneum completely surrounds the central cavity in the adult. Examples: earthworm, Phylum Annelida; dog, Phylum Chordata. A **coelom** is an important feature of most animals, although in some animals, such as molluscs and insects, it becomes secondarily reduced.

The coelom represents an important evolutionary advance among animals, and this cavity has several important functions. For example, the coelom:

1. allows for the expansion and movements of internal organs;

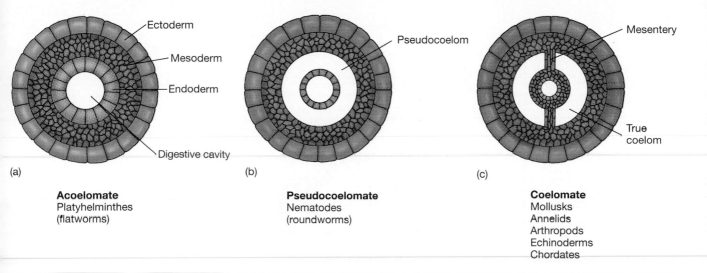

(a)

**Acoelomate**
Platyhelminthes
(flatworms)

(b)

**Pseudocoelomate**
Nematodes
(roundworms)

(c)

**Coelomate**
Mollusks
Annelids
Arthropods
Echinoderms
Chordates

**FIGURE 8.1** Animal body types.

2. permits the lengthening and regional specialization of the digestive tract;
3. facilitates the exchange of gases, nutrients, and waste products;
4. provides storage space for gametes; and
5. serves as a hydrostatic skeleton in soft-bodied forms.

## Segmentation

Some animals exhibit a **serial repetition** of body parts similar to a line of boxcars in a freight train. Such serial repetition of parts is called segmentation, or **metamerism.** The individual parts are called segments, or metameres.

Segmentation is most obvious in three phyla—Annelida, Arthropoda, and Chordata—although several other phyla show varying degrees of segmentation. In the annelids and arthropods, segmentation is readily apparent both externally and internally. In chordates, such as a fish, dog, or human, the evidence of segmentation is mostly internal. A study of the vertebral column, muscles, and nerves clearly reveals the segmented nature of their bodies. Segmentation also is clearly evident during the embryonic development of these groups.

## Cephalization

Most animals have a definite front end (**anterior**) and a rear end (**posterior**). This differentiation of the anterior/posterior body axis and the related concentration of nervous and sensory structures to form a head at the anterior end is called cephalization. Externally we recognize a head by the presence of eyes, ears, nose, antennae, and other features. These are sense organs (or contain sense organs) that provide an animal with information about the environment immediately ahead of it. Usually, but not always, the mouth is located on the head since much of an animal's activity is directed toward locating and obtaining an adequate supply of appropriate food. Internally the head also contains a concentration of nervous tissue, organized into a brain in higher animals, that receives and processes the sensory information received by the sense organs. Proximity of the brain to the sense organs speeds reception and processing of sensory information often essential for survival. Thus it is easy to understand why so many animals exhibit a well-developed head!

Many multicellular animals have a well-developed head, and the degree of cephalization tends to increase with the complexity of other organs and organ systems, particularly among active and motile animals. More complex vertebrates such as frogs, birds, and mammals show the highest degree of cephalization. Sedentary or sessile (attached) animals such as oysters, barnacles, and sea urchins show little or no signs of cephalization. Also, burrowing animals such as earthworms, shipworms, and burrowing snakes show less cephalization than their relatives that live active lives above ground. *How might you logically explain the lack of cephalization in such animals in light of what you know about the processes of animal evolution?*

## Hints for Dissection

Many of the animals to be studied in later exercises will require dissection for the study of internal structures. To gain the most from these studies, it is important that you follow directions carefully and that you develop good dissection skills. You must take special care in each dissection to avoid damaging important structures before you have had an opportunity to complete your study.

Some important anatomical terms are illustrated in figure 8.2 and are defined in the Key Terms section at the end of this chapter. Some understanding of basic anatomy is essential for success in zoology, regardless of your future specialization. You should review figure 8.2 and the list of key terms to reinforce your understanding of anatomy and proper terminology as you proceed through the remaining exercises in this book.

The chief objective in the dissection of a specimen is to expose body parts for the study of their structure and their relationships to other parts. Therefore, it is important to proceed

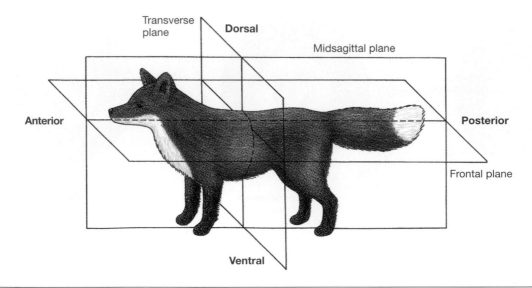

**FIGURE 8.2**  Basic anatomical directions and planes.

carefully with each dissection and to avoid cutting or re-moving anything, unless so instructed in the printed directions or by your laboratory instructor. Read through the instructions completely before you begin each dissection. This will save you time later and will often prevent disappointing results.

Body parts should be parted carefully along natural lines of separation and attachment wherever possible. This can often be done best (without cutting) with forceps, probes, the handle of a scalpel, or with a finger.

Good quality dissection instruments are essential to good laboratory work. Your dissection kit should include the following items:

Scissors, fine points (or one fine point and one blunt point), about 4–6 inches long

Forceps, fine points, about 4–5 inches long

Scalpel with replaceable blade

Extra scalpel blades

Two teasing needles

Probe, about 6 inches long

Plastic centimeter ruler

Dropping pipette

Scissors should be constructed of first-quality chrome or stainless steel and have the blades joined by a screw rather than a rivet. Some common dissecting instruments are illustrated in figure 8.3.

Dissecting instruments should be kept clean and sharp at all times. A small oilstone and a piece of emery cloth or fine sandpaper should be used to keep cutting edges and needle points sharp. Two types of scalpel are in common use. One type has a permanent blade that can be resharpened, and the second type has a separate handle and replaceable blades that are not usually sharpened. Blades on scalpels with replaceable blades should be replaced often to ensure a good edge. Sharp scalpels and scissors allow clean cuts and mini-mize damage to adjacent tissues. Always wash and fully dry your dissecting instruments with a paper towel after each use. With proper care, a good set of instruments will last for many years.

The scalpel is used for making incisions in the body wall and occasionally for sectioning interior structures. Hold the scalpel upright with the handle nearly perpendicular to the surface to be cut (figure 8.4) and make a clean forward incision while supporting surrounding tissues with your fingers. Be conscious of internal structures and avoid cutting too deeply. If you have to saw with the scalpel to cut the tissue, your blade is too dull. Replace disposable blades or sharpen your scalpel regularly. Used blades should be discarded only in the sharps disposal container provided in the laboratory.

Forceps and dissecting needles should be used for loosening, lifting, and moving various structures to facilitate their study and to expose underlying parts. It is important to gain some idea of the thickness of the covering or surface layer to be cut and its relationship to other important structures nearby before starting your dissection. Consult any available charts, photographs, models, and demonstration dissections of the animal to be studied. A few minutes spent in such preliminary orientation will often prevent later disappointment. Also remember that your laboratory work may be evaluated by your instructor partly on the basis of your care and skill in dissecting.

Read the dissection instructions in the laboratory manual and follow them carefully. Always remember that **direct study** of internal anatomy from your dissections is the primary objective. The illustrations in this manual are intended to **help** you in your dissections and study but are **not intended to substitute** for the study of real specimens. Refer to the illustrations frequently, but focus your study primarily on the animal, not on the pictures.

In some of the exercises, you are also asked to make drawings of the animals you study and dissect. Take care to be accurate and neat in your drawings so that they will be

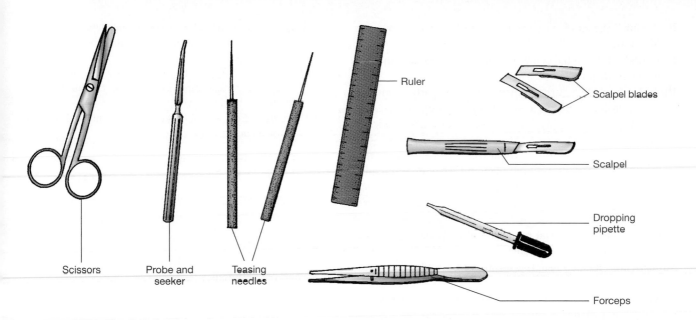

FIGURE 8.3   Dissecting instruments.

Ruler

Scalpel blades

Scalpel

Dropping pipette

Scissors

Probe and seeker

Teasing needles

Forceps

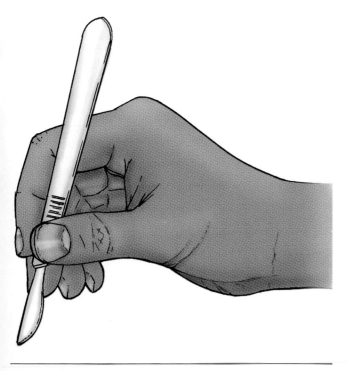

FIGURE 8.4   Proper use of the scalpel.

helpful to you for later study and review. Remember that your laboratory notes and drawings serve a purpose similar to your lecture notes. They provide a record to help you to recall actual structures during your later study and reviews. They help you to retain and to recall mental pictures of the things actually seen through the microscope and during your dissections. Include a magnification scale in your drawings to help you remember size and dimensions.

You should also use the Notes and Sketches section at the end of each chapter for any additional things that you observe or need to remember.

## Key Terms

**Aboral**   away from or opposite to the mouth.

**Anal**   toward the anus or away from the mouth.

**Asymmetry**   an irregular arrangement of body parts; without a central point, axis, or plane of symmetry.

**Bilateral symmetry**   an arrangement of body parts on opposite sides of a central plane (midsagittal plane), which divides the body into two symmetrical halves (mirror images).

**Caudal**   toward the tail or tail end; the opposite of cephalic.

**Cephalic**   of or pertaining to the head; the opposite of caudal.

**Cephalization**   the concentration of nervous and sensory structures to form a head at the anterior end.

**Cranial**   relating to the skull or cranium.

**Cross section**   sections of the body cut on any transverse plane; such sections are perpendicular to the sagittal and frontal planes.

**Deep**   pertaining to structures away from the surface of the body; the opposite of superficial.

**Distal**   away from the center or point of attachment; the opposite of proximal.

**Dorsal**   relating to the back or upper surface; the opposite of ventral.

**Frontal plane**   plane parallel to the dorsal and ventral surfaces of the body, which bisects a bilaterally symmetrical animal into upper and lower halves.

**Lateral**   toward the side; the opposite of medial.

**Longitudinal**   lengthwise; parallel to the long axis.

**Medial**   toward the sagittal plane or center of the body; the opposite of lateral.

**Median**   located in or near the sagittal plane.

**Oral**   toward the mouth.

**Peripheral**   toward the outer surface.

**Posterior**   the hind part (rear) of the body; the opposite of anterior.

**Proximal**   toward the center or point of attachment; the opposite of distal.

**Radial symmetry**   arrangement of body parts symmetrically around a central axis; any plane through the central axis divides the body into symmetrical halves (mirror images).

**Sagittal plane**   any longitudinal plane passing from the head to tail. The midsagittal plane bisects a bilateral animal into two symmetrical halves (mirror images). All longitudinal planes parallel to the midsagittal plane are parasagittal planes.

**Segmentation**   the serial repetition of body parts into distinct segments or metameres.

**Superficial**   located near the surface of the body; the opposite of deep.

**Transverse plane**   any plane perpendicular to the sagittal and frontal planes. Sections of the body cut on a transverse plane are called cross sections.

**Ventral**   relating to the belly or underside; the opposite of dorsal.

## Internet Resources

Visit the zoology website at http://www.mhhe.com/zoology to find live Internet links for each of the references listed below.

1. Wandtafeln (wall charts) of Rudolph Leuckart. Includes images of these remarkable charts, which are a unique teaching aid in the study of zoology. Classic old-fashioned art.

2. System Comparison Table. A lengthy table that compares various systems of the earthworm, the frog, the snake, the shark, the perch, the pigeon, and the pig.

3. University of Minnesota Phylum Comparison Table. I would suggest that you print this out and fill it in during the following labs.

4. Diversity of Life: A Definition of Terms. This will be useful throughout the course.

## Critical Thinking Questions

1. Discuss the role cephalization may have played in the evolution of complex animals.

2. Describe the coelom and explain how the coelom may have contributed to the evolution of more complex animals.

3. Compare the following symmetry types: asymmetry, radial symmetry, and bilateral symmetry. Might there be some advantages of one over the other?

4. Discuss how higher grades of tissue organization, coelom formation, and segmentation have contributed to the great success of the more complex animals.

5. Discuss the human body in terms of anterior–posterior; cephalic–caudal; distal–proximal; dorsal–ventral; etc. Why is knowing precise "directions" so important in dissection?

## NOTES AND SKETCHES

# CHAPTER 9

## Platyhelminthes

## OBJECTIVES

After completing the laboratory work in this chapter, you should be able to perform the following tasks:

1. Briefly outline the characteristics of the Phylum Platyhelminthes and identify the major advances in organization over the Phylum Cnidaria.
2. List and briefly characterize each of the four classes of the Phylum Platyhelminthes.
3. Describe the behavior of a free-living flatworm (planarian) such as *Dugesia* or *Planaria* and relate this behavior to the function of its sense organs.
4. Discuss the structure of the epidermis of a planarian and explain the location and function of cilia, rhabdites, and gland cells.
5. Identify the major structures that can be seen in microscopic cross sections at various levels of the body of a planarian (for example, anterior, pharyngeal region, and posterior).
6. Explain the organization of the reproductive system of a planarian and identify its principal reproductive organs.
7. Discuss the general morphology of the trematode *Clonorchis* and compare it with a free-living turbellarian such as *Planaria* or *Dugesia*.
8. Describe the life cycle of *Clonorchis* and identify the principal stages in microscopic preparations.
9. Describe the general morphology of the liver fluke *Fasciola* and identify its principal organs.
10. Identify in microscope slides the scolex, proglottids, and strobila of a tapeworm and explain the basic organization of a tapeworm.
11. Describe the life cycle of a tapeworm such as *Taenia* or *Dipylidium,* and identify its principal stages in microscopic preparations.

## Introduction

The Platyhelminthes, or flatworms, are soft, wormlike animals with flattened, elongated bodies. They exhibit several important structural advances over the cnidarians, including **three distinct tissue layers (triploblastic construction), bilateral symmetry,** and several well-developed **organ systems.** Approximately 13,000 species of flatworms have been described.

The body parts of bilaterally symmetrical animals are arranged along a midsagittal plane, so that the left and right sides are approximately mirror images of each other (see figure 8.2 in the last chapter). This type of symmetry is characteristic of most higher metazoans, except for adult echinoderms. Triploblastic body construction with **endoderm, ectoderm,** and **mesoderm** tissues is also characteristic of all higher metazoans from Platyhelminthes to Chordata. Each of these embryonic tissue layers develops into specific types of adult structures as discussed in Chapter 8.

Flatworms lack the large central body cavity found in most advanced animals. Instead, the interior of flatworms is typically filled with loosely packed **parenchyma** tissue with irregular spaces between the cells and clumps of cells. Since flatworms have no circulatory system or heart, body fluids percolate through these irregular interior spaces to bring nutrients and oxygen to the cells and remove wastes from them. The body fluids are moved in part by muscular contractions. This type of organization without a central body cavity is called **acoelomate construction.**

Free-living flatworms (figure 9.1) have well-developed digestive, excretory, reproductive, nervous, and muscular systems, but, as noted above, they have no circulatory system or central body cavity. Some flatworms exhibit spiral cleavage and determinate development that suggest evolutionary affinities with the molluscs, annelids, and arthropods—animals that make up the group known as the protostomes.

Most flatworms, however, have become parasites or micropredators and show many adaptations for parasitism or micropredation, which include reduction and/or modification of several of their organ systems. These flatworms tend to show fewer of their ancestral characteristics. Their modifications—due to their specialized mode of life—tend to obscure any clue to their possible evolutionary relationships.

## Classification

The current classification of the phylum includes four classes: Turbellaria, Trematoda, Monogenea, and Cestoda. For many years, flatworms were divided into three classes: The Class Trematoda included both the **monogenetic** (one host) and the **digenetic** (two hosts) trematodes, which were considered orders. More recently, taxonomists have considered the differences between these two large and important groups of parasitic flatworms to be sufficiently important that they should be considered separate classes.

### Class Turbellaria (Free-living Flatworms)

Mainly **free-living flatworms** with a dorsoventrally flattened body covered by a ciliated epidermis. Mouth usually ventral, leading into a gastrovascular cavity with a single opening; anus lacking. Freshwater and marine forms. Examples: *Dugesia, Bipalium, Mesostoma, Polychoerus, Leptoplana.*

### Class Trematoda (Flukes)

Parasitic flatworms with a body covered by an external tegument (cuticle) secreted by underlying cells. Ovoid body typically with one anterior and one midventral sucker for attachment to the host. Internal parasites with complex life cycles, usually involving several successive larval stages and two or more hosts. Examples: *Clonorchis* (human liver fluke), *Fasciola* (sheep liver fluke), and *Schistosoma* (human blood fluke).

### Class Monogenea (Flukes)

Parasitic flatworms with an external tegument, with a large posterior sucker for attachment and an anterior sucker reduced or absent, life cycle with a single host. Most are external parasites of fish. Examples: *Gyrodactylus, Polystoma.*

### Class Cestoda (Tapeworms)

Internal parasites with specialized scolex with hooks and/or suckers for attachment to host; body divided transversely into a series of similar proglottids; thick external tegument; mouth and digestive tract absent. Usually with a complex life cycle involving successive larval stages and alternate hosts. Examples: *Dipylidium caninum* (dog tapeworm), *Taenia pisiformis* (dog and cat tapeworm), and *Dibothriocephalus* (fish tapeworm).

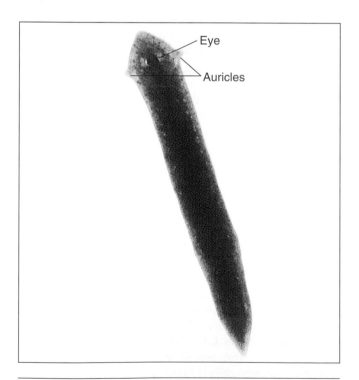

**FIGURE 9.1** Brown planaria, living.
Courtesy of Carolina Biological Supply Company, Burlington, NC.

## Materials List

**Living specimens**
- *Dugesia* or *Planaria*
- *Stenostomum*
- *Cercariae*

**Preserved specimens**
- Whole tapeworms

**Prepared microscope slides**
- *Planaria*, whole mount, representative cross sections
- *Stenostomum*, whole mount
- *Clonorchis*, whole mount
- *Fasciola*, whole mount
- *Dipylidium*, whole mount
- *Taenia pisiformis*, scolex and representative sections
- *Taenia solium*, scolex and representative sections (demonstration)
- *Clonorchis*, miracidia, sporocysts, redia, cercaria, metacercaria (demonstrations)
- *Dibothriocephalus latus*, scolex and representative sections (demonstration)
- Onchosphere (six-hooked) larva (demonstration)
- Cysticercus larva (demonstration)

**Plastic mounts**
- *Dipylidium*, whole mount (demonstration)
- *Taenia pisiformis*, representative sections (demonstration)

**Miscellaneous**
- Methyl cellulose solution
- Beef liver
- Methylene blue solution (1%)

# Free-living Flatworms: Class Turbellaria

## A Planarian: *Dugesia*

Most members of the Class Turbellaria are marine. However, freshwater members of this class, commonly referred to as planarians, can be found in many springs, brooks, ponds, and lakes. These flatworms are typically grey, brown, or black and usually move about on the bottom or on submerged sticks, stones, or plants in shallow water. Many species are negatively phototrophic and are found on the underside of stones or leaves. Common North American genera include *Dugesia*, *Phagocota*, and *Polycelis*.

### Behavior and External Anatomy

Observe a living specimen and note its general size and shape. *Is the worm uniformly pigmented, or does it show some distinctive pattern of pigmentation?* Locate the anterior **head,** the **eyes (how many on your specimen?),** and the **auricles** (lateral projections from the head; absent from some species).

The eyes serve as light receptors but are not image forming. The auricles are well-equipped with touch and chemical receptors. The head region also contains a concentration of **nerve ganglia,** which function in the processing of sensory information and thus serve as a primitive "brain."

Most behavior in turbellarians appears to result from simple trial and error, although some experimental studies have indicated that flatworm behavior can be modified to some extent by prior experience (i.e., they "learn").

Observe the smooth, gliding locomotion of the worm. This form of locomotion is due to the action of cilia on the ventral surface of the body, coordinated with rhythmic muscular contractions of the body. Note the behavior of the head and the auricles during locomotion. *How is this behavior related to the function of sensory and nervous structures of the head?* Gently touch the "head" of the worm with a clean dissecting needle. *How does the worm react?* Touch other body regions in a similar way and compare the reaction. Turn the worm over on its dorsal side. *How does it react? Can you relate your observations of the worm's behavior to the structure of the head and the concentration of sense organs and nervous elements there? Can you make any conclusions about the possible advantages to an animal of having an anterior head with a concentration of sense organs and nervous tissue?*

### Internal Anatomy

◆ To study the anatomy further (figure 9.2), obtain a microscope slide with a stained whole mount of a planarian.

Review the structures previously noted in the living specimen and also observe the **three-branched gastrovascular cavity.** Note the one anterior and two posterior branches of the cavity and the many smaller lateral branches, or diverticula.

◆ Study also a microscope slide with cross sections through the anterior, pharyngeal, and posterior portions of the body (figure 9.3). In the section from the anterior portion of the body, locate (1) the **epidermis,** the external layer of cells surrounding the body; (2) the large, vacuolated cells of the **gastrodermis** lining the digestive tract; (3) the layers of **longitudinal** and **circular muscles** lying just inside the epidermis; (4) the large, irregularly shaped cells of the **parenchyma** tissue, which fills most of the interior space of the body; and (5) the two large **ventral nerve cords** (figure 9.4). The ventral nerve cords are connected at regular intervals by transverse nerves, giving it a ladderlike appearance (figure 9.3*c*).

Careful observation will reveal that the cells of the ventral epidermis are ciliated, but those of the dorsal epithelium are not. Mucus and other types of **gland cells** are present in the epidermis and in the underlying mesenchyme. Many of the epidermal cells contain densely staining **rhabdites.** Rhabdites are small, rodlike bodies whose function is not yet fully understood. There is some evidence that they are discharged if the worm is attacked and swell up to form a slimy coat for defense. Some of the gland cells have long ducts that extend to the surface. These gland cells produce mucus for lubrication and other sticky materials for adhesion, capturing prey, and other functions. Among the parenchyma tissue, you should also

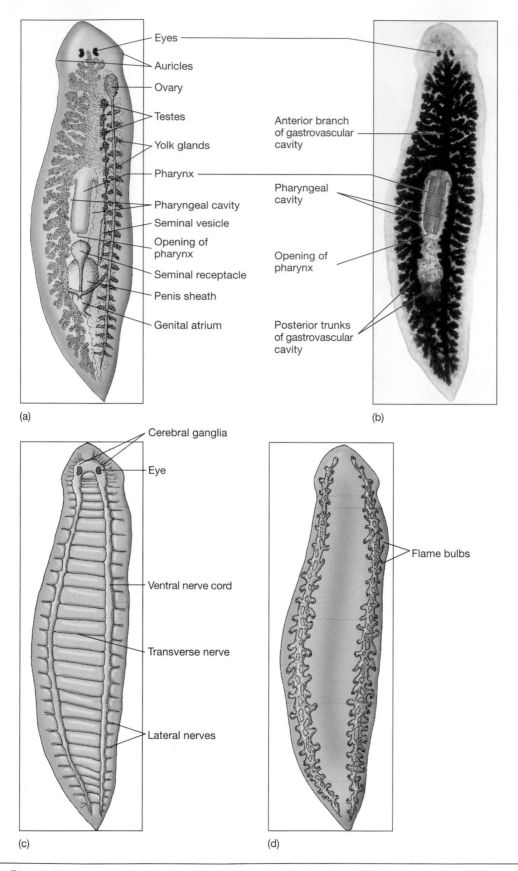

**FIGURE 9.2**  *Planaria.* (*a*) Reproductive system. (*b*) Microscopic whole mount. (*c*) Nervous system. (*d*) Excretory system.
(*b*) Photo courtesy of Carolina Biological Supply Company, Burlington, NC.

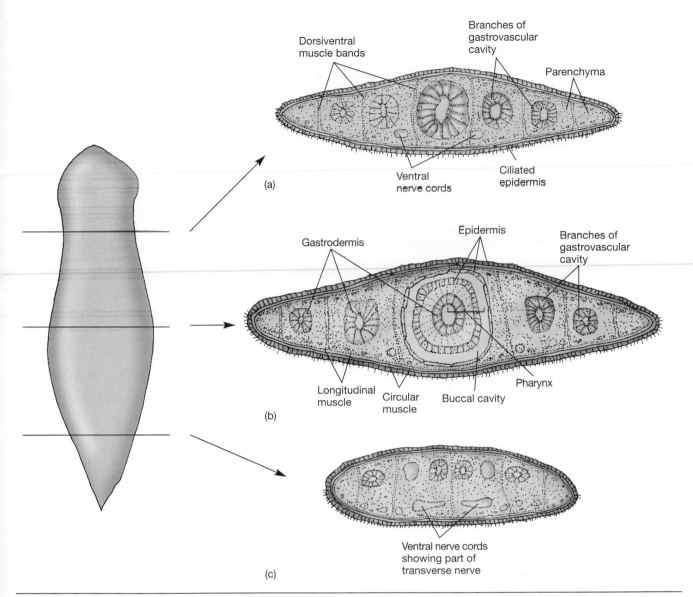

**FIGURE 9.3** *Planaria*. Cross sections through three body regions: (*a*) anterior region, (*b*) pharyngeal region, and (*c*) posterior region.

find several small branches of the gastrovascular cavity surrounded by large vacuolated gastrodermal cells.

In the middle section through the buccal cavity of the planarian, find the large muscular **pharynx** lying within the **buccal cavity.** The pharynx has powerful muscles that allow it to be extended from the buccal cavity through the **ventral mouth** into the host's body to suck up fluids and soft tissues. Within the pharynx, locate several types of muscles: inner and outer layers of **circular muscle,** inner and outer layers of **longitudinal muscle,** and bands of **radial muscle** that extend across the pharynx from the outside to the central lumen. Find the **ciliated epithelium** on the outside of the pharynx. Many **gland cells** that produce mucus and proteolytic enzymes can also be found within the pharynx.

◆ Study the section through the posterior portion of the body and note the several branches of the **gastrovascular cavity,** the ventral **nerve cords,** the circular and longitudinal **muscles,** and the **epidermis.**

## Reproduction and Excretion

The reproductive system of *Dugesia* and other freshwater flatworms is small and difficult to observe except in special microscopic preparations. The reproductive organs are shown in figure 9.2, but most of them probably will not be visible in your slide. Because each worm has both male and female sex organs, planaria are termed **monoecious.**

Although planarians are monoecious, they are not normally self-fertilizing. In sexual reproduction, sperm is transferred from the male system of one worm by the male copulatory organ, the **penis,** to the **seminal receptacle** of the partner. The sperm subsequently moves to the oviduct of the female reproductive system where fertilization occurs. The fertilized eggs are later deposited outside the body in cocoons where they develop directly into young worms (figure 9.5).

In many species of planarians, however, the most common form of reproduction is **asexual.** A worm separates into two parts, and each part regenerates the missing structures. Planarians have great powers of regeneration, and even relatively

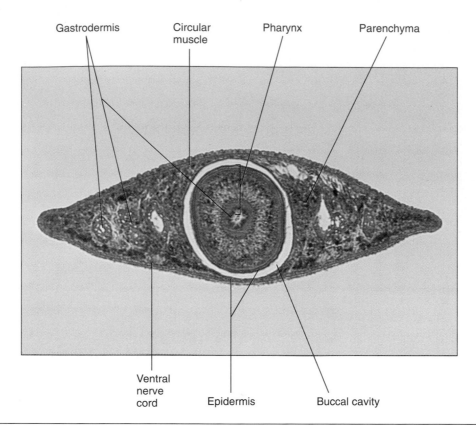

Gastrodermis    Circular    Pharynx    Parenchyma
                muscle

Ventral    Epidermis    Buccal cavity
nerve
cord

**FIGURE 9.4** *Planaria*. Cross section through pharynx.
Courtesy of Carolina Biological Supply Company, Burlington, NC.

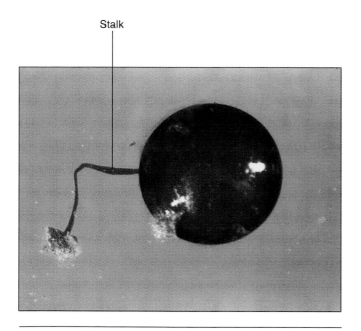

Stalk

**FIGURE 9.5** *Planaria*. Cocoon.
Courtesy of Carolina Biological Supply Company, Burlington, NC.

small parts of a worm can develop into a complete animal. If time and materials permit, your instructor may be able to help you set up an experiment with regeneration in planarians.

The excretory/osmoregulatory system of *Dugesia* and other planarians is not well-developed and probably represents an ancestral condition. It consists of a system of **flame bulbs** (a type of protonephridia) interconnected by a system of collecting ducts leading to a posterior excretory pore. The excretory structures are difficult to observe, except in special microscopic preparations. The flame bulbs appear to function mainly in **osmoregulation.** The body fluids and cellular contents are hypertonic to the environment (contain more dissolved salts, etc.); thus, a planarian must constantly eliminate excess water from the body. Nitrogenous wastes, resulting from the breakdown of proteins and other nitrogenous matter in the food, are excreted, mainly as ammonia ($NH_3$), directly from the body cells.

### Feeding and Digestion

Planarians, such as *Dugesia,* are chiefly carnivores and typically feed on protozoans and small animals, such as rotifers and small crustaceans. Food is ingested by the protrusible pharynx, which is extended through the midventral mouth while feeding. Do not confuse the opening of the muscular pharynx with the actual midventral mouth opening. Proteolytic enzymes, secreted from glands near the tip of the pharynx, aid in the penetration of a prey organism, such as a crustacean. The contents of the prey (a *Daphnia,* for example) can then be sucked into the muscular pharynx and passed into the gastrovascular cavity. The digestive system of a planarian is a gastrovascular cavity with a single opening that serves as both the entrance for food and the exit for waste materials.

- Place a small piece of fresh beef or pork liver in a dish containing one or more planarians. Observe their feeding behavior. Note the ventral **mouth** and the extension of the protrusible **pharynx** through the mouth when food is located. *How does the worm locate the food? What sensory structures may be involved?*

    *How does this basic organization of the digestive system compare with that of a cnidarian or a higher animal, such as an earthworm or a frog? Which type of system would be more efficient? Why?*

Digestion in a planarian is both **extracellular** (outside of the digestive cells) and **intracellular** (inside of the digestive cells). Digestive enzymes are secreted by **gland cells** in the gastrodermis that lines the gastrovascular cavity to assist in the breakdown of food materials. Later, small bits of food are engulfed by **phagocytic cells** in the gastrovascular lining. *How many cell types can you identify in the gastrodermis in your cross section?*

### Muscular System

Other structures you should identify in the cross sections include the **longitudinal** and **circular muscle layers** just beneath the epidermis. *Which of these layers lies closer to the epidermis? How can you relate these muscle layers to the locomotion that you observed in the living worms? Contraction of which layer would increase body length? How would this be helpful in locomotion?*

- Locate also in the cross sections the **dorsiventral muscle bands.** *What is their function?* Find the loosely packed parenchyma cells that fill most of the interior spaces. *How are interior cells nourished? How are wastes removed from them?* Near the ventral epidermis, find the **two ventral nerve cords.**

## The Flukes: Class Trematoda

### The Human Liver Fluke: *Clonorchis (Opisthorchis) sinensis*

#### Class Trematoda

Members of the Class Trematoda are all **endoparasites** and have well-developed **suckers** for attachment—one located in the region of the mouth, and one located on the midventral surface. The outer covering of the trematode body is highly modified and lacks cilia. The outer layer, the **tegument** (formerly called the cuticle), is a syncytial extension of underlying cells embedded in the body wall. Electron microscope studies have revealed that the tegument has a complex structure. Tapeworms (Class Cestoda) also have a tegument with a similar structure.

The tegument serves an active role both in protecting trematodes from the digestive enzymes of the hosts and in the uptake of nutrients from the host gut. The tegument is an excellent example of morphological and physiological adaptation of a parasite for its very special mode of life.

*Clonorchis sinensis* (figure 9.6), sometimes also called *Opisthorchis sinensis,* is a common and important human

parasite in parts of the world, particularly the Orient. Like many other trematodes, this species has a complex life cycle involving several hosts and a series of larval stages.

- Obtain a prepared microscope slide with a stained whole mount of an adult fluke and observe its size, shape, and general morphology under your stereoscopic microscope. Observe the **oral sucker** surrounding the mouth at the anterior end.

Behind the mouth is a muscular **pharynx,** a short **esophagus,** and two **intestinal caeca.** Note that the digestive tract has a single opening, the **mouth,** and thus represents an **incomplete digestive system.** A bilobed **cerebral ganglion** ("brain") lies on the dorsal side of the pharynx (small and difficult to see in most slides). Near the branching of the intestinal caeca, note the **posterior sucker.** Just posterior to this sucker, along the midline of the body, is the long, coiled **uterus** containing many eggs.

Posterior to the uterus lie the many-branched **testes** where the sperm are produced. Observe the many small **yolk glands** along the lateral margin of the body in its midregion. The yolk glands connect with the ovary by means of two delicate **yolk ducts.** The **ovary** is a single, small structure located near the center of the body. It is connected with the **seminal receptacle,** which serves to store sperm received during copulation.

- Study figure 9.7 and the demonstration materials provided to illustrate the life cycle of *Clonorchis.*

Note that the life cycle includes parasitic stages in three different hosts: *human, snail,* and *fish.* In order to survive, a parasite with such a complex life cycle including several hosts must have an effective means of transfer from one host to the next. Unless the proper host is available at the appropriate time, the life cycle cannot continue, and the parasite will die. This fact is used as the basis for the control of many parasitic diseases of humans and animals, such as malaria and schistosomiasis.

The adult liver fluke lives in the bile duct of humans and of other carnivorous animals. The host in which the adult (sexually mature) stage of a parasite resides is designated as the **definitive host.** All other hosts in the life cycle are termed **intermediate hosts.**

Human infections of *Clonorchis* occur from eating raw fish. Adult worms live in the bile ducts of the liver of the human host, and fertilized eggs are released into the bile duct. The eggs pass into the small intestine and are later voided in the feces of the human host. If the feces get into water, the eggs may be eaten by certain species of snails (**first intermediate host**). Inside the digestive system of the snail, the egg hatches into a larval form called a **miracidium.** The miracidium lives in the tissues of the snail, passing through several other larval stages (**sporocyst, redia,** and **cercaria**) and reproducing asexually to produce thousands of new larvae. The last larval stage, the cercaria, escapes from the snail and swims in the water until it contacts the **second intermediate host,** specific species of fish. When the fish is contacted, the cercaria burrow through the skin, shed their tails, and encyst to form still another stage, the **metacercaria.**

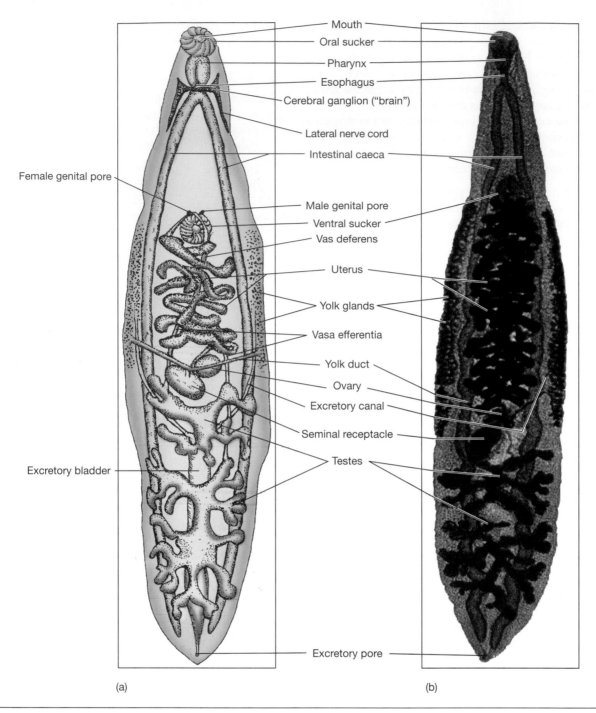

Mouth
Oral sucker
Pharynx
Esophagus
Cerebral ganglion ("brain")

Lateral nerve cord

Intestinal caeca

Female genital pore

Male genital pore
Ventral sucker
Vas deferens

Uterus

Yolk glands

Vasa efferentia

Yolk duct
Ovary
Excretory canal
Seminal receptacle

Testes

Excretory bladder

Excretory pore

(a)                                    (b)

**FIGURE 9.6** *Clonorchis*, whole mount.
(b) Photo courtesy of Carolina Biological Supply Company, Burlington, NC.

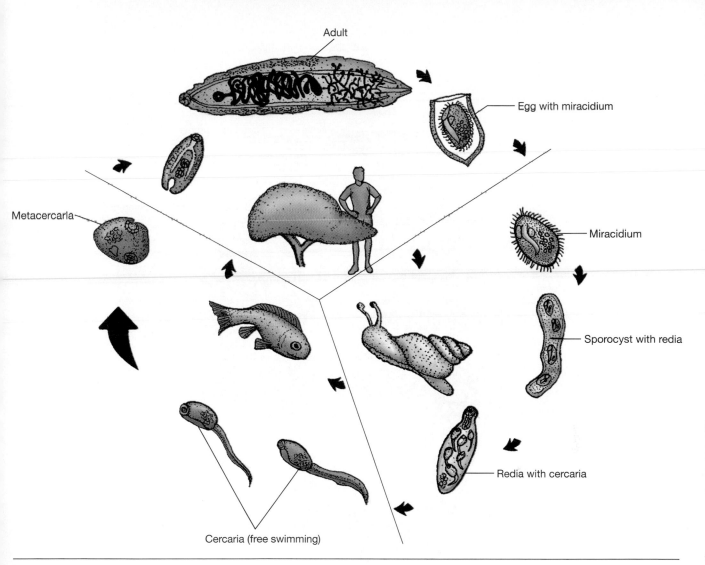

Adult

Egg with miracidium

Metacercaria

Miracidium

Sporocyst with redia

Redia with cercaria

Cercaria (free swimming)

**FIGURE 9.7** *Clonorchis*, life cycle.

If raw or improperly cooked fish containing metacercaria is eaten by a human or another appropriate definitive host, the cyst walls are digested, and the metacercaria are released. Subsequently, they migrate into the bile ducts of the liver where they develop into adult flukes to complete the life cycle.

## Cercaria Larvae

Living cercaria larvae of trematodes are easy to obtain and study in the laboratory. Many aquatic snails serve as intermediate hosts of trematodes, and specimens collected in lakes, ponds, and streams can serve as a good source of cercaria larvae. Following collection, snails should be washed and placed in wide-mouth bottles. Examine the containers several times each day because the cercariae of some species are shed only at specific times. Hold, the containers in a bright light in front of a dark background. Cercariae will appear as small white swimming objects. The emerging cercariae stream out in a white cloud from heavily infected snails.

Alternatively, infected snails (with directions for obtaining the cercariae) can be obtained from a biological supply house.

◆ Capture cercariae with a capillary pipette and make a wet mount on a clean microscope slide after adding a drop of 1% methylene blue or other appropriate stain. Draw most of the water from under the coverslip with a piece of filter paper to compress the cercariae enough so that they are held in place but not crushed under the weight of the coverslip.

## The Sheep Liver Fluke: *Fasciola hepatica*

### Class Trematoda

This large trematode is similar in structure to *Clonorchis*, although it is considerably larger, and its reproductive system is more complex (figure 9.8). *Fasciola hepatica* is a common parasite of sheep and cattle, but it also occasionally parasitizes other mammals, including humans. The adult

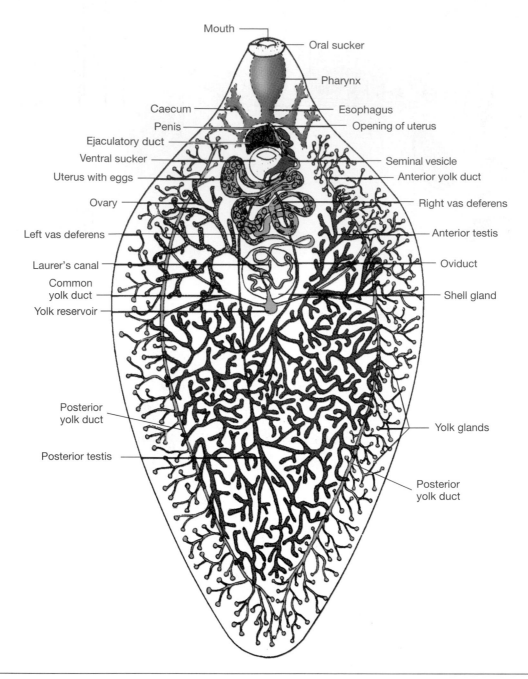

Mouth
Oral sucker
Pharynx
Caecum
Esophagus
Penis
Opening of uterus
Ejaculatory duct
Ventral sucker
Seminal vesicle
Uterus with eggs
Anterior yolk duct
Ovary
Right vas deferens
Left vas deferens
Anterior testis
Laurer's canal
Oviduct
Common
yolk duct
Shell gland
Yolk reservoir
Posterior
yolk duct
Yolk glands
Posterior testis
Posterior
yolk duct

**FIGURE 9.8**  *Fasciola*, whole mount.

flukes live mainly in the bile ducts of the mammalian host and cause the breakdown of the adjacent liver tissue, producing the disease called "liver rot."

The life cycle of *Fasciola hepatica* is of historic importance since it was the first to be worked out for a digenetic (two-host) trematode (figure 9.9). Adult worms live in the bile duct of a sheep and produce many **eggs** that pass from the bile duct to the intestine and are deposited with the feces. In water or warm, moist conditions, the **miracidia larvae** develop within the eggs and escape. These microscopic ciliated larvae must burrow into the appropriate species of snail within a few hours to continue

their development. Unless they find an appropriate host, the miracidia die.

In the appropriate snail, which serves as the **intermediate host,** the miracidia develop into **sporocysts.** Within the sporocysts, several germ cells develop into the next larval stage, the **redia** (plural: rediae). The rediae also reproduce asexually and produce still more rediae. The last generation of rediae produces another larval stage, the **cercaria** (plural: cercariae). The cercariae escape from the snail and become free-swimming, until they reach some aquatic plants. Here they settle, lose their tails, and encyst as **metacercariae.** The encysted larvae can survive for several weeks on the

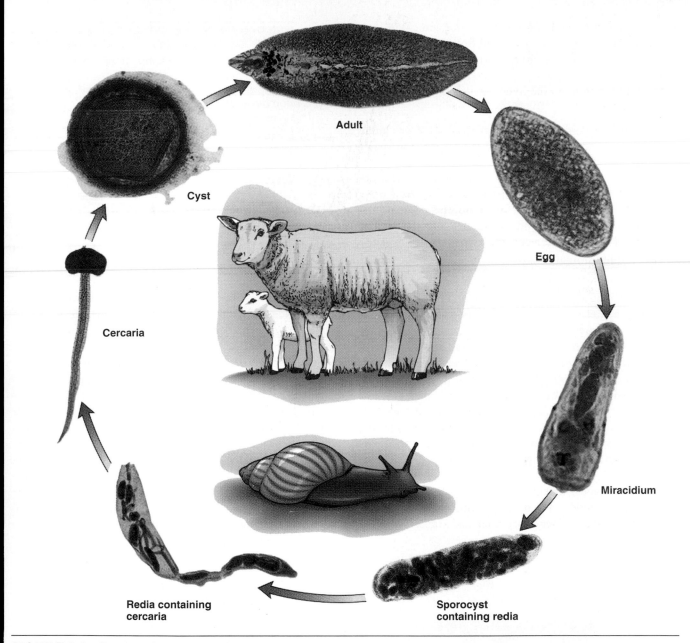

**Adult**

**Cyst**

**Egg**

**Cercaria**

**Miracidium**

**Redia containing cercaria**

**Sporocyst containing redia**

**FIGURE 9.9** *Fasciola*. Sheep liver fluke, life cycle.
Photos courtesy of Carolina Biological Supply Company, Burlington, NC.

plants, and if the plant and metacercariae are eaten by a sheep or other appropriate final or **definitive host,** the cyst wall is digested, and the larvae burrow through the intestinal wall to the body cavity and travel to the liver. During their migration through the liver, the young adults feed on the liver tissue.

◆ Observe the demonstrations of the various stages in the life cycle and try to understand the sequence of stages with the aid of figure 9.9.

Obtain a slide with a stained whole mount of an adult *Fasciola* and note its general shape. *How does its shape differ from that of* Clonorchis? Locate the **anterior sucker** around the mouth and the nearby **ventral sucker.** Between the two suckers, find the **genital pore** (figure 9.10).

The digestive system includes the anterior **mouth,** a muscular **pharynx,** a short **esophagus,** and two branches of the **intestine,** each of which has many smaller lateral branches, the **intestinal caeca** (singular: caecum). *What would you expect to be the main purpose of the intestinal caeca?* Both male and female reproductive systems are present as in *Clonorchis*. With the aid of figure 9.8, locate the principal organs of each system.

The female organs (mainly found in the anterior half of the body) include the following: **ovary, yolk glands, yolk ducts, yolk reservoir, shell gland, oviduct, uterus** with eggs, and the **opening of the uterus** just inside the genital pore.

The male organs include the following: **testes** (one anterior and one posterior), **vas deferens** *(how many?),* **seminal vesicle** *(how many?),* **ejaculatory duct,** and a muscular **penis.**

## Demonstrations

1. Slides with adult stages of some other representative trematodes
2. Slides representing the stages in the life cycle of *Clonorchis* and/or *Fasciola*
3. Living cercaria

# The Tapeworms: Class Cestoda

## The Dog Tapeworm: *Dipylidium caninum*

### Class Cestoda

Tapeworms are highly specialized internal parasites and show several important adaptations for their parasitic mode of life. The adult worms inhabit the intestines of various species of vertebrate animals, and the larvae live in the tissues of some alternate host. In general, the life cycles of tapeworms, or cestodes, are less complicated than those of the trematodes.

The flat, ribbonlike body of a tapeworm is typically divided into many sections called **proglottids,** and the body is divided into three major regions: an anterior **scolex,** a specialized holdfast organ with suckers and/or hooks (figure 9.11); followed by a narrow **neck,** which contains the budding zone where, at the posterior end, new proglottids are produced asexually; and the **strobila,** the rest of the long body, usually consisting of many maturing and ripe proglottids. Although the tapeworms superficially appear to be segmented, the proglottids are not generally believed to represent true body segments because of the way in which

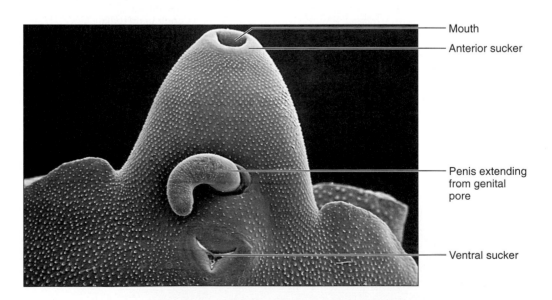

**FIGURE 9.10** *Fasciola*. Ventral surface of anterior end. Note sculpturing on tegument. Magnification 28×.
Scanning electron micrograph by Louis de Vos.

they are formed and because each proglottid is a complete reproductive unit within itself. Many zoologists therefore view the body of a tapeworm as comparable to a colony, or a chain of individuals, rather than as being actually segmented.

◆ Obtain a prepared microscope slide or a plastic mount with the scolex and some representative proglottids of the dog tapeworm, *Dipylidium caninum* (figures 9.12 and 9.13). Locate and study the following structures: the scolex, the neck, and the strobila. On the scolex, find the **rostellum,** the raised tip of the scolex, which bears several rows of **hooks,** and four lateral **suckers.** The hooks and suckers aid in attachment to the intestinal wall of the host. Note the absence of a mouth, pharynx, and digestive system. *Can you explain how a tapeworm can survive without these structures?* Ancestral flatworms almost certainly did have a mouth and a digestive system. *How would you explain their ab-*

*sence in tapeworms descended from ancestors that had such structures?*

◆ *Where would you find the youngest proglottids in an intact tapeworm? The oldest?* Study several proglottids in different stages of development and observe the different degrees of elaboration of the reproductive system. Identify immature, mature, and ripe proglottids (figure 9.11).

Select a mature proglottid for more detailed study and identify the **genital pore, vagina, oviduct, yolk glands, uterus, testes, vasa deferentia** (singular: vas deferens), **excretory canals,** and **longitudinal nerve cords.** Note the complete absence of any digestive structures in the tapeworm. *How do you suppose tapeworms obtain their nourishment?*

Also study some of the ripe proglottids and observe the many ovarian capsules containing several eggs or embryos. *Is the number of eggs per capsule always the same?*

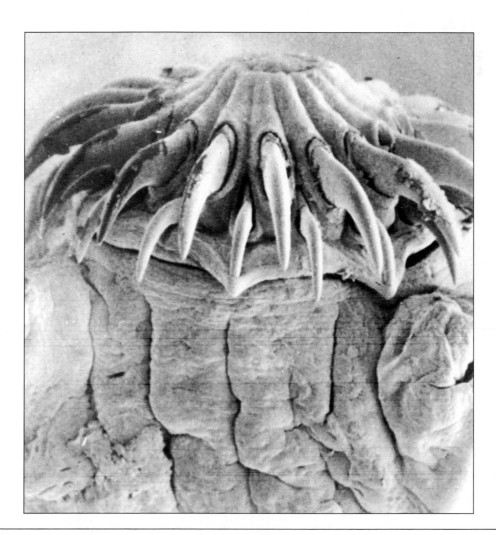

**FIGURE 9.11** *Taenia taeniaeformis*, from cat. Magnification 300×.
Scanning electron micrograph by Fred H. Whitaker.

Platyhelminthes **139**

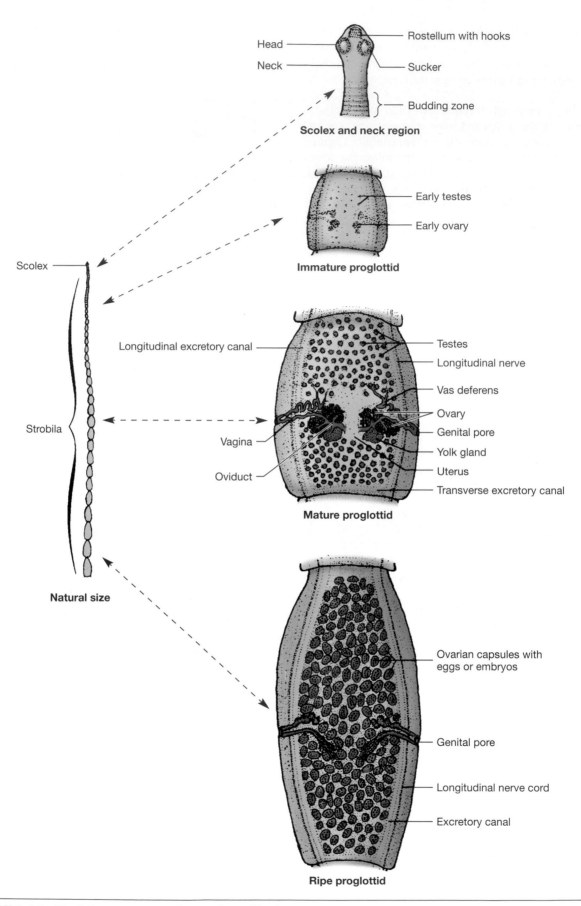

**FIGURE 9.12** *Dipylidium*, representative sections of body.

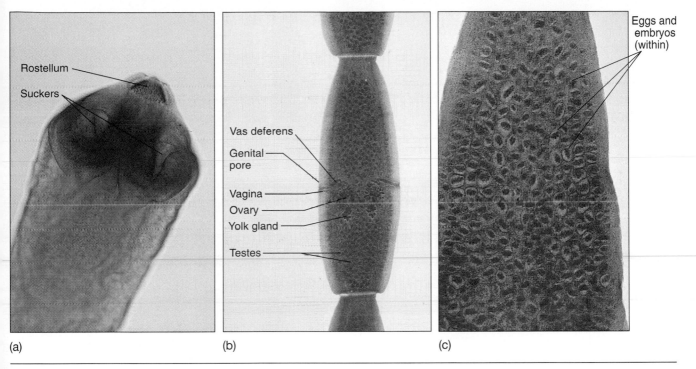

Rostellum

Suckers

Vas deferens

Genital pore

Vagina

Ovary

Yolk gland

Testes

Eggs and embryos (within)

(a)      (b)      (c)

**FIGURE 9.13** *Dipylidium canium.* (a) Scolex, (b) mature proglottid, and (c) portion of ripe proglottid.
Courtesy of Carolina Biological Supply Company, Burlington, NC.

Locate some of the transitional proglottids between the mature and ripe proglottids to observe the progressive **atrophy** of the reproductive organs. *Which organs disappear first? Which persist the longest? How do these observations relate to mode of reproduction of this tapeworm?*

The life cycle of *Dipylidium* includes two distinctive larval stages, a six-hooked **onchosphere larva** and a **cysticercoid larva** ("bladder worm"). Ripe proglottids containing eggs and embryos pass out of the host's intestine via the feces. The eggs are ingested by flea larvae and hatch into onchosphere larvae inside the intestinal wall of this intermediate host. Later the eggs develop into cysticercoid larvae as the fleas become adults. A dog or cat may become infected by nipping a flea. Children sometimes become infected by being licked by a dog or by ingesting eggs deposited in the soil.

### The Dog and Cat Tapeworm: *Taenia pisiformis*

#### Class Cestoda
*Taenia pisiformis* (figures 9.14 and 9.15) is another commonly studied tapeworm. Adult tapeworms of this species occur in the small intestine of dogs and cats. The larval stages are found in the liver and mesenteries of rabbits.

◆ Obtain a microscope slide and/or plastic mount with a scolex and representative proglottids of *Taenia pisiformis*. Identify the **scolex, neck, hooks,** and **lateral suckers.** Observe that *Taenia pisiformis*, like other tape-

worms, has no mouth or digestive system. *How does this tapeworm obtain its nourishment?*

Locate and identify **immature, mature,** and **ripe proglottids.** Select a mature proglottid on your slide and study its internal structure. Find the **uterus, yolk glands, oviduct, vagina, seminal receptacle,** and **genital pore** comprising the female reproductive system. Among the male reproductive structures, locate the **testes, vas efferens, vas deferens,** and the copulatory organ, the **cirrus.** Also find the lateral **excretory canals.** Two **longitudinal nerve cords** should be visible lateral to the excretory canals.

## Adaptations for Parasitism

Tapeworms provide a good example of the adaptations of animals for a special mode of life. Tapeworms exhibit several structural and physiological features that increase their chances of survival as internal parasites of vertebrates. List six different adaptations related to their parasitic mode of life that you have been able to observe or to identify from your reading.

1. _____
2. _____
3. _____
4. _____
5. _____
6. _____

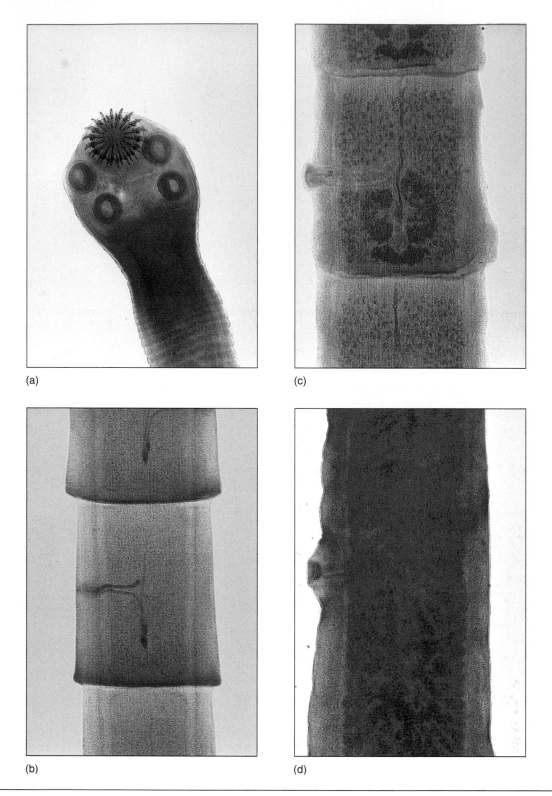

(a)

(b)

(c)

(d)

**FIGURE 9.14** *Taenia pisiformis*, representative sections. (*a*) Scolex, (*b*) immature proglottid, (*c*) mature proglottid, and (*d*) ripe proglottid.

Courtesy of Carolina Biological Supply Company, Burlington, NC.

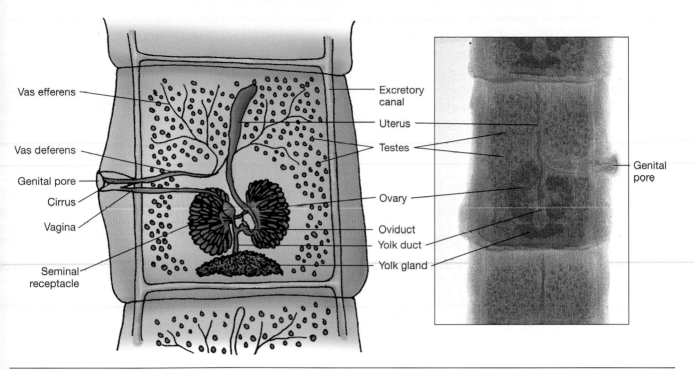

**FIGURE 9.15** *Taenia pisiformis*, mature proglottid.
Photo courtesy of Carolina Biological Supply Company, Burlington, NC.

Labels (left diagram): Vas efferens, Vas deferens, Genital pore, Cirrus, Vagina, Seminal receptacle

Labels (center): Excretory canal, Uterus, Testes, Ovary, Oviduct, Yolk duct, Yolk gland

Labels (right photo): Genital pore

## Demonstrations

1. Microscopic preparations of other tapeworms such as *Taenia solium* (pork tapeworm), *Taeniarhynchus saginatus* (beef tapeworm), and *Dibothriocephalus latus* (broad or fish tapeworm).
2. Onchosphere (six-hooked) larva.
3. Cysticercus larva.
4. Preserved whole tapeworms.

## Key Terms

**Bilateral symmetry**   body form with parts arranged symmetrically along a midsagittal plane.

**Cercaria (plural: cercariae)**   motile, free-swimming, tadpolelike larval stage in the life cycle of certain trematodes. Body resembles miniature fluke.

**Definitive host**   final host in the life cycle of a parasite with more than one host; host in which the sexually mature parasite develops.

**Intermediate host**   host in the life cycle of a parasite that bears a larval or immature stage of that parasite.

**Miracidium**   ciliated larval stage in the life cycle of many trematodes. Develops from the fertilized egg and gives rise to the sporocyst stage.

**Proglottid**   reproductive body division of a tapeworm. Since each proglottid contains a complete set of male and female organs, these units are often considered analogous to a complete individual, and thus the tapeworm as a colony.

**Redia**   larval stage in the life cycle of certain trematodes. Formed by the sporocyst and produces many cercariae.

**Scolex (plural: scoleces or scolices)**   specialized holdfast organ of the cestodes (tapeworms); with hooks and/or suckers for attachment.

**Sporocyst**   saclike larval stage in the life cycle of some trematodes. Develops from a miracidium and produces several rediae.

**Strobila**   the major portion of the body of a tapeworm excluding the scolex. Includes all of the proglottids.

**Tegument**   living syncytial protective outer covering of the cestode and trematode body; formerly called the cuticle; secreted by large underlying cells with numerous tubular cytoplasmic connections with the tegument.

**Triploblastic construction**   body type with three distinct tissue layers: endoderm, mesoderm, and ectoderm. Characteristic of flatworms and all higher metazoan phyla.

## Internet Resources

Visit the zoology website at http://www.mhhe.com/zoology to find live Internet links of each of the references listed below.

1. Animal Diversity Web, University of Michigan. Phylum Platyhelminthes. Information on platyhelminths, links to all classes.
2. Platyhelminthes page.
3. Platyhelminthes. Arizona's Tree of Life Web Page. Pictures, characteristics, phylogenetic relationships, references on flatworms. A picture of a flatworm, and a link to trematodes (aspidogastreans).
4. National Center for Infectious Diseases. This CDC site has many links to information on bacterial, viral, protozoan, and worm-related diseases (primarily affecting humans).

## Critical Thinking Questions

1. Compare the Phylum Platyhelminthes structurally with the Phylum Cnidaria. What are the significant changes in symmetry, tissues formation, and life cycles found in the Cnidaria?

2. Discuss the importance of a triploblastic germ tissue construction. Of what evolutionary significance is the mesoderm?

3. The parasitic flatworms have unusual life cycles and hosts. Speculate how these unusual life cycles might have evolved.

4. Why is encystment important to survival in many parasitic flatworms? Do you suppose that encystment in the different parasitic groups arose once or many times over the course of evolution? Support your answer.

5. Discuss the most important adaptations observed in flatworms. What adaptations are most important for endoparasites? Which are most important for ectoparasites?

## Suggested Readings

Buchsbaum, R., M. Buchsbaum, J. Pearse, and V. Pearse. 1987. *Living Invertebrates*. Pacific Grove, CA: The Boxwood Press.

Croll, N.A. 1966. *Ecology of Parasites*. Cambridge, MA: Harvard University Press.

Lemly, A.D., and G.W. Esch. 1984. Effects of the trematode *Uvulifer ambloplitis* on juvenile bluegill sunfish. *Journal of Parasitology* 70(4):475–92.

Noble, E.R., and G.A. Noble. 1982. *Parasitology: The Biology of Animal Parasites,* 5th ed. Philadelphia: Lea & Fibiger.

Olsen, W.O. 1962. *Animal Parasites: Their Biology and Lifecycles*. Edina, MN: Burgess Publishing Co.

Olsen, W.O. 1974. *Animal Parasites*. Baltimore, MD: University Park Press, pp. 199, 228–31.

Pappas, P.W. 1983. *Biology of the Eucestoda*. London: Academic Press, pp. 28–35.

## NOTES AND SKETCHES

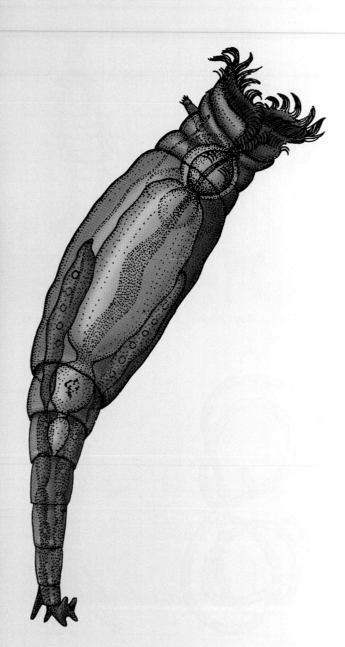

## OBJECTIVES

After completing the laboratory work in this chapter, you should be able to perform the following tasks:

1. List and briefly describe six of the phyla presently grouped together as pseudocoelomate animals.
2. Distinguish between a coelom and a pseudocoelom. Explain the significance of each.
3. Describe the external morphology of the roundworm *Ascaris* and locate the chief external features on a preserved specimen. Distinguish between mature male and female specimens.
4. Describe the male and female reproductive systems of *Ascaris* and identify the principal reproductive organs in dissected specimens and on microscope slides.
5. Name three parasitic roundworms, explain their importance in human or animal health, and briefly describe their life cycles.
6. Describe the anatomy of a rotifer, such as *Philodina,* and locate its principal organs in a specimen or on a drawing.
7. Name three other pseudocoelomate animals and briefly explain their economic, ecological, and scientific importance.
8. Compare the body organization of a pseudocoelomate animal (such as a roundworm) to that of an acoelomate animal (such as a flatworm). List the major advances shown by the pseudocoelomates over the organization of the acoelomates.
9. Summarize the arguments for and against considering the several groups of pseudocoelomate animals as members of a single phylum. *Why have there been differences of opinion among zoologists on this issue?*

## Introduction

The roundworms, Phylum Nematoda, are often studied together with several other groups of pseudocoelomate animals. Although these animal groups in the past have been lumped together in the single large and diverse Phylum Aschelminthes, they are now usually considered separate phyla because of their many differences.

The most prominent similarity among these groups is the **pseudocoelom,** a central body cavity lying between the endoderm and mesoderm layers. The pseudocoelom develops from the embryonic blastocoel, and although these animals are triploblastic, their central body cavity develops quite differently from those animals with a true coelom.

Other common features said to be shared among many of these pseudocoelomate animals include a **cylindrical body** with an external **cuticle, microscopic size,** a **complete digestive tract** with both a mouth and an anus, a **protonephridial excretory system,** and a **fixed number of cells** (or nuclei) in the adult. This latter condition results from the cessation of mitoses in the adult, which limits the number of cells (and nuclei) present in each mature individual.

Numerous exceptions to these "common" features have been found and have led to the present uncertainty about the true evolutionary relationships among these animal groups. For instance, recent electron microscopic investigations have shown that several of these groups have a much reduced body cavity, and, in some cases, this cavity may not actually be a pseudocoelom. Much more study is needed on these enigmatic animals.

Eight phyla are now commonly included in this grouping of pseudocoelomates, including the **Nematoda, Rotifera, Gastrotricha, Kinorhyncha, Priapulida, Loricifera, Nematomorpha,** and **Acanthocephala** (figure 10.1). Sometimes two other phyla, the Entoprocta and the Gnathostomulida, are also included in this group, but their relationships to the other groups are even less clear. Most of the pseudocoelomate animals are microscopic; only the

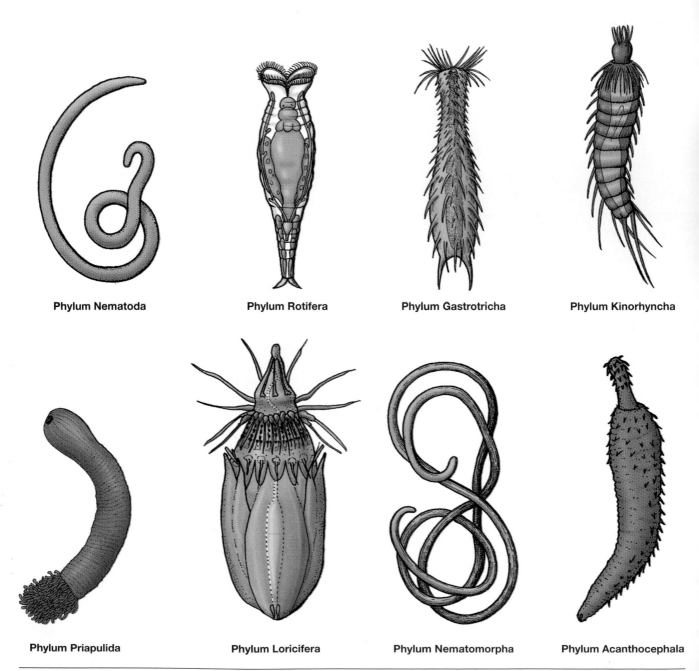

Phylum Nematoda　　　Phylum Rotifera　　　Phylum Gastrotricha　　　Phylum Kinorhyncha

Phylum Priapulida　　　Phylum Loricifera　　　Phylum Nematomorpha　　　Phylum Acanthocephala

**FIGURE 10.1** Principal phyla of pseudocoelomate animals.

Nematoda, Nematomorpha, Acanthocephala, and the Priapulida include species of macroscopic size.

In this exercise, we will study some representatives of two of these phyla, the Nematoda and the Rotifera, to illustrate some of the important features of the organization of pseudocoelomate animals and the biology of a few representatives.

# Classification

The eight phyla currently grouped among the pseudocoelomates are briefly described here.

## Phylum Nematoda (Nemathelminthes or Roundworms)

Free-living and parasitic worms with an **elongate cylindrical body tapered at both ends,** a **triradiate pharynx,** and a **specialized excretory system.**

## Phylum Rotifera (Rotifers)

Microscopic animals, found primarily in freshwater habitats, with an anterior ciliated locomotory and feeding organ, the **corona,** and a specialized internal grinding organ, the **mastax.**

## Phylum Gastrotricha (Gastrotrichs)

Microscopic aquatic animals with an **external cuticle** covered by scales or spines, and **patches or tracts of cilia** on the epidermis of the ventral surface.

## Phylum Kinorhyncha (Kinorhynchs)

Microscopic marine animals with strong **superficial segmentation** (segmented cuticle), a **retractile head with spines,** and **lateral spines** along the body.

## Phylum Priapulida (Priapulids)

Burrowing marine worms, usually microscopic to 13 cm in length, with a **spiny, retractable proboscis** (prosoma), a **warty cuticle,** superficial segmentation, and usually with one or two terminal **caudal appendages** of unknown function.

## Phylum Loricifera (Loriciferans)

Microscopic marine animals found in marine sediments. Bilaterally symmetrical with a **retractable spiny head** and a **lorica** (exoskeleton) of six cuticular plates covering the abdomen. Additional rows of **spines** or **stylets** encircle the anterior end of the body.

## Phylum Nematomorpha (Gordiacea or Horsehair Worms)

Slender, elongate worms with a **cylindrical body rounded at the ends,** a **reduced digestive tract,** and **larvae that parasitize insects** or other arthropods.

## Phylum Acanthocephala (Spiny-headed Worms)

Parasitic worms with adults living in the intestines of vertebrates and larvae in arthropod hosts; **anterior proboscis with hooks** for attachment to the gut wall of the host; **no digestive tract.**

| Materials List |
| --- |

**Living specimens**
  *Anguilla aceti*
  *Philodina*
  *Caenorhabditis elegans*
**Preserved specimens**
  *Ascaris*
**Microscope slides**
  *Ascaris,* cross sections of male and female
  *Ancylostoma,* whole mount
  *Enterobius,* pinworm, whole mount
  *Chiloplacus,* free-living nematode (demonstration)
  *Trichinella,* whole mount (demonstration)
  Representative Rotifera (demonstration)
**Plastic mounts**
  *Ascaris,* male and female worms (demonstration)
  *Macracanthorhynchus,* acanthocephalan, parasite of swine (demonstration)
  *Ancylostoma,* hookworm (demonstration)
**Miscellaneous**
  Nile blue solution

# Phylum Nematoda

## A Parasitic Roundworm: *Ascaris lumbricoides*

### External Anatomy

*Ascaris lumbricoides* (figure 10.2) is a large roundworm parasite of humans and swine. The adult worms range in length from about 15–40 cm. Living specimens are yellow or pink, but preserved specimens usually range from pink to brown.

◆ Study a preserved specimen to determine some of the characteristic features of roundworms, members of the Phylum Nematoda.

Observe the elongate cylindrical body that tapers at both ends. The anterior end is more pointed than the posterior, which tends to be slightly blunt. The triangular **mouth,** surrounded by three distinct **lips,** is located at the anterior end of the body (figure 10.3). Under your dissecting microscope, observe the three lips and the mouth. Also locate the **anus,** which appears as a transverse slit on the ventral side of the body near the posterior end. The ventral location of the anus is a convenient means of distinguishing the dorsal and ventral surfaces of the worm. You should be able to find four longitudinal lines extending lengthwise along the body wall: one dorsal, one ventral, and two lateral in position. These

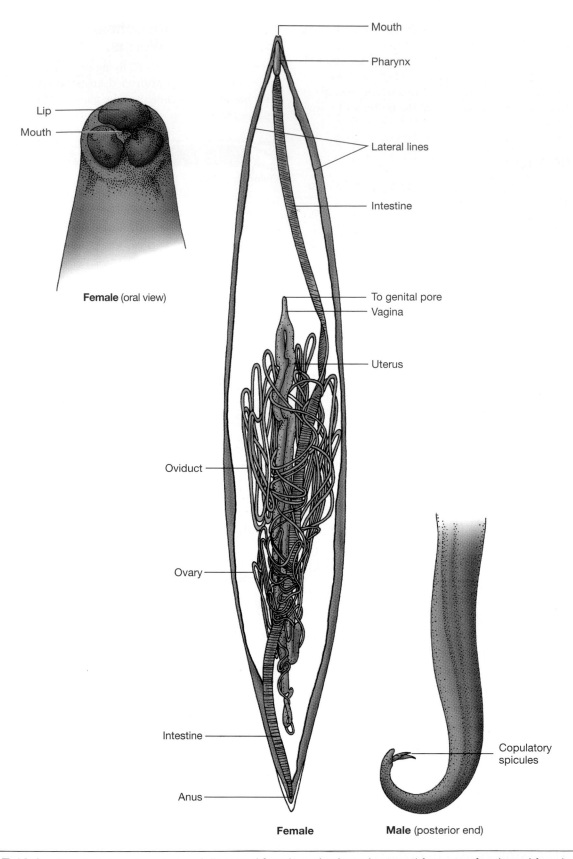

Lip
Mouth
**Female** (oral view)

Mouth
Pharynx

Lateral lines

Intestine

To genital pore
Vagina

Uterus

Oviduct

Ovary

Intestine

Anus

**Female**

Copulatory
spicules

**Male** (posterior end)

**FIGURE 10.2** *Ascaris,* internal anatomy of dissected female and selected external features of male and female.

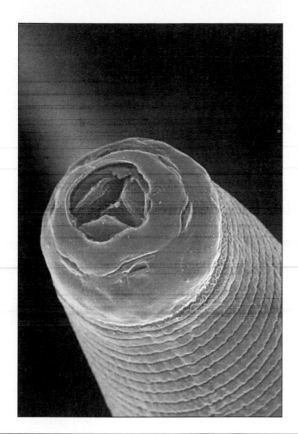

**FIGURE 10.3** Nematode. Mouth, lips, and anterior end showing triangular mouth surrounded by three teeth as in *Ascaris.* Magnification 5,200×.
Scanning electron micrograph by Louis de Vos.

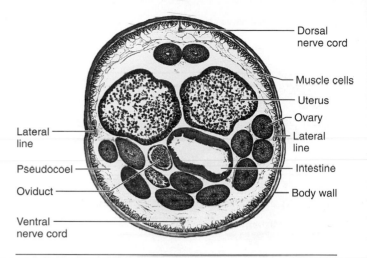

**FIGURE 10.4** *Ascaris,* cross section, female.
Courtesy of Carolina Biological Supply Company, Burlington, NC.

▲ **Caution:** Although the *Ascaris* specimens you will receive in the laboratory have been preserved in formalin, the eggs found in the female specimens are enclosed in a very resistant "shell." They may not all be killed, even after several weeks in formalin. Although acquiring an infection in this way is a remote possibility, you should always follow good laboratory safety practices. Always wear dissecting gloves and avoid putting your hands in your mouth after handling live or preserved animals. You should wash your hands after you complete your work.

longitudinal lines are more apparent in living specimens than in preserved specimens, and their significance will be better understood after your study of the microscopic anatomy of the worm later in this exercise.

A further examination of the external features of your specimen should reveal its sex. Most species of nematodes are **dioecious** (sexes are separate), and the males and females usually exhibit external morphological differences. Male specimens of *Ascaris lumbricoides* typically exhibit a curved posterior end and a pair of **copulatory spicules** extending from the anus (figure 10.2). Female worms have a ventral **genital pore** about one-third the distance from the anterior end. *From your observation of the location of the copulatory spicules in the male, what can you conclude about the location of the male genital opening?* Examine several specimens of *Ascaris lumbricoides* to determine whether you can find any other morphological differences between the sexes.

### Internal Anatomy

◆ After you have completed your study of the external features of *Ascaris,* carefully dissect the worm by making a longitudinal slit down its dorsal side. Take care with your incision to avoid damaging the internal organs. Pin down the body wall on each side of the worm in a wax-bottom dissecting pan to expose fully the internal structures, and cover your specimen with a little water to keep the internal organs moist and flexible.

Observe the long, ribbonlike **intestine** extending most of the length of the body and the numerous tangled threadlike coils surrounding the middle part of the intestine (figure 10.2). The female reproductive system is Y-shaped, and each branch of the Y consists of a long, threadlike **ovary** connected with a larger **oviduct** and a still larger **uterus.** The distinction among these three organs is not apparent externally but is easy to observe in microscopic cross sections. The two uteri join to form a single midventral muscular tube, the **vagina,** which opens to the exterior via the female genital pore, or **vulva,** observed previously.

The male reproductive system is not branched, as in the female reproductive system, and it consists of a single, continuous tubular structure. This continuous structure includes a threadlike **testis** connected with a larger-diameter **vas deferens,** or sperm duct, a still larger diameter **seminal vesicle,** and a short, muscular **ejaculatory duct,** which empties into the **cloaca** at the posterior end of the digestive tract.

The digestive system consists of the ribbonlike **intestine,** which is attached anteriorly to a short muscular **pharynx** into which the mouth opens, and posteriorly to the **cloaca.**

### Cross Sections

Prepared microscope slides with cross sections of male and female worms illustrate other aspects of nematode structure (figures 10.4 and 10.5).

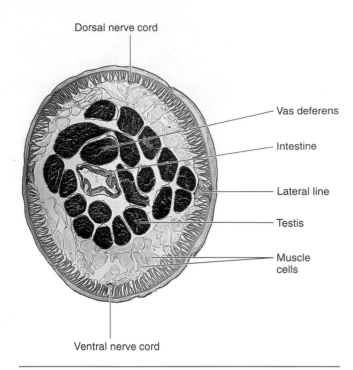

Dorsal nerve cord

Vas deferens

Intestine

Lateral line

Testis

Muscle cells

Ventral nerve cord

**FIGURE 10.5** *Ascaris,* cross section, male.

Courtesy of Carolina Biological Supply Company, Burlington, NC.

◆ Examine a cross section of a female worm and identify the thick **cuticle** covering the body and the underlying layer of **hypodermis,** which secretes the cuticle. Observe the many **longitudinal muscle bands;** there are no circular muscles present in roundworms.

This anatomical feature produces the peculiar whiplike movements often exhibited by living roundworms. Free-living nematodes are sometimes called "whipworms." Also associated with the body wall are the **dorsal** and **ventral nerve cords** and the two **lateral lines.** Located within each of the lateral lines is an **excretory canal,** part of the highly specialized excretory system of the nematodes.

Nematodes lack protonephridia but do have a unique system of gland cells connected to a canal system that empties through a ventral excretory pore. This system, called the **renette gland,** is located ventrally near the pharynx. The function of this unique excretory system is not well understood, but *Ascaris* is known to excrete nitrogenous waste in the form both of ammonia and urea.

Observe the thin-walled intestine and note that the intestinal wall consists of a single layer of large **gastrodermal cells.** The body cavity of *Ascaris* is a **pseudocoelom,** since it lacks a mesodermal layer around the gut.

In the cross section of the worm, the female reproductive system should be represented by numerous round or oval structures of various sizes. The **ovaries** are small in diameter and bear some resemblance to tiny wagon wheels. A layer of columnar ovarian cells surrounds a central core, or **rachis.** The **oviducts** are slightly larger in diameter than the ovaries and have a central opening, or lumen, for the passage of eggs. If the section was cut through the region of the vagina, you could observe a single, large median structure,

usually filled with eggs. Sections posterior to the vagina region usually show two large uteri, also filled with eggs.

Cross sections of male specimens (figure 10.5) exhibit features similar to female specimens except for the reproductive structures. The reproductive system is usually represented by several sections through the **testes,** which are small round structures containing many spermatogonia. Several sections through the larger **vas deferens** should also be present in the section. You should find numerous spermatocytes in the vas deferens of mature male specimens. A large **seminal vesicle** containing mature spermatozoa should also be present.

### Life Cycle

*Ascaris lumbricoides* has a direct life cycle with no intermediate hosts. Mature female worms in the intestine of the host produce eggs that are expelled in the feces. The fertilized eggs develop into larvae in warm, moist soil. If eggs containing larvae (figure 10.6) are ingested by an appropriate host, development may continue. Hatching occurs in the small intestine, and the larvae emerge, burrow through the intestinal wall, and are carried to the liver of the host via the hepatic portal vein. From the liver, the larvae migrate to the lungs. Later they move up through the bronchi and trachea and reach the oral cavity. From there, the worms are swallowed and are carried to the small intestine. Sexual maturity is reached within about two months, when fertilized eggs begin to be released. Fertilization occurs after copulation between male and female worms in the host's intestine.

## A Parasitic Roundworm: *Trichinella spiralis*

Another common parasitic nematode is *Trichinella spiralis,* the trichina worm. *Trichinella* is the causative agent of the disease **trichinosis** and occurs commonly in humans, swine, bears, rats, and other mammals. The disease is usually contracted by humans by eating raw or inadequately cooked pork or bear meat.

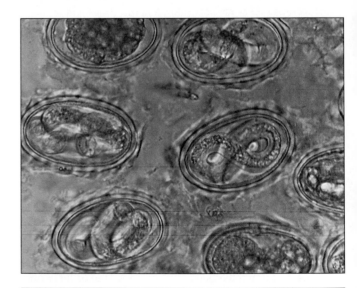

**FIGURE 10.6** *Ascaris.* Eggs with developing larval worms.

Courtesy of Carolina Biological Supply Company, Burlington, NC.

◆ Observe a microscope slide showing encysted worms in the skeletal muscles of a pig or a rat (figure 10.7).

Encysted worms in infected meat may be released through the digestion of the cyst walls by enzymes in the human digestive tract. The juvenile worms mature sexually in about two to three days and then mate. The females burrow into the wall of the small intestine and shortly after begin to produce living offspring. A single female may produce up to 1,500 young in her three-month life span. These juvenile worms enter the host's lymphatic system and are distributed through the body via the circulatory system. The young worms burrow mainly into the active skeletal muscles, where they grow to about 1 mm in length and become enclosed in cysts. Although such encysted worms cannot continue to develop unless the infected tissue is eaten by another suitable host, the encysted worms may remain alive within the cysts for several years.

*Trichinella* infections in humans often result in mild muscle aches that are mistaken for the flu. Such infections commonly go undiagnosed and sometimes are discovered only during an autopsy. Severe pain and serious damage can result, however, from the burrowing of adult and juvenile worms through the tissues of the body, and rare infections of the brain can result in death.

## Hookworms: *Ancylostoma duodenale* and *Necator americanus*

*Ancylostoma duodenale* and *Necator americanus* are the two most common and most serious human hookworms. *A. duodenale* is predominantly a northern species common in Europe, North Africa, northern China, and Japan, while *N. americanus* is primarily a tropical species occurring in many

warmer areas of the world. At one time *N. americanus* was a very serious public health menace in the southeastern United States. Improved sanitation and control measures in the past few decades, however, have greatly reduced the severity of the hookworm problem.

Both species are principally human parasites, but are sometimes found in other animal species. Both species exhibit specialized **teeth** and/or **cutting plates** around the mouth (figure 10.8), which aid in the attachment of the adult worms to the wall of the small intestine. The body of *N. americanus* is smaller and more slender than that of *A. duodenale*.

◆ Study the demonstration slides of hookworms.

### Life Cycle

The life cycles of the two species are basically similar. The adults live in the small intestine, where they attach to the intestinal mucosa and feed on blood and fluids from tissues. Adult females produce large numbers of eggs each day (5,000–20,000). The eggs are released in the host feces, and the embryonated eggs develop and hatch in the dung. The young, free-living larvae feed on bacteria and other debris, grow, molt several times, and reach the infective stage in about five days. During this time, they live in the upper layer of soil and frequently migrate to the surface.

They are attracted to heat and are stimulated by physical contact. These stimuli facilitate contact and penetration of exposed human skin—such as that of a bare foot. Wearing shoes, therefore, instead of walking barefoot, is a very effective defense against hookworm infection. Often simple sanitation measures, like wearing shoes and taking care in the disposal of human feces, can be among the most effective protection against many types of parasitic infections.

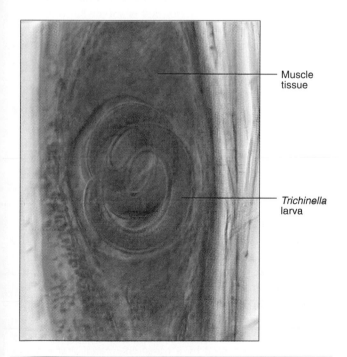

**FIGURE 10.7** *Trichinella.* Larval worm, encysted in muscle tissue.
Courtesy of Carolina Biological Supply Company, Burlington, NC.

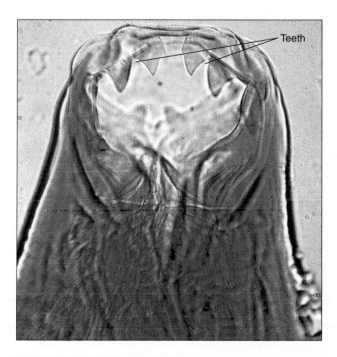

**FIGURE 10.8** *Ancylostoma,* hookworm, mouth and buccal cavity showing teeth.
Courtesy of Carolina Biological Supply Company, Burlington, NC.

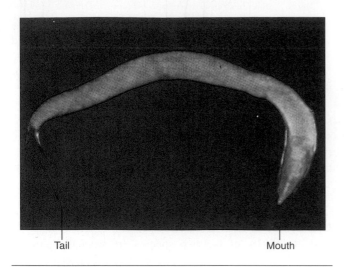

Tail        Mouth

**FIGURE 10.9** *Enterobius,* pinworm.
Courtesy of Carolina Biological Supply Company, Burlington, NC.

After contact with exposed human skin, the larvae burrow through the skin until a blood or lymph vessel is reached. In the bloodstream, they are carried to the heart and then to the lungs. There they burrow into the alveoli. From the lungs, they are passed up the bronchi and trachea to the throat. From the throat, they may be swallowed and pass through the digestive tract to the small intestine, where they undergo a final molt and attach to the mucosa. Sexual maturity is reached in about six weeks.

## Pinworm: *Enterobius vermicularis*

The most common nematode parasite of humans in the United States is the pinworm, *Enterobius vermicularis* (figure 10.9). One epidemiological study showed that as many as one-third of North American preschool children may be infected. Infections of adults are much less common.

◆ Obtain a microscope slide with a stained whole mount of an adult *Enterobius* and observe its size, shape, and internal organs.

Adult female pinworms are 8–13 mm long and have a pointed tail. They have enormous reproductive potential; a single female can lay more than 16,000 eggs in her lifetime.

Male and female adult worms live in the host intestine where they attach to the walls. Gravid females migrate posteriorly to the anus from which they often emerge at night to deposit their eggs on the surrounding skin. Infections and re-infections usually arise from ingestion of eggs transferred from contaminated bedding or from fingernails contaminated from scratching the anal area.

## Free-living Nematodes

### Caenorhabditis: *An Important Research Animal*

The free-living soil nematode, *Caenorhabditis elegans,* has become an important animal used in many research studies during the last 20 years (figure 10.10). It is a very useful animal

for research studies in genetics, development, and cell biology. *C. elegans* is a valuable animal for research studies because of its short life span of about 3.5 days, its transparent eggs and body, its simple anatomy, a fixed number of cells in the adult, and its ease of laboratory culture.

Embryonic cleavage in *C. elegans* is precise and development is strongly determinate. Since the fate of each embryonic cell is fixed very early in development, the species is very useful in studies on the role of genes in development. In this species, more than 500 specific genes have been identified and their roles investigated. Recently, the entire sequence of nucleotides in the genome *C. elegans* was determined.

Many studies have followed the development of specific cells and the formation of their products, such as specific muscle proteins, in the larvae and adults of *C. elegans*. Other studies have focused on the interactions of cells during development. The worms are easily cultured in the laboratory in petri dishes and feed on the bacterium *Escherichia coli*. In nature, *C. elegans* feeds on soil bacteria.

If specimens of *C. elegans* are not available, *Rhabditis,* a related soil nematode similar in structure and readily available from most biological supply companies, can be substituted for this exercise.

◆ Prepare a wet mount with a few specimens of *C. elegans* and observe the slide under your compound microscope. If the worms are too active for good observation, briefly hold the slide over a warm incandescent lamp or pass it quickly two or three times through the flame of an alcohol lamp or Bunsen burner to kill the worms. Be careful not to cook them!

The body of *C. elegans* resembles a flexible cylinder tapered at both ends; the anterior end is somewhat blunt, and the posterior end terminates in a pointed tail. The entire body is covered with a collagenous **cuticle** secreted by an underlying layer of **hypodermal cells.** Beneath the hypodermal layer is a layer of **longitudinal muscles** arranged in a series of muscle bands. There are no circular muscles. This arrangement of muscles results in the characteristic whiplike movements of the worms as they move along the substrate. The central cavity is a **pseudocoel** and is derived from the embryonic blastocoel (see Chapter 8). Within the pseudocoel is the digestive tract, which has no muscle layers or other mesodermally derived tissues separating it from the pseudocoel.

The animals have a relatively simple histological structure with few types of specialized cells and tissues. Most of the internal organ systems are tubular and are suspended in the fluid-filled pseudocoel. There are no circulatory or gas exchange organs. The excretory system of nematodes is different from those found in other animals, and its functions are not well understood. In *C. elegans,* the **excretory system** consists of a pair of lateral canals that run along the body wall and are connected by a transverse canal that empties through a ventral excretory pore in the pharyngeal region.

The **digestive system** begins with an anterior mouth surrounded by three pairs of lips. The lips lead into the short

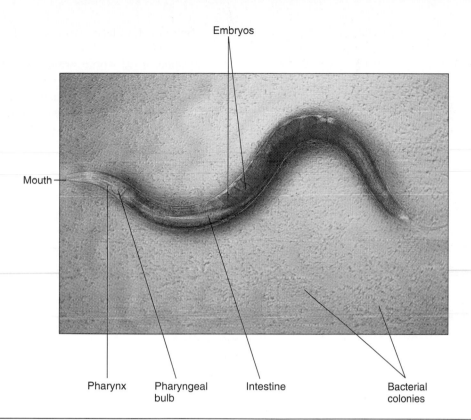

Embryos

Mouth

Pharynx

Pharyngeal
bulb

Intestine

Bacterial
colonies

**FIGURE 10.10** *Caenorhabditis elegans.*
Courtesy of Carolina Biological Supply Company, Burlington, NC.

**pharynx,** which terminates in a muscular **pharyngeal bulb.** Posterior to the esophagus is the long **intestine,** which terminates in the **rectum.** The rectum empties through the ventral, subterminal **anus.**

The **nervous system** of *C. elegans* is simple, basically resembling that of *Ascaris,* and consists of an **anterior nerve ring** that surrounds the pharynx, a **ventral nerve cord,** and a smaller **posterior nerve ring.** Several sensory and motor neurons connect with this central nervous system. Detailed studies of the nervous system of *C. elegans* have revealed that the entire nervous system of an adult of this species consists of fewer than 300 neurons.

*C. elegans,* like many other free-living nematodes, is predominantly **hermaphroditic.** Most individuals have two reflexed **gonads** that form a loop within the pseudocoel. In each tubular gonad is an **ovary** that extends anteriorly before looping back posteriorly and connecting with a larger **oviduct.** Locate the oviduct with large maturing eggs (and embryos) near the midventral opening, the **vulva,** through which the embryos are released. Sperm are produced in the small **testis** also located in each tubular gonad. The sperm are stored in a **spermatheca.** Fertilization takes place within the oviduct after release of sperm from the spermatheca, and the embryos within the transparent eggs usually undergo several cleavages before their release. Development is rapid and usually involves four larval stages preceding the adult. The collagenous cuticle is shed and secreted anew after each larval stage.

Males are produced only rarely in this species, so **self-fertilization** in hermaphroditic individuals is the most common strategy of reproduction. Even this low frequency of males, however, is believed to be important in providing additional genetic variability to enhance the survival of the species.

## The Vinegar Eel: *Anguillula aceti*

The vinegar eel (figure 10.11), *Anguillula aceti* (also sometimes called *Turbatrix aceti*), is a tiny, free-living nematode often found (in times past) in unpasteurized vinegar. Today most vinegar sold in stores is pasteurized and has added preservatives to prevent the growth of vinegar eels and other undesirable flora and fauna. Vinegar eels are harmless, however, and would pose no health threat if accidentally ingested—they would add a little protein to your diet!

The worms are most abundant in the bottom sediments of unpasteurized vinegar and other fermented fruit juices. Vinegar eels thrive in such acidic conditions and feed on the yeast and bacteria growing in the sediments. This combination of yeast and bacteria is often called the "mother" of the vinegar.

From a biological point of view, "mother" of the vinegar is not a bad term for these sediments. *Can you explain why this might be an appropriate name from what you know of the process of fermentation?*

◆ Examine a sample from the culture of the worms and observe their active, whiplike swimming movements. *How can you relate their mode of locomotion to what you have learned about the anatomy of the nematodes? What kind of muscles are involved?*

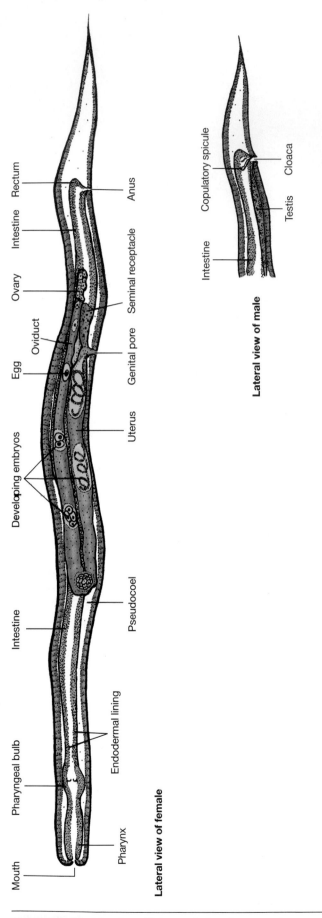

**Rectum** · **Intestine** · **Ovary** · **Oviduct** · **Egg** · **Developing embryos** · **Intestine** · **Pharyngeal bulb** · **Mouth**

Anus · Seminal receptacle · Genital pore · Uterus · Pseudocoel · Endodermal lining · Pharynx

**Lateral view of female**

Copulatory spicule · Intestine · Testis · Cloaca

**Lateral view of male**

**FIGURE 10.11** *Anguillula aceti,* the vinegar eel.

Select a few large worms for further study and place them in a small drop of vinegar on a clean microscope slide. If the worms are too active for study, they may be killed by briefly warming the slide over an incandescent lamp or a small flame. The specimens may then be observed unstained, or they may be lightly stained with hematoxylin or Nile blue sulphate in 70% alcohol.

Observe the blunt anterior end with its terminal mouth and the pointed posterior end. ***Where is the location of the anus?*** Identify the straight digestive tract consisting of the **mouth, pharynx, pharyngeal bulb** containing a three-part pharyngeal valve, an **intestine,** a thin-walled **rectum,** and a subterminal **anus.** In male specimens, the rectum opens into a **cloaca,** which in turn opens through the anus.

Select a large female worm for the study of the reproductive system. Identify the **female genital pore** on the ventral side of the body. This genital opening serves both for receiving sperm during copulation and for the release of the offspring, which are born live in this species. Extending posteriorly from the genital opening is a blind sac, the **seminal receptacle,** believed to serve for the temporary storage of sperm after copulation. Anterior to the genital pore is a single large **uterus,** which may contain several developing young. At its anterior end (about one-third the distance from the anterior tip of the body), the uterus is attached to a smaller **oviduct,** which bends posteriorly and connects with the threadlike **ovary,** where the eggs are produced.

Males may be recognized by their **copulatory spicules** and by the different structure of their reproductive system. Observe the opening of the **genital pore** as it enters the **cloaca,** the small spherical spermatozoa in the **vas deferens,** and the filamentous **testis,** which may extend almost to the middle third of the body, where it terminates in a slight enlargement. Observe a demonstration of a male worm that has been treated with strong formalin—this treatment causes the copulatory spicules to project from the anus.

## Collecting Free-living Nematodes

Many species of free-living roundworms live in the soil, in decaying vegetation, and in many aquatic habitats, such as lakes, ponds, and streams. You can easily collect living aquatic nematodes for study in the laboratory by gathering small bits of decaying plant material from the edge or bottom of a pond or stream. Place the plant material in a small culture dish and gently tease it apart with dissecting needles and/or forceps.

You should easily identify the nematodes by their shape and characteristic whiplike movements. A stereoscopic microscope equipped with a substage mirror that allows viewing with transmitted light from the bottom works best with such material.

Soil-dwelling nematodes can also be obtained easily for laboratory study. Collect some samples of black or dark soil with ample organic material ("black dirt") and place small bits of the soil in water in a culture dish and examine in the same manner as described above.

Pieces of raw or boiled potato can also be used for bait to attract soil nematodes. Place the potato pieces in moist soil in the garden or a flower bed, under a rock or board, or beneath a log in the forest for a few days. After collecting the bait, you can immediately tease apart the potato pieces to search for nematodes, or you can incubate the pieces for a few days in a sterile petri dish or test tube in a warm place. The latter procedure may permit you to find eggs and developing larvae in the bait.

## Some Other Important Nematodes

**Rhabditis maupasi,** a small nematode found in the soil that often parasitizes earthworms.

**Wuchereria bancrofti,** a filarial worm and important human parasite in many warm parts of the world. Causative agent of the disease elephantiasis, which is characterized by a massive accumulation of fluids after the blockage of lymphatic circulation by the worms, causing great enlargement of the limbs or other parts of the body.

**Heterodera schactii,** the sugar beet nematode. A specialized plant-eating nematode that feeds on the root of the sugar beet. An aggressive agricultural pest of great economic importance.

# Phylum Rotifera

## A Rotifer: *Philodina*

Rotifers are common and abundant freshwater animals. They are important constituents of the plankton of lakes, ponds, and streams, and are significant as a food source for many species of fish and filter-feeding animals.

*Philodina* (figure 10.12) is a common North American rotifer that illustrates many of the typical features of the Phylum Rotifera.

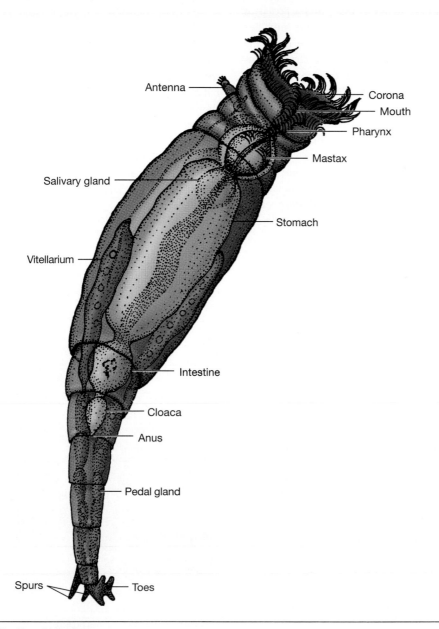

**FIGURE 10.12** Rotifer. *Philodina.*

◆ Obtain a living specimen of *Philodina* and prepare a wet mount on a microscope slide to study its behavior and appearance.

Living specimens are superior to fixed and stained specimens for the study of general anatomy of rotifers because of the serious contraction and distortion almost always encountered during their fixation and staining.

Observe the elongate cylindrical body, which can be divided into three general regions—the **head,** the **trunk,** and a posterior **foot.** Superficially, the body is divided into several segments, usually about 16. The rotifers may not have true segmentation, but the cuticle covering the body is often divided into a number of superficial segments. Observe the telescoping of the segments when the animal retracts its head. Note also the large ciliated corona at the anterior end. Most rotifers have a conspicuous **corona,** which serves both for locomotion and for feeding. In *Philodina,* the corona consists of two large ciliated **trochal discs** and a posterior band of cilia, the **cingulum.** Special **retractor muscles** serve to retract the corona. Observe also, on the head, the dorsal fingerlike process, the **rostrum.** The rostrum is believed to be a sensory structure. Locate the long, tapering **foot** with two **spurs** and four retractable **toes** at the posterior end of the body.

The body wall is covered externally by a thin **cuticle** secreted by a syncytial **hypodermis.** Cell membranes are generally lacking in the organs and tissues of adult rotifers, and specific organs typically have a fixed number of nuclei. Although there are no definite muscle layers associated with the body wall of rotifers, several distinct **muscle bands** connect various parts of the body. As in members of related phyla, the central body cavity of rotifers is a **pseudocoelom.**

The **mouth** is a funnel-shaped opening located at the base of the corona. It receives food particles swept down by ciliary currents created by the corona. Connecting with the mouth is a muscular **pharynx,** which encloses a conspicuous grinding apparatus, the **mastax.** The mastax is a distinctive feature of the rotifers and can be easily identified in a living rotifer because of its active, almost constant movement. In *Philodina,* the mastax is specialized for grinding plankton, periphyton, and detritus, but other types of mastax are found in different species of rotifers, and their structure appears to be closely related to the food habits of the various species.

Posterior to the pharynx is a short **esophagus** surrounded by large **salivary glands.** The esophagus opens into a large, thick-walled **stomach** where digested food is absorbed by the gastrodermal lining. Behind the stomach is a short **intestine** leading into the **cloaca,** which in turn empties through the **anus.**

Most of the reproductive system of *Philodina* can be seen only in special preparations, but a pair of large, yolk-forming glands, the **vitellaria,** are readily visible in living specimens. Observe the two large vitellaria in the posterior part of the trunk region of your specimen. ***How many transparent nuclei can you identify?*** The vitellaria are syncytial like most other rotifer organs and have a fixed number of

nuclei. Other parts of the reproductive system include two small anterior **ovaries** (difficult to see), which connect via small ducts to the large vitellaria, and two tiny oviducts that lead from the vitellaria to the cloaca.

Males are unknown in the group of rotifers to which *Philodina* belongs (Order Bdelloidea), and the eggs produced develop exclusively by parthenogenesis. Other species of rotifers, however, do exhibit normal sexual reproduction.

## Collecting Rotifers

Rotifers are easily collected in plankton and benthic (bottom) samples from ponds and lakes. Collect a plankton sample from a local pond or lake with a plankton net and examine portions of your sample in a cavity slide or in a watch glass under a stereomicroscope. Rotifers are best observed using transmitted or oblique lighting.

Bottom samples will yield other species of rotifers that live in association with the substrate rather than forming part of the plankton; some species of rotifers swim among the algae attached to stones, twigs, aquatic plants, and other submerged objects (figure 10.13). Other species of rotifers, called sessile rotifers, live attached to the substrate. Some of these sessile rotifers form delicate tubes from which they protrude while feeding and into which they can retract when disturbed or threatened.

By examining samples of plankton as well as material from bottom samples, you can observe some of the varied ways that rotifers have adapted for different modes of life in freshwater habitats.

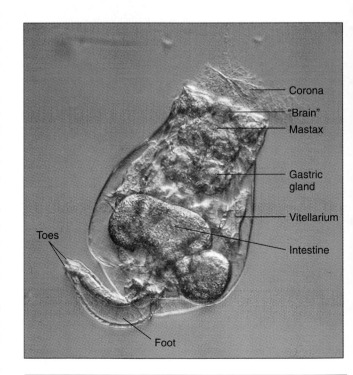

**FIGURE 10.13** Living rotifer.
Courtesy of Carolina Biological Supply Company, Burlington, NC.

## Some Other Common Rotifers

**Epiphanes senta,** a common, relatively large rotifer often cited in textbooks as a representative rotifer. Exhibits a single median ovary and vitellarium; males are small and simplified in structure.

**Asplanchna,** this genus includes several species of planktonic rotifers with a saclike body, an incomplete digestive tract, and a well-developed corona.

## Demonstration

Specimens of representative pseudocoelomates

## Key Terms

**Cloaca** a chamber at the posterior end of the digestive tract in certain types of animals that serves to receive wastes from the digestive tract, gametes from the reproductive tract, and/or wastes from the excretory system.

**Corona** a specialized organ at the anterior end of rotifers, usually ciliated, which typically serves for locomotion and in food gathering. The circular motion of the coronal cilia in many species of rotifers provides the basis for the common name, "wheel animals," often given to rotifers.

**Cuticle** a thick, noncellular protective covering secreted by an underlying layer of epithelial cells, the hypodermis.

**Hypodermis** a type of tissue made up of epidermal cells that secrete an overlying cuticle.

**Mastax** a specialized grinding organ found in the pharyngeal region of rotifers.

**Pseudocoelom** a type of internal body cavity that lacks a mesodermal lining around the digestive tract. It arises embryonically as a remnant of the blastocoel.

**Trichinosis** a parasitic disease of humans and certain other carnivorous mammals caused by the nematode parasite *Trichinella spiralis*. Usually acquired by eating rare or uncooked pork or bear meat.

## Internet Resources

Visit the zoology website at http://www.mhhe.com/zoology to find live Internet links of each of the references listed below.

1. Animal Diversity Web, University of Michigan. Phylum Nematoda. Information on nematodes ranging from the dog intestinal roundworm to the nematode that causes onchocerciasis.

2. Nematode Dissection Pages.
   - Nematodes
   - Nematoda
3. *Caenorhabditis elegans* WWW server. This site contains information on this nematode, links, meetings, the genome project, and a literature search.
4. Center for Disease Control. The CDC site that takes you to information on infectious pseudocoelomates, such as intestinal nematodes, hookworms, and dracunculiasis.

## Critical Thinking Questions

1. Distinguish between a pseudocoelom and a coelom. Is this distinction just a matter of "hair-splitting," or is there some evolutionary significance to this division?

2. Outline the life cycles of three different parasitic roundworms and discuss their medical and economic importance.

3. Discuss the arguments for and against considering the various groups of pseudocoelomate animals as members of a single phylum. Why do you suppose there is such a difference of opinion among systematists?

4. How important is sexual dimorphism in the evolution of more complex animals? (Recall that virtually all higher metazoans exhibit some form of sexual dimorphism.)

5. Is the total sequencing of the genome of *Caenorhabditis elegans* of scientific interest only? Is there any way you can see to apply this information—perhaps by comparing the genome sequence of *C. elegans* to those of other animals?

## Suggested Readings

Brusca, R.C., and G.J. Brusca. 1990. *Invertebrates*. Sunderland, MA: Sinauer Associates.

Gaugler, R., and H.K. Kaya, eds. 1990. *Entomopathogenic Nematodes in Biological Control*. Boca Raton, FL: CRC Press.

Meglitsch, P.A., and F.R. Schram. 1991. *Invertebrate Zoology*. 3d ed. New York: Oxford University Press.

# NOTES AND SKETCHES

# CHAPTER 11

## Mollusca

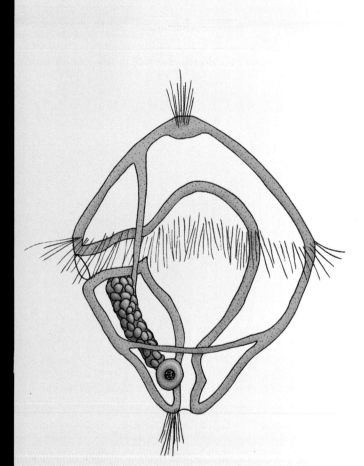

## OBJECTIVES

After completing the laboratory work in this chapter, you should be able to perform the following tasks:

1. List and briefly characterize the seven classes of living molluscs. Give an example of each class.
2. Identify the principal internal organs on a specimen of freshwater mussel and briefly explain the function of each organ.
3. Explain the pattern of water flow through a freshwater mussel and its importance.
4. Explain the reproductive and life cycles of a freshwater mussel.
5. Describe the organization of the garden snail *Helix* and identify its principal features on a dissected specimen.
6. Identify the chief external features of the squid and show four morphological features illustrating adaptation for its specific mode of life.
7. Explain the concept of adaptive radiation and how the molluscs demonstrate this concept.

## Introduction

Molluscs, a large and diverse group of animals, include chitons, clams, oysters, snails, slugs, squids, nautili, and octopi. Estimates of the number of living molluscs range as high as 150,000 species, and some 45,000 fossil species have been described. This is an ancient phylum of animals with an excellent fossil record extending back to the Cambrian era. They are among the most diverse and successful of animal phyla.

Molluscs provide an excellent example of **adaptive radiation,** with a great diversity of forms that have become adapted for virtually every type of freshwater and marine habitat as well as numerous terrestrial habitats. Molluscs are one of the few animal phyla to successfully adapt to the rigors of the terrestrial environment.

Members of this phylum typically have soft, unsegmented bodies, which usually are enclosed, wholly or in part, by a thin fleshy layer derived from the dorsal body wall, the **mantle.** The folds of the mantle form a mantle cavity that typically encloses the gills. The mantle usually secretes a hard **shell.** In some of the more specialized molluscs, however, the shell has been lost or reduced, or has become embedded in the soft tissue.

Molluscs are **triploblastic coelomate** animals, as are the annelids and all other higher metazoan animals, including the chordates. The **coelom** is the principal internal body cavity in most of these groups and is completely lined by mesodermal tissue (muscle, connective tissue, etc.). The coelom arises during embryonic development as a cavity within the developing mesoderm. (See Chapter 8.) In the molluscs, the coelom is reduced, but the body wall is thick and muscular.

In most higher metazoans, the coelom expands greatly to become the principal internal cavity and serves many important functions. For example, it provides space for the development of internal organs, serves in the temporary storage of metabolic wastes, provides space for the temporary storage of gametes, and provides a hydrostatic (fluid) skeleton to facilitate the movements and burrowing activities of soft-bodied animals.

In modern molluscs, however, the coelom is reduced largely to the cavities surrounding the heart, gonads, and excretory organs (nephridia). However, abundant evidence indicates that molluscan ancestors had a more prominent and spacious coelom.

The molluscan body typically consists of three major parts: an anterior **head,** a ventral **foot,** and a dorsal **visceral mass.** These basic parts are variously modified in different molluscs; these morphological variations clearly illustrate the remarkable diversity of form that can be achieved by alterations on a relatively simple body plan.

Because of the great diversity of form in the Phylum Mollusca, there is no "typical" mollusc. After you complete your study of the principal representative of this phylum (the freshwater mussel), it is important that you make a careful study of the demonstration material to gain a better appreciation of the other kinds of molluscs. We will study representatives of three of the classes of molluscs in more detail in this chapter.

## Classification

This great diversity of form among the mollusca—an excellent example of **adaptive radiation**—is clearly represented by the morphological diversity within the seven classes of the phylum. Figure 11.1 illustrates six of the major classes of living molluscs.

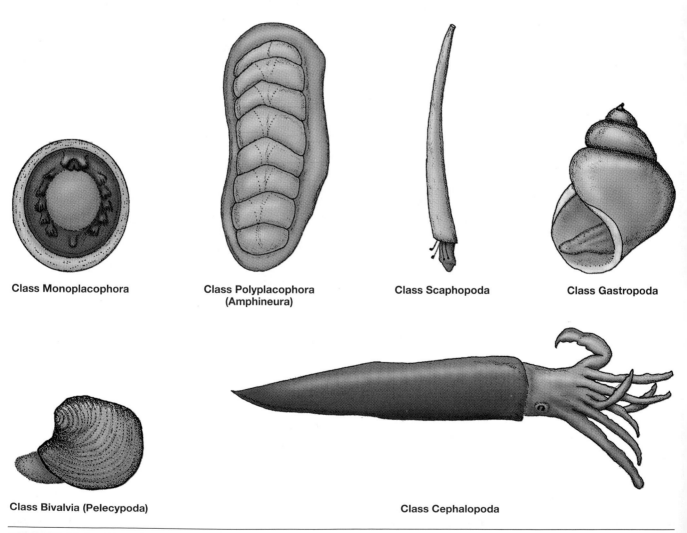

Class Monoplacophora

Class Polyplacophora (Amphineura)

Class Scaphopoda

Class Gastropoda

Class Bivalvia (Pelecypoda)

Class Cephalopoda

**FIGURE 11.1** Six major classes of living molluscs.

## Class Monoplacophora

Primitive molluscs with a conical, one-piece shell similar to the shell of a limpet. Embryonic development shows a close relationship between molluscs and the segmented annelids. Several organ systems also show definite evidence of segmentation or at least the linear repetition of parts. Some investigators, however, believe this "segmentation" is convergent to that exhibited by annelids, and therefore is really "pseudosegmentation." Known only from fossil forms and from a few specimens dredged in recent years from great depths in the sea at a few locations. All marine. Example: *Neopilina galathea* from deep ocean waters off the west coast of Mexico.

## Class Polyplacophora (Amphineura)

The chitons. Body oval shaped with eight dorsal calcareous plates and a large, flat foot. Algal feeders, mainly in the marine intertidal zone, but found at all depths of the ocean down to 4,000 meters. Examples: *Chaetopleura, Chiton, Cryptochiton.*

## Class Aplacophora

The solenogasters. A small group of peculiar wormlike marine molluscs with no head or shell, and with a reduced foot; with an anterior mouth and a posterior anus. The body surface is covered by a cuticle with imbedded spicules. Some recent authorities suggest splitting the Aplacophora into two classes, the Solenogastres and the Caudofoveata. Examples: *Neomenia, Crystallophrisson (Chaetoderma).*

## Class Scaphopoda

The tooth shells or tusk shells. Body elongate dorsoventrally and encased in a tapered, tubular, one-piece shell open at both ends. The proboscis-like head sports slender tentacles. Burrowing marine molluscs found in sandy and muddy sea bottoms. Scaphopod shells were once used by several groups (or tribes) of Native Americans on the North American Pacific coast as money. Example: *Dentalium.*

## Class Gastropoda

The snails, slugs, whelks, and limpets. Animals with a long, flat foot; a distinct head with eyes and tentacles; and a dorsal visceral mass usually housed in a spiral shell. Interestingly, several ancient groups of now-extinct gastropods were bilaterally symmetrical with coiling only in a single plane. The asymmetrical growth of the visceral mass and the overlying mantle is responsible for the spiral shape of the shell. Gastropods comprise the largest and most successful class of molluscs. About 20,000 fossil species and 35,000 extant species are known, making the gastropods the most diverse of all molluscan classes. Ecologically, they are the most versatile molluscs with freshwater, marine, and terrestrial species. Some gastropods are carnivores, some are herbivores, and still others are parasites. Nudibranchs (commonly called sea slugs) are an order of colorful marine gastropods without a shell. They are often found on rocky shores, in sea-grass beds, and pilings. They are carnivores that feed on sponges, hydrozoans, and bryozoans. Their bright colors serve as warning coloration to repel predators. Examples: *Helix* (European garden snail), *Limax* (common slug), *Littorina* (periwinkle), *Busycon* (whelk), *Aplysia* (nudibranch), *Physa* and *Lymnea* (freshwater snails).

## Class Bivalvia (Pelecypoda)

The clams, mussels, oysters, and scallops. Many bivalves are commercially important. Body laterally compressed, small foot, no head, body contained in a bivalve (two-piece) shell hinged on the dorsal side. Most pelecypods are adapted for a sedentary life in marine or freshwater habitats. The mantle is composed of two large tissue folds; it encloses a large mantle cavity and secretes the hinged, bivalve shell. Typically they are filter feeders and have specialized gills that serve to trap suspended food particles. Examples: *Anadonta* and *Unio* (freshwater mussels), *Mercenaria* (hardshell clam), *Crassostrea* (an oyster), *Mytilus* (marine mussel), *Pecten* and *Aquipecten* (scallops).

## Class Cephalopoda

The squids, cuttlefish, nautili, and octopi. Cephalopods are the most advanced molluscs and possess a large head with conspicuous eyes; mouth surrounded by eight or ten (or more) fleshy arms or tentacles; elongate body; shell often internal, reduced, or absent. There are three major groups: the nautiloids, the ammonoids, and the coleoids. Cephalopods range in size from 2 cm to nearly 20 meters in length. They are typically active marine animals, preying on various fish, molluscs, arthropods, and worms. Examples: *Loligo* (squid), *Octopus, Nautilus* (figure 11.2).

**FIGURE 11.2** *Nautilus.* Chambered nautilus. Shell bisected to show internal chambers.
Courtesy of Carolina Biological Supply Company, Burlington, NC.

## Materials List

**Living specimens**
  *Physa* or similar freshwater snail
  Representative molluscs (demonstration)
**Preserved specimens**
  Freshwater mussel
  *Loligo*
    *Loligo,* liquid mount of dissection
  *Helix*
    *Helix,* plastic or liquid mount of dissection
  Dextral (right-handed) and sinestral (left-handed)
    gastropod shells (demonstration)
  Dried pen of squid (demonstration)
  *Nautilus* shell (demonstration)
  Representative molluscs (demonstration)
**Prepared microscope slides**
  Mussel shell, cross section (demonstration)
  Glochidium larva (demonstration)
  Radula of gastropod (demonstration)
**Chemicals**
  Carmine or carbon powder (for demonstration of ciliary
    action)
**Audiovisual materials**
  Anatomy of the Freshwater Mussel

## Demonstration

Living and preserved representatives of various classes of molluscs

# A Freshwater Mussel

## Class Bivalvia (Pelecypoda)

Freshwater mussels are sedentary animals that live in or on the bottoms of lakes, ponds, and streams. The particular species available for laboratory study vary from time to time and from place to place, and may include any of several large species in the genera *Anodonta, Lampsilis, Elliptio,* and *Quadrula.* The basic anatomy of the hardshell marine clam *Mercenaria* (formerly called *Venus*) is also very similar to the following description and can be used instead of the freshwater mussel.

◆ Place a mussel in a dissecting pan and note the hard **bivalve shell.**

The two valves of the bivalve shell are joined along the dorsal surfaces by an elastic **hinge ligament.** The exterior surface of the valves is covered by a dark, horny material, the **periostracum.** Observe the concentric **lines of growth** on the exterior surface of the shell; these lines are formed as the mantle secretes new material at the edge of the growing shell.

The shell actually consists of three layers, the exterior periostracum, a middle **prismatic layer,** and an inner **nacreous layer.** The exterior periostracum is made up of a

structural protein, conchiolin, which retards dissolution of the shell by the slightly acidic waters in which freshwater mussels are typically found. The middle prismatic layer consists of crystalline calcium carbonate ($CaCO_3$) and provides strength. The inner nacreous layer ("mother-of-pearl") consists of numerous fine layers of $CaCO_3$ and is iridescent. Near the anterior end of each valve is a raised portion, the **umbo,** which represents the oldest part of the valve. At the edge of the shell, between the valves, you should be able to observe two openings between the edges of the mantle (figure 11.3): the (lower) **incurrent siphon** and the (upper) **excurrent siphon.**

In figure 11.3, locate the two large muscles that hold the valves together, the **anterior adductor** and **posterior adductor muscles.** They work antagonistically to the **hinge ligament,** a strong chitinous structure on the dorsal edge of the valves that tends to keep them open. You must cut through the anterior and posterior adductor muscles (carefully) to access the interior organs.

◆ Check to see if your mussel has a wooden peg inserted to keep the valves spread apart. If not, it will be necessary for you to pry open the valves by inserting the handle of your forceps or the handle of your scalpel between the valves along the ventral edge. (**Be careful of the blade!**) Twist the handle, and when the valves are sufficiently separated, place a wedge between them so that they are separated about 6–12 mm. Carefully insert your scalpel, blade first, into the space between the left valve and the closely adhering mantle in the region just below the anterior adductor muscle (figure 11.3). Keep the blade close against the shell, loosen the mantle from the valve, and cut through the large anterior adductor muscle. Repeat the procedure at the posterior end and cut the posterior adductor muscle. Now carefully lift the loosened left valve, separating the mantle from it as you lift. **Warning:** The heart is located in the pericardial cavity near the dorsal side of the shell; take care to avoid damage to this region.

## Feeding, Digestion, and Respiration

The thin layer of tissue that lines each valve is the **mantle.** It attaches to the inside of each valve along the **pallial line,** which is located about half of the distance from the edge of the valve to the center. Enclosed within the mantle is a space, the **mantle cavity,** which encloses the other organs. Along the edge of the mantle, identify the ventral **incurrent siphon,** with **sensory papillae** lining its borders, and the dorsal **excurrent siphon.** Locate the muscular **foot,** the **gills,** the **labial palps,** and the **mouth** located at the base of the palps (figure 11.4).

The large gills play an important role in both respiration and feeding. Each gill consists of a double fold (**lamella**) of tissue suspended in the mantle cavity (figure 11.5). The lamellae of each gill (left and right) join ventrally to form a **food groove** and dorsally to form a **suprabranchial chamber,** which carries water posteriorly to the excurrent siphon. The gills contain blood vessels and obtain some

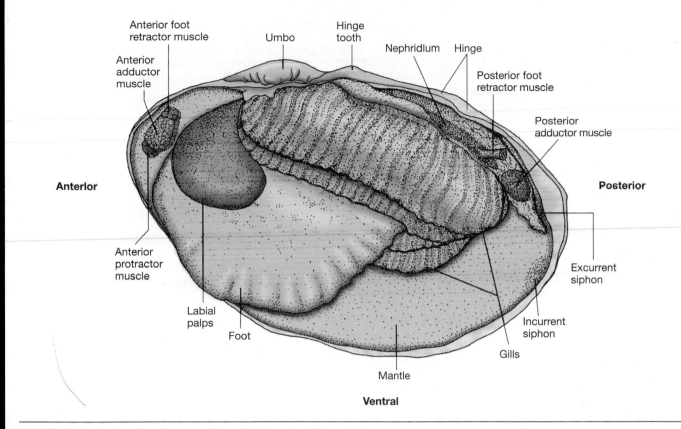

Dorsal

Anterior foot
retractor muscle

Anterior
adductor
muscle

Umbo

Hinge
tooth

Nephridium

Hinge

Posterior foot
retractor muscle

Posterior
adductor muscle

Anterior

Posterior

Anterior
protractor
muscle

Excurrent
siphon

Labial
palps

Foot

Incurrent
siphon

Gills

Mantle

Ventral

**FIGURE 11.3**  Freshwater mussel, partly dissected, lateral view.

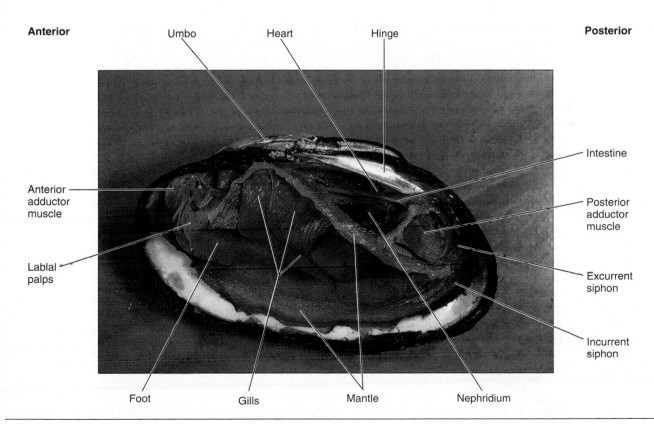

Anterior

Umbo

Heart

Hinge

Posterior

Anterior
adductor
muscle

Intestine

Posterior
adductor
muscle

Labial
palps

Excurrent
siphon

Incurrent
siphon

Foot

Gills

Mantle

Nephridium

**FIGURE 11.4**  Freshwater mussel, part of mantle removed.
Photograph by Ken Taylor.

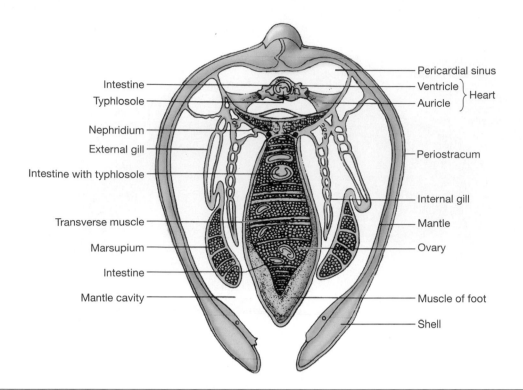

Intestine
Typhlosole
Nephridium
External gill
Intestine with typhlosole
Transverse muscle
Marsupium
Intestine
Mantle cavity

Pericardial sinus
Ventricle ⎱
Auricle ⎰ Heart
Periostracum
Internal gill
Mantle
Ovary
Muscle of foot
Shell

**FIGURE 11.5**   Freshwater mussel, cross section through heart region.

oxygen from the incoming water currents. The mantle is also highly vascularized and serves as another respiratory organ.

Cilia on the surface of the mantle and gills beat in one direction to create water currents in the mantle cavity. This ciliary action draws water into the incurrent siphon. Food particles contained in the incoming water are filtered from the water as it passes through the gills, and are trapped in mucus secreted by glands in the gill tissue. The trapped food particles are collected in the food tubes of the gills and transported anteriorly by coordinated ciliary movements to the labial palps and into the mouth. Along this route, nonfood particles are sorted out and eliminated.

◆   If live mussels or clams are available in the laboratory, you can observe the coordinated ciliary movements on the surface of the gills by placing a few particles of carbon powder (or carmine powder) on the moistened surface of a gill. Make certain that the surface of the gills is moist (not wet) before you apply the carbon, and take care to apply only a **few particles** of carbon as a marker. (Too much carbon will coat the whole surface of the gills and obscure action of the cilia.) When working with live mussels or clams, be sure to add water to your specimen frequently to keep the soft internal tissues moist and flexible. The soft tissues dry out rapidly when they are exposed to the air.

If the gills of your mussel appear fat or swollen, they may contain eggs or larvae. This is because the gills of female freshwater clams also serve as a brood pouch or **marsupium** (figure 11.5). Reproduction and the life history of the freshwater clam will be considered later in this exercise.

Most parts of the digestive system, including the **esophagus, stomach, digestive gland,** and **intestine,** are located within the foot and visceral mass (figure 11.6). To study the various structures enclosed within the **visceral mass,** you will need to cut along the ventral surface of the foot and bisect it into right and left halves. After you have bisected the foot, you should be able to locate the digestive structures mentioned previously and the yellowish **gonad** tissue that surrounds a portion of the intestine.

Find the mouth, which is behind the anterior adductor muscle. Food particles, collected on the gills and transported to the labial palps, pass through the mouth and esophagus and into the stomach, where the process of digestion begins. The stomach floor is folded into numerous ciliary grooves that aid in sorting food particles from sediment and other nondigestible particles. Such sorting begins when particulate matter is trapped on the gills and it continues enroute to the stomach. Particles rejected in the stomach are passed into the intestine for elimination. The intestine traverses the entire length of the pericardial cavity (a remnant of the embryonic coelom).

Digestive enzymes are released into the stomach by the **crystalline style,** a gelatinous rod that extends into the stomach and releases enzymes as it is rotated by the action of certain stomach muscles. Other enzymes are secreted by the digestive gland. Digestion is both extracellular and intracellular. Small food particles are engulfed by phagocytic cells in the digestive gland, and digestion is completed intracellularly. The greenish pulpy tissue surrounding the stomach is the digestive gland. Undigested material passes through the intestine to the **rectum,** exits through the rectal tube or **anus** (which lies above the posterior adductor muscle), and is flushed from the mantle cavity via the excurrent siphon.

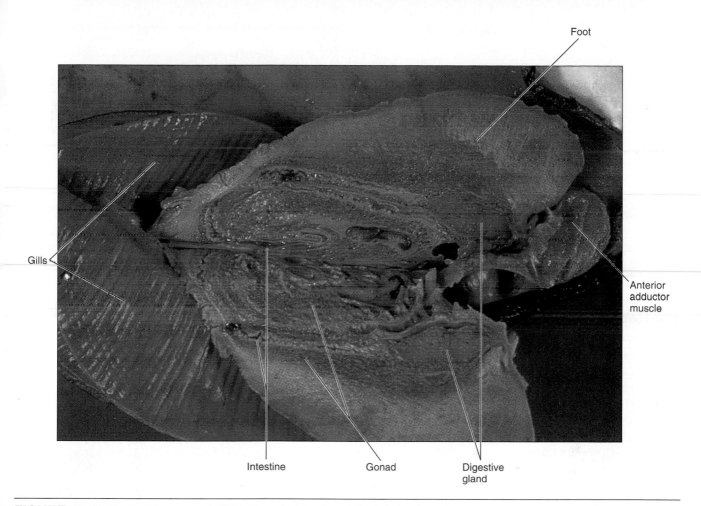

Foot

Gills

Anterior
adductor
muscle

Intestine       Gonad       Digestive
gland

**FIGURE 11.6**  Freshwater mussel. Foot dissected to show internal structures.
Photograph by Ken Taylor.

◆ Review the location and function of each part of the digestive system and observe how water currents and the gills play a central role in both digestion and respiration. To make sure you understand the processes of feeding and digestion, trace the path of some food and nonfood particles from its entrance through the incurrent siphon to the ejection of the undigested material from the excurrent siphon. Be sure that you know the role of each structure along this path.

## Muscles

You previously located two large and important muscles, the anterior and posterior adductor muscles. Three other muscles facilitate movements of the foot. The **anterior protractor** aids in extending the foot, and the **anterior foot retractor** draws the foot into the shell. These two muscles are located near the mouth and the anterior adductor muscle. Often they are partially fused with the anterior adductor muscle and may be difficult to distinguish from it.

Another muscle, the **posterior foot retractor,** is found near the posterior end of the shell, slightly dorsal and anterior to the posterior adductor muscle. The posterior foot retractor draws the foot toward the posterior of the shell.

## Circulation

The heart is located in the **pericardial cavity** found near the dorsal surface (figure 11.4). The pericardial sinus is enclosed by a thin membrane, the **pericardium.** The pericardial cavity represents the reduced coelom of the mussel. Look within the pericardial cavity and locate the muscular **ventricle** surrounding the intestine. Attached to it are two thin-walled **auricles.** Two major blood vessels carry blood away from the heart: the **anterior aorta** supplies the foot and most of the viscera, and the **posterior aorta** supplies the rectum and the mantle.

The circulatory system of the mussel is designated an **open system.** This means that the blood is not confined to a definite system of closed vessels, but that in the body tissues blood passes into large spaces or sinuses (figure 11.7). From the visceral organs, blood passes to the nephridia for the removal of metabolic wastes and then to the gills for gas exchange before returning to the heart via the veins. The mantle also serves as a respiratory organ, and oxygen-rich blood from the mantle returns directly to the heart. Dissolved in the nearly colorless blood is **hemocyanin,** a pale blue-green pigment that aids in carrying oxygen.

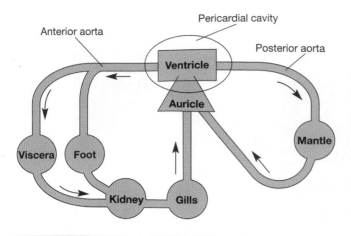

**FIGURE 11.7** Pattern of open circulation in a mussel.

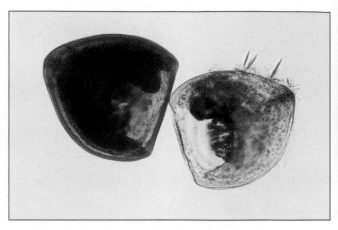

**FIGURE 11.8** Glochidia larvae of freshwater mussel.
Courtesy of Carolina Biological Supply Company, Burlington, NC.

## Excretion, Osmoregulation, and Reproduction

Ventral to the heart and embedded in the mantle tissue is a mass of brownish or greenish tissue, the **nephridia.** The nephridia are the excretory organs, or "kidney," of the mussel. They remove nitrogenous and other wastes from the blood. In freshwater mussels, the nephridia also provide an osmoregulatory function by excreting large amounts of water to maintain proper water balance in the body tissues.

Most bivalves are dioecious: in contrast, however, the sexes are separate in freshwater mussels. The male and female gonads are generally quite similar in appearance, so determination of the sex of a specimen requires microscopic examination of gonad tissue. The gonads consist of a yellowish mass of tissue and are located within the foot, surrounding a portion of the intestine (figure 11.6).

Freshwater mussels of the type most commonly used for laboratory dissection have a very specialized mode of reproduction not typical of other molluscs. Mature females can be identified easily by the presence of eggs and young larvae contained in the marsupium, a specialized portion of the gills that serves as a brood pouch (figure 11.5).

The ripe eggs are released, and they pass into the suprabranchial chamber, where they are fertilized by sperm brought in by water currents. The fertilized eggs attach to the gills, which enlarge to form the **marsupia** (figure 11.5). The eggs develop into tiny bivalve larvae, the **glochidia** (figure 11.8), which are released from the mussel to spend a portion of their lives (10 to 30 days) as external parasites on certain species of host fish. The growing glochidia induce the host to form a cyst around the larva. The mature glochidia measure only 0.5 to 5 mm in length, depending on the species of mussel. Later, the parasitic larvae detach from the host fish and move to the bottom, where they become free-living and develop into mature mussels (see Demonstrations).

◆ If you have a mature female with a marsupium, cut away a portion of the gill and remove a few of the eggs or larvae. Examine the contents under a stereoscopic microscope. *Can you identify the stages of development represented among the eggs and/or larvae?*

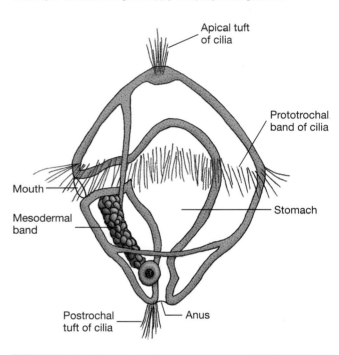

**FIGURE 11.9** Trochophore larva.

Most marine pelecypods, including mussels, clams, and oysters, shed their gametes into the water. Here fertilization takes place, which leads to the formation and development of a free-swimming larva. A **trochophore larva** is typical of many marine molluscs and annelids (figure 11.9). Later in development the trochophore transforms into a **veliger larva** (figure 11.10) with a large, bilobed flap of ciliated tissue, the velum, which enables it to swim; a distinct head with tentacles and eyes; and a larval shell into which the head and velum can be retracted. Veliger larvae are found in the life cycle of many gastropod, pelecypod, and scaphopod molluscs, indicating their close evolutionary ancestry.

## Nervous Coordination

Although the mussels lack a differentiated head, the basic plan of their nervous system is similar to that of other molluscs and consists of three pairs of ganglia and their connecting nerves.

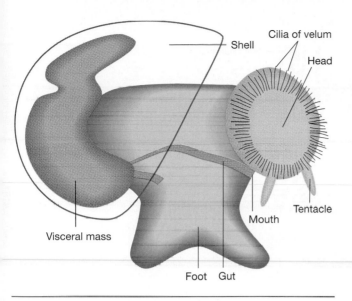

**FIGURE 11.10** Schematic of veliger larva.

A pair of **cerebropleural ganglia** (right and left) is located just posterior to the anterior adductor muscle; the two **visceral ganglia** lie just ventral to the posterior adductor muscle; and the **pedal ganglia** are embedded in the visceral tissue within the muscular foot.

## Demonstrations

1. Living bivalves in aquaria to show action of siphons
2. Microscopic section of mussel shell
3. Glochidium (larva)
4. Living or preserved specimens of other common pelecypods

## *Helix:* The Garden Snail (Demonstration)

### Class Gastropoda

*Helix* (figure 11.11) is a large terrestrial pulmonate ("land snail") often called the garden snail or the edible snail. It is the snail (escargot) found in gourmet food shops and served in restaurants. Study a specimen that has been relaxed before preservation and identify its fleshy anterior **head** with two pairs of **tentacles,** one pair of **eyes** at the tip of the second pair of tentacles, and a ventral **mouth.** Posterior to the head is the large, muscular **foot.** Above the foot is the **visceral mass** covered by the coiled calcareous shell. When the animal is threatened or disturbed, the head and foot can be retracted into the dorsal shell. Some snails (not *Helix*) have a chitinous plate, the operculum, that covers the opening when the head and foot are withdrawn into the shell, thereby providing additional protection from predators such as birds.

The anatomy of *Helix* and most other gastropods is difficult to work out because the adult body is coiled and has lost much of the bilateral symmetry of its internal organs.

Study a demonstration of a dissected snail to observe the principal internal organs (figure 11.12).

The digestive system consists of the ventral **mouth;** a muscular **pharynx** with a dorsal chitinous **jaw** and a ventral chitinous **radula;** a tubular **esophagus;** a large, thin-walled **crop** for storage; and a rounded **stomach.** The **radula** is used to shred the plant material on which *Helix* feeds. The radula of the *Helix,* like that of many other gastropods, consists of numerous chitinous teeth attached to a muscular organ that can move the radula back and forth against food materials (figure 11.13). The shredded particles are then ingested.

Food passes through the mouth and on to the stomach where it is digested. A **digestive gland** located in the apex of the shell releases enzymes into the stomach. Connected to the stomach is a long, coiled intestine terminating in the anus, which empties through the fleshy mantle on the right side of the shell.

Respiration in *Helix* is accomplished by a highly vascularized portion of the mantle that serves as a **"lung."** The circulatory system consists of a **heart** with one **auricle** and one **ventricle.** Blood is pumped from the ventricle through arteries to the principal organs, and then drains into **blood sinuses** (spaces) from which it flows to the vascularized lung portion of the mantle. A series of pulmonary vessels collect oxygenated blood from the lung and carry it to the heart.

*Helix* has no real brain, but the nervous system consists of four pairs of **ganglia** connected by nerves with each other and with major organs of the body. The principal sense organs are the two **eyes,** an **olfactory organ,** and two **statocysts,** which serve as balancing organs.

*Helix* is monoecious. A single, unpaired **gonad,** located near the apex of the shell, produces both eggs and sperm. Mating between two snails results in the exchange of sperm

**FIGURE 11.11** *Helix,* garden snail, living.
Courtesy of Carolina Biological Supply Company, Burlington, NC.

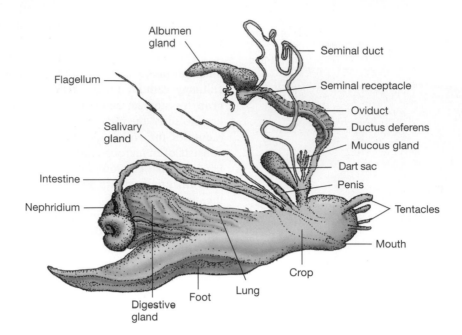

FIGURE 11.12 *Helix,* the garden snail, internal organs.

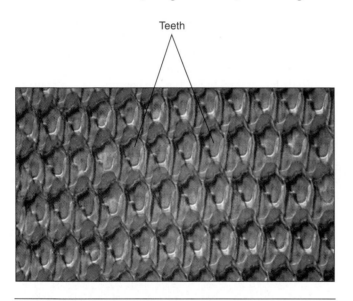

FIGURE 11.13 Radula of snail, stained preparation. Magnification 660×.

Courtesy of Carolina Biological Supply Company, Burlington, NC.

and cross fertilization. Later, each of the snails deposits eggs in a gelatinous mass. Development is direct into miniature snails without larval stages.

## Demonstrations

1. Microscope slide of radula
2. Examples of dextral and sinestral shells
3. Examples of gastropods with reduced, internal, or no shell (slugs, nudibranchs, etc.)
4. Living or preserved specimens of other gastropods

# A Squid: *Loligo*

## Class Cephalopoda

Squid are large, swift-moving, and highly specialized molluscs (figure 11.14). They exhibit a well-developed nervous system, complex sense organs, and sophisticated patterns of behavior. They are good representatives of the Class Cephalopoda, the most highly evolved class of molluscs. The name *cephalopoda* means "head-foot" and is quite appropriate because the foot, one of the basic anatomical elements of molluscs, is highly modified in this class and actually serves as a head. The "head-foot" contains several large nerve ganglia, which coordinate nerve impulses, as well as several sense organs that provide important sensory information.

Squid are often abundant in the surface waters of the oceans, where they may form large schools. Dozens or hundreds of squid can be captured together in trawl nets cast overboard in coastal waters. Other species of squid inhabit the ocean depths, but all species are marine.

Several species of the genus *Loligo* are found in the coastal waters of the Atlantic and Pacific Oceans and in the Gulf of Mexico. *Loligo pealeii* is a common small Atlantic squid; *Loligo opalescens* is a small squid abundant in Pacific coastal waters. Squids in the genus *Loligo* are typically pelagic. They consume large quantities of fish by first biting off their heads, and then quickly chopping up the rest of the body with their sharp, parrotlike beaks. *Architeuthis,* the giant squid found in deeper waters of the Atlantic Ocean, is the largest known invertebrate animal, with some individuals reaching a length of 15–20 meters. It is a common prey item of the sperm whale.

## External Anatomy

◆ Examine a preserved squid to identify its major external features. Living squid in an aquarium are interesting to

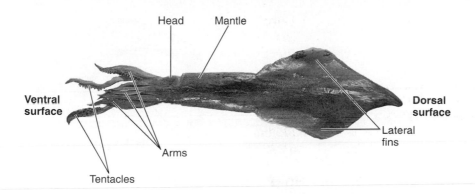

Head    Mantle

Ventral
surface

Dorsal
surface

Lateral
fins

Arms

Tentacles

**FIGURE 11.14**   Squid, *Loligo.*
Courtesy of Carolina Biological Supply Company, Burlington, NC.

observe for swimming, feeding, and other types of behavior, but are difficult to obtain and keep at inland locations.

On the preserved specimen observe the **head, tentacles,** and two large **eyes.** The eyes are equipped with a pupil, iris, cornea, lens, and retina and are capable of forming clear images; they are remarkably similar to vertebrate eyes in many respects. This is an interesting and important example of **convergent evolution,** in which two organisms with quite different evolutionary histories (squid and humans) have evolved quite similar structures. Recently, however, researchers discovered the Pax 6 gene, which regulates eye development in animals as diverse as flies, mice, and cephalopods. This suggests that eyes in virtually all animals may share a common genetic origin. Therefore, the eyes of invertebrates and vertebrates may be more homologous—at least genetically—than we formerly believed.

Note that the body of the squid is slender and tapered, which facilitates its swift movement through the water. The orientation of the squid is rather unusual, since morphologically the head and tentacles represent the **ventral surface** of the body, and the pointed end opposite the head represents the **dorsal surface** (figure 11.14). The funnel, or **siphon,** is located on the **posterior surface.** While swimming, a squid actually travels with its dorsal surface forward and its anterior surface up!

Squid swim by a type of jet propulsion in which water is squirted out of the funnel. *Loligo* alternately contracts its antagonistic circular and radial muscles to pump water in and out of the mantle cavity. Since the funnel is under muscular control, the direction of water ejection and thus the direction of movement can be swiftly changed. Movements of the arms and the two lateral fins also aid in steering the animal. The cephalopods are the fastest moving and most mobile of all molluscs.

Locate the four pairs of **arms** on the head of your specimen and two pairs of elongated **tentacles.** The tentacles are retractile and play an important role in feeding. Study one arm and observe that it bears two rows of **suckers** (figure 11.15). *Are the suckers all the same size? Which suckers are largest? Which are smallest?*

Each sucker is made up of a rounded cup attached to the arm by a stalk or **pedicle.** Remove a sucker from one of the arms and observe it under a stereoscopic microscope. Identify the **chitinous ring** with teeth, which supports the edge of the cup, and the small **piston** at the base of the cup. Note that the suckers on the two tentacles are found only near the distal ends.

A special modification of the suckers on one of the arms of mature males results in a reduction in the size of the suckers and an increase in the length of the pedicle supporting the terminal suckers (figure 11.15). This modified arm is called a **hectocotyl** and serves in the transfer of sperm bundles (**spermatophores**) to the female in mating. Such a modification of an anatomical structure by one sex (males in this case) and not the other is an example of **sexual dimorphism.**

Locate the muscular membrane at the base of the arms and tentacles. In the center of the membrane is the **mouth** opening (figure 11.15). Inside the mouth, find the two chitinous **beaks,** which are used to seize and tear prey. Prey organisms are captured by swift movements of tentacles, which seize the prey, hold it fast with the suckers, and draw it toward the mouth with the aid of the other arms. Poison glands provide secretions, which aid in subduing active prey.

The fleshy mantle surrounds the remainder of the body and encloses the internal organs. The ventral edge of the mantle adjacent to the eyes is called the **collar.** Locate the **funnel** extending from the posterior side of the collar. On the anterior side of the collar, you should find a projection of the mantle that represents the ventral tip of a translucent chitinous structure called the **pen.** In the squid, the skeleton is reduced to the pen plus several cartilages, (which protect the **cephalic ganglia** or "brain" within the head-foot), and other cartilages embedded in the collar region of the mantle and funnel.

Locate also the two large **lateral fins** extending from the mantle. As noted previously, the fins are movable and serve as steering aids during swimming. In the mantle epithelium are numerous **chromatophores,** specialized pigment-containing cells that allow the squid to change the color of its integument. Each chromatophore consists of a small sac of a single pigment (e.g., red, blue, yellow) surrounded by an elastic membrane attached to many small muscles. Contraction of these tiny muscles causes the pigment to expand and

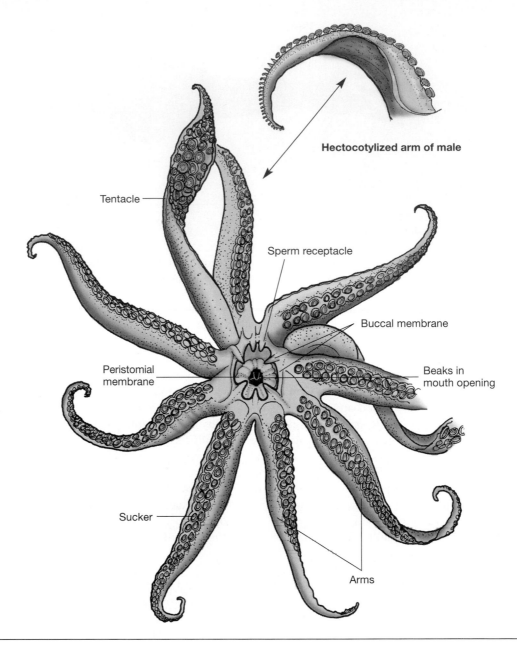

Hectocotylized arm of male

Tentacle

Sperm receptacle

Buccal membrane

Peristomial membrane

Beaks in mouth opening

Sucker

Arms

**FIGURE 11.15**  Squid, ventral view of mouth area.

add to the general coloration of the squid. A chromatophore whose pigment is not expanded contributes little to the apparent color of the squid.

Since squid have several types of chromatophores, each with a different colored pigment, many shades of color can be achieved. This aids greatly in the camouflage and protective coloration of the squid. Live squid show almost constant changes of color because of the continuous activity of the chromatophores, which are under nervous (and muscular) control.

## Internal Anatomy (Demonstration)

Study a demonstration of a dissected specimen to observe the principal organs (figure 11.16). Posterior to the head is the **funnel** connected with two large **funnel retractor**

**muscles.** Dorsal to the funnel, locate the **rectum** and **ink sac.** A median **kidney** surrounds the **intestine.** Two **branchial hearts** pump blood to the two **gills.** Anterior to the paired branchial hearts is the **intestinal caecum.** The chitinous pen is attached to the anterior wall of the mantle cavity and provides support for the body.

## Reproduction

Male squid have a single elongated testis in the posterior end of the coelom adjacent to the intestinal caecum. Sperm are released into the coelom, pass through the **sperm duct,** and are bundled into **spermatophores**—complex structures containing many individual spermatozoa. Later the spermatophores are stored in an anterior **spermatophore sac.** During copulation, spermatophores are picked up by the

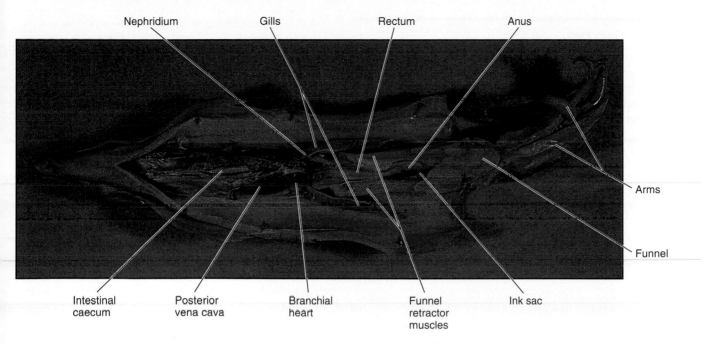

Nephridium      Gills      Rectum      Anus

Arms

Funnel

Intestinal caecum    Posterior vena cava    Branchial heart    Funnel retractor muscles    Ink sac

**FIGURE 11.16** Squid, dissection of male, posterior view.
Photograph by Ken Taylor.

specialized hectocotyl arm and are transferred to the mantle cavity or to the seminal receptacle near the base of the arms of a female.

The most prominent reproductive organs in a mature female are two large, white **nidamental glands.** Eggs are formed in a posterior **ovary,** shed into the mantle cavity, and pass into the **oviduct.** An expanded portion of the oviduct is modified to form an **oviducal gland,** which secretes the eggshell. Later the eggs pass on through the oviduct and are released into the mantle cavity through an opening near the nidamental glands. These glands secrete gelatinous capsules that surround the eggs after they are released.

# Octopus (Demonstration)

## Class Cephalopoda

The body of an octopus is globular and saclike with well-developed eyes and a complex brain (figure 11.17). Octopi usually crawl about the sea bottom using their eight flexible arms equipped with numerous suction cups used for locomotion and to grasp prey. When startled, they can also swim by jets of water forced through the funnel from the mantle cavity.

Octopi are carnivores and feed on molluscs by drilling holes in the shells with their radula and injecting a poison to kill their prey. They also have a powerful chitinous beak that is used to shred tissue or attack enemies.

**Demonstrations**

1. Dried pen of squid to illustrate internal skeleton
2. Preserved or plastic mount of octopus
3. Shell of chambered nautilus

Arms

Eye

Dorsal surface

Suckers

Mouth

Ventral surface

**FIGURE 11.17** Octopus.
Courtesy of Carolina Biological Supply Company, Burlington, NC.

## Key Terms

**Adaptive radiation**   evolutionary diversification of a group of animals sharing common ancestry to fill numerous ecological niches.

**Chromatophores**   specialized pigment-containing cells found in the integument of cephalopod molluscs and many crustaceans (Phylum Arthropoda).

**Coelom**   central body cavity lined by mesodermal tissues; develops from spaces arising within the mesoderm during embryonic development. Characteristic of annelids, molluscs, and all higher metazoans, although modern molluscs have a secondarily reduced coelom.

**Convergent evolution**   the evolution of two apparently similar structures within two or more groups of animals having different ancestral lines.

**Foot**   in molluscs, the foot is a large ventral mass of muscular tissue; usually functions in locomotion; sometimes modified for other functions.

**Glochidium larva** (plural: glochidia)   specialized larval form in the life cycle of certain freshwater bivalve molluscs; enclosed in a transparent bivalve shell, these larvae spend part of their life cycle as parasites on the gills of freshwater fish. Later they detach and develop into free-living adult mussels.

**Hemocyanin**   a blue copper-containing respiratory pigment in the blood of molluscs and certain other invertebrate animals. Carries oxygen to the body cells.

**Mantle**   a thin, outer layer of tissue largely enveloping the bodies of most molluscs. Usually secretes a shell.

**Marsupium**   a modified portion of the gills in females of certain freshwater clams and mussels; serves as a brood pouch for eggs and young larvae.

**Sexual dimorphism**   condition in which male and female members exhibit a different form.

**Veliger larva**   typical larval form of many marine molluscs, with a prominent bilobed flap of ciliated tissue used in swimming; the flap can be retracted into the transparent shell. Also with a distinct head and tentacles.

## Internet Resources

Visit the zoology website at http://www.mhhe.com/zoology to find live Internet links of each of the references listed below.

1. Animal Diversity Web, University of Michigan. Phylum Mollusca. Information on molluscs, and links to all seven classes of molluscs. Some images may not be displayed due to copyright restrictions.
   Links to individual classes:
   • Aplacophora
   • Monoplacophora
   • Polyplacophora
   • Scaphopoda
   • Gastropoda
   • Bivalvia
   • Cephalopoda
2. Dissection Aid. Dissection guides and lesson plans to dissect a clam, a squid, an earthworm, a frog, and a heart and a kidney of a vertebrate for home-schooled students.

3. Dissection of the Squid. Terrific photos of external and internal organs (loads slowly); some pictures are labeled, others are not. Separate pictures of various aspects of external anatomy, individual pictures of various internal organs or organ systems.
4. Squid Dissection. This lesson plan was actually designed for students in grade 2, but there are some interesting facts!
5. Clam Dissection. University of Minnesota. Nice photograph of internal anatomy, but not labeled.
6. Mollusca. Arizona's Tree of Life web page. Pictures, characteristics, phylogenetic relationships, references on molluscs. Pictures and references, and links to polyplacophorans, gastropods, bivalves, and cephalopods. Most sites have reference lists.

## Critical Thinking Questions

1. Discuss what is meant when one says that the molluscs provide an excellent example of adaptive radiation. First define *adaptive radiation* and then give specific molluscan examples to support your view.

2. What is the significance of knowing that molluscs are triploblastic coelomate animals? With which group(s) of animals does this fact suggest kinship and common evolutionary descent?

3. The coelom in molluscs is reduced to cavities surrounding the heart, gonads, and excretory organs. Does this indicate that the coelom has been reduced from that of an ancestral form that had a fully developed coelom, or does it mean that this coelom is primitive and never fully evolved? Give reasons for your point of view. Perhaps there is another viewpoint or two that might be considered?

4. Humans are dioecious and also exhibit sexual dimorphism. Discuss the various groups of molluscs and indicate which ones are dioecious and also exhibit sexual dimorphism. Do you think there is anything in common between those groups that are sexually dimorphic?

5. Prior to a few years ago, researchers believed that eyes evolved independently in the animal kingdom perhaps 12 different times. However, the discovery of the Pax 6 gene, which regulates eye development in all animals studied to date, from molluscs to primates, suggests that the eyes of all animals may share a common genetic origin. Find evidence to support the idea that the eyes of animals are not convergent (and therefore analogous) but are homologous and share a common ancestor.

6. Do you think there may be an evolutionary or physiological reason to explain why hemocyanin (a blue copper-containing compound) is the respiratory pigment of molluscs and other

invertebrates instead of hemoglobin (an iron-based pigment) common in vertebrates?

7. Discuss the importance of the similarity of ciliated larvae found in many invertebrate marine groups. What does this imply? The common evolutionary origin of these larvae, or simply the fact that cilia are useful in marine environments? Do you believe the cilia borne by these larvae are analogous or homologous?

## Suggested Readings

Abbott, R.T., and S.P. Dance. 1986. *Compendium of Seashells,* 3d printing, revised. Melbourne, Florida, and Burlington, MA: American Malacologists, Inc.

Brown, D.S. 1994. *Freshwater Snails of Africa and their Medical Importance,* 2d ed. London; Bristol, PA: Taylor & Francis.

Gilbert, D.L., Adelman, W.J. Jr., and J.M. Arnold (editors). 1990. *Squid as Experimental Animals.* New York: Plenum Press.

Lane, F.W. 1960. *Kingdom of the Octopus; The Life History of the Cephalopoda.* New York: Sheridan House.

The Veliger (international journal of mollusks). B. Roth, editor. 745 Cole Street, San Francisco, California 94117.

Wells, M.J. 1978. *Octopus: Physiology and Behavior of an Advanced Invertebrate.* London: Chapman and Hall. New York: Halsted Press.

## NOTES AND SKETCHES

# NOTES AND SKETCHES

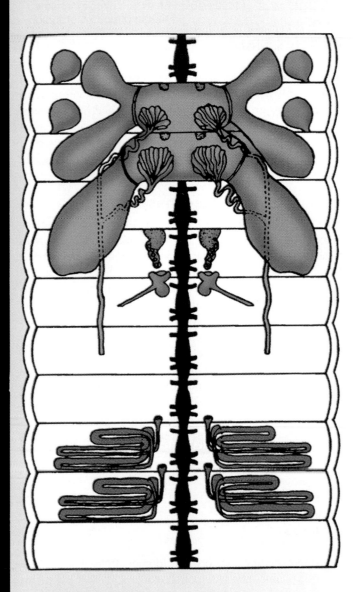

After completing the laboratory work in this chapter, you should be able to perform the following tasks:

1. Briefly outline the characteristics of the Phylum Annelida and cite five significant advances of this phylum over the Phylum Platyhelminthes and the Phylum Nematoda.
2. List and briefly characterize each of the three classes of annelids. Cite an example of each class.
3. Identify the principal external features of the polychaete worm *Nereis* and explain the functions of these features.
4. Describe the internal structure of *Nereis* as a representative polychaete and identify the principal structures in a microscopic cross section or drawing.
5. Explain the difference between a coelom and a pseudocoelom.
6. Identify the parts of the digestive system in a dissected earthworm and give the function of each part.
7. Describe the basic pattern of circulation and identify the chief parts of the circulatory system of an earthworm.
8. Identify the main reproductive organs and discuss reproduction in earthworms.
9. Describe the external anatomy of a leech and explain how it differs from that of an earthworm.

## Introduction

The Phylum Annelida includes more than 9,000 species of segmented worms—animals whose cylindrical, elongate bodies are composed of a longitudinal series of segments (also called somites or metameres) (figure 12.1). In addition to their conspicuous segmentation, the annelids also exhibit several other important structural advances over the flatworms and roundworms. Some important advances include the following: a spacious coelom, a closed circulatory system, an efficient excretory system, a highly developed muscular system, a well-developed central nervous system, and an increased concentration of nerve centers (ganglia) and sensory organs at the anterior end of the animal (cephalization).

Annelids are the first animals we shall study that exhibit a well-developed coelom. Although molluscs are also coelomate

**FIGURE 12.1** Earthworm. Notice the external segmentation and the lack of cephalization.
Courtesy of Carolina Biological Supply Company, Burlington, NC.

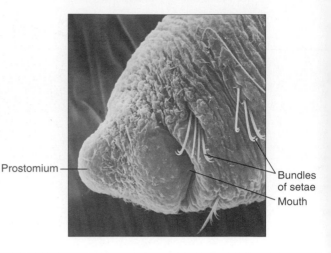

**FIGURE 12.2** Mouth region of an aquatic oligochaete showing characteristic arrangement and gross morphology of setae. Magnification 450×.
Scanning electron micrograph by Thomas Bouillon.

animals, the coelom of molluscs is reduced to the cavities in which the heart (pericardial cavity), nephridia, and gonads are located. The annelids, in contrast, have a large coelomic cavity in which most of the internal organs are suspended. Annelid worms illustrate well the basic body organization of a coelomate animal and the relationship of the coelomic cavity, its peritoneal lining of mesodermal origin, and the internal organs.

## Classification

The phylum is divided into three classes, chiefly on the basis of **segmentation,** the distribution of setae, and the presence or absence of a clitellum. Setae are thin, chitinous bristles or rods that are secreted by certain cells in the body wall. Setae are used in locomotion and, in many annelids, also aid in feeding. The number, morphology, and distribution of setae is often important in the classification of annelids (figure 12.2).

The clitellum is a glandular swelling produced by certain segments. The clitellum may be permanent or formed seasonally; it secretes a cocoon into which eggs are deposited for fertilization and later incubation.

### Class Polychaeta (Polychaete Worms)

These annelid worms typically have many segments, lateral appendages (parapodia) with many setae, and a well-developed head region, but lack a clitellum and permanent gonads, and are usually dioecious and often have a trochophore larva. Most species are marine. This class also includes the archiannelids, a group of small marine worms formerly considered by many biologists to be a separate class. Examples: *Nereis* (formerly *Neanthes,* the clamworm); *Arenicola* (lug worm); *Chaetopterus; Hydroides* (feather-duster or fan worm); *Diopatra* (plume worm); *Polygordius* (an archiannelid).

### Class Oligochaeta (Bristleworms)

This class includes the earthworms and many species of related freshwater worms that have many segments but few setae per segment, form a clitellum, but lack parapodia and a

differentiated head region. They are generally monoecious, with permanent gonads and direct development. Oligochaetes are mostly found in the soil and in freshwater habitats. Examples: *Lumbricus* and *Allolobophora* (earthworms); *Eisenia* (dung worm); *Tubifex* (sewage worm); *Aeolosoma; Chaetogaster.* The last three genera are aquatic and have many common freshwater representatives.

### Class Hirudinea (Leeches)

This group includes the leeches, a relatively small class (about 500 species) of specialized annelids found mainly in freshwater habitats. A few species are also terrestrial. Leeches have dorsoventrally flattened bodies, anterior and posterior suckers, thick muscular bodies, no head or tentacles at their anterior ends, and lack parapodia and setae. The number of segments is limited (34), and a seasonal clitellum is formed during the reproductive season. Examples: *Hirudo, Glossiphonia, Haemadipsa.*

Three representatives of this phylum have been selected for special study to illustrate the principal features of the Annelida: a clamworm, an earthworm, and a leech.

| Materials List |
|---|

**Living specimens**
    Earthworm
    Leech
**Preserved specimens**
    *Nereis*
    Leech
    Representative polychaetes (demonstration)
    Polychaete tubes (demonstration)
    Earthworm cocoons (demonstration)
    Representative oligochaetes (demonstration)
    Representative leeches (demonstration)
    Leech, dissected to show internal anatomy
        (demonstration)

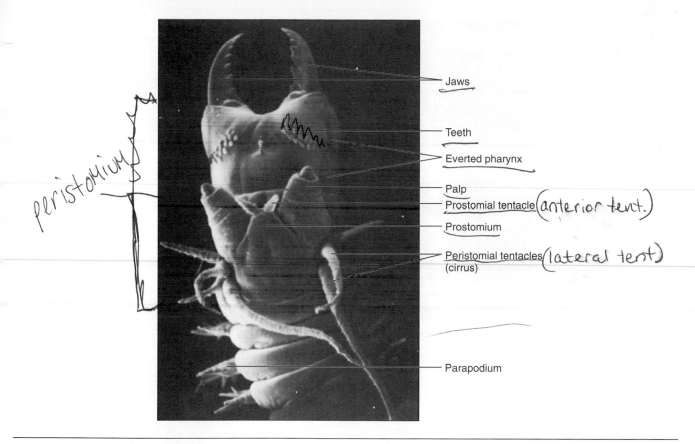

Handwritten annotations: "peristomium" (left margin), "anterior tent.", "lateral tent."

Labels on figure: Jaws, Teeth, Everted pharynx, Palp, Prostomial tentacle, Prostomium, Peristomial tentacles (cirrus), Parapodium

**FIGURE 12.3** *Nereis,* head region, dorsal view.
Scanning electron micrograph by Betsy Brown.

**Prepared microscope slides**
    Earthworm, cross section
    Trochophore larva (demonstration)
**Miscellaneous supplies**
    Dissecting pans
    Pins
**Audiovisual materials**
    Chart of earthworm anatomy
    Anatomy of the Earthworm video

# A Marine Annelid: *Nereis virens*

## Class Polychaeta

The sandworm or clamworm, *Nereis virens* (figures 12.3 and 12.4), is a large marine annelid commonly found in many areas along the Atlantic coast of North America. *Nereis* usually lives in burrows in the sand or in mud bottoms of the shallow coastal waters during the day and emerges at night to search for food.

### External Anatomy

◆ Select a preserved specimen of *Nereis* and observe the numerous body segments. Internally, the segments are separated by thin sheets of tissue called **septa.** Septa are often invaded by muscles, allowing for controlled body movement along the length of the coelom. Both septa and their mesenteries may be partially opened, allowing

for communication between the coelomic compartments. The internal structure of *Nereis* is generally similar to that of the earthworm, which you will study later.

All the segments of *Nereis* except the first and the last bear a pair of lateral **parapodia** (figure 12.5). Observe the demonstration slide of a cross section of *Nereis* to study the structure of these appendages. Note that the parapodia are **biramous** (two-branched), with a dorsal **notopodium** and a ventral **neuropodium.** Each main branch bears bundles of chitinous bristles, or **setae,** and also has several smaller side branches. The parapodia play an important role both in locomotion (swimming and creeping along the sea bottom) and in respiration. *What aspects of their structure appear to be advantageous for these functions?*

Examine the well-developed head region made up of the **peristomium,** the first complete body segment that encircles the mouth, and the **prostomium,** a triangular mass of tissue located on top of the peristomium (figure 12.3). Several sensory structures are found on the prostomium, including two fleshy **palps** attached to the sides of the prostomium, two pairs of pigmented **eyes** on the dorsal surface, and two **anterior tentacles.** Four pairs of **lateral tentacles** (or cirri) are also found attached to the peristomium.

During feeding, the pharynx may be everted through the mouth, exposing two powerful **jaws** and many small **teeth** on the wall of the pharynx (figures 12.3 and 12.4). The jaws aid in seizing prey, and the pharyngeal teeth help

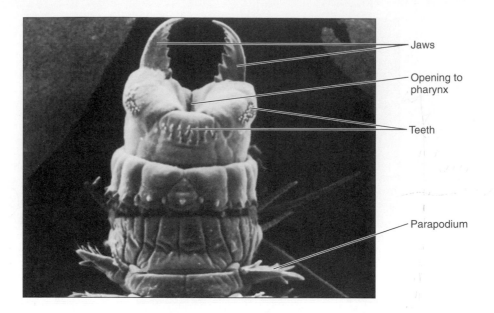

**FIGURE 12.4** *Nereis,* head region, ventral view.
Scanning electron micrograph by Betsy Brown.

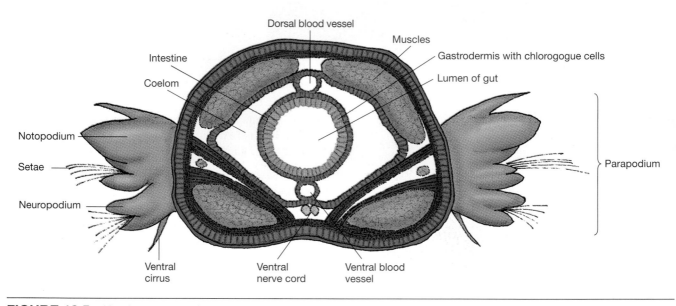

**FIGURE 12.5** *Nereis,* cross section.

hold the prey as the everted pharynx is withdrawn back into the mouth.

The sexes are separate in *Nereis,* as in most polychaete worms, although there are no permanent gonads. The gametes (eggs and sperm) are formed in the body wall at certain times, pass into the coelom, and then are released into the sea. There the eggs may be fertilized and develop into a **trochophore larva** (figure 11.9). Other larval stages follow before metamorphosis into the adult worm.

Many species of polychaete worms show marked structural, physiological, and behavioral changes during the breeding season. Some species aggregate in large swarms and rise to the surface before releasing their gametes into the

sea. This special behavior represents an important adaptation that serves to increase the likelihood of fertilization.

The trochophore larva formed by many marine polychaetes bears a close resemblance to the larvae of many marine molluscs. This similarity of trochophore larvae is one of the important pieces of evidence suggesting a close evolutionary relationship between these two phyla. Similarly, the resemblance of the trochophore larvae of annelids and molluscs to the marine larvae of several other invertebrate phyla, in addition to other similarities in their patterns of embryonic and larval development, suggests other important evolutionary relationships with groups such as the flatworms and the arthropods.

## Cross Section

Obtain a microscope slide of *Nereis* with a stained cross section of the midsection to help you understand the internal organization of a polychaete. First, locate the body wall with its external **cuticle** covering the surface of the body (figure 12.5). Beneath the cuticle is a layer of **hypodermal cells** that secrete the cuticle. Identify the **circular** and **longitudinal muscle** layers inside the hypodermis. Between these muscle layers, find the thin **peritoneum** that lines the large central cavity, the **coelom.**

◆ *What are the functions of the coelom? Why is it important to the evolution of higher animals?* Review the discussion of the coelom and its functions in Chapter 8, if needed.

Locate the large circular **intestine** in the middle of the coelom. Observe the large **gastrodermal cells** lining the inside of the intestine. These cells absorb nutrients from the food digested in the lumen of the intestine. Also identify the circular and longitudinal muscle layers and the outer thin peritoneum in the intestinal wall.

You should also be able to identify the **dorsal** and **ventral blood vessels** and the **ventral nerve cord** in your cross section. Depending on where your cross section was cut, you may be able to find some portions of the parapodia and fibers of the **oblique muscles** connecting the midventral line with the dorsal margin of the parapodia. These muscles aid in locomotion as well as other movements of the body. If

your section was cut through the parapodia, you may also be able to identify the **setal sac** in which some of the parapodial setae are anchored and, adjacent to the setal sac, some **nephridial tissue.** The excretory organs of the polychaetes consist of many pairs of **nephridia** located in most body segments.

Later in this exercise you will have an opportunity to compare the internal structures of this polychaete to that of an oligochaete when you study cross sections of an earthworm.

## Other Polychaetes

Polychaete worms are often abundant in the shallow coastal waters of the sea. This class includes many varied forms, some of which form permanent tubes that they secrete or build from various materials. These species are often called sedentary polychaetes. Errant polychaetes do not form permanent tubes. Observe the demonstrations of various living and preserved polychaetes available in the laboratory. Three common and well-known species are illustrated in figures 12.6, 12.7, and 12.8.

*Hydroides* (figure 12.6) secretes a permanent calcareous tube on the surface of shells and rocks. The tube is open at the anterior end and tapers at the closed posterior end. While feeding, the worm extends a feathery crown of anterior appendages (extensions of the prostomium) from the open anterior end of its calcareous tube. Because of these feathery appendages, which are often brightly colored, *Hydroides* and other members of the Family Serpulidae are called feather-duster or fan worms.

*Diopatra,* the plume worm (figure 12.7), is a polychaete common in intertidal areas on the Atlantic coast of the United States. The body of the plume worm is iridescent and bears several pairs of bright red (blood-filled) gills. *Diopatra* forms a vertical, parchmentlike tube that may reach 3 feet in length. Most of the tube is buried in the sediment, but 2–3

**FIGURE 12.6** *Hydroides,* a tube-dwelling polychaete. Notice "feather-duster" everting.
Photograph by Charles Wyttenbach.

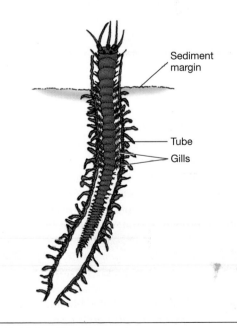

Sediment margin

Tube
Gills

**FIGURE 12.7** *Diopatra,* a plume worm.

**FIGURE 12.8** *Arenicola,* a lug worm. These sedentary polychaetes burrow into the sand and feed on detritus.

inches usually extend above the sediment. The tube is camouflaged by bits of shell, sea grass, and other debris.

*Arenicola,* the lug worm (figure 12.8), is a burrowing polychaete that lives in a **U**-shaped tube lined with mucus. In its habits, *Arenicola* resembles a seagoing earthworm. The worm burrows in shallow marine sediments by extending its proboscis into the sand or mud and retracting it filled with sand. The sand contains detritus and small organisms on which the worm feeds as the sand passes through the digestive tract. A burrowing *Arenicola* works from 5 to 8 hours a day, resting every few minutes. Where this worm is abundant, it moves about 1,900 tons of sand per acre to the surface every year. Within several years, the worms redistribute virtually all of the upper 60 cm of ocean bottom in the littoral zone. *Arenicola* shows several external structural adaptations related to its peculiar burrowing habits, including reduced parapodia, small gills, a large bulbous proboscis, and the lack of tentacles and other head appendages.

## Demonstrations

1. Representative species of Polychaeta, such as *Amphitrite, Aphrodite, Arenicola, Chaetopterus, Hydroides, Sabella, Diopatra,* and *Polygordius*
2. Examples of different kinds of tubes formed by polychaetes
3. Microscope slide of trochophore larva

# The Earthworm

## Class Oligochaeta

Earthworms are terrestrial annelids that typically live in tunnels they burrow in the soil. Internally, they demonstrate many of the typical features of the Annelida, and they are the most commonly studied representatives of the phylum. The external features of earthworms, however, exhibit several modifications probably related to their burrowing habits and underground life. Most apparent of these external adaptations is the absence of a distinct head and the lack of parapodia.

The traditional earthworm for classroom dissection is *Lumbricus terrestris,* upon which this and most other textbook descriptions are based. *Lumbricus terrestris* is a species

common in Europe and is known as the night crawler in northeastern United States and in eastern Canada. It is found only rarely south of the Pleistocene glacier boundaries.

Many of the earthworms supplied for dissection by North American supply houses in recent years, however, have been collected in other areas and are different species. These species often differ in the specific location of reproductive structures and pores and other minor anatomical details. In most respects, however, the anatomy of most species of earthworms should correspond to the description of *Lumbricus terrestris* described here.

## External Anatomy

◆ Obtain an anesthetized or freshly killed specimen of a large species of earthworm and examine the major external features. Preserved specimens may also be studied, but fresh specimens are far superior because the organs of preserved specimens are often fragile and their colors faded, making them difficult to identify.

Observe the long, cylindrical body, which tapers at each end. The anterior end can be identified most readily by the **clitellum,** a conspicuous swollen region including several segments located near the anterior end. The clitellum secretes material that forms the cocoon in which fertilized eggs are incubated (figure 12.9).

Note the absence of a distinct head in the earthworm and the lack of external sense organs at the anterior end. *How can you relate these adaptations to the burrowing habits of the earthworm?*

The mouth is located in the first anterior segment, and a small, rounded projection, the **prostomium,** overhangs the mouth. The **anus** is located in the last segment.

Observe the darker and more rounded dorsal surface of the worm. Locate the **setae,** small chitinous bristles located in each segment except the first and last. To find the setae, rub your finger lightly back and forth along the sides of the worm. Observe how the setae protrude slightly from the body surface and catch your finger—in one direction only!

**FIGURE 12.9** Earthworm cocoons.
Courtesy of Carolina Biological Supply Company, Burlington, NC.

*In which direction do they catch your finger? What does this tell you about the orientation of the setae? How would this aid in forward movements?*

The setae are located in four distinct rows. Two pairs of setae are located in rows on the ventral surface, and two pairs are located on the sides (figure 12.10).

Note the iridescent **cuticle,** a protective body covering that reduces water loss. The iridescence results from the many small striations in the surface of the cuticle. These striations cause the cuticle to act as a diffraction grating to separate incident light into separate wavelengths and reinforce the wavelengths of those colors that produce the iridescence.

Also located on the surface of the body are several openings from the excretory and reproductive systems. A pair of small **excretory pores,** the nephridiopores, is found on the ventral surface of each segment except the first three or four and the last. Openings from the reproductive system include the following: a pair of **male pores** on the ventral surface of segment XV surrounded by swollen liplike structures, a small pair of **female pores** on the ventral surface of segment XIV, and openings to two pairs of **seminal receptacles** located laterally in the grooves between segments IX–X and X–XI. (Note: segments are usually designated by Roman numerals.) The use of a stereoscopic (dissecting) microscope will help to locate these tiny external openings to the excretory and reproductive systems of the earthworm.

## Internal Anatomy

◆ Place your earthworm in a wax-bottom dissecting pan and carefully pin the specimen down, dorsal side up, with one pin through the prostomium and another near the posterior end. Study figure 12.14 and note the thin body wall and the closeness of the underlying parts before you begin your dissection. With the tips of your scissors, slightly puncture the body wall to one side of the dorsal midline and carefully make a longitudinal incision from behind the clitellum, continuing it anteriorly to the mouth. Cut to the side of the large dorsal blood vessel and avoid severing it. Take special care, also, to avoid damage to the brain, which is located near the mouth.

Figure 12.11 shows a dissected earthworm and its internal organs. Observe the large body cavity, the **coelom,** that is divided into many distinct compartments by the **septa.** The septa are membranous, transverse partitions that divide the body of the earthworm internally into many segments.

Carefully separate the body wall from the internal structures and pin down the body wall on each side. Place the first pins on opposite sides of segment XV, easily located because of the presence of the large male pores. Insert the pins through the tissue and into the wax bottom of the pan at a 45 degree angle away from the worm to provide working space and a clear view of the internal organs. Place additional pins through each fifth segment (for example, segments V, X, XV, XX, XXV, XXX, and so on) to serve as convenient landmarks to identify the location of various structures. Cover the specimen with about 1 centimeter of water to keep the internal organs moist and flexible.

### Digestive System

The digestive tract is a straight tube extending from the mouth to the anus. Examine your specimen and locate the **mouth** at the anterior end. The mouth opens into the small **buccal cavity** (figure 12.12). Immediately behind the buccal cavity is the muscular **pharynx.** The pharynx is attached to the body wall by numerous threadlike **dilator muscles** that can produce a sucking action to draw food materials through the mouth into the buccal cavity and pharynx. The tubular,

Dorsal setae

Ventral setae

**FIGURE 12.10** Earthworm portion of body wall showing lateral rows of setae. Magnification 50×.
Scanning electron micrograph by Louis de Vos.

thin-walled **esophagus** extends from segments VI to XIII and passes the food posteriorly to the **crop.**

Associated with the lateral wall of the esophagus of earthworms are two or more pairs of **calciferous glands.** These glands develop as evaginations of the esophageal wall and regulate the levels of calcium and carbonate ions ($Ca^{++}$ and $CO_3^=$) and the pH of the blood.

Food is stored in the thin-walled, extensible crop before passing into the muscular **gizzard,** where it is ground into smaller pieces and passed to the **intestine** to be digested and absorbed. Undigested materials are passed to the **anus** for elimination.

Much of the intestine, as well as the dorsal blood vessel, is covered by a yellowish layer of **chlorogogue tissue.** The chlorogogue tissue does not appear to be directly involved in digestion, but it does store glycogen and lipids and has other functions analogous to those of the liver in vertebrate animals, such as providing chemical defense for the body. It also deaminates proteins and releases ammonia and urea, thereby serving an excretory function.

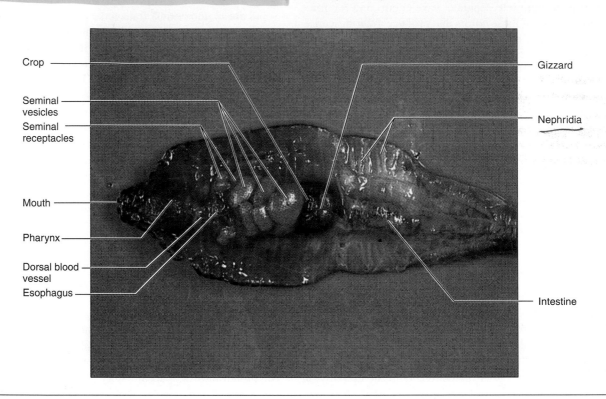

**FIGURE 12.11** Earthworm, dissected, dorsal view.
Photograph by Ken Taylor.

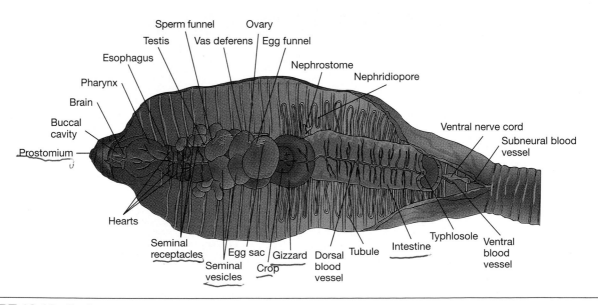

**FIGURE 12.12** Earthworm, internal organs.

## Circulatory System

The earthworm has a *closed* circulatory system, which contains red blood. The red color is due to the respiratory pigment, **hemoglobin,** dissolved in the **plasma,** or fluid portion, of the blood. In addition to the plasma, the blood also contains numerous colorless **blood cells,** or corpuscles.

Locate the large **dorsal blood vessel** lying just above the digestive tract, and the pair of **parietal vessels** extending laterally from the dorsal blood vessel in each segment. In segments VII to XI, inclusive, the parietal vessels are enlarged and much more muscular. These five pairs of enlarged **aortic arches** are often called "hearts." Their contractions aid in propelling the blood through the circulatory system.

The yellowish tissue that covers the surface of the large blood vessels consists of **chlorogogue cells.** Some of the other major blood vessels include the **ventral vessel,** lying beneath the digestive tract; the **subneural vessel,** lying beneath the ventral nerve cord; and two **lateral neural vessels,** one on each side of the ventral nerve cord (see figure 12.14).

In an anesthetized worm, waves of muscular contraction can be readily observed along the dorsal vessel and the five pairs of "hearts." *In which direction does the blood flow through these vessels?* Valves in the dorsal vessel and in the hearts prevent backflow and aid in the efficient circulation of the blood.

## Excretory System

Each segment, except the first three or four and the last, contains a pair of tubular excretory structures, the **nephridia.** Each nephridium is located partly in two adjacent segments. Consult figures 12.12 and 12.13. Locate the funnel-shaped opening into the coelom, the **nephrostome,** which projects through the septum into the next anterior segment. Also locate the **coiled tubular portion,** which empties through a ventral opening in the body wall, the **nephridiopore.**

The **calciferous glands,** formed as evaginations of the wall of the esophagus, also serve as excretory organs. They were described previously during our discussion of the digestive system.

## Reproductive System

◆ Cut across the digestive tract about 12 mm behind the gizzard and carefully free it with the associated blood vessels from the underlying structures up to the level of segment IV.

The earthworm is monoecious: each individual has both male and female sex organs (figure 12.13). Despite the monoecious condition, however, cross fertilization between different individuals is necessary for sexual reproduction in earthworms.

Several of the male reproductive organs are larger and more conspicuous than the female organs. Locate the following male reproductive structures and note the segments in which they are located: (1) three pairs of large **seminal vesicles** (storage sacs in which the sperm mature and are stored until copulation); (2) two pairs of small **testes** embed-

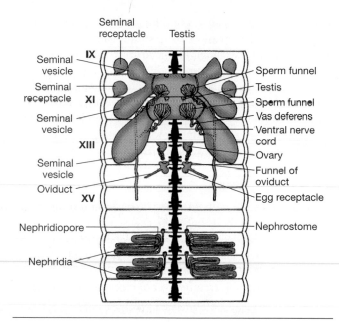

**FIGURE 12.13** Earthworm, schematic of reproductive and excretory systems.

ded within the tissues of the seminal vesicles; (3) two pairs of small ciliated **sperm funnels** (one pair on each side), which connect via short tubules to (4) two **vasa deferentia** (sperm ducts), which lead to the male genital pores on segment XV.

◆ Make a wet mount of a small piece of one of the seminal vesicles, gently squash the tissue, and examine under high power of your compound microscope. You should be able to observe many spermatozoa and perhaps specimens of the protozoan parasite *Monocystis,* a common parasite of annelids and arthropods. Many earthworms are infested with this apicomplexan parasite, which thrives in the tissues of the seminal vesicles.

The female reproductive structures are small and difficult to locate in most specimens. They include a pair of **ovaries** in segment XII, which discharge mature ova into the coelom. The ova are collected in two ciliated **egg funnels** and passed through the oviducts to the **female genital openings** on segment XIV. You should be able to locate two pairs of **seminal receptacles** in segments IX and X, where sperm is received during copulation. Sperm stored in the seminal receptacles is later used in fertilization of eggs released into the cocoon.

## Nervous System

The nervous system of the earthworm consists of the following principal components: (1) an anterior pair of large **suprapharyngeal ganglia** (the "brain"); (2) two **circumpharyngeal connectives;** (3) a pair of smaller **subpharyngeal ganglia;** and (4) a **ventral nerve cord** with an enlarged ganglion in each segment (figure 12.13). Locate several nerves extending from the suprapharyngeal and subpharyngeal ganglia to the anterior segments. Three pairs of **lateral nerves**—two from the ganglion itself and one anterior to the ganglion—lead from the ventral nerve cord in each posterior segment.

So-called giant fiber cells are common but highly variable within the "brain" or the ganglia of the ventral nerve cord. These giant fibers are important for the rapid conduction of impulses used, for example, to generate startle reactions to predators.

## Cross Sections

◆ Obtain a slide with a stained cross section of an earthworm and study its histological structure (figure 12.14). Observe the body wall and its composition. Locate the outer **cuticle** and the underlying **hypodermis**. *What is the nature of the cuticle and how is it formed?* Beneath the hypodermis is a thin layer of **circular muscles** and a thicker layer of **longitudinal muscles.**

Observe the thin layer of flattened cells, the **peritoneum,** at the inner margin of the longitudinal muscle layer. Covering the outer surface of the intestine (and some of the larger blood vessels as noted earlier) is a specialized type of peritoneum, the **chlorogogue.** The cells of this layer are believed to function like those of the vertebrate liver in intermediary metabolism, in the removal of waste products, and also in the synthesis of hemoglobin, the red oxygen-carrying pigment in the blood. The body cavity of the earthworm is a true **coelom.** *What makes this a coelom?*

You may be able to observe within the coelom the **nephridia** lateral to the intestine. The nephridia in cross section may appear to be irregularly shaped since they are often cut through the coiled portions of the tubule (figure 12.15).

The wall of the intestine consists of several layers of cells, including an outer **chlorogogue,** a thin layer of **longitudinal muscle** and **circular muscle,** and a layer of **endodermal epithelium** lining the cavity, or lumen, of the gut. Dorsally, observe the pronounced fold of tissue, the **typhlosole,** which extends into the lumen and increases the surface area of the intestine available for absorption.

Observe also the **dorsal** and **ventral blood vessels** and the **ventral nerve cord.** On each side of the nerve cord is a **lateral neural blood vessel,** and beneath the nerve cord is the **subneural blood vessel.**

## Demonstrations

1. Charts and models to illustrate earthworm anatomy
2. Earthworm cocoons
3. Behavior of living earthworms (locomotion, reactions to stimuli, etc.)
4. Examples of other Oligochaetes, such as *Tubifex, Aeolosoma, Enchytraeus, Stylaria,* and *Chaetogaster*

# Leeches
## Class Hirudinea

Although they are commonly called "bloodsuckers," many of the 500-plus species of leeches (figure 12.16) are actually scavengers or predators on small invertebrates, rather than ectoparasites that feed on the blood of some vertebrate host to which they periodically attach.

The large European leech *Hirudo medicinalis* described in most textbooks was used in the early days of medical practice for bloodletting, since it was once believed that

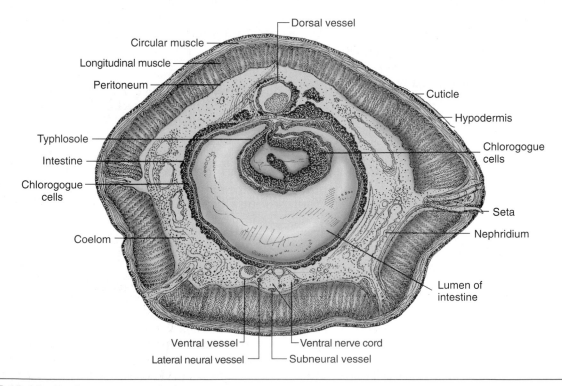

**FIGURE 12.14** Earthworm, cross section.

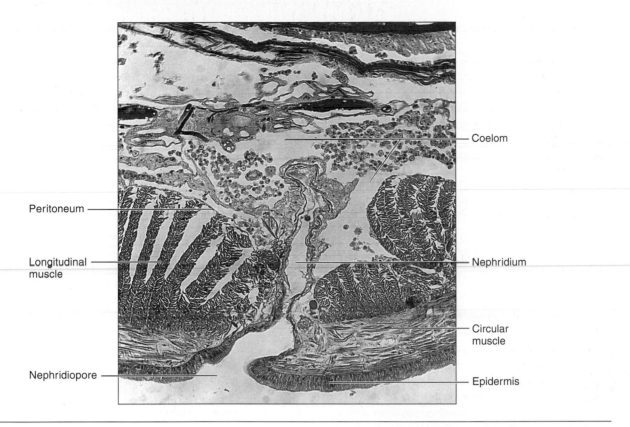

Peritoneum

Longitudinal muscle

Nephridiopore

Coelom

Nephridium

Circular muscle

Epidermis

**FIGURE 12.15**   Earthworm, portion of cross section showing opening of nephridiopore through body wall.
Courtesy of Carolina Biological Supply Company, Burlington, NC.

many diseases and bodily disorders were caused by an accumulation of excess blood. For this reason, this species is often called the "medicinal leech." Leeches are now being used after microsurgery (e.g., for the reattachment of severed fingers, ears, etc.) to reduce the swelling due to venous insufficiency.

Most leeches live in freshwater streams, ponds, and lakes. There are also some marine species and a few terrestrial forms. Leeches are relatively large annelids, and the adults of most species reach about 2–5 cm in length. Among their distinctive features are large **anterior and posterior suckers,** a **fixed number of segments,** and male and female **copulatory organs.** However, they also lack appendages and setae, which distinguishes them from the oligochaetes and the polychaetes. The leeches most often available for laboratory study in the United States are large aquatic species of the genera *Haemopis, Malacobdella,* or *Placobdella.*

## External Anatomy

◆ Examine a preserved leech and observe the elongate, flattened body. The body is flattened dorsoventrally and usually tapers toward the anterior end. Note the absence of setae and lateral appendages. Locate the numerous rings or **annuli** that circle the body.

Leeches are segmented both internally and externally; however, unlike the oligochaetes and polychaetes, leeches have a limited number of body segments (34). Also, the external annuli are more numerous than the internal segments; various internal segments bear from 1–5 annuli each.

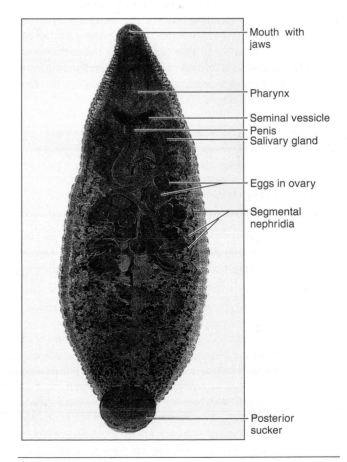

Mouth with jaws

Pharynx

Seminal vessicle
Penis
Salivary gland

Eggs in ovary

Segmental nephridia

Posterior sucker

**FIGURE 12.16**   Leech, stained whole mount.
Courtesy of Carolina Biological Supply Company, Burlington, NC.

Identify the large **anterior sucker** and the smaller **posterior sucker.** The suckers serve both for attachment to the host and for locomotion. Leeches move in a looping fashion by attaching to the substrate by the anterior sucker, contracting the body, attaching by the posterior sucker, loosening the anterior sucker, and extending the body forward. Aquatic leeches can also swim by means of undulatory motions of the body resulting from rhythmic contractions of their muscular bodies. Some are graceful swimmers.

Locate the **mouth** in the center of the anterior sucker. At the anterior margin of the posterior sucker find the **anus** on the middorsal surface.

## Internal Anatomy (Demonstration)

◆ Study a demonstration of a dissected leech to observe its internal anatomy.

The principal internal features include a complete digestive system with a **mouth** containing teeth, a muscular **pharynx,** a large **crop** (that serves to store blood in parasitic leeches), a small **stomach,** a narrow tubular **intestine,** and an enlarged **rectum** that opens dorsally via the **anus** located just anterior to the posterior sucker.

Leeches also have a closed circulatory system, several pairs of segmental nephridia that serve as excretory organs, and permanent gonads. The **reduced coelom** contains one pair of ovaries and several pairs of testes. Although leeches are monoecious (hermaphroditic), reproduction involves copulation and reciprocal fertilization between different individuals. The fertilized eggs are deposited in a cocoon secreted by the clitellum formed by an enlargement of segments IX–XI. The eggs develop directly into miniature leeches; there are no larval stages.

### Demonstrations

1. Preserved specimens of several different leeches, such as *Haemopis, Malacobdella, Placobdella,* and *Hirudo*
2. Dissected leech to illustrate internal anatomy

## Key Terms

**Biramous** having two branches or rami; as the parapodium of a polychaete worm with a dorsal notopodium and a ventral neuropodium.

**Clitellum** a thickened, glandular portion of certain midbody segments of many oligochaetes and hirudineans; characteristic of sexually mature individuals. Forms a cocoon in which eggs are deposited and incubated.

**Coelom** a central body cavity fully lined by peritoneum, an epithelial lining of mesodermal origin. It forms the principal body cavity of annelids and most higher phyla, although in some groups the coelom is secondarily reduced.

**Nephridium** (plural: nephridia) a tubular excretory organ found in annelids and other invertebrate phyla; may function both in osmoregulation and in the removal of nitrogenous and other body wastes.

**Parapodium** (plural: parapodia) a lateral appendage typical of polychaete worms; always occurs in pairs on opposite sides of the body; usually with many setae arranged in bundles.

**Segmentation** having a body consisting of many similar units or subdivisions arranged along the anterior-posterior axis. Each unit is called a segment, somite, or metamere. Also called metamerism.

**Septum** (plural: septa) a thin wall of tissue separating two adjacent segments or masses of tissue.

**Seta** (plural: setae) a chitinous bristle or rod secreted by certain epidermal cells of oligochaetes and polychaetes. Setae are also found in certain arthropods and other invertebrates.

**Trochophore larva** a marine larval form characteristic of many polychaete worms and certain other invertebrate groups; similar larvae are formed by molluscs and certain other invertebrate phyla. Body is pear-shaped with one or more bands of cilia around the body, tufts of cilia at the anterior and posterior ends, and a complete digestive tract.

## Internet Resources

Visit the zoology website at http://www.mhhe.com/zoology to find live Internet links of each of the references listed below.

1. Animal Diversity Web, University of Michigan. Phylum Annelida.
   • Polychaeta
   • Oligochaeta
   • Hirudinea
2. Earthworm dissection. Nice photos, both labeled and unlabeled.
3. University of Minnesota—Phylum Annelida. Information and links on annelids.
4. All About Earthworms. Some interesting facts, anatomical structure, and two very short videos are at this site.
5. Earthworm Dissection. This site leads you to dissection guides.
6. Introduction to the Annelida. University of California at Berkeley Museum of Paleontology. This site provides an introduction to the annelids. It contains information on annelids in the fossil record and annelid life histories, systematics, and morphology, as well as many links.
7. Annelida. Arizona's Tree of Life web page. Pictures, characteristics, phylogenetic relationships, references on annelids. Pictures and references on annelids.

## Critical Thinking Questions

1. Discuss the significant advances (if any) of the Phylum Annelida over the Phylum Platyhelminthes and the Phylum Nematoda. How do you know these are indeed advances? How do you measure organismal success? By the complexity of the organism? By the species diversity of the group? By

the complexity of the life cycle? By the populations of its member species? Is one organism "more advanced" than another simply because it is more complex? Do complex species survive better than simpler species?

2. Compare the coelom of the molluscs with the coelom of the annelids. Of what advantage is an enterocoel when it is obvious that both groups are very successful? In fact, it may be argued that the mollusc's reduced coelom is an advantage and part of the reason for its diversity and adaptive radiation. What do you think?

3. Of what importance is increasing "cephalization"? Was this a significant evolutionary event (to have neural and sensory structures congregate in the anterior of the body)?

4. Segmentation really becomes dominant (some investigators say it appears for the first time) in the annelids. Of what importance is segmentation? Examine how segmentation might have contributed to the evolution of appendages and locomotion as well as the ultimate specialization of these appendages.

5. Parasites are generally considered to be organisms that live *within* or *on* another organism. Some investigators disagree that leeches are parasites, and instead call them "micropredators" because they obtain a blood meal and then fall off the host. How would you classify leeches?

6. Compare the trochophore larva with the larva of other marine invertebrate groups. Do you see any similarities? If so, what are they? Give an explanation for these similarities.

7. Discuss the ecological importance of oligochaetes in general. What do you suppose would happen in many ecosystems if earthworms were to suffer serious population declines? Support your predictions.

## Suggested Readings

Dales, R.P. 1967. *Annelids.* 2d ed. London: The Hutchinson Publishing Group.

Jamieson, B.G.M. 1981. *The Ultrastructure of the Oligochaeta.* New York: Academic Press.

Laverack, M.S. 1963. *The Physiology of Earthworms.* New York: Pergamon Press, MacMillan Co.

Mann, K.H. 1962. *Leeches (Hirudinea), Their Structure, Physiology, Ecology, and Embryology.* New York: Pergamon Press.

Meglitsch, P.A. and F.R. Schram. 1991. *Invertebrate Zoology.* 3d ed. New York: Oxford University Press.

Mill, P.J. (editor). 1978. *Physiology of Annelids.* London: New York: Academic Press.

Sawyer, R.T. 1984. *Leech Biology and Behavior.* 3 vols. New York: Oxford University Press.

## NOTES AND SKETCHES

# NOTES AND SKETCHES

After completing the laboratory work in this chapter, you should be able to perform the following tasks:

1. List the four subphyla of arthropods. Briefly characterize each subphylum and cite examples.
2. Identify the principal external features of the horseshoe crab, *Limulus;* briefly describe its method of feeding and locomotion.
3. Identify the chief external features of a spider on a preserved specimen.
4. Compare the basic morphological organization of a spider with that of the horseshoe crab, a crayfish, and a grasshopper.
5. Identify the chief external features of a crayfish. Compare and contrast the morphology of the crayfish with that of the horseshoe crab.
6. Discuss the concept of serial homology and give examples using the appendages of a crayfish.
7. Describe the respiratory system of a crayfish and explain what appendages are involved in ventilation or gas exchange.
8. Describe the circulatory system of a crayfish and compare it with that of an earthworm.
9. Identify the main morphological features of *Daphnia* and explain how this animal is adapted for its planktonic life.
10. Describe the feeding mechanism of *Daphnia* and identify the chief structures involved.
11. Identify the principal external features of the cockroach *Periplaneta americana* using a specimen or a drawing.
12. Identify and explain the function of each of the principal internal organs of the cockroach *Periplaneta americana*.
13. Identify the major external features of the grasshopper *Romalea* and explain how they represent and illustrate the basic organization of an insect.
14. Describe the principal types of insect metamorphoses and development; identify the stages in complete metamorphosis.

## Introduction

The arthropods represent the most diverse, abundant, and successful of the animal phyla. About 1 million species of arthropods have now been described with some recent estimates of species diversity ranging between 10 and 30 million species. Members of this phylum dwell in almost every type of habitat. In addition to the many kinds of aquatic arthropods, there are also many terrestrial species and numerous species of flying insects—the only invertebrates that have evolved a capability for flight. The body of arthropods is covered externally with a chitinous exoskeleton and, in varying degrees, is segmented both internally and externally. The jointed appendages are among the most apparent and versatile features of the arthropods, and they have been adapted for walking, running, digging, swimming, jumping, feeding, reproduction, sound production, sensory perception, and defense.

# Phylum Arthropoda

## Major Features

Among the important major features of the Phylum Arthropoda are the following: (1) a hardened, chitinous exoskeleton; (2) paired, jointed appendages, often modified for various functions; (3) a segmented body divided into two or three functional regions, such as head and trunk, cephalothorax and abdomen, prosoma and opisthosoma, or head, thorax, and abdomen; (4) a reduced coelom; (5) a complete digestive tract with mouthparts developed from modified anterior appendages; (6) an open venous system but often a closed arterial system with a dorsal heart, arteries, and open spaces (sinuses) in the tissues that serve to collect blood prior to its return to the heart; (7) a highly organized brain and a nervous system with paired ventral ganglia and nerve cords; and (8) complex sense organs (such as compound eyes) and behavior.

## Exoskeleton

The arthropod exoskeleton is a dominant feature of the phylum and has clearly influenced the evolution of other organ systems. The exoskeleton typically consists of a series of hardened plates, called **sclerites.** Between adjacent sclerites are thin, pliable regions that permit the sclerites to move relative to each other. The simplest pattern is that of a series of segmentally arranged dorsal plates, **tergites** or **terga** (singular: tergum), and corresponding ventral plates, **sternites** or **sterna** (singular: sternum) connected laterally by flexible pleural membranes. Between successive dorsal and ventral plates is a flexible nonsclerotized intersegmental membrane. The basal portion of an appendage, the precoxa, arises from the lateral pleural membrane between the tergum and sternum. Further support for the appendage may be provided by development of additional pleural sclerites (pleurites) anterior and posterior to its attachment to the body.

In different groups of arthropods, this basic pattern of the exoskeleton has been variously modified. Sclerites may become fused together, enlarged, or reduced, or they may become modified as a result of selective forces in response to changes in locomotion or changes in mode of life. Also, secondary segmentation may occur when, as a result of evolution, the original segmental boundaries are obscured and secondary segments are formed in new locations. Such

flexibility in the exoskeleton and other body systems has certainly contributed to the great success of the arthropods.

**Chitin** is a major component of the exoskeleton and consists of a large polysaccharide bound with a protein to form a complex glycoprotein, N-acetylglucosamine. The arthropod cuticle is made up of many layers of chitin, which give it strength and flexibility. Chitin is found in numerous groups of animals in addition to arthropods. Arthropod exoskeletons are often strengthened by the addition of calcium compounds and other substances that harden or **sclerotize** portions of the body covering.

In this exercise, you will have an opportunity to study the representatives of three of the four major groups (subphyla) of arthropods: the **chelicerates** (horseshoe crab, spider); the **crustaceans** (crayfish, water flea); and the **uniramians** (cockroach, grasshopper, honeybee).

# Classification

The classification of the arthropods is complicated and unsettled even today because the phylum is so large and diverse. We can present only a brief summary of some of the major classes. Recent studies have led to some changes in the classification of the phylum. Most authorities now recognize four major groups of arthropods—the trilobites, the chelicerates, the crustaceans, and the uniramians, but this current classification is by no means cast in stone.

## Subphylum Trilobita (Trilobitomorpha)

**The Trilobites** (figure 13.1). Primitive marine arthropods with a flattened, ovoid body divided into three parts, an anterior head (cephalon), a middle thorax, and a posterior pygidium ("tail"). Two longitudinal grooves divide the body into three lobes. One pair of simple antennae on the head; all other appendages biramous and similar. All extinct.

## Subphylum Chelicerata

**The Chelicerates.** Body with distinct **prosoma** and **opisthosoma,** or prosoma and opisthosoma broadly fused. Antennae absent. Six pairs of uniramous appendages, including one pair of chelicerae (specialized pincerlike appendages modified for feeding). Chelicerates do *not* have mandibles.

**FIGURE 13.1** Fossil trilobite, portion of head surface broken away. Most trilobite fossils are shed exoskeletons that are filled and fossilized to create "casts."

### Class Merostomata (Water Scorpions and Horseshoe Crabs)

Marine animals with abdominal book gills and one pair of lateral compound eyes. Well-developed cephalothorax and abdomen. Examples: *Eurypterus* (water scorpion, fossil species only), *Limulus* (*Xiphosura*) (horseshoe crab).

### Class Pycnogonida (Sea Spiders)

Small, marine, spiderlike chelicerates; well-developed cephalothorax and reduced abdomen. Example: *Pycnogonum*.

### Class Arachnida (Scorpions and relatives, Spiders, Harvestmen Ticks, and Mites)

Mainly terrestrial animals; no marine species, although a few inhabit aquatic areas; with six pairs of appendages. From the anterior, the first pair of appendages are chelicerae with claw or fang; the second pair are pedipalps (variously modified for sensory perception, sperm transfer, grasping); the remaining four pairs are walking legs. Head and thoracic areas fused to form cephalothorax (= prosoma). Prosoma and opisthosoma distinct and narrowly connected by a pedicel, or else broadly fused. Examples: *Centruroides* (scorpion), *Argiope* (spider), *Dermacentor* (tick), *Trombicula* (mite).

## Subphylum Crustacea

**The Crustaceans.** Mainly aquatic animals with cephalothorax (fused head and thorax) usually covered by a dorsal carapace; hardened chitinous exoskeleton; biramous appendages. Head with one pair of antennae, one pair of biramous antennules, one pair of mandibles, and two pairs of maxillae (accessory mouthparts).

### Class Branchiopoda (Branchiopods)

Small crustaceans with trunk appendages flattened and leaflike, often modified for filter-feeding. Commonly live in temporary ponds or pools, mainly freshwater. Examples: *Daphnia* (water flea), *Artemia* (brine shrimp).

### Class Cirripedia (Barnacles)

Adults sessile and modified for sessile or parasitic existence; carapace covered with calcareous plates; six pairs of thoracic appendages (cirri) modified for feeding. All marine. Examples: *Balanus* (acorn barnacle), *Lepas* (goose barnacle).

### Class Malacostraca

Large crustaceans with trunk consisting of fourteen segments: a thorax with eight segments, an abdomen with six segments, and a posterior telson. All segments typically bear appendages, first antennae, walking legs, and anterior abdominal appendages biramous; first one, two, or three pairs of thoracic appendages modified as maxillipeds. Examples: *Homarus* (lobster), *Penaeus* (shrimp), *Procambarus* (crayfish), *Callinectes* (blue crab).

## Subphylum Uniramia

**The Uniramians.** Mainly terrestrial animals with uniramous (unbranched) appendages; head with one pair of antennae, one pair of mandibles, and one or two pairs of maxillae. Respiration via spiracles leading to tracheal tubules; excretion by Malpighian tubules.

### Class Diplopoda (Millipedes)

Elongate cylindrical body with distinct head and many similar trunk segments; one pair of short antennae; two pairs of walking legs per trunk segment. Example: *Narceus* (*Spirobolus*).

### Class Chilopoda (Centipedes)

Elongate, flattened body with many similar segments; one pair of antennae; each trunk segment with one pair of walking legs. Examples: *Lithobius, Scolopendra.*

### Class Pauropoda (Pauropods)

Tiny cylindrical body; one pair of triramous (three-branched) antennae; no eyes; trunk with nine or 10 pairs of walking legs. Example: *Pauropus.*

### Class Symphyla (Garden Centipede)

Mostly minute, slender, elongate bodies with 14 segments; one pair of long antennae; one pair of spinnerets on last segment; 12 pairs of walking legs. No eyes. Example: *Scutigerella.*

### Class Insecta (Insects)

Body with three regions (tagmata; tagma=sing): head, thorax, and abdomen; one pair of uniramous antennae; three pairs of walking legs and often two pairs of wings in adults. Examples: *Romalea* (grasshopper), *Musca* (housefly), *Apis* (honeybee), *Pulex* (flea).

## Materials List

**Living specimens**
Crayfish (*Procambarus*, etc.)
*Daphnia*
Spider (*Argiope* or similar)

**Preserved specimens**
*Limulus*
*Romalea*
*Argiope*
Crayfish, female with eggs (demonstration)
Representative crustaceans (demonstration)

**Prepared microscope slides**
Crayfish, compound eye (demonstration)
Crayfish, gill, cross section (demonstration)
Insect cuticle (demonstration)
*Argiope,* walking leg (demonstration)
Mites and ticks

**Plastic mounts**
*Argiope*
*Peripatus*
*Lepas*
*Balanus*
Crayfish appendages
Millipede
Centipede
Honeybee
Insect life cycle

**Audiovisual materials**
Anatomy of the Crayfish video

## Subphylum Chelicerata

### The Horseshoe Crab: *Limulus*

The horseshoe crab, *Limulus (Xiphosura) polyphemus* (figure 13.2), is common in the shallow waters and sandy shores of the Atlantic coast of North America from Canada to Mexico. It reaches a length at maturity of 0.5 meter. It is one of the few surviving members of an ancient group of chelicerate arthropods whose ancestors date back at least 200 million years.

◆ Study a preserved or dried specimen of *Limulus* and observe the tough, leathery **carapace** that covers the **prosoma** (fused head and thorax) (figure 13.3).

Behind the horseshoe-shaped cephalothorax is a tapering, hexagonal **opisthosoma** (abdomen) and a long posterior **telson.** On the dorsal surface of the carapace, identify the two lateral **compound eyes** and the two **simple eyes** located on opposite sides of a small anterior spine. Along the sides of the abdomen, find six pairs of **movable spines.**

On the ventral surface of the prosoma (figure 13.4), identify the **seven pairs of appendages:** an anterior pair of **chelicerae** located in front of the mouth, four pairs of **chelate** (pincer) **legs,** and a longer sixth pair of appendages with specialized tips for cleaning the gills and pushing the carapace through the sea bottom sediments. Behind the sixth pair of appendages, locate a seventh pair of small, spiny **chilaria.** On the opisthosoma, find the six pairs of broad, flat structures, one pair per opisthosomal segment. The first pair of plates forms the **genital operculum,** which bears a pair of genital pores through which the gametes are shed. Behind the second to sixth pair of opisthosomal appendages are the many thin, leaflike, highly vascularized folds of the **book gills.** When submerged, these opisthosomal folds beat almost constantly to provide a continuous flow of seawater over the respiratory surfaces of the book gills.

The **mouth** is located ventrally near the center of the base of the walking legs. The spiny **gnathobases** at the base of the legs shred the food and push it toward the mouth.

*Limulus* feeds on worms, molluscs, and other small invertebrates in shallow ocean sediments. While *Limulus* burrows in the mud, it captures prey with its chelicerae. The

(a)

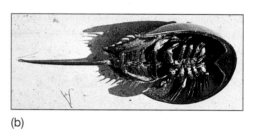
(b)

**FIGURE 13.2** Horseshoe crab, (*a*) dorsal view, (*b*) ventral view.
Courtesy of Carolina Biological Supply Company, Burlington, NC.

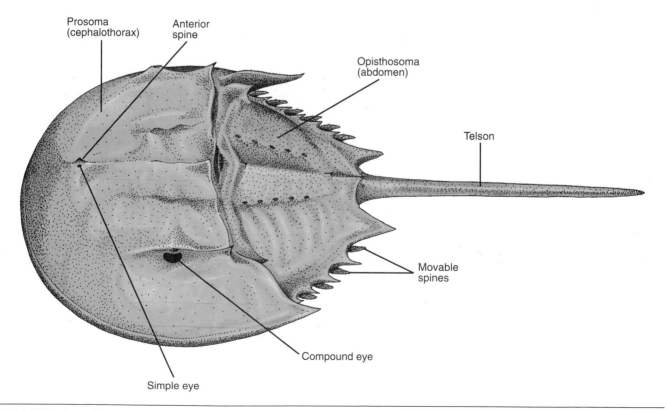

**FIGURE 13.3** Horseshoe crab, dorsal view.

crab plows through the sand by arching its back and pushing against the sand with the sixth pair of appendages and the telson. Awkward as it may sound, this crude form of locomotion and feeding has been effective for some 200 million years. *Limulus* is also able to swim short distances by vigorous beating of the abdominal platelike appendages.

## A Spider: *Argiope*

Spiders, belonging to the Order Araneida (= Araneae), are an ancient group, with some fossils dating back over 300 million years. With two exceptions—Antarctica and the oceans below the intertidal zone—spiders can be found throughout the world. Of an estimated world total of 50,000 species, about 35,000 have been named and described. Spiders are often very abundant: for example, grassy fields may contain over two million spiders per hectare. All spiders are carnivorous, and thus are important for controlling insect pests. The common garden spider (*Argiope* sp.) serves well as an example of this important group of chelicerate arthropods (figure 13.5).

◆ Obtain a preserved or plastic embedded specimen of the garden spider *Argiope* and study it under your dissecting microscope.

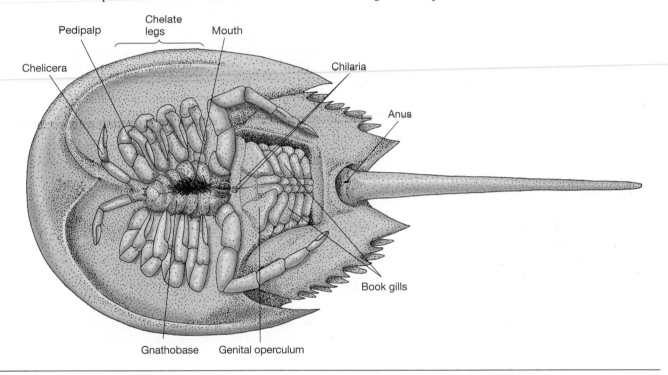

**FIGURE 13.4** Horseshoe crab, *Limulus,* ventral view.

**FIGURE 13.5** Garden spider, *Argiope,* ventral view.
Courtesy of Carolina Biological Supply Company, Burlington, NC.

Arthropoda and Onychophora     **193**

The anterior region from which the legs project is called the **prosoma** (= cephalothorax) and is composed of the head and thorax fused together. The posterior part is called the **opisthosoma** (= abdomen). The two regions are connected by a narrow **pedicel.** The pedicel is seen most easily by gently bending the spider to expose the area between the overlapping opisthosoma and the prosoma. If your specimen is of a genus other than *Argiope,* the body shape and size may vary, but the anatomical parts will be similar.

When spiders die, their legs close over the ventral side. The reason for this is that spiders do not have extensor muscles for their legs; they have only flexor muscles. Living spiders extend their legs by using blood as a hydraulic fluid. As the spider dies, the haemolymph is withdrawn into the body, and the flexor muscles contract, bringing the legs under the body. A tarantula leg has 31 muscles; most of them are flexors.

Externally, the body of a spider is covered by a **chitinous exoskeleton,** which consists of numerous hardened plates called **sclerites** (including, for example, carapace, sternum). The sclerites are separated by sutures or thin membranous areas that are not sclerotized.

The body wall is made up of three main layers: an outer **cuticle;** an underlying **hypodermis,** which secretes the cuticle; and a thin, noncellular **basement membrane.** Insects have a body wall with a similar structure. The sclerites provide body rigidity, offer sites for muscle attachment, prevent desiccation, and protect the animal from predators.

Male and female spiders are usually very different in appearance from one another. The females may grow to be ten or more times larger than the males and may be very differently marked and colored. Such difference between the male and female is called **sexual dimorphism.**

The head region of the spider bears the eyes and mouthparts (figure 13.6). Most spiders have eight eyes (four pairs), but some have fewer or even none at all. Garden spiders have eight simple eyes called **ocelli,** visible at the anterior end of the carapace. The **carapace** is the portion of the exoskeleton covering the prosoma (fused head and thorax). Vision plays a relatively minor role in the behavior of most spiders; they rely more on tactile and chemical senses than vision to locate their prey. Eyes often aid in detecting subtle changes in light intensity and motion; but, with the exception of hunting spiders (such as the wolf, crab, and jumping spiders), the eyes of spiders probably do not form actual images, such as those formed by the eyes of vertebrates.

The area immediately below the eyes and down to the edge of the carapace is called the **clypeus.** Extending downward from below the clypeus is a pair of **chelicerae.** The chelicerae consist of a thick basal segment called the **paturon** and a small distal segment called the **fang** (figure 13.7). The fang articulates with the paturon. It is with the fang that a spider bites and poisons its prey.

Lateral to and slightly behind the chelicerae is a pair of **pedipalps** (= palps) that resemble small legs. The pedipalps are an important aspect of sexual dimorphism in spiders. (If your specimen is a male, the distal end of the pedipalp is enlarged to form the **copulatory organ.**) Now, turn the spider ventral side up and gently pry the legs away from the underside of the spider. Pedipalps consist of only six segments (figure 13.8a and b). The basal segment (coxa) of the pedipalp is expanded to form an accessory mouthpart called the **maxilla.** The maxilla often has a brush of hairs at its distal end called the **scopula.** The scopula aids in sponging up fluids from the prey as the spider feeds. The maxillae are manipulated to squeeze the fluids from the prey.

◆ Examine a walking leg from your specimen and in a prepared microscope slide. Identify the parts as shown in figure 13.8c. Compare the segments of male and female pedipalps (figure 13.8a and b) with those of a

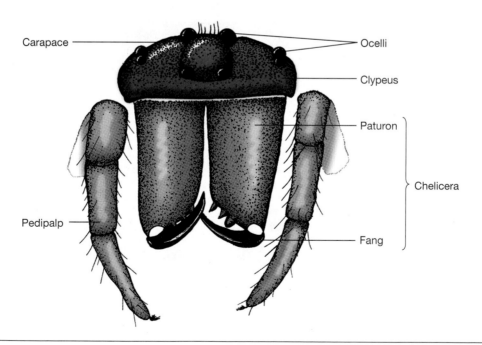

**FIGURE 13.6**  Garden spider, *Argiope,* anterior view of head.

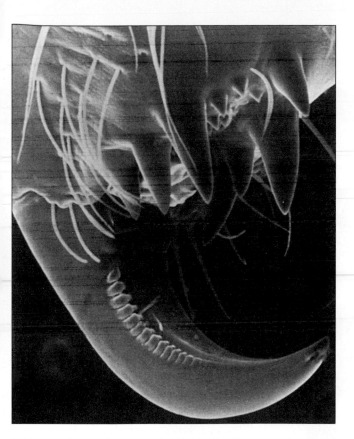

**FIGURE 13.7** Fang of spider.
Scanning electron micrograph courtesy of North Carolina State University.

walking leg. *How do these appendages illustrate adaptation? Which segments of the pedipalps show the highest degree of specialization?*

The **labium** (figure 13.9) is located on the ventral surface between two endites (that form the anterior maxillae) and immediately anterior to the sternum. The labium is sternite No. 2, or the ventral plate of the second fused segment.

The mouth (not visible) is located between the area formed by the endites of the pedipalp, the labium, and chelicerae. Note that there are no mandibles with which to chew the food. Spiders can ingest only liquid foods. When spiders "eat" their prey, they first suck out the blood. Then, through the wound, they pour in digestive juices (the chief enzymes are proteases and lipases) that convert the prey's inner tissues to a soup. The soup is then sucked out by the spider, leaving the hollow exoskeleton of the prey.

The **sternum** is the large plate on the ventral surface of the prosoma (figure 13.9). It is surrounded laterally by the coxae of the four pairs of walking legs. The sternum represents the fused sternites of segments Nos. 3–6.

The **pedicel** connects the prosoma to the opisthosoma and represents the seventh body segment. Except in the very primitive spiders (Suborder Liphistiomorpha), none of the original opisthosomal segmentation is evident in spiders. From histological studies, we know that the opisthosoma in spiders represents segments 7 to 17, and that the spinnerets come from segments 9 and 10, and the anus from segment 18.

On the ventral, anterior portion of the opisthosoma, find the distinct groove that runs from one side to the other (figure 13.9). This is the **epigastric furrow.** At each end of the epigastric furrow is a **lung slit,** marking the entrance to a **book lung.** Most of the air that larger spiders breathe enters through the lung slits. On the central, anterior edge of the epigastric furrow are the **gonopores.** If your specimen is a female, there will be a large, intricate, sclerotized plate extending posteriorly, the **epigynum.** If your specimen is a male, a small plate may or may not be visible. Male spiders do not have a penis. Instead, the palpal organ of the pedipalp is enlarged and serves as the **copulatory organ** (penis analogue) (figure 13.8*b*). Some workers believe that the copulatory organ may have evolved from the tarsal claw. The female epigynum and male palps are very important taxonomic features.

Near the ventral, posterior end of the opisthosoma, locate the three pairs of **spinnerets** (figure 13.9). Two pairs, the anteriors and posteriors, are larger and distinct; the anterior median pair is smaller and tucked in between the larger pairs. The single **anal tubercle** is located at the central posterior edge of the spinneret cluster. Between the anterior bases of the anterior spinnerets, find the small organ called the **colulus.** It is considered to be a small, vestigial spinneret. Immediately anterior to the spinnerets, in a groove, is an extremely small **spiracle,** which serves as the entrance to the tracheal system.

Turn the spider over and observe the dorsal side. On the central, slightly posterior part of the carapace is a depression that represents the **cervical groove** (figure 13.10). Aside from marking the boundary between the cephalic and thoracic regions of the prosoma, it is also an **apodeme,** serving as a muscle attachment site on the inside of the carapace.

Gently manipulate the spider so that the pedicel can be seen from above. On the dorsal surface of the pedicel is a small pair of plates called the **larum** (may be difficult to see).

There are no external, dorsal features worthy of note on the opisthosoma of spiders. However, in the central, dorsal anterior region is the cardiac area, beneath which lies the longitudinal **heart** that pumps blood anteriorly. Pale markings on the opisthosoma result from the deposition of the nucleotide guanine, an excretory product of spiders.

## Demonstrations

1. Living *Argiope* or other spider in terrarium
2. Plastic mount of *Argiope*

# Subphylum Crustacea, Class Branchiopoda

## A Water Flea: *Daphnia*

The water flea, *Daphnia* (figure 13.11), is a common microscopic crustacean found in many bodies of fresh water. A few marine Cladocera are known, but nearly all branchiopods are freshwater species. Its structure is simpler than that of the crayfish, and because of its transparency and small size, living *Daphnia* can be studied easily in the laboratory. *Daphnia* and

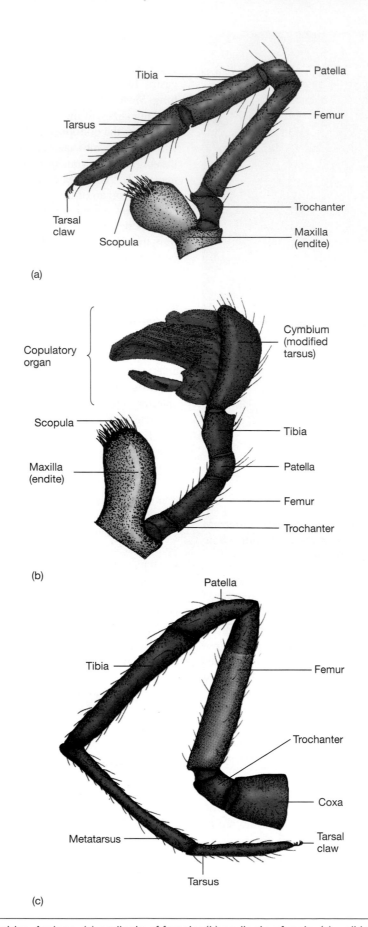

**FIGURE 13.8** Garden spider, *Argiope,* (a) pedipalp of female; (b) pedipalp of male; (c) walking leg.

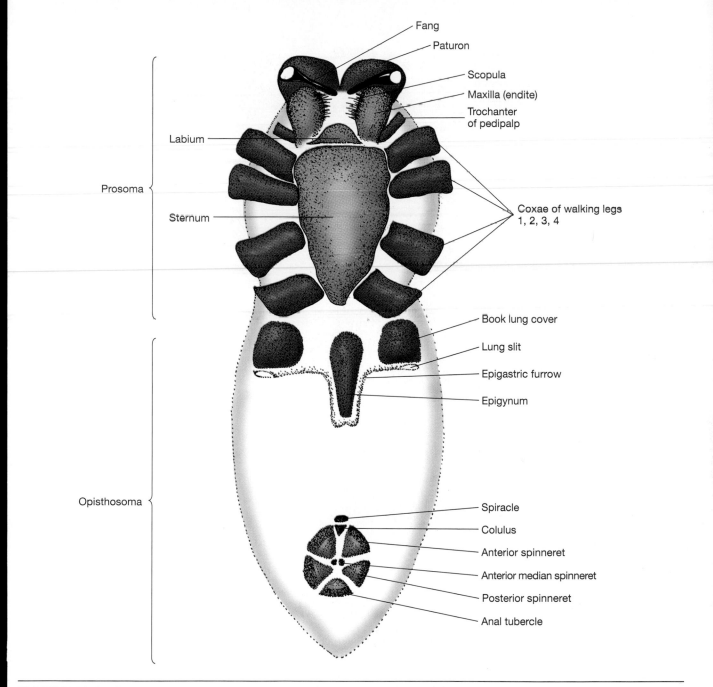

**FIGURE 13.9** Garden spider, *Argiope,* ventral view, most parts of appendages removed.

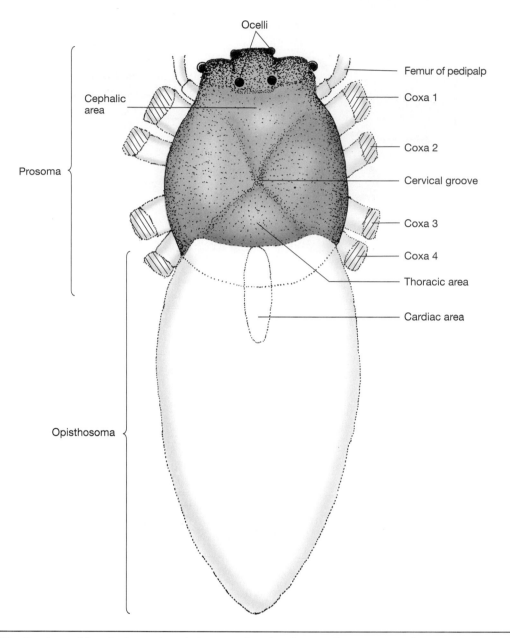

Ocelli

Cephalic area

Prosoma

Opisthosoma

Femur of pedipalp

Coxa 1

Coxa 2

Cervical groove

Coxa 3

Coxa 4

Thoracic area

Cardiac area

**FIGURE 13.10** Garden spider, *Argiope,* dorsal view, most parts of appendages removed.

other members of the Class Branchiopoda of the Subphylum Crustacea are popularly known as "water fleas" because of their characteristic swimming movements. These animals constitute an important element in the diet of many fish and other larger aquatic animals. The *Daphnia* you will observe are classified in the Order Diplostraca, Suborder Cladocera. They can tolerate a wide range of pH and salinity and are important fauna of temporary ponds, especially those without predators.

◆ Make a wet mount of a living *Daphnia* on a microscope slide and observe its characteristic shape.

### External Anatomy
*Daphnia* swims by rapid movements of its two large **antennae** and feeds on microscopic food particles filtered from the water by the complex movements of five pairs of

thoracic appendages, which bear many setae. Food particles removed from the water are collected in a **median ventral groove** located at the base of the legs and are passed forward to the mouthparts where larger particles are macerated by the sclerotized (hardened) **mandibles** before passing into the mouth (figure 13.12).

The body of *Daphnia* is laterally compressed and not obviously segmented. It consists of three main regions or **tagmata:** an anterior head, a large thorax, and a smaller postabdomen. A true abdomen is absent in the Cladocera. The thorax is covered by a part of the chitinous exoskeleton called the **carapace,** which has a bivalved appearance, but which is fused dorsally and open ventrally. The compact head is fully enclosed in the exoskeleton (that is, it does not open ventrally) and is bent downward. Under certain conditions, some Cladocera exhibit seasonal changes in the

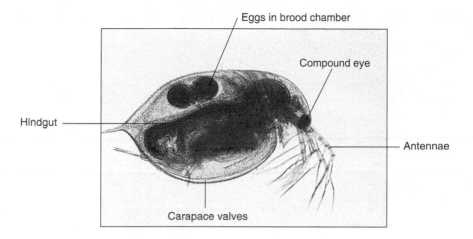

Eggs in brood chamber

Compound eye

Hindgut

Antennae

Carapace valves

**FIGURE 13.11** *Daphnia,* living.
Courtesy of Carolina Biological Supply Company, Burlington, NC.

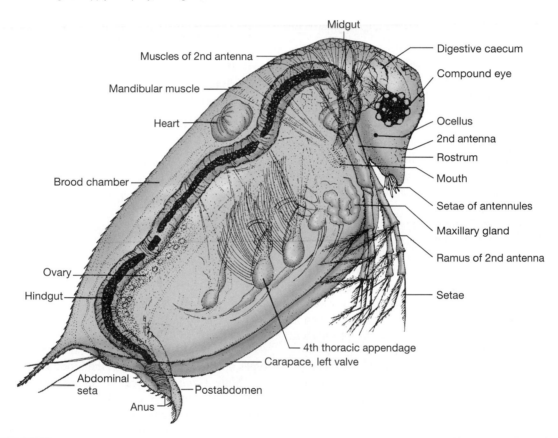

Midgut

Muscles of 2nd antenna

Mandibular muscle

Heart

Brood chamber

Ovary

Hindgut

Abdominal seta

Anus

Postabdomen

4th thoracic appendage

Carapace, left valve

Setae

Ramus of 2nd antenna

Maxillary gland

Setae of antennules

Mouth

Rostrum

2nd antenna

Ocellus

Compound eye

Digestive caecum

**FIGURE 13.12** *Daphnia,* lateral view.

morphology of the head by forming a dorsal elongation or "helmet." (Certain other parts of the body may be modified also.) This kind of morphological variation is called **cyclomorphosis,** and there is evidence that it is influenced by temperature, turbulence of the water, heredity, and other factors, such as the presence of predators.

The most conspicuous structure on the head is the single large compound eye with a central mass of pigmented granules surrounded by numerous hyaline lenses (figure 13.13). A single small *ocellus* (simple eye) lies posterior to the compound eye.

Locate the first and second antennae. The first antennae, or antennules, are small and unsegmented, and are located on the ventral side of the head near the posterior margin. They bear chemical sense receptors called **chemoreceptors** (specialized setae). The tapering projection of the head between the antennules is the **rostrum.**

The second antennae are very large structures attached laterally near the posterior margin of the head. Each second antenna consists of a stout, unjointed **basal segment,** and segmented **dorsal** and **ventral rami.** Each ramus consists of

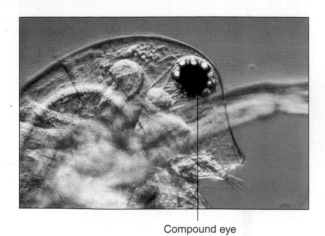

Compound eye

**FIGURE 13.13** *Daphnia,* compound eye.
Courtesy of Carolina Biological Supply Company, Burlington, NC.

several segments and bears numerous **plumose** (featherlike) **setae.** The second antennae are moved by a set of strong muscles that originate on the dorsal side of the carapace near the junction of the head and the thorax. Ventrally, near the same junction, the mouth may be found. It is surrounded by six mouthparts: (1) a dorsal median **labrum** (upper lip); (2) a pair of stout **mandibles;** (3) a pair of small pointed **maxillae,** which are used to push food between the mandibles; and (4) a ventral median **labium** (lower lip).

The **thorax** consists of six segments, and all but the fifth segment bear paired appendages. These five pairs of lobed, leaflike (foliaceous) appendages are basically biramous, but this condition is masked by their high degree of specialization. All of the thoracic appendages bear many hairs and setae used to strain food particles from the water.

As previously stated, a true abdomen is absent in *Daphnia,* but there is a definite **postabdomen** located at the posterior end of the body. It is usually held bent forward under the thorax so that the dorsal side of the postabdomen is directed downward. The postabdomen bears no appendages, but it has two rows of spines, and it terminates in two long abdominal setae. It is effectively equipped with musculature and appears to function mainly in cleaning debris from the thoracic legs, although it may also aid locomotion.

### Internal Anatomy

The complicated musculature of *Daphnia* tends to obscure some of the smaller anatomical details, but the major elements of most of the internal systems can be seen in a whole specimen. Study as many as possible of the following features on your specimen.

1. **Digestive system**—Relatively unspecialized, and consisting of the **mouth, foregut** (esophagus) with a cuticular lining, **midgut** (stomach-intestine), and a **hindgut** (rectum), also with a cuticular lining. Digestion takes place principally in the midgut region, which does not have a cuticular lining. Located anteriorly (in the head region above the compound eyes) are two digestive caeca, which probably secrete digestive enzymes.

2. **Circulatory system**—A simple oval **heart** lies behind the head and dorsal to the gut. Blood enters through two lateral **ostia** and leaves via an anterior opening. There are no definite blood vessels, and the blood is roughly channeled through the large central **hemocoel** by a series of thin mesenteries, and by muscular contraction.

3. **Respiratory system**—Cutaneous or cuticular. Exchange of gases occurs through the body wall, especially through that portion lining the inner surfaces of the valves, and it also occurs through the surfaces of the legs.

4. **Excretory system**—Two looped **maxillary glands** located near the anterior end of the valves are thought to be excretory in function.

5. **Nervous system**—Double **ventral nerve cord,** a few **ganglia,** paired **nerves,** and a **brain** located between the foregut and the compound eyes.

6. **Reproductive system**—Two elongated **ovaries** lie lateral to the midgut in the female *Daphnia.* Posteriorly, each ovary is connected to the **brood chamber** (a cavity dorsal to the body proper and located between the valves of the carapace; closed posteriorly by dorsal processes of the postabdomen) by a thin **oviduct.** In the male, two **testes** occupy the same location but are smaller in size and are continued posteriorly as **sperm ducts,** which follow the gut and open on the postabdomen near the anus. In some species of Cladocera, a part of the postabdomen of the male is modified to form a copulatory organ. The second antennae are large and inflated.

Under favorable conditions, a female *Daphnia* lays thin-shelled, diploid eggs, which contain little yolk and develop parthenogenetically. Under adverse conditions, however, the females produce thick-shelled, resistant eggs with large yolk reserves. The latter type of eggs are haploid and require fertilization by a male for their development. Both types of eggs are deposited in the brood chamber.

Male *Daphnia* appear in natural populations only at certain times of the year and often constitute only a small proportion of the total population. The specific factors responsible for the production of males are not known, despite much investigation, but it appears that male production can be induced by a combination of stressful environmental factors: crowding, decreasing food supply, and water temperature. These environmental factors appear to affect the metabolism of the female in such a way that parthenogenetic male eggs rather than parthenogenetic female eggs are released into the brood chamber.

There are no larval stages in the development of the Cladocera (a situation not at all typical of Crustacea!), but four different periods can be recognized in the life history of a cladoceran: egg, juvenile, adolescent, and adult.

Parthenogenetic eggs develop in the brood chamber into a stage already having the adult form before their release (first instar young). Subsequent development continues through a series of molts in which the carapace and other chitinous portions are lost and the animal increases in size. The average life span covers about 17 **instars** (instar = intermolt stage).

Resistant eggs are retained in the brood chamber until the next molt. A portion of the brood chamber surrounding them is modified into a thick-walled protective case called an **ephippium.** At the next molt of the parent, the ephippium and the eggs it contains are shed.

# Subphylum Crustacea, Class Malacostraca

## The Crayfish: *Procambarus*

The crayfish is a large aquatic arthropod, which effectively illustrates many of the basic characteristics of the phylum. Large southern crayfish (*Procambarus*) or western crayfish (*Astacus*) are especially suitable for laboratory study because of their large size, but members of other common genera, such as *Cambarus* and *Orconectes,* can also be used. The American lobster *Homarus* is very similar in structure to the crayfish and may also be used when available.

### External Anatomy

◆ Obtain an anesthetized or freshly killed crayfish and study the general organization of the body (figures 13.14 and 13.15). Observe that the body is divided into two major regions: an anterior **cephalothorax,** which is made up of a fused head and thorax, and a posterior **abdomen.**

Extending anteriorly from the cephalothorax between the compound eyes is a pointed **rostrum.** Find the **cervical groove,** which represents the line of fusion between the head and the thorax. The hard outer covering (exoskeleton) of the cephalothorax is the carapace; note that the carapace and the chitinous exoskeleton on the dorsal side of the abdomen are hardened by the deposition of mineral salts (except in newly molted specimens). The crayfish body is made up of a total of nineteen segments. Segments 1–13 make up the cephalothorax (five in the head and eight in the thorax), and segments 14–19 constitute the abdomen.

Observe the two large **compound eyes** on movable eyestalks, two large **antennae** with a large scalelike exopodite (sometimes called second antennae), and two smaller biramous **antennules** (sometimes called first antennae). The pointed rostrum projects forward between the eyes. The **mouth** is ventral, surrounded and largely concealed by several pairs of modified appendages that serve as mouthparts (these will be studied later). Locate the five pairs of large **walking legs** and five pairs of small abdominal appendages or **swimmerets.** The first pair of walking legs bear large pincers or **chelae** and thus they are called **chelipeds** ("pincer legs"). *How many of the other pairs of walking legs also bear chelae?* On the last abdominal segment is a pair of large flattened lateral appendages, the **uropods.** Find the **anus** on the ventral side of the **telson,** a medial extension of the last abdominal segment (not an appendage).

Crayfish are dioecious, and each sex exhibits distinctive secondary sexual characteristics. Females have a broad abdomen, rudimentary appendages (swimmerets) on the anterior abdominal segments, a **seminal receptacle** on the ventral surface of the exoskeleton between the fourth and fifth pairs of walking legs, and two female sex openings located at the bases of the third pair of walking legs. Males have a narrower abdomen, enlarged swimmerets (**gonopods**) on the first abdominal segments (modified for the transfer of sperm to the seminal

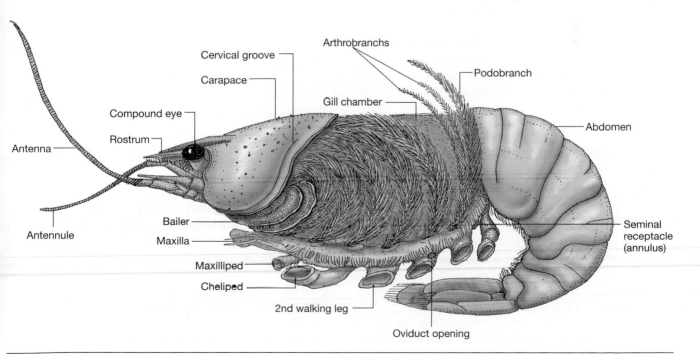

**FIGURE 13.14**   Crayfish, lateral view, portion of carapace removed to show gills.

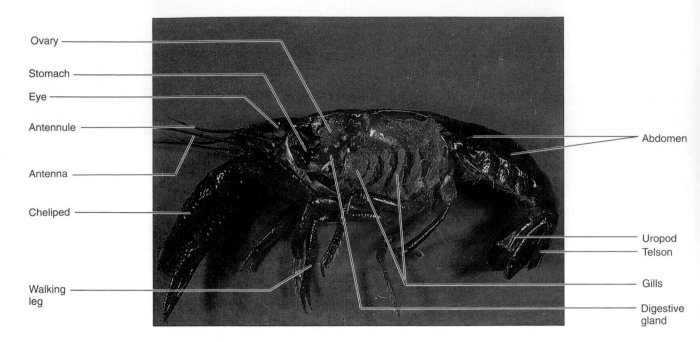

Ovary
Stomach
Eye
Antennule
Antenna
Cheliped
Walking leg

Abdomen
Uropod
Telson
Gills
Digestive gland

**FIGURE 13.15**   Crayfish, lateral view, part of carapace removed.
Photograph by Ken Taylor.

receptacle of the female), and openings of the vasa deferentia (male sex ducts) at the base of each fifth walking leg.

◆ Determine the sex of the specimen you are examining and also observe specimens of the opposite sex.

*Appendages*

In early ancestors of the arthropods, the appendages of all body segments were biramous and similar on all segments. During the evolution of arthropods, various segments and the attached appendages have become adapted to serve in sensory perception, defense, swimming, walking, grasping, sperm transfer, and so forth. Since the body segments and appendages arise from similar embryonic tissue during development, they are said to be homologous structures. The concept of **homology** is an important principle in zoology, and the study of structural homologies has provided much valuable evidence about the evolutionary history and relationships of animals.

The paired appendages of the crayfish provide an excellent illustration of **serial homology,** the adaptation of a longitudinal series of originally similar structures to carry on different functions. The appendages of primitive arthropods probably were **biramous** (two-branched). This is the simplest type of arthropod appendage and is most nearly approximated in the crayfish by the swimmerets of the abdominal segments.

◆ Remove an appendage from the third abdominal segment by cutting it free near its attachment, and observe that it consists of a basal portion, the **protopodite,** and two terminal branches, a lateral (outer) **exopodite** and a medial (inner) **endopodite.**

Carefully remove the appendages from one side of the body, starting with the first walking leg (cheliped). Next, remove the appendages forward of this cheliped,

followed by those appendages behind the cheliped, and arrange them in order on a sheet of paper. Study each appendage with the aid of figure 13.16 and the following descriptions.

1. **Antennule** (first antenna)    base composed of three segments, followed by two long, many-jointed filaments. This appendage is probably not truly biramous. The slender inner branch arises from the base of the outer thicker filament (expodite) and is, therefore, probably not homologous with an endopodite. The three basal segments are homologous with the protopodite. Within the basal segment of each antennule is a **statocyst,** a saclike sense organ that opens dorsally via a small pore.

2. **Antenna** (second antenna)    protopodite of two segments; endopodite long, many-jointed; exopodite a short scale.

3. **Mandible**    basal segment of protopodite greatly enlarged as functional jaw; second segment small and forming the base of the palpus; endopodite, the two small distal segments of the palpus; exopodite secondarily lost.

**Note:** The metastoma, or bifurcated "lower lip" between the mandible and the first maxilla, is a leaflike blade that fits closely over the convex surface of the mandible. It is not a true appendage.

4. **First maxilla**    protopodite of two flat, leaflike segments; endopodite of two segments, the distal one narrow and pointed; exopodite secondarily lost.

5. **Second maxilla**    protopodite of two flat, bilobed segments; endopodite, a small pointed segment;

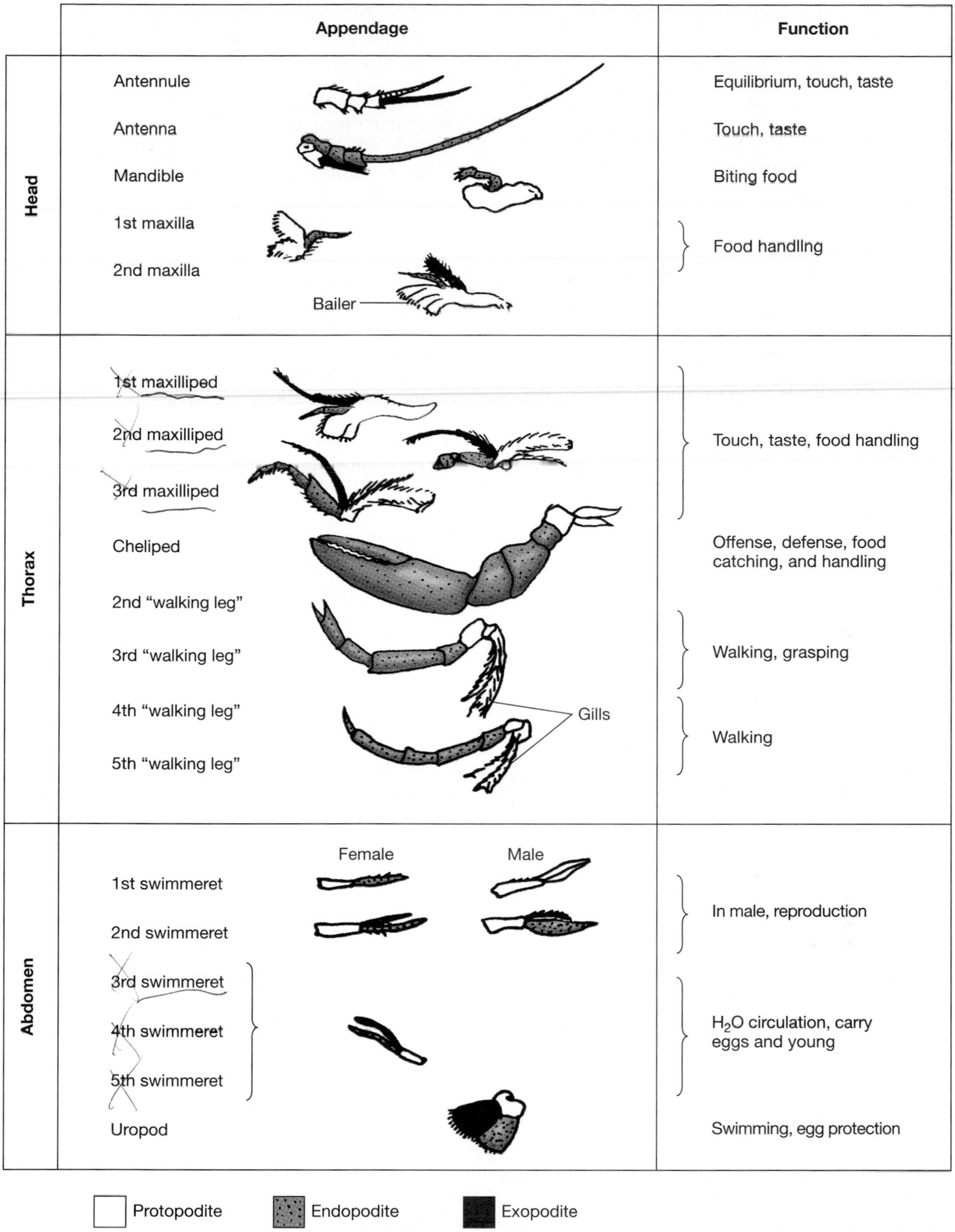

| | Appendage | Function |
|---|---|---|
| **Head** | Antennule | Equilibrium, touch, taste |
| | Antenna | Touch, taste |
| | Mandible | Biting food |
| | 1st maxilla | } Food handling |
| | 2nd maxilla | |
| | Bailer | |
| **Thorax** | 1st maxilliped | } Touch, taste, food handling |
| | 2nd maxilliped | |
| | 3rd maxilliped | |
| | Cheliped | Offense, defense, food catching, and handling |
| | 2nd "walking leg" | } Walking, grasping |
| | 3rd "walking leg" | |
| | 4th "walking leg" | } Walking |
| | 5th "walking leg" | |
| | | Gills |
| **Abdomen** | Female      Male | |
| | 1st swimmeret | } In male, reproduction |
| | 2nd swimmeret | |
| | 3rd swimmeret | } $H_2O$ circulation, carry eggs and young |
| | 4th swimmeret | |
| | 5th swimmeret | |
| | Uropod | Swimming, egg protection |

☐ Protopodite   ▦ Endopodite   ■ Exopodite

**FIGURE 13.16**   Functions of crayfish appendages.

protopodite forming part of a large elongated plate, the **bailer.**

6. **First maxilliped**   protopodite of two flat segments extending inward; endopodite and exopodite are both present.

7. **Second and third maxillipeds**   protopodite of two segments; endopodite of five distinct segments; exopodite relatively thin.

8. Five walking legs or **pereiopods**   protopodite of two segments; endopodite of five joints as in the second and third maxillipeds; exopodite lost. In the large chelipeds (first walking legs), the second segment of the protopodite and the first segments of the endopodite are fused together.

9. **First and second swimmerets**   the first is rudimentary in the female; in the male, the protopodite and the endopodite are fused, and the entire structure is modified for transferring spermatozoa to the seminal receptacle of the female; exopodite is secondarily lost.

10. **Third to fifth** (second to fifth in the female) **abdominal appendages** (swimmerets) protopodite of two segments; endopodite and exopodite flat and filamentous.

11. **Sixth abdominal appendage** (uropod)   greatly enlarged and modified to form (with the telson) the tail fan, which is used for swimming backward.

### Respiratory System

◆ Locate the gills within the **gill chambers** at each side of the carapace (figures 13.14 and 13.15). The lateral flaps of the carapace, covering the gills, are termed **branchiostegites.** Remove the left branchiostegite carefully by cutting away the carapace with your scissors to expose the gills; make your cut carefully to avoid damage to the underlying gills.

The gills are feathery projections of the body wall that contain blood channels. Water is pumped through the gill chamber by the action of the **bailer,** a paddlelike projection of the second maxilla. Oxygen dissolved in the water diffuses across the thin walls of the gills and combines with the blood in the gills. Carbon dioxide diffuses across the walls of the gills in the opposite direction.

Observe that the gills occur in distinct longitudinal rows. *How many rows of gills are there in the specimen? The outer row is attached to the base of certain of the appendages. Which ones? These outer gills are the podobranchs ("foot gills"). How many podobranchs do you find in the specimen?* Separate the gills carefully with your probe or dissecting needle, and locate the inner row(s) of gills. These inner gills are the **arthrobranchs** ("joint gills") and are attached to the chitinous membrane, which joins the appendages to the thorax (figure 13.14). *How many rows of arthrobranchs do you find in the specimen?*

◆ Remove a gill with your scissors by cutting it free near its point of attachment and place it in a watch glass filled with water. Observe the numerous gill filaments

arranged along a central axis. Also study a demonstration slide of a crayfish gill and note the afferent and efferent blood vessels in the central axis.

### Internal Anatomy

◆ Remove the remainder of the carapace up to the rostrum by cutting forward from the rear of the carapace up to a level just behind the eyes. Leave the eyes, rostrum, and adjacent anterior portion of the carapace intact. With your scalpel, carefully separate the hard carapace from the thin, soft underlying layer of **hypodermis** as you cut away the carapace. The hypodermis is the tissue that secretes the carapace and other parts of the exoskeleton. Next, remove the remaining gills on the left side by cutting them near their points of attachment, and cut away the chitinous membrane underlying the gills to expose the internal organs as shown in figure 13.15.

*Circulatory System.*   Locate the small membranous **heart** just posterior to the stomach (figure 13.17). If you are dissecting a preserved specimen, you may find the heart filled with colored latex.

Identify the **pericardial sinus** (the space in which the heart lies), the **ostia** (openings in the wall of the heart), and the delicate **arteries** attached to the heart. Because of their small size, the arteries are often difficult to locate. Carefully search among the internal organs and try to locate the following arteries and determine which structures they supply with blood: (1) **ophthalmic artery,** (2) two **antennary arteries,** (3) two **hepatic arteries,** (4) **dorsal abdominal artery,** and (5) **sternal artery.** The circulatory system of the crayfish is an **open system;** blood is pumped from the heart through the major arteries that branch into smaller arteries and supply blood to the various organs of the body. From the arteries, the blood flows into open spaces, or **sinuses,** within and between the organs and then into a large **sternal sinus** along the floor of the thorax. Channels from the sternal sinus lead to the gills where the blood is oxygenated before returning to the pericardial sinus and into the heart via the ostia.

Crayfish blood consists of a nearly **colorless plasma** containing a dissolved respiratory pigment **hemocyanin** (a complex copper-containing protein) and numerous amoeboid blood cells.

*Muscular System.*   First locate some of the principal muscles of the crayfish. These include the two longitudinal bands of **abdominal extensor muscles** that run along the dorsal side of the thorax and extend back along the dorsal surface of the abdomen (figure 13.18). *What is their function?* Find the large **abdominal flexor muscles** in the abdomen lying below the extensor muscles. These large muscles serve to bend or flex the abdomen; hence, they provide the force for the quick backward thrust and propulsion of the animal when it is alarmed.

The **gastric muscles** attach the stomach to the inner wall of the carapace, and two large **mandibular** muscles **originate** (fixed end of the muscle) on the carapace and **insert** (movable end of the muscle) on the mandibles. Contraction of these muscles causes the crushing or grinding

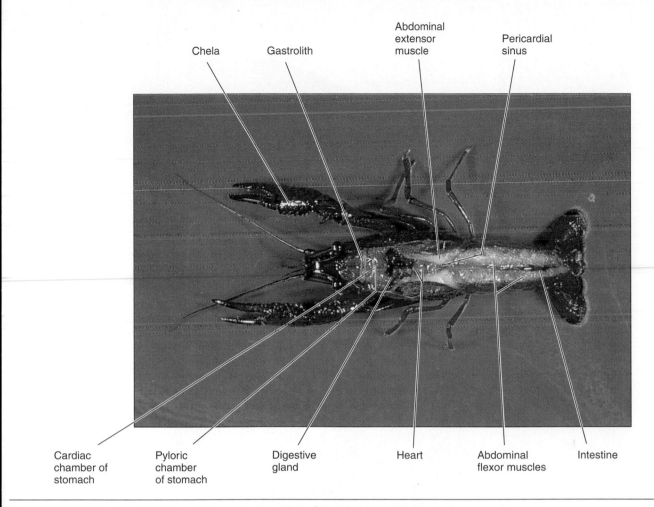

Chela  Gastrolith  Abdominal extensor muscle  Pericardial sinus

Cardiac chamber of stomach  Pyloric chamber of stomach  Digestive gland  Heart  Abdominal flexor muscles  Intestine

**FIGURE 13.17**  Crayfish, dorsal view, upper portion of exoskeleton removed.
Photograph by Ken Taylor.

action of the mandibles. Each of the other appendages is also equipped with musculature: **flexor muscles** serve to draw the appendage closer to the body or to its point of attachment, and **extensor muscles** serve to extend, or straighten out, the appendage.

*Female Reproductive System.*  Locate the paired **ovaries** lying ventral and lateral to the pericardial sinus (figure 13.18) and trace the delicate **oviducts** to their openings at the base of the third walking legs. *Are there eggs in the ovaries? What color are they?* Locate again the **seminal receptacle** where sperm is deposited during copulation. Observe the demonstration of a female "in berry" with clusters of eggs attached to the bristles on the swimmerets.

*Male Reproductive System.*  Find the paired **testes** beneath the pericardial cavity, and trace the **vasa deferentia** to their openings. *Where are the male openings located?* Compare the first abdominal appendages of the male specimen to those of a female specimen.

*Digestive System.*  The **mouth** is obscured by the several oral appendages and can be located by moving these appendages aside with your forceps. Locate, behind the mouth,

the tubular **esophagus** leading to the **stomach** just behind the rostrum (figures 13.17, 13.18, and 13.19). The stomach is divided into a large anterior **cardiac chamber** and a smaller posterior **pyloric chamber.**

◆ Cut open the stomach and study the **gastric mill,** a grinding apparatus consisting of three chitinous teeth. Observe in the pyloric chamber the **chitinous bristles,** which serve as a strainer to prevent large chunks of food from passing unground into the intestine. You may also find calcareous **gastroliths** attached to the walls of the stomach.

These calcareous bodies serve as stores of calcium salts, and they appear to play an important role in the calcification of the exoskeleton after molting. Locate the large digestive gland, or **hepatopancreas.** *What is its function?* Trace the **intestine,** behind the **stomach,** into the abdomen. Find the blind, saclike **intestinal caecum** in the abdomen and the terminal **anus.**

*Excretory System.*  The excretory structures of the crayfish are two large **green glands** located ventrally in the head region near the base of the antennae. These glands are usually dark red or brown (not green!) in preserved specimens.

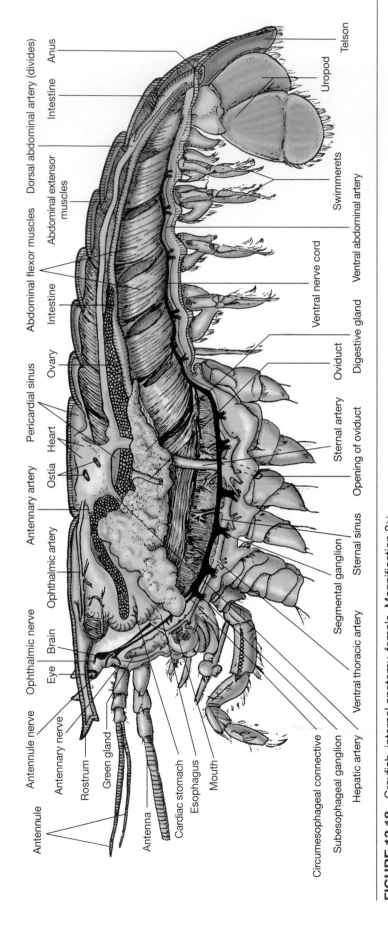

**FIGURE 13.18** Crayfish, internal anatomy, female. Magnification 2×.

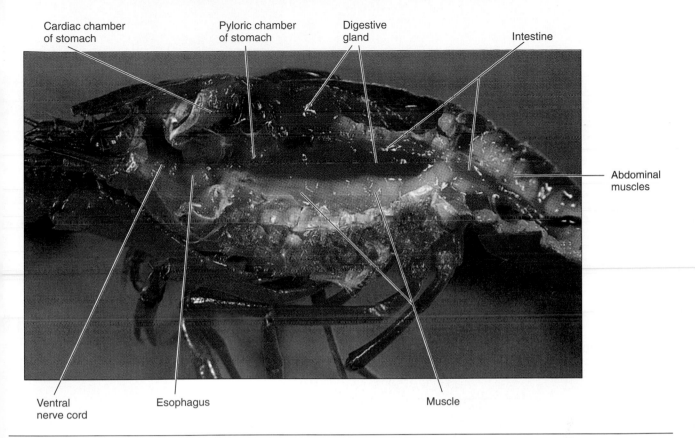

Cardiac chamber of stomach • Pyloric chamber of stomach • Digestive gland • Intestine • Abdominal muscles • Ventral nerve cord • Esophagus • Muscle

**FIGURE 13.19**  Crayfish, lateral view, gills removed to show internal organs. Magnification 2×.
Photograph by Ken Taylor.

◆ Find the two **excretory pores** at the bases of the antennae through which the nitrogenous wastes are discharged to the exterior.

### Nervous System and Sense Organs.

◆ Carefully remove the organs of the reproductive, digestive, and circulatory systems in the head and thoracic regions to expose the ventral surface of the cephalothorax. Also remove the muscles and intestine from the abdominal region, carefully cut away the skeletal plates that cover the anterior part of the nervous system, and locate the paired **ventral nerve cord** (figure 13.20). Observe the **segmental ganglia** and their **paired lateral nerves.** Trace the nerve cord forward to the thorax and carefully cut away the skeleton that surrounds the nerve cord in this region. Follow the nerve cord anteriorly and locate the **supraesophageal ganglion** (the "brain"), the subesophageal ganglion, and the circumesophageal connectives, which run laterally and connect the supra- and subesophageal ganglia.

Also find the several pairs of nerves that lead from the brain to the eyes, antennules, antennae, and mouthparts.

The most prominent sense organs of the crayfish are the **compound eyes** and the **statocysts,** both located previously. The crayfish also has numerous **sensory hairs** (tactile receptors and mechanoreceptors) found on various appendages

that are sensitive to touch. Special **chemoreceptors** on the antennae, antennules, and mouthparts also provide senses of taste and smell.

The statocysts are a pair of small sacs located in the basal segments of the antennules. They have a cuticular lining bearing sensory hairs. A sand grain within the sac, the **statolith,** stimulates the sensory hairs and thus provides the animal with a sense of equilibrium. The lining of the statocysts and the statoliths are lost and replaced at each molt.

## Demonstrations

1. Compound eye (microscope slide)
2. Section of crayfish gill (microscope slide)
3. Female crayfish with eggs attached to swimmerets
4. Live crayfish in aquaria for study of behavior
5. Preserved specimens of several other types of freshwater and marine Crustacea

The large compound eyes are made up of many individual units, the **ommatidia.**

◆ Remove one of the compound eyes and examine its surface under your stereoscopic microscope. The surface of the eye is covered with a transparent **cornea** secreted by underlying cells. Observe that the cornea is divided into many sections, or **facets.**

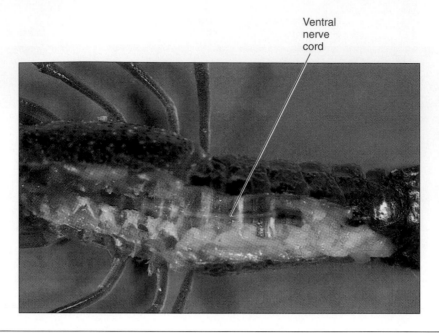

**FIGURE 13.20** Crayfish, internal organs removed to show ventral nerve cord.
Photograph by Ken Taylor.

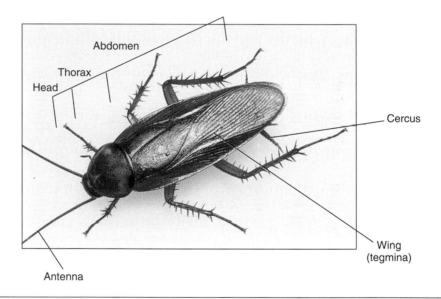

**FIGURE 13.21** Cockroach, *Periplaneta*.
Courtesy of Carolina Biological Supply Company, Burlington, NC.

Each facet of the cornea represents the outer end of an ommatidium. The compound eye probably forms a mosaic light-and-dark impression of its surroundings. Because of its structure, the compound eye is well-adapted for detecting movements.

## Subphylum Uniramia, Class Insecta

Insects constitute the largest, most diverse, and most wide-spread class of arthropods, and indeed of all the multicellular animals. More than one million species of insects have been described, and several thousand new species are added to the list each year. The insects are easily distinguished from the other classes of arthropods by a combination of the following characteristics: (1) **chitinous exoskeleton;** (2) **body divided into three distinct regions** called **tagmata** (head, thorax, and abdomen); (3) **one pair of uniramous antennae;** (4) **three pairs of jointed legs;** (5) and, usually, **two pairs of wings.**

In this exercise, we will examine two common insects as examples of this diverse group of animals, the cockroach *Periplaneta americana* and the grasshopper *Romalea microptera*. These two representatives should give you a good introduction to the basic organization of the insects.

### A Cockroach: *Periplaneta americana*

*Periplaneta americana* is the largest of the three most common cockroaches in North America and Europe (figure 13.21). It is called the American cockroach even though the species appears to have originated in Africa and was subsequently introduced into all parts of the world. *P. americana* is commonly found in restaurants, grocery stores, bakeries, homes, and other places where food is prepared or stored. This cockroach has developed a relationship, though not always a harmonious one, with humans during the past few centuries, and this species has spread worldwide.

Some authors place cockroaches in the Order Orthoptera along with the grasshoppers, crickets, mantids, and walking sticks. Other authors place them with the mantids in the Order Dictyoptera or in a separate Order Blattaria (or Blattodea). In this, as in many other cases, expert taxonomists do not always agree!

Mature specimens of *P. americana* range from 30–45 mm in length, and males tend to be slightly larger in size than females. Both sexes of this species have well-developed wings; males occasionally fly or glide through the air, although they are most often seen crawling along some surface. They tend to thrive in dark, moist places with an available food supply. They are negatively phototrophic, avoid well-lit areas, and are most active at night. Cockroaches are omnivorous and have been reported to feed on fruit, fish, peanuts, flour, paste, paper, cloth, dead insects, and many other objects.

### External Anatomy

Obtain an anesthetized or freshly killed cockroach and study its external anatomy. Preserved specimens may also be used for external anatomy, but fresh specimens are far superior for the study of internal structures. Specimens can either be anesthetized with carbon dioxide or immobilized by cold torpor simply by placing them in a closed container in a refrigerator for 20–30 minutes. Specimens killed by freezing are also suitable for dissection.

Observe the chitinous **exoskeleton** that covers the body. The exoskeleton is made up of many hardened plates, or sclerites, that are connected by thin membranes giving some

flexibility to the exoskeleton. Observe that the body is divided into three regions: an anterior **head,** a middle **thorax,** and a posterior **abdomen** (figure 13.21). Each region consists of several segments, although many of the segments, particularly those making up the head, are fused together.

Identify the two large **compound eyes** and the two long, many-segmented **antennae** on the head. Adjacent to the base of each antenna is a white spot, believed to be the vestige of an **ocellus** or simple eye found in certain other insects. Viewed from the front, the head is roughly pear-shaped; the dorsal portion, which bears the kidney-shaped eyes and antennae, is broader than the ventral portion, which bears the mouthparts (figure 13.22).

The **mouthparts** are modified appendages that aid in grasping and macerating food. Locate the medial **labrum** (not a modified appendage), or front lip. Lateral and posterior to the labrum are the paired **mandibles.** Posterior to the mandibles are the paired **maxillae.** Each maxilla bears a maxillary palp that aids in manipulating food. The medial **labium,** or posterior lip, which consists of several fused parts, also has palps.

By lifting the labrum with your dissecting needle, locate the hardened mandibles that aid in chewing or macerating food for passage to the mouth. Between the mandibles is the **hypopharynx,** which extends upward into the preoral cavity; at its base is the mouth. A duct from the paired salivary glands empties at the base of the hypopharynx.

The head is joined to the thorax by a short, flexible **cervical region.** The thorax consists of three segments and bears two pairs of wings and three pairs of legs. The anterior thoracic segment (*prothorax*) never bears wings and is longer than the next two segments, each of which bears a pair of wings. The **anterior wings** borne by the second thoracic segment (*mesothorax*) are thickened and help protect the membranous **second pair of wings** that arise from the third thoracic segment (*metathorax*).

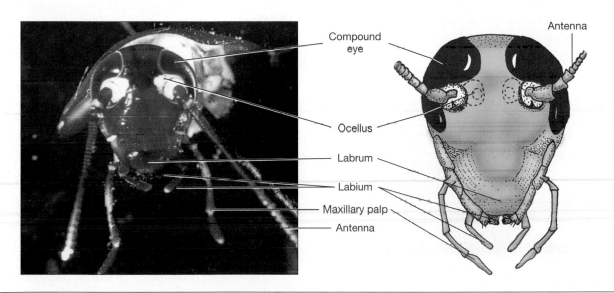

**FIGURE 13.22** Cockroach, anterior view of head.
Courtesy of Carolina Biological Supply Company, Burlington, NC.

Arthropoda and Onychophora **209**

Observe the numerous **veins** in the wings. The pattern of wing venation is specific and is characteristic of each species of cockroach; therefore, it often serves as an important taxonomic characteristic.

Two pairs of **spiracles**—small slitlike openings into the tracheal system—are found in the membranes connecting the first and second and the second and third thoracic segments. Locate the spiracles on one side near the dorsal surface of the intersegmental membrane.

The abdomen consists of 11 segments, each with distinct dorsal and lateral plates called tergites and sternites, respectively. A collection of these tergites is called a *tergum* and a collection of sternites is called a *sternum*. Each segment is joined to the next by a flexible **intersegmental membrane.** Several of the posterior segments have become fused so that some of the dorsal and ventral plates are not clearly identifiable. Each of the first eight segments bears a pair of oval spiracles. The **anus** opens in an unpigmented area on the last segment, dorsal to the external genitalia.

Cockroaches are sexually dimorphic. Sex can be determined in *P. americana* by observing the number of projections from the last abdominal segment. Males have four projections from this segment, two lateral **cerci** and two medial **styli** (figure 13.23). Females have two cerci but lack the styli.

### Internal Anatomy

Start your dissection by removing the legs and wings near their bases with your dissecting scissors. Next, cut carefully through the terga of the thoracic and abdominal segments except the last abdominal segment. Make a shallow incision near the lateral margins of the terga. Be careful to avoid damaging internal organs as you cut through the terga. Do not cut into the head and last abdominal tergum. After you have completed these lateral incisions, pin the insect with its ventral surface down in a dissecting dish with a wax-covered bottom. Cover the insect with water or a saline solution (0.75% NaCl) or 70% alcohol to keep its internal organs soft and pliable. Place the pins along the lateral margin of the insect beyond the incisions in the terga. Angle the heads of the pins outward to give you a clear view of the cockroach.

Beginning at the posterior end and working forward, carefully remove the thoracic and abdominal terga that have been cut. Use your forceps and a dissecting needle (or an in-

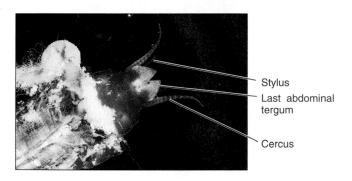

**FIGURE 13.23** Cockroach, posterior end of male abdomen.
Courtesy of Carolina Biological Supply Company, Burlington, NC.

sect pin) to separate the terga from the internal tissues. Take special care to separate the transparent dorsal blood vessel that runs along the dorsal midline just below the terga. Observe and cut the strong **dorsoventral muscles** attached to the thoracic terga. *What do you think the function of these muscles might be? How might movements of the ventral body wall be important to the metabolism of an active insect?*

After removing the terga, locate the dorsal blood vessel again; the portion of this vessel in the abdominal region is called the **heart.** Blood enters the heart through paired **ostia** (slitlike valves) and flows anteriorly toward the head through the aorta. Blood passes from the aorta in the head and flows over organs in the **hemocoel** as it moves posteriorly through the body to the heart.

Running along the dorsal blood vessel are silvery **tracheae** and **air sacs** of the respiratory system. Much of the internal space of the abdomen is filled with a white, spongy **fat body.** The fat body is an important organ, somewhat analogous in function to the vertebrate liver. It is the major site of intermediary metabolism, and it also stores fats, amino acids, and carbohydrates.

***Digestive System.*** Carefully separate the heart, tracheae, fat body, and associated tissues to locate the digestive tract, which extends through the center of the abdomen. Adjacent to the digestive tract are the more transparent reproductive organs; be careful not to damage or remove them accidentally while removing parts of the fat body to locate the digestive tract.

Follow the digestive tract anteriorly and find the translucent **salivary glands** lateral to the digestive tract in the thorax. These glands secrete **amylase,** an enzyme that digests starch. The salivary glands empty through ducts leading to the base of the hypopharynx. Follow the digestive tract to its junction at the head and cut the digestive tract and the salivary glands at this junction. Free the digestive tract from the surrounding tracheae and membranes and pin the anterior end of the digestive tract at an angle with the body so that you can examine it in more detail (figures 13.24 and 13.25).

As in other insects, the digestive tract of the cockroach is divided into three regions, a chitinized **foregut,** a **midgut,** and a chitinized **hindgut.** The foregut extends from the mouth via a narrow **esophagus** to an expanded **crop.** The crop serves mainly to store food. Posterior to the crop is a muscular bulb, the **proventriculus,** which works like a gizzard to break up solid food. A **stomodual valve** at the posterior end of the proventriculus regulates the passage of food material into the midgut.

Digestion and absorption take place mainly in the midgut and in several fingerlike **gastric caeca** that extend out from the anterior end of the midgut. Several enzymes are secreted in the midgut and the caeca. The **pyloric valve** at the posterior end of the midgut controls the passage of materials into the hindgut.

The hindgut is composed of a large, dark-colored anterior **ileum;** a thin, light-colored **colon;** and a **rectum** found in the last abdominal segment (figure 13.24). Many thin,

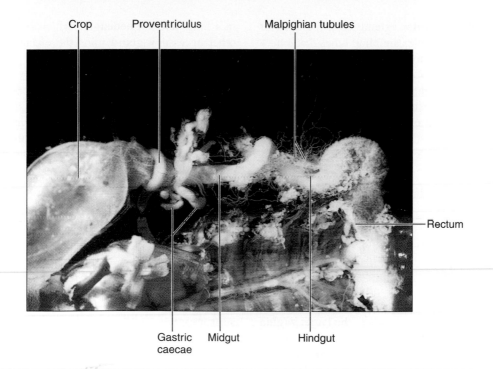

FIGURE 13.24   Cockroach, digestive system, posterior portion.
Courtesy of Carolina Biological Supply Company, Burlington, NC.

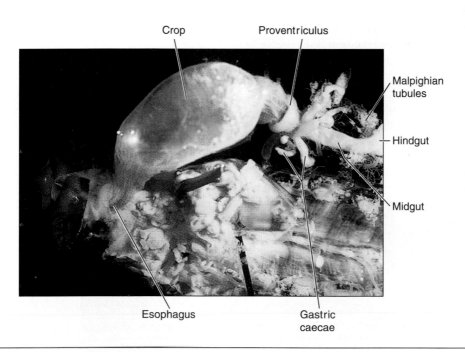

FIGURE 13.25   Cockroach, digestive system, anterior portion.
Courtesy of Carolina Biological Supply Company, Burlington, NC.

translucent **Malpighian tubules** extend from the anterior end of the hindgut. These tubules are blind tubes that empty into the gut. They serve to remove metabolic waste materials and to regulate the concentration of various ions in body fluids. Undigested materials entering the hindgut from the midgut pass through the ileum and colon to the posterior rectum. The rectum is made up of six **rectal pads,** as well as other muscle and connective tissue, and serves to reabsorb water and to store waste materials temporarily before discharge through the anus.

***Reproductive System.*** Cut through the rectum and remove the digestive tract to study the reproductive system. If you have a gravid female, you will find large **ovaries** with developing **eggs** filling much of the abdominal space. A transparent tube, the **lateral oviduct,** extends posteriorly from each ovary and runs toward the ventral midline where it joins the other lateral oviduct to form the **common oviduct.** See figures 13.26 and 13.28.

The short common oviduct extends to the larger **vagina** or genital chamber near the posterior of the insect. Dorsal and posterior to the vagina are two white **accessory glands.** Anterior to the accessory glands is a small **spermatheca** and an elongate **spermathecal gland.**

In a male specimen, you will find a cluster of tubules near the base of the last tergite. These tubules are part of the **accessory glands,** which surround the paired **testes.** By carefully separating the tissues of the accessory glands, you may be able to locate the **testes** and the **seminal vesicles** in which the sperm are stored in bundles of **spermatophores** (figures 13.27 and 13.28).

From the seminal vesicles, an **ejaculatory duct** runs posteriorly through a mass of muscles and sclerites near the posterior end of the insect. The ejaculatory duct terminates in a long, hooked **aedeagus,** which serves as a penis. Another **accessory gland** joins the ejaculatory duct near the base of the aedeagus.

Male *P. americana* have a complex copulatory apparatus consisting of several elongate processes equipped with hooks and barbs that aid in positioning the male and female genitalia during copulation. During mating, a spermatophore (discrete package of sperm and nutrients for the female) is transferred from the male and passes into the female genital chamber where it is attached to the inside of the chamber near the spermathecal opening.

***Nervous System.*** Remove the remaining tissue of the fat body in the median ventral portion of the body to locate the translucent **ventral nerve cord** that runs along the ventral midline (figures 13.29 and 13.30). Observe the **segmental ganglia** and the peripheral nerves extending laterally from each of the ganglia. In the head near the base of the antennae are three pairs of ganglia fused together to form one large **supraesophageal ganglion,** which serves as the brain of the cockroach. The supraesophageal ganglion innervates the eyes, antennae, and other sense organs of the head. It connects to the anterior end of the ventral nerve cord by a pair of **circumesophageal connectives.** Ventral to the digestive tract at the anterior end of the ventral nerve cord is the **subesophageal ganglion,** also made up of three pairs of fused ganglia. This ganglion innervates the mouthparts of the cockroach.

## A Grasshopper: *Romalea microptera*

Grasshoppers, along with katydids, locusts, and crickets, are members of a relatively unspecialized order of insects (Order Orthoptera). Specimens of the large, black, lubber grasshopper are usually provided for laboratory study

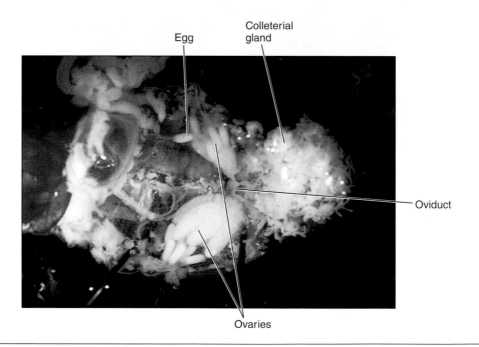

Egg Colleterial gland Oviduct Ovaries

**FIGURE 13.26** Cockroach, female reproductive system.
Courtesy of Carolina Biological Supply Company, Burlington, NC.

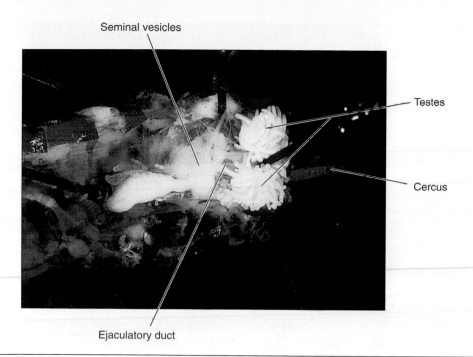

Seminal vesicles

Testes

Cercus

Ejaculatory duct

**FIGURE 13.27** Cockroach, male reproductive system.
Courtesy of Carolina Biological Supply Company, Burlington, NC.

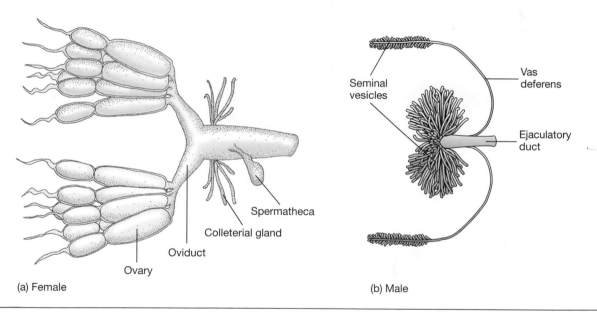

Spermatheca

Colleterial gland

Oviduct

Ovary

(a) Female

Seminal
vesicles

Vas
deferens

Ejaculatory
duct

(b) Male

**FIGURE 13.28** Cockroach, reproductive system, (*a*) female; (*b*) male.

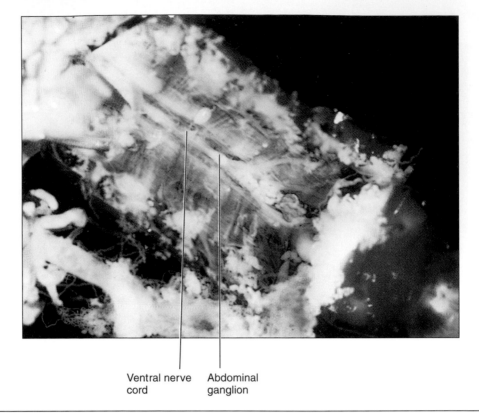

Ventral nerve    Abdominal
cord             ganglion

**FIGURE 13.29**  Cockroach, central nervous system.
Courtesy of Carolina Biological Supply Company, Burlington, NC.

Supraesophageal ganglion ("brain")

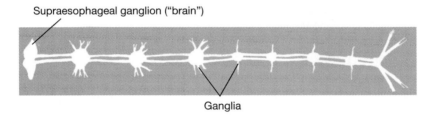

Ganglia

**FIGURE 13.30**  Cockroach, nervous system showing supraesophageal ganglion ("brain") and parallel ventral nerve cord and associated ganglia.

(figure 13.31), but the following information generally applies to other common species of grasshoppers as well.

### External Anatomy

◆ Obtain a preserved grasshopper and study its external features.

Externally, the body of a grasshopper is covered by a **chitinous exoskeleton** consisting of numerous hardened plates, or **sclerites,** separated by sutures, or thin membranous areas. The latter provide the flexibility necessary for movements of the body segments and the appendages. The body wall is made up of three principal layers—an outer **cuticle;** an underlying **hypodermis,** which secretes the cuticle; and a thin, noncellular **basement membrane.** The cuticle is a complex structure consisting of three layers: a very thin outer epicuticle, an exocuticle, and an innermost endocuticle lying directly upon the hypodermis. The endocuticle and exocuticle are composed largely of chitin, a

nitrogenous polysaccharide (**N-acetylglucosamine),** while the epicuticle is nonchitinous and consists chiefly of a network of fibrous proteins impregnated with wax. Study the demonstration provided in the laboratory, showing the microscopic structure of insect cuticle.

The head of the grasshopper (figure 13.32) bears the eyes, antennae, and mouthparts. Two large **compound eyes** occupy prominent sites on opposite sides of the head, and three **ocelli** (simple eyes) form an inverted triangle on the front of the head between the compound eyes. The two many-jointed **antennae** arising anterior to the compound eyes serve as organs of touch, smell, and taste. The **mouthparts** of a grasshopper are relatively unspecialized in comparison with those of most other kinds of insects and are of the **chewing type.** They consist of seven parts: a dorsal **labrum** or upper lip; a pair of sclerotized (hardened) **mandibles;** a pair of **maxillae,** which manipulate the food; a broad median **labium** or lower lip; and, in the center of these

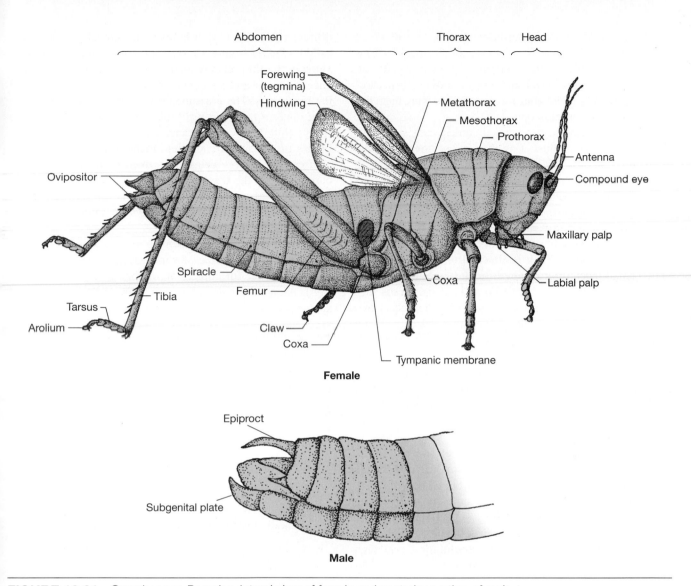

**Female**

Abdomen · Thorax · Head

Forewing (tegmina)
Hindwing
Metathorax
Mesothorax
Prothorax
Antenna
Compound eye
Maxillary palp
Labial palp
Coxa
Tympanic membrane
Coxa
Claw
Femur
Spiracle
Tibia
Tarsus
Arolium
Ovipositor

**Male**

Epiproct
Subgenital plate

**FIGURE 13.31**  Grasshopper, *Romalea,* lateral view of female and posterior portion of male.

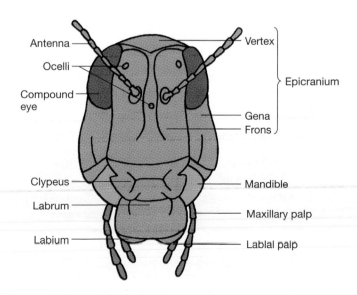

Antenna
Ocelli
Compound eye
Clypeus
Labrum
Labium
Vertex
Epicranium
Gena
Frons
Mandible
Maxillary palp
Lablal palp

**FIGURE 13.32**  Grasshopper, head, frontal view.

Arthropoda and Onychophora

mouthparts, a cylindrical **hypopharynx** at the base of which the salivary glands empty. The labium and the maxillae bear sensory palps, which project ventrally. The mouthparts can be most easily observed and studied by removing them carefully from the head with your forceps and studying them under a stereoscopic microscope.

The thorax, like the head and abdomen, consists of several fused segments, illustrating the multisegmented ancestry of the insects. In the thorax, there are three segments, a large anterior **prothorax,** followed by two smaller segments, the **mesothorax** and the **metathorax.** Each thoracic segment bears one pair of legs. Only the mesothorax and the metathorax bear a pair of wings.

The legs are clearly jointed and consist of five main segments: a short **coxa,** which is attached to the ventral body wall; a small **trochanter** (often difficult to see) fused with a stout **femur;** a slender, spiny **tibia;** and a distal **tarsus** subdivided into five **tarsomeres,** the last of which bears two terminal **claws.** Between the claws is a fleshy pad, the **arolium,** which provides a grip on smooth surfaces. All three pairs of legs are used by the grasshopper for walking and climbing, but the third pair (the metathoracic legs) is specially modified for leaping. They are equipped with an enlarged femur containing strong voluntary muscles and with an elongated tibia to provide additional leverage. Another modification of the metathoracic legs can be seen in male specimens of *Romalea* and its close relatives. The inner surface of the femur bears a row of small spines that is rubbed against the lower edge of the front wing to produce the sounds that comprise the characteristic song of the males. Certain other species of grasshoppers produce their song by rubbing together their forewing and hindwing. This type of sound production is called **stridulation,** and such sounds can be produced by insects of other orders by rubbing other parts of their bodies together.

The long, narrow **forewings** are borne on the mesothorax and serve as a covering for the hindwings when at rest. These leathery wings—found commonly in members of the orthopteroid orders (grasshoppers, cockroaches, etc.)—are called *tegmina* (figure 13.31). Observe the numerous **wing veins** in the thin, membranous **hindwings.** Note the difference in texture and pigmentation between the forewings and the hindwings. The grasshopper wing develops as a saclike outgrowth of the body wall; this outgrowth later flattens to form a thin double membrane, and the two opposing membranes fuse together enclosing tracheae, nerves, and blood sinuses. The hollow veins are formed by thickening of the cuticle around the **tracheae** (air tubes) or along the sinuses. The pattern of wing venation in insects is very precise and species-specific and therefore plays an important role in classification.

The cylindrical abdomen of *Romalea* consists of 11 segments, the last three of which are reduced in size and modified either for copulation or for egg laying (oviposition). Along the sides of certain thoracic and abdominal segments are 10 pairs of tiny respiratory openings, the **spiracles,** which open into the system of **tracheal tubules.** The first abdominal segment also bears on each side a large, oval **tympanic membrane,** which serves as an organ of hearing. There is one pair of **sensory cerci** on the eleventh segment.

### Internal Anatomy

Freshly killed or anaesthetized specimens are most suitable for the study of internal anatomy, although well-preserved specimens can also be used with some success. The internal tissues and organs are difficult to preserve, however, because of the chitinous exoskeleton of the grasshopper, which limits penetration of the preservative fluids. Study figure 13.33 for orientation before starting your dissection.

◆ Remove the wings, and starting at the posterior end of the body, cut through the body wall along each side just above the spiracles. Remove the freed dorsal wall and pin back the sides of the remaining body wall in your dissecting pan. Cover your dissection with water to keep the internal organs moist and pliable.

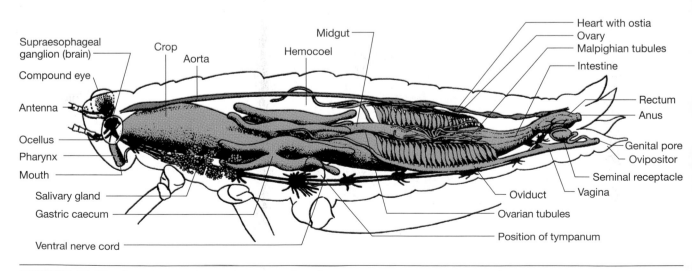

**FIGURE 13.33** Grasshopper, internal anatomy, female.

Observe the **hemocoel,** the tubular **heart** with **ostia,** and the various parts of the digestive tract. The digestive tract of insects consists of three main divisions: **foregut, midgut,** and **hindgut.** These divisions are often functionally subdivided for specific functions in particular insects.

The **Malpighian tubules** are outgrowths of the digestive tract that originate at the junction of the midgut and hindgut. They have an excretory function. Other internal organs include the **tracheal tubules** and the **gonads.** *Is your specimen a male or female?* Observe the texture and shape of the gonads and also the posterior segments of the abdomen to determine the sex. Along the ventral surface, find the **salivary glands** and the **nerve cord.** Note the **segmental ganglia** and the numerous **lateral nerves.** Observe the powerful muscles of the thorax associated with the legs and wings.

## Demonstrations

1. Plastic mount of cockroach life cycle
2. Microscopic mounts of grasshopper mouthparts
3. Microscope slide showing spiracle
4. Life cycle stages (larvae, pupae)
5. Microscope slide with section of cuticle

# Insect Metamorphosis

Many insects undergo marked changes of body form during their postembryonic life so that the immature stages bear little resemblance to the adult stages. The processes of these changes in form are called **metamorphosis.** We can distinguish three general types of postembryonic development common in the insects:

1. **No metamorphosis.** (egg → young adult) Some primitive insects (**ametabolous insects**) exhibit no metamorphosis but hatch directly from the egg in a form resembling miniature adults. Examples of this type of development are provided by the silverfish, or bristletails (Order Thysanura), and the springtails (Order Collembola).
2. **Incomplete or gradual metamorphosis.** (egg → nymph → adult) Young grasshoppers emerge from the egg generally resembling the adult but with a disproportionately large head and without wings. The emergent form is the **nymph,** and transformation to the adult form involves a series of five molts during which the wing pads gradually develop into functional wings, and the head-body proportions approach the adult condition. Insects that exhibit incomplete metamorphosis are called **hemimetabolous insects.** In addition to grasshoppers, this type of metamorphosis is exhibited by several other groups of insects, including the cicadas (Order Homoptera), walking sticks, mantises, cockroaches (Order Orthoptera), termites (Order Isoptera), and the true bugs (Order Hemiptera).

Some aquatic hemimetabolous (technically called paurometabolous or gradual metamorphosis) insects have immature forms (**naiads**) that generally resemble the adults but are adapted for aquatic life. The naiads often have external gills, and they lack wings. They, too, undergo a series of molts and gradually assume the adult form. Dragonflies (Order Odonata), stone flies (Order Plecoptera), and mayflies (Order Ephemeroptera) are examples.

3. **Complete metamorphosis.** (egg → larva → pupa → adult) Most insects do undergo striking changes in form during their postembryonic life and are called holometabolous insects. The characteristic series of stages in metamorphosis is as follows: **egg, larva, pupa,** and **adult.** The life cycle of the fruit fly *Drosophila,* a member of the Order Diptera, is illustrated in figure 13.34.

The egg hatches into a segmented, wormlike larva. The larva feeds, grows, and undergoes several molts. After the final molt, the larva transforms into the pupa—a stationary, nonfeeding stage. Although, externally, the pupa appears to be inactive, internally, radical changes are taking place. Most of the larval tissues are reabsorbed, and the adult organs are newly formed from special groups of embryonic cells called **imaginal discs.** These groups of embryonic undifferentiated cells are retained through the larval period and do not resume their development until the pupal stage. When the adult is fully formed within the pupal case, the case is ruptured, and the adult emerges. Insects demonstrating complete metamorphosis include the beetles (Order Coleoptera), the flies (Order Diptera), the bees and wasps (Order Hymenoptera), and the butterflies and moths (Order Lepidoptera).

◆ Study the demonstration materials on insect metamorphosis provided in the laboratory. The grasshopper affords an excellent illustration of hemimetabolous metamorphosis. Observe the eggs and several nymphal stages, and compare each of them with the adult. Note especially the changes in the head-body proportions and the gradual development of the wing pads.

Study complete metamorphosis in a beetle (such as a June beetle, a Mexican bean beetle, or a meal worm *Tenebrio*) and in the silkworm moth (*Bombyx*), or a similar moth. Observe each of the principal stages—the egg, the wormlike larva, the pupa, and the adult. On figure 13.35, draw the series of stages for one of the insects studied.

## Phylum Onychophora

A small and unusual group of animals, the Phylum Onychophora, shares characteristics with both the annelids and the arthropods (figure 13.36). Some investigators suggest that the onychophorans may have descended from an ancestral

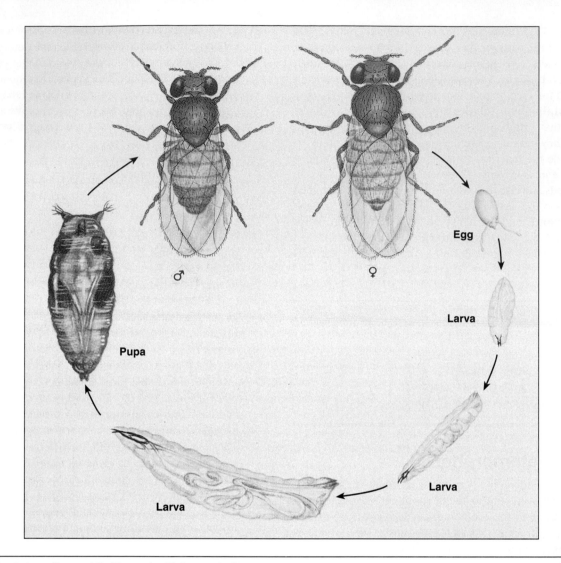

**FIGURE 13.34** *Drosophila* life cycle. Holometabolism.
Courtesy of Carolina Biological Supply Company, Burlington, NC.

**FIGURE 13.35** Student drawing of insect metamorphosis stages.

segmented group that also gave rise to both the extant arthropods and the annelids.

The Onychophora is an ancient group that arose in the Paleozoic era and apparently has changed little since the Cambrian period. The belief of most authorities is that long ago the Onychophora evolved separately from the annelids and the arthropods, and that perhaps the three phyla evolved independently from some common ancestral stock.

The onychophorans are a small group of about 80 species of small (1.5 to 15 cm long) wormlike animals found in tropical rain forests and other damp or wet habitats in tropical and subtropical areas. Distribution is widespread, but scattered and discontinuous. *Peripatus* is the best-known genus. The caterpillarlike onychophorans exhibit several morphological characteristics similar to those of the annelids and the arthropods. In addition, however, investigators have found that they also have several uniquely derived features.

**FIGURE 13.36** *Peripatus.* An onychophoran. Approximately life size.
Courtesy of Carolina Biological Supply Company, Burlington, NC.

## External Anatomy

◆ Examine a preserved or plastic-embedded specimen of *Peripatus* and observe its general appearance.

The body has a velvety surface that is covered by a thin, chitinous **cuticle.** The body is divided into an anterior **head** and a posterior **trunk. Segmentation** is not apparent from the dorsal side, but note the numerous paired, segmentally arranged legs. The short, stumpy legs bear terminal **claws.**

Study the anterior head with one pair of segmented **antennae,** one pair of **simple eyes** (ocelli), and a pair of **oral papillae** just anterior to the **ventral mouth.** Inside the mouth is a pair of sclerotized, hooklike **mandibles** used to hold and to tear prey.

## Internal Anatomy

Internally the onychophorans also show a strange mixture of annelid and arthropod features (figure 13.37). The body cavity is a **hemocoel,** which contains a simple **tubular digestive tract,** two anterior **slime glands,** a pair of **ventral nerve cords,** the **reproductive organs,** and other internal organs. Respiration is accomplished by a branched **tracheal system** with numerous **spiracles** opening to the outside. The **open circulatory system** consists of a dorsal tubular heart with a pair of openings (ostia) in each segment. Blood flows out the ostia into several **blood sinuses** and back to the heart. A pair of **segmental nephridia** with ciliated funnels open through pores at the base of each leg.

Among the annelid characteristics shown by onychophorans are the muscular body wall, segmental nephridia with ciliated openings, and nonjointed legs. Arthropodlike features include the chitinous cuticle, a reduced coelom, an open circulatory system with a large interior hemocoel, and jaws formed from modified appendages.

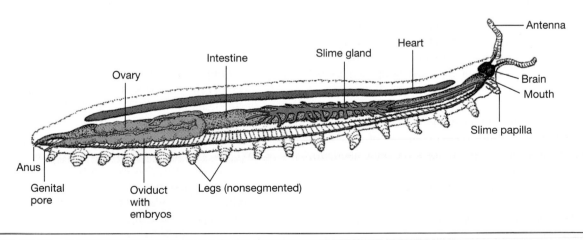

**FIGURE 13.37** *Peripatus,* internal anatomy. Approximately 3× life size.

## Key Terms

**Biramous appendage**   a two-branched appendage characteristic of the crustaceans and certain other invertebrate groups. Consists of a basal portion (protopodite) and two distal branches or rami. The medial (interior) branch is the endopodite, and the lateral (exterior) branch is the exopodite.

**Book gills**   respiratory organs consisting of many thin, highly vascularized sheets of tissue. Found in *Limulus*.

**Cephalothorax**   fused head and thorax regions typical of the Crustacea and the Arachnida.

**Chelicera** (plural: chelicerae)   pincerlike first appendage characteristic of the Chelicerata.

**Cheliped**   one of the first pair of walking legs in the crayfish, lobster, and many other large crustacea. Equipped with a terminal pincerlike claw called a chela.

**Chitin**   a complex structural carbohydrate that forms a chief component of the exoskeleton of many arthropods and other invertebrates. N-acetylglucosamine.

**Compound eye**   a type of image-forming photoreceptor with many individual units, the ommatidia. Typical of insects but also found in many crustaceans.

**Endopodite**   the medial branch (ramus) of the distal portion of a biramous appendage.

**Exopodite**   the lateral (exterior) branch (ramus) of the distal portion of a biramous appendage.

**Exoskeleton**   an external supporting structure or support for the body. Found in arthropods and many other kinds of animals.

**Extensor muscle**   a muscle that extends or straightens out an appendage or other body part.

**Flexor muscle**   a muscle that flexes or bends an appendage or other body part toward the body.

**Green glands**   type of excretory organ found in the crayfish and many other crustacea. Located near the base of the antennae.

**Hepatopancreas**   spongy internal organ of many arthropods that secretes digestive enzymes, absorbs food, and stores food reserves. Formed as an outgrowth of the midgut.

**Larva**   an independent, immature stage of an animal morphologically dissimilar to the adult.

**Metamorphosis**   a change from one form to another in the developmental history of an animal.

**Nymph**   an immature stage in the life history of insects exhibiting incomplete metamorphosis (hemimetabolism). Morphologically similar to the adult.

**Ocellus** (plural: ocelli)   a simple eye found in many invertebrates.

**Protopodite**   the basal portion of a biramous appendage in arthropods.

**Pupa**   a metamorphic stage following the larva stage and preceding the adult stage in insects with holometabolism.

**Serial homology**   the adaptation of a longitudinal series of originally similar organs to perform different functions, as in the appendages of the crayfish and many other arthropods.

**Sexual dimorphism**   significant structural, physiological, and behavioral differences between the male and female members of a species.

**Statocyst**   organ of equilibrium found in certain invertebrates, including crustaceans.

**Statolith**   a sand grain or calcareous granule within a statocyst.

## Internet Resources

Visit the zoology website at http://www.mhhe.com/zoology to find live Internet links of each of the references listed below.

1. Animal Diversity Web, University of Michigan. Phylum Arthropoda. Information about the phylum in general, as well as links to the various classes:
   - Chelicerata
   - Crustacea
   - Uniramia
2. Insect Lithographs. Cool old lithographs of insects—about 60 years old.
3. Insects on WWW. More links than you could ever have time to peruse, from Virginia Tech, Colorado State University, and Iowa State University.
   - Virginia Tech
   - Colorado State
   - Iowa State
4. Arthropods. University of Minnesota. Information about arthropods, and links to various other related sites, including the crayfish dissection home page, and a cockroach dissection home page.
5. Arthropoda. Arizona's Tree of Life web page. Pictures, characteristics, phylogenetic relationships, references on arthropods. Links to specific arthropod groups.
6. BIOSIS Crustacea. This site provides links to information on the paleontology, taxonomy, life history, and ecology of members of most crustacean groups.

## Critical Thinking Questions

1. Why do you suppose the arthropods are the most diverse and among the most abundant of all groups of organisms? Think in terms of cephalization, the specialization of the arthropodan appendages, segmentation, exoskeleton, generally high fecundity (fertility), and the range of food sources utilized.

2. Why do you suppose there are no truly marine insects? (It is true that some gerrids are surface dwellers thousands of miles from shore, but there are no insects truly adapted to marine environments.) Explain your reasoning.

3. Examine the structural and chemical similarities of chitin and cellulose. (You may have to use an introductory biology textbook for this purpose.) Do you see any similarities? What are they? Is this surprising to you, or would you expect it based on evolutionary theory?

4. If crustaceans have modified legs for mouthpart appendages, this must mean that the original segmentation of the head region was composed of more than one segment. Count the mouthparts and estimate how many segments made up the "ancestral" anterior region of crustaceans (this also works for insects).

5. What is the significance of tagmatization (the amalgamation of body segments into distinct sections such as head, thorax, and abdomen)? Discuss the evolutionary importance of tagmatization.

6. Why is sclerotization so important to the exoskeleton of insects and crustaceans? What advantages does sclerotization provide?

## Suggested Readings

Berenbaum, M.R. 1995. *Bugs in the System.* Reading, MA, Addison-Wesley Publishing Co.

Borror, D.J., Triplehorn, C.A., and N.F. Johnson. 1997. *Introduction to the Study of Insects.* 6th ed. Philadelphia: Saunders College Publishing.

Chapman, R.F. 1999. *The Insects: Structure and Function.* 4th ed. Cambridge: Cambridge University Press.

Fitzpatrick, J.F. Jr., Bamrick, J. Cawley, E.T., and W.G. Jaques. 1983. *How to Know the Freshwater Crustacea.* Dubuque, IA: William C. Brown Publishers.

Foelix, R.F. 1982. *Biology of Spiders.* Cambridge, MA: Harvard University Press.

Kaston, B.J. 1978. *How to Know the Spiders.* 3rd ed. Dubuque, IA: William C. Brown Publishers.

Ruppert, E.E. and R.D. Barnes. 1994. *Invertebrate Zoology.* 6th ed. Philadelphia: Saunders College Publishing.

Schram, F.R. 1986. *Crustacea.* New York: Oxford University Press.

# NOTES AND SKETCHES

# NOTES AND SKETCHES

# CHAPTER 14

## Echinodermata

### OBJECTIVES

After completing the laboratory work in this chapter, you should be able to perform the following tasks:

1. List and briefly characterize the five classes of living echinoderms.
2. Cite five significant differences between the echinoderms and each of the following phyla studied previously: Arthropoda, Annelida, and Mollusca.
3. Identify the principal external features of a starfish. Explain the function of the madreporite, pedicellariae, papulae, and tube feet.
4. Identify the parts of the digestive tract of a starfish and explain the function of each part.
5. Identify the parts of the water vascular system of a starfish and give the function of each part.
6. Identify the main external features of a sea urchin.
7. Compare the basic organization of a starfish, a sea urchin, and a sea cucumber.

## Introduction

The Phylum Echinodermata is a unique group of spiny-skinned marine animals, which includes the sea stars or starfish, echinoids, feather stars, sea cucumbers, brittle stars, and sea lilies (figure 14.1). Echinoderms are exclusively marine and are unable to regulate the ionic composition of their body fluids. Nonetheless, echinoderms are widespread in the seas, occurring in shallow waters and in the depths of the sea in all parts of the world.

Echinoderms have inhabited the seas since the early Cambrian era, over 600 million years ago. They flourished in the seas at several times in the past, and many more fossil species have been described (some 20,000) than living species (about 6,000).

Adult echinoderms are easily recognized by their obvious five-part (pentamerous) **radial symmetry,** spines, and calcareous **endoskeleton** made up of many small plates, which may be separate (as in the sea cucumbers) or fused into a rigid framework or **test** (as in a sea urchin or sand dollar). In many of the echinoderms, these mesodermally derived skeletal plates bear protruding spines. Echinoderms have a large enterocoelous coelom lined with a ciliated peritoneum, a portion of which forms into the unique **water vascular system** during embryonic development.

Class Asteroidea

Class Holothuroidea

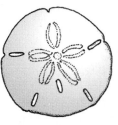

Class Echinoidea

Class Ophiuroidea

Class Crinoidea

**FIGURE 14.1**   Major classes of living echinoderms.

The water vascular system is of great importance to the echinoderms and plays a vital role in their locomotion, attachment, gas exchange, food handling, and sensory perception.

The echinoderms are of special interest because of their close evolutionary relationship to the Phylum Chordata. Strong embryological evidence suggests a close link between the echinoderms and the chordates. This evidence includes their radial indeterminate cleavage, enterocoelous coelom formation, and their endodermal endoskeleton, all of which are chordatelike characteristics. Like chordates, echinoderms are deuterostomes.

# Classification

Some disagreement exists among specialists on the classification of the Phylum Echinodermata, but most recent workers have adopted the following classification of living echinoderms, with three subphyla and five classes represented among the present North American fauna.

## Subphylum Echinozoa

Nonsessile bottom-dwelling (benthic) animals with a rounded or disc-shaped body without arms.

### Class Echinoidea

Sea urchins, heart urchins, and sand dollars. Echinoderms with a rigid test of fused skeletal plates. Body covered with movable spines. Five rows of tube feet (bearing suckers) around the test. Examples: *Arbacia, Lytechinus, Strongylocentrotus* (all sea urchins); *Echinarachnius, Mellita,* and *Dendraster* (sand dollars).

### Class Holothuroidea

Sea cucumbers. Nonsessile soft-bodied animals having a flexible body wall with many tiny, embedded calcareous ossicles; no spines or arms. Body elongated in the oral-aboral axis to a cucumber or picklelike form; a whorl of tentacles surround the mouth. Usually with five or more series of tube feet arranged in longitudinal rows. Examples: *Thyone, Cucumaria, Stichopus.*

## Subphylum Crinozoa

Lilylike or cup-shaped body with branching arms; some species attach to the substrate during all or part of life by a stem; stalk or cirri present; oral surface directed upward.

### Class Crinoidea

Sea lilies, basket stars. Flowerlike echinoderms with a central calyx and five (or multiples of five) branching arms. Ciliated ambulacral groove along the oral surface of the arms serves for food gathering. Some species attach to the sea bottom by a stalk; others free-swimming. Examples: *Cerocrinus* (crinoid), *Hathrometra* (= *Antedon*) (feather star).

## Subphylum Asterozoa

Nonsessile, star-shaped echinoderms, with radial arms and oral surface directed downward.

## Class Asteroidea

Starfish or sea stars. Animals with a star-shaped body, typically with five arms and a flexible skeleton with many calcareous plates. The arms are continuous with and not clearly distinct from the central disc. With ambulacral grooves on the oral side of each arm; tube feet bearing suckers. Examples: *Asterias, Pisaster, Astropecten.*

## Class Ophiuroidea

Brittle stars. Star-shaped echinoderms with arms distinct from the central disc; no ambulacral grooves along the arms; tube feet absent or reduced to sensory organs. Examples: *Amphipholis, Ophiothrix,* and *Ophioderma* (brittle stars); and *Gorgonocephalus* (Pacific basket star).

## Materials List

**Preserved specimens**
  *Asterias*
  *Arbacia*
  *Cucumaria* or *Thyone*
  Aristotle's Lantern (demonstration)
  Starfish, dried to show ossicles (demonstration)
  Representative echinoderms (demonstration)
**Prepared microscope slides**
  Starfish, pedicellaria (demonstration)
  Starfish, dermal branchiae (demonstration)
  Starfish, bipinnaria larva (demonstration)
  Starfish, metamorphosis (demonstration)
**Audiovisual materials**
  Anatomy of the Starfish video

## The Common Starfish: *Asterias*

### Class Asteroidea

*Asterias forbesi* is a common starfish (figure 14.2) found along the Atlantic coast of the United States. Obtain a living or preserved specimen of *Asterias* and observe its radial symmetry. The radial symmetry of adult echinoderms appears to be a secondary condition, since echinoderm larvae are bilaterally symmetrical. Observe a demonstration slide of the bipinnaria larva of the starfish. *In what ways does the radial symmetry of the echinoderms differ from that of the coelenterates?*

The common starfish along the Pacific coast of the United States is *Pisaster,* whose anatomy is similar to that of *Asterias* and can be used in this exercise where available.

### External Anatomy

Specimens preserved for laboratory study often are injected with a colored material to facilitate the study of the water vascular system, including the external tube feet. Keep your specimen wet by adding a little water to the dissecting pan. Note the **central disc** on the upper, or **aboral,** side; the five rays or **arms;** and the **madreporite**—a light-colored circular area near the edge of the disc at the junction of two rays. The madreporite is a calcareous disc that serves as the entrance to the water vascular system. The two arms adjacent to the madreporite comprise the **bivium;** the other three arms make up the **trivium.** The arm opposite the madreporite is referred to as the **anterior** arm.

Note the many spines scattered over the surface of the arms and the central disc. Among the spines there are also many small pincerlike structures located near the bases of

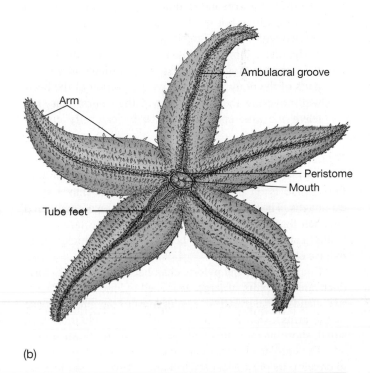

**FIGURE 14.2** Starfish, (*a*) aboral view, (*b*) oral view.
Photograph courtesy of Carolina Biological Supply Company, Burlington, NC.

the spines (see demonstration slide). These structures are the **pedicellariae.**

◆ Carefully scrape a small area of the arm with your scalpel, add the scrapings to a drop of water, and observe the material under the low power of your microscope. *Can you identify any pedicellariae among the scrapings? From their structure, what do you think the function of the pedicellariae might be?*

Also among the spines are numerous **dermal branchiae,** soft, hollow projections of the body wall used for gas exchange (see demonstration slide). Observe the skeletal calcareous plates, or **ossicles,** on a dried starfish specimen.

Note on the oral side of the starfish: the **mouth,** guarded by specialized **oral spines** and surrounded by a soft membrane, the **peristome;** the five **ambulacral grooves,** extending from the mouth and along the middle of each arm; and the numerous **tube feet** in the grooves, extending from the water vascular system. There is also a pigmented **eyespot** and a **sensory tentacle** at the tip of each ray, but these latter structures are usually difficult to find in preserved specimens.

### Internal Anatomy

◆ Carefully examine figures 14.3 and 14.4 for orientation before starting your dissection. Cut off about one-half inch from the tip of the anterior arm. Then carefully cut along each side of the anterior arm to its junction with the central disc along each side and across the top of the arm at the margin of the disc. Observe the hard, calcareous plates, or ossicles, of the skeleton as you cut. Remove the portion of the body wall covering the aboral surface of the arm and examine the large **coelom** within the arm, which houses the internal organs. The coelom is lined with a ciliated **peritoneum,** as mentioned previously, and is normally filled with coelomic fluid. This fluid carries oxygen and dissolved nutrients to various parts of the body. Next, remove the portion of the body wall covering the aboral side of the central disc, but leave the madreporite in place by carefully cutting around it.

### Digestive System

The starfish has a complete digestive system. The mouth of the starfish is on the oral surface and opens into a short esophagus. The esophagus leads to a large, two-chambered stomach that fills most of the central disc. The large, thin-walled cardiac chamber lies below a smaller, thick-walled pyloric chamber.

Emptying into the pyloric chamber through five pyloric ducts are five pairs of large, branched pyloric caeca (digestive glands). The pyloric caeca fill most of the interior space in the arms. Also leading from the pyloric chamber is a small, short intestine that empties through the aboral anus.

The cardiac chamber can be everted through the mouth to envelop its prey. Most starfish are carnivores and feed on many kinds of invertebrates, including clams, oysters,

**FIGURE 14.3** Starfish, partly dissected, aboral view.
Photograph by Ken Taylor.

gastropods, crustaceans, polychaetes, other echinoderms, and small fish. Prey is often engulfed whole, including shell or exoskeleton, and the undigestible parts are later regurgitated. Typically, food is digested in the cardiac chamber of the stomach. Digestive enzymes are secreted principally by the pyloric caeca, which also play a major role in absorption and storage of food materials.

### Water Vascular System

The **water vascular system** (figure 14.5) is a unique feature of the echinoderms. It develops embryologically as a specialized part of the coelom and is lined with a **ciliated epithelium.** The water vascular system has several important functions, including locomotion, feeding, gas exchange, excretion, and sensory perception. The canals of this system are filled with a water vascular fluid that has a chemical composition resembling seawater plus some soluble proteins and a higher concentration of potassium ions. Suspended in the water vascular fluid are numerous **coelomocytes,** amoeboid cells that serve in defense, excretion, and perhaps in other roles as well.

Starfish have a well-developed water vascular system. Among its principal parts is the external **madreporite,** seen previously on the aboral surface, which marks the entrance to a short **stone canal** (figure 14.5). The madreporite is

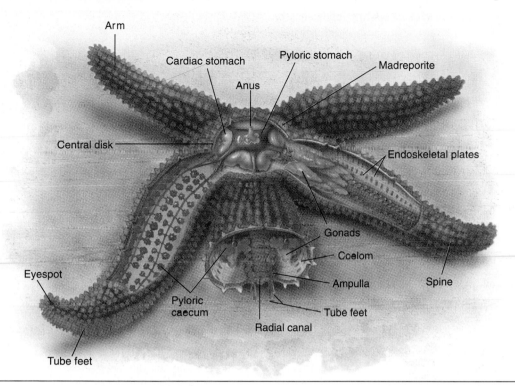

**FIGURE 14.4** Starfish, internal anatomy, aboral view. Approximately life size.

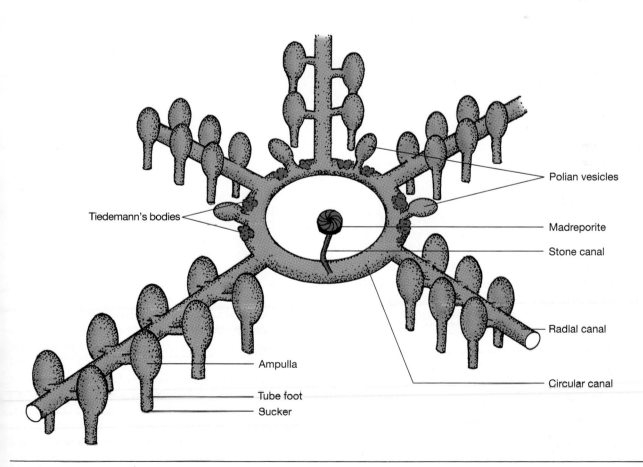

**FIGURE 14.5** Water vascular system of a starfish.

covered by the thin, ciliated epidermis of the body surface, and is believed to function as a pressure regulator for the water vascular system. The stone canal gets its name from the calcareous deposits surrounding it. The stone canal leads to the **circular canal,** which encircles the mouth. Attached to the circular canal are one or more muscular sacs, the **Polian vesicles,** and four or five pairs of irregularly shaped **Tiedemann's bodies.** The Polian vesicles are believed to serve as fluid reservoirs and pressure regulators that maintain proper hydrostatic pressure within the water vascular system. The Tiedemann's bodies produce coelomocytes and may function in the chemical regulation of the coelomic fluid. **Coelomocytes** are amoeboid cells that circulate through the coelomic spaces and canals and ingest foreign materials. There is evidence from some echinoderms that coelomocytes may also be involved in nutrient transport.

Five **radial canals** lead from the circular canal along the top of the **ambulacral groove** leading into each arm. Many short **lateral canals** connect the radial canals with each pair of **tube feet.** Each tube foot consists of a bulblike **ampulla** attached to each suckerlike tube foot.

Within each lateral canal is a valve that closes to allow local pressure changes in the individual tube feet. Contraction of the circular muscles in the wall of the ampulla increases the internal pressure and extends the foot. Contraction of longitudinal muscles in the wall of the tube foot withdraws the foot. These pressure changes within the water vascular system (plus nervous control from the central nervous system) allow coordinated movements of the tube feet and, thus, coordinated locomotion of a starfish.

### Nervous System
The starfish has a very simple nervous system. A circular **nerve ring** surrounds the mouth, and a **radial nerve** extends from the nerve ring into each arm. A simple light-sensitive **eyespot** at the tip of each arm is the only differentiated sense organ; however, sensory cells are scattered throughout the epidermis.

### Gas Exchange and Excretion
Gas exchange is carried on through the thin walls of the **dermal branchiae** extending from the body surface and also through the walls of the tube feet. Excretion of nitrogenous wastes is accomplished mainly by diffusion of ammonia and urea from the gas exchange surfaces and the body wall. Coelomocytes engulf insoluble particulate matter and carry these wastes to the surface where they are discharged. Echinoderms have no differentiated excretory organs.

### Reproductive System
The reproductive system consists of a pair of **gonads** located near the base of each arm (figure 14.4). The sexes are separate in sea stars, but cannot be distinguished except by microscopic examination of the gonad contents.

◆ Take a small amount of tissue from one of the gonads and macerate it in a drop of water on a clean microscopic slide. *Can you identify any large oocytes (eggs) or any flagellated sperm?*

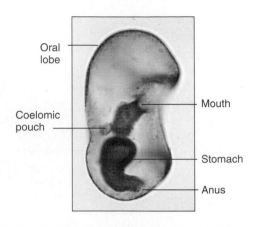

**FIGURE 14.6** Bipinnaria larva, side view. See also figure 4.4.
Courtesy of Carolina Biological Supply Company, Burlington, NC.

From each gonad a duct leads to an external pore on the aboral surface. Eggs and sperm are released by mature sea stars into the seawater, where fertilization occurs, and the zygotes develop into free-swimming **bipinnaria larvae** (figure 14.6).

# A Sea Urchin

## Class Echinoidea
Sea urchins are echinoderms with a **rigid endoskeleton** or **test,** consisting of many fused plates or ossicles. They lack arms and typically are globular or ovoid in shape; their bodies are covered with movable **spines** (figure 14.7). Among the spines are many tube feet and numerous **pedicellariae,** which serve to keep the test clean of debris and fouling organisms.

◆ Study a living or preserved specimen of *Arbacia* or a similar sea urchin. Identify the rigid **test, spines, dermal branchiae,** and **pedicellariae.** Also locate the five series of **tube feet** along five symmetrically placed **ambulacral regions** around the globular body. In a living specimen, the tube feet can extend well beyond the spines.

On the oral side of the urchin (flattened surface), find the **mouth** surrounded by five **calcareous teeth** and a circular oral membrane, the **peristome.** Five pairs of peristomial gills are located around the peristome. The teeth are connected internally to a complex chewing apparatus, called **Aristotle's Lantern,** which anchors the teeth and enables them to scrape algae from submerged rocks along the shore where they live. Although some sea urchins are scavengers, most species are principally herbivorous, and many species feed on algae from the sea floor; they are sometimes called the grazers of the sea. Their relatives, the sand dollars and heart urchins, are adapted for life on soft sea bottoms in which they burrow in search of food and shelter.

**FIGURE 14.7**  Sea urchin, *Arbacia*.
Courtesy of Carolina Biological Supply Company, Burlington, NC.

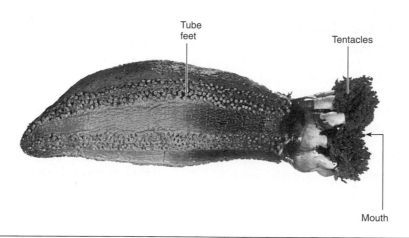

Tube feet

Tentacles

Mouth

**FIGURE 14.8**  Sea cucumber, *Cucumaria*. Notice elongate oral-aboral shape.
Courtesy of Carolina Biological Supply Company, Burlington, NC.

## Some Common Echinoids

*Arbacia punctulata*—purple sea urchin found along the Atlantic coast from Cape Cod to Florida.

*Strongylocentrotus drobachiensis*—green sea urchin found in colder waters north of Cape Cod along the western Atlantic, in northern Europe, and from Washington to Alaska on our Pacific coast.

*Strongylocentrotus purpuratus*—the Pacific purple sea urchin found on surf-swept rocks from California to Alaska.

*Echinarachnius parma*—Northern Atlantic sand dollar found on sandy areas on the sea bottom from Labrador to New Jersey, and also in the Northern Pacific from British Columbia across to Japan.

*Mellita quinquiesperforata*—Keyhole sand dollar abundant from Cape Hatteras to the Caribbean.

# A Sea Cucumber

## Class Holothuroidea

Sea cucumbers are soft-bodied echinoderms with many tiny plates or ossicles embedded in their **leathery body wall.** They are principally filter feeders, collecting plankton with their tentacles, or deposit feeders, collecting food materials mixed with the bottom sediments.

◆ Study a living or preserved *Cucumaria* (figure 14.8) or *Thyone* and identify the **oral tentacles,** the external **tube feet,** and the **leathery body wall.**

## Some Common Holothuroidea

*Thyone briareus*—burrowing sea cucumber found in shallow waters on soft, sandy, or muddy sea bottoms from Cape Cod to the Gulf of Mexico. Tube feet scattered over body surface.

*Cucumaria frondosa*—deeper water, North Atlantic species in which the tube feet are arranged along five distinct ambulacral areas.

*Leptosynapta tenuis*—long, wormlike sea cucumber, modified for burrowing in soft bottom sediments. Lacks regular tube feet.

## Demonstrations

1. Pedicellariae (microscope slide)
2. Dermal branchiae (microscope slide)
3. Bipinnaria larva (microscope slide)
4. Starfish metamorphosis (microscope slide or color transparency series)
5. Preserved or dried Aristotle's Lantern from sea urchin
6. Dried starfish to show ossicles
7. Preserved specimens representing other types of echinoderms, such as crinoids, basket stars, and brittle stars

## Key Terms

**Dermal branchiae**   hollow projections of the body wall that function in gas exchange in starfish and other asteroids.

**Pedicellariae**   small, pincerlike structures found on the surface of many starfish (Class Asteroidea), sea urchins (Class Echinoidea), and brittle stars (Class Ophiuroidea). Serve to keep the body surface free from debris and fouling organisms.

**Radial symmetry**   type of body organization in which all parts of the body are arranged symmetrically around a central axis.

**Test**   a rigid endoskeleton composed of many fused calcareous plates such as exhibited by the sea urchins.

**Water vascular system**   a unique organ system found only in the Phylum Echinodermata. Functions in feeding, attachment, locomotion, gas exchange, and sensory perception. Formed embryologically as a specialized portion of the coelom.

## Internet Resources

Visit the zoology website at http://www.mhhe.com/zoology to find live Internet links of each of the references listed below.

1. Animal Diversity Web, University of Michigan. Phylum Echinodermata. Information about echinoderms, with links to various groups of echinoderms. Nice pictures: check out the magnificent urchin, *Astrophyga magnifica.* It's obvious why it was named!
2. Starfish. External and internal anatomy.
3. Echinoderms. University of Minnesota. Information about echinoderms, and a link to the sea star dissection home page.
4. Introduction to the Echinoderms. University of California at Berkeley Museum of Paleontology. This site provides information on the echinoderm fossil record, life histories, systematics, and morphology. It also provides a great number of links to sites that focus on each of the echinoderm classes.

## Critical Thinking Questions

1. Compare the structural similarities and differences among the Arthropoda, Annelida, and Mollusca. In what important ways do they differ from Echinodermata? What does this suggest about the ancestry of these phyla?

2. Explain how the water vascular system functions. Identify the madreporite, stone canal, circular canals, Tiedemann's bodies, papulae, and tube feet and explain their interactions.

3. The echinoderms are deuterostomes. What significance might this have in the evolutionary sense considering that chordates are also deuterostomes but the "lower" invertebrates are not?

4. Examine the variations in the form of echinoderms, keeping in mind the elongation of the oral-aboral surface (e.g., the Holothuroidea). Can you imagine how a bilateral larvae could give rise to a radial creature, and then in the Holothuroidea evolve an almost elongated, biradial-bilateral body form?

5. Why do you suppose the echinoderms are so well-represented in the fossil record?

## Suggested Readings

Abbott, D.P. 1987. *Observing Marine Invertebrates: Drawings from the Laboratory.* Stanford, CA: Stanford University Press.

Binyon, J. 1972. *Physiology of Echinoderms.* Oxford: Pergamon Press.

Lawrence, J.M. 1987. *A Functional Biology of Echinoderms.* Baltimore: Johns Hopkins University Press.

Nichols, D. 1969. *Echinoderms,* 4th ed. London: Hutchinson University Library.

# NOTES AND SKETCHES

# NOTES AND SKETCHES

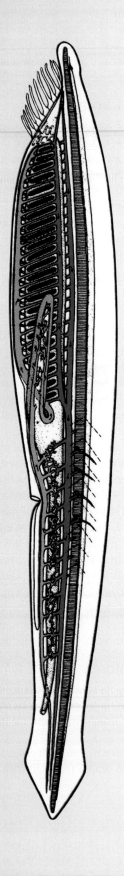

After completing the laboratory work in this chapter, you should be able to perform the following tasks:

1. Describe the life cycle of a tunicate (Subphylum Urochordata) and identify the principal features of a tunicate tadpole larva. Explain the morphological and evolutionary significance of the larva.
2. Identify the main organs in an adult tunicate and discuss the adaptations of the adult for its sessile mode of life.
3. Explain the feeding mechanism of a tunicate and identify the principal structures involved.
4. Identify the main organs in a lancelet and discuss the significance of its basic chordate features.
5. Compare the organization of a tunicate and a lancelet and explain their main similarities and differences.

## Introduction

The Phylum Chordata includes a remarkable range of forms, varying from relatively simple marine animals to highly specialized birds and mammals. Despite the wide range of diversity in the phylum, however, all members exhibit four fundamental chordate characteristics during some stage in their life histories. These distinctive chordate features are (1) a dorsal hollow nerve cord; (2) a notochord—a dorsal supporting rod; (3) paired pharyngeal gill slits; and (4) a postanal tail. The chordates comprise a large, diverse, and distinct phylum of animals that have successfully populated the land, the waters, and the air. This large and important phylum is commonly divided into two groups and three subphyla.

## Classification

### Group Acrania (Protochordata)

Chordates without a cranium or braincase.

#### Subphylum Urochordata (Tunicates or Sea Squirts)

Animals with a well-developed notochord and dorsal nerve cord in the free-swimming larva; specialized adults: sessile or planktonic, and lacking a notochord and dorsal nerve cord. Examples: *Molgula* (sea grape), *Styela, Amaroucium* (sea pork).

### Subphylum Cephalochordata (Lancelets)

Elongate, fishlike chordates with a persistent notochord and dorsal nerve cord. Example: *Branchiostoma (Amphioxus).*

## Group Craniata

Chordates with a cranium enclosing the brain and sense organs of the head; exhibit substantial cephalization.

### Subphylum Vertebrata (Vertebrates)

Chordates with a backbone, skull, brain, and kidneys.

***Class Agnatha (Lampreys and Hagfish).*** Fishlike vertebrates without jaws or appendages. Example: *Petromyzon* (lamprey), *Myxine* (hagfish).

***Superclass Elasmobranchiomorphii (Class Chondrichthyes) (Sharks and Rays).*** Cartilaginous fishes, usually with external placoid scales. Examples: *Squalus* (shark), *Raja* (skate), *Chimaera* (ratfish).

***Class Osteichthyes (Bony Fishes).*** Fishes with skeleton primarily of bone, usually with swim bladder or lungs. Examples: *Amia* (bowfin), *Perca* (perch), *Lepistoseus* (gar), *Micropterus* (bass).

***Class Amphibia (Salamanders, Frogs, and Toads).*** Four-limbed vertebrates (tetrapods) with a soft, moist skin; three-chambered heart; eggs enclosed in gelatinous covering; fertilization and development usually restricted to fresh water. Adults usually aquatic or semiaquatic. Examples: *Rana* (frog), *Bufo* (toad), *Ambystoma* (salamander).

***Class Reptilia (Lizards, Snakes, Turtles, and Alligators).*** Four-limbed vertebrates with a dry, cornified skin; eggs enclosed in a protective shell resistant to drying (amniote egg); most with three-chambered heart. Examples: *Anolis* (lizard), *Chrysemys* (turtle), *Crotalus* (rattlesnake), *Alligator.*

***Class Aves (Birds).*** Winged vertebrates with feathers and constant high body temperature. Examples: *Sturnus* (starling), *Cyanocitta* (blue jay), *Corvus* (crow).

***Class Mammalia (Mammals).*** Warm-blooded vertebrates; body covered with hair; young nourished by milk produced by female. Examples: *Homo* (human), *Sus* (pig), *Equus* (horse), *Canis* (dog).

In this chapter we will study representatives of the first two subphyla, the urochordates and the cephalochordates.

## Materials List

Preserved specimens
  *Molgula*
  *Branchiostoma*
  Representative tunicates
Prepared microscope slides
  *Amaroucium,* tadpole larva
  *Branchiostoma,* whole mount
  *Branchiostoma,* cross section

# Subphylum Urochordata, Tunicates or Sea Squirts

Urochordates are interesting, important, and unusual animals. They are interesting and important because they represent the simplest living chordates. About 1,250 species are currently recognized. They are unusual because adult urochordates became adapted to a special mode of life as sedentary, filter-feeding, marine animals, and, in the process, have lost two of the four basic chordate features. Larval urochordates, however, have retained all four chordate features and clearly establish the urochordates as legitimate ancestors of the higher chordates, including fishes, birds, mammals, and humans.

## The Tunicate Larva

The larval stages of tunicates are of special importance because they clearly exhibit the four fundamental chordate characteristics: **tubular nerve cord, notochord, pharyngeal gill slits,** and **postanal tail.** The first two of these features are lost in the adult tunicate, presumably because of the specialization of the adult due to its sessile mode of life.

Tunicate larvae are often called **"tadpole larvae"** because of their superficial resemblance to the tadpole larvae of amphibians.

◆ Study a prepared microscope slide with a whole mount of the larva of *Amaroucium,* or a similar tunicate larva, and observe its characteristic form (figure 15.1).

Note the **muscular tail** and thickened body. At the anterior end of the body, find the adhesive papillae with which the larva attaches to the substrate prior to metamorphosing into a sessile adult. Near the dorsal surface is a **sensory vesicle** with two conspicuously pigmented sense organs—a light-sensory **ocellus,** or eyespot, and a **statolith** that serves as a balancing organ.

Anterior to the darkly pigmented sensory vesicle, locate the **incurrent siphon** through which water is pumped into the pharynx. The **excurrent siphon** is found slightly posterior to the sensory vesicle, near the attachment of the tail. Locate also the **gill slits** in the wall of the pharynx.

The **notochord,** the **dorsal nerve cord,** and conspicuous **muscle bands** are present in the tail. During metamorphosis, the larva settles to the bottom, attaches by means of a secretion of glands in the adhesive papillae, and partly reabsorbs its tail. The notochord, nerve cord, and muscles are broken down, and several new organs appear in the body of the tunicate to complete the transformation into the adult. The life cycle of a tunicate is illustrated in figure 15.1.

## The Adult Tunicate

Adult tunicates may be solitary or colonial, sessile or planktonic. The most familiar ones, however, are the sessile forms found attached to rocks, jetties, and pilings in shallow seas (figure 15.2).

The typical structure of an adult tunicate is well-illustrated by *Molgula,* the sea grape, found in many places along the Atlantic coast.

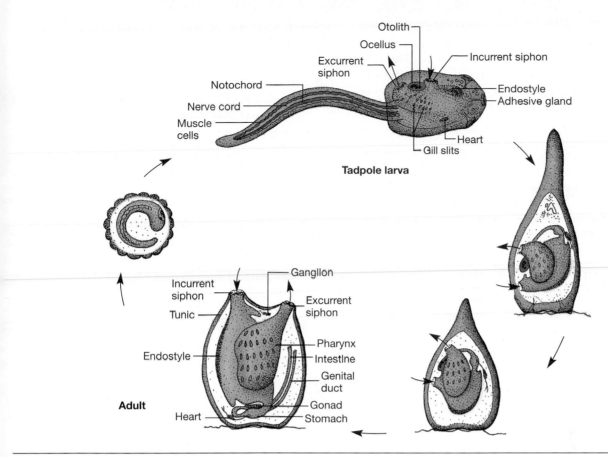

**FIGURE 15.1** Tunicate, life cycle. Stages not to scale.

**FIGURE 15.2** A colonial tunicate.
Photograph courtesy of Carolina Biological Supply Company, Burlington, NC.

◆ Study the demonstration of a partially dissected specimen of *Molgula* and note the tough, fibrous outer covering called the **tunic** (figure 15.3). This outer covering is unusual because it is composed partly of cellulose, a carbohydrate normally produced only by plants.

Projecting from the globular body are two siphons: the incurrent siphon is anterior, and its opening is divided into six lobes. The more posterior excurrent siphon has a square open-

ing. Locate the numerous gill slits within the pharynx. Ventral to the pharynx are the short esophagus, the bulbous stomach, and the beginning of the intestine. The intestine continues dorsally and ends at the anus just below the excurrent siphon.

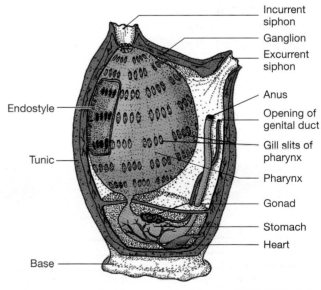

**FIGURE 15.3** Tunicate, adult. Part of the tunic and pharynx removed.

*Molgula,* like most other tunicates, is a specialized filter feeder. Water is drawn in through the incurrent siphon by the action of cilia lining the internal chambers. The water is drawn into the large pharynx, and minute organisms suspended in the water (plankton) are trapped in the mucus on the pharyngeal walls as the water passes out through the many gill slits.

Special ciliary currents collect the food within the pharynx and pass it into the esophagus and on to the stomach. Water leaving the pharynx passes into the atrium, a large cavity surrounding the pharynx, and from the atrium the water flows out through the excurrent siphon.

The **endostyle,** a ciliated groove located on one side of the large pharynx, is of special importance. The endostyle secretes a stream of mucus that is carried along by the motion of the cilia to form a sheet that traps small food particles filtered from the incoming seawater. The mucous sheet with its captured food particles is then passed to the esophagus and into the stomach and through the digestive system. The endostyle has been found to be the forerunner of the vertebrate thyroid gland.

The circulatory system in *Molgula,* as in most tunicates, consists of a ventral heart located near the stomach. Attached to the heart are two large vessels that carry blood to other organs. The tunicate heart is unique in that it provides two-way propulsion. The heart first pumps blood in one direction, then reverses and pumps blood in the opposite direction.

In the adult tunicate, the dorsal hollow nerve cord and the notochord are absent. Their loss is generally believed to represent a specialization of the adult tunicate for its peculiar sessile mode of life.

# A Lancelet: *Branchiostoma*

## Subphylum Cephalochordata

The Subphylum Cephalochordata consists of a few species of small fishlike animals called lancelets (figure 15.4), which live in shallow seas in many parts of the world. In certain areas, the lancelets are sufficiently abundant that they are used as human food. The common American lancelets are traditionally called *Amphioxus* although most specialists now use the generic name *Branchiostoma.*

### External Anatomy

◆ Study first a preserved specimen and observe its general form and external features. Then obtain a prepared microscope slide with a stained whole mount of a lancelet to study its internal anatomy.

Mature lancelets are usually between five and eight mm in length, but smaller, immature forms are normally used in making microscopic whole mounts.

In the preserved specimen, note the slender, elongate shape of the animal, the absence of a distinct head, and the lack of paired fins or limbs.

◆ Handle the specimen with care and do not dissect it. Return it intact to the proper container when you have completed your study.

Refer to figure 15.4 and identify the anterior **rostrum** and the **oral hood** bordered by a fringe of ciliated **oral cirri** enclosing the large **vestibule.** The **mouth** is an opening in a membrane, the **velum,** located at the rear of the vestibule. Surrounding the mouth are several **velar tentacles.** *How many?* Cilia arranged in bands along the walls of the vestibule form the "wheel organ," which generates water currents and carries seawater containing suspended food organisms into the mouth.

### Internal Anatomy

Behind the mouth is a large pharynx with many diagonal **gill slits** on each side. Between the gill slits are **gill bars,** each supported by a thin cartilaginous rod (figure 15.5). The pharynx plays an important role both in feeding and in gas exchange, but its primary function is in feeding. The gas exchange function is clearly secondary. Posterior to the pharynx is a straight intestine, which ends at the subterminal anus. A slender pocket, the **midgut caecum** (believed by some workers to be homologous with the vertebrate liver), opens on the ventral side of the intestine

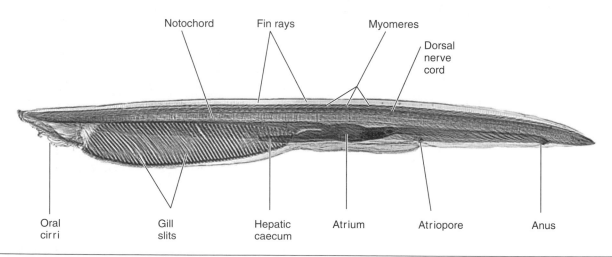

Notochord    Fin rays    Myomeres    Dorsal nerve cord

Oral cirri    Gill slits    Hepatic caecum    Atrium    Atriopore    Anus

**FIGURE 15.4** Lancelet, whole mount. Magnification 3×.
Courtesy of Carolina Biological Supply Company, Burlington, NC.

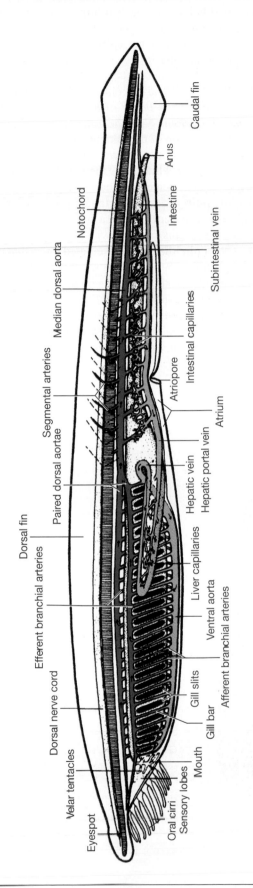

Labels on the figure (top to bottom / left to right):

Caudal fin
Anus
Notochord
Intestine
Median dorsal aorta
Subintestinal vein
Intestinal capillaries
Segmental arteries
Atriopore
Paired dorsal aortae
Atrium
Efferent branchial arteries
Hepatic vein
Hepatic portal vein
Dorsal fin
Liver capillaries
Ventral aorta
Dorsal nerve cord
Gill slits
Afferent branchial arteries
Gill bar
Velar tentacles
Mouth
Eyespot
Oral cirri
Sensory lobes

**FIGURE 15.5** Lancelet, circulatory system. Small arrows in blood vessels show direction of blood flow.

near the junction of the pharynx and intestine and extends forward.

Locate the **dorsal, caudal** (tail), and **ventral fins.** Note the short **fin rays** composed of connective tissue within the fins. Observe the **atriopore,** a midventral opening located anterior to the ventral fin. Water taken into the pharynx passes out through the gill slits into the **atrium** and out of the atrium via the atriopore (figure 15.4).

Lancelets, like the tunicates studied earlier, are filter feeders. Seawater containing planktonic organisms is drawn by ciliary currents through the mouth into the pharynx where food particles are trapped by mucous secretions. The mucus, containing trapped food particles, is swept posteriorly to the intestine, and the water passes out of the pharynx through the lateral gill slits. Gas exchange also occurs as the water passes through the gill slits and past the gill bars, which are highly vascularized with blood vessels.

Note the conspicuous V-shaped structures along each side of the body; these are the muscle segments, or **myomeres.** Contraction of these muscles produces a side-to-side lateral bending of the body, which aids the lancelet in swimming and burrowing in the bottom sediments where it commonly dwells.

Locate the dorsal **notochord,** which extends longitudinally just dorsal to the pharynx and intestine. The notochord is a thin, cartilaginous rod surrounded by a sheet of connective tissue. The contraction of the myomeres against the rigidity of the notochord produces the lateral swimming movements of the body that propel the lancelet forward. Find the **dorsal nerve cord** just ventral to the dorsal fin and dorsal to the notochord. At its anterior end is a slight enlargement, the **cerebral vesicle,** a very primitive sort of "brain."

The circulatory system of *Branchiostoma* consists of a network of elastic vessels similar to that found in the higher chordates, but it lacks a distinct heart. It is difficult to study the details of the circulatory system except by the dissection of specially prepared specimens, but portions of the circulatory system can be observed in the microscopic cross sections to be studied later. Figure 15.5 shows the principal blood vessels and the general pattern of circulation in a lancelet.

Blood from the digestive tract is collected by the **subintestinal vein,** which leads to the **hepatic portal vein,** which, in turn, carries the blood to the midgut caecum. The **hepatic vein** leaves the liver and leads to the **ventral aorta** below the pharynx. Numerous **afferent branchial arteries** (each with a contractile bulb at its base) branch from the ventral aorta and carry the blood upward to the gill bars where it is oxygenated. Pulsations of the ventral aorta and of the enlargements at the bases of the afferent branchial arteries appear to aid in pumping the blood through the system. From the gills, the blood is transported by the **efferent branchial arteries** and is collected in the **paired dorsal aortas** (right and left) above the gills. The two dorsal aortas (figure 15.5) join posteriorly to form a single **median dorsal aorta** just behind the pharynx. This latter vessel carries oxygenated blood posteriorly to the body tissues and the intestine to complete the circuit.

## Cross Sections

Prepared microscopic cross sections will help greatly to supplement your observations of the preserved specimens and whole mounts, and will improve your understanding of the anatomy of *Branchiostoma*.

◆ Study several cross sections from different regions of the body and attempt to identify as many of the internal and external structures as possible.

The following description should aid you in identifying various structures in the cross section. It is based upon a cross section through the pharyngeal region as illustrated in figure 15.6.

Observe the two-layered **skin,** with an outer **epidermis** consisting of a single layer of columnar epithelium, and an underlying **dermis,** a thin layer of gelatinous connective tissue. Find the **dorsal fin** and **fin ray,** the **myomeres** bounded by the myosepta of connective tissue, the **notochord,** and the **dorsal nerve cord** above it. Locate the **central canal** within the dorsal nerve cord. Observe also the nerve cells and fibers of the dorsal nerve cord and the **spinal nerves** (not present in every section because of variations in the plane of sectioning).

Observe the laterally compressed pharynx and the numerous **gill slits** between the **gill bars.** The ventral ciliated groove is the **endostyle,** or hypobranchial groove, and the dorsal ciliated groove is the **epipharyngeal groove.** Both play an important role in the trapping and transporting of food particles.

The chamber around the pharynx is the **atrium.** *What tissue layer forms the lining of the atrium? What tissue layer lines the coelom?* Note the **gonads** extending into the atrium and the section of the midgut caecum also extending into it. The paired cavities above and lateral to the pharynx are portions of the **coelom,** as is the small cavity found below the endostyle. Locate the **blood vessels** in your section, including the **paired dorsal aortae** above the pharynx and the **single median ventral aorta** below it. Also, find the **hepatic vein** or veins closely associated with the midgut caecum. The **nephridia** are small ciliated ducts that connect the dorsal portions of the coelom with the atrium. The two **metapleural folds** at the two sides on the ventral surface of the body should be apparent.

Compare the structures observed in the cross section through the pharyngeal region with a cross section through a more posterior region, as illustrated in figure 15.7.

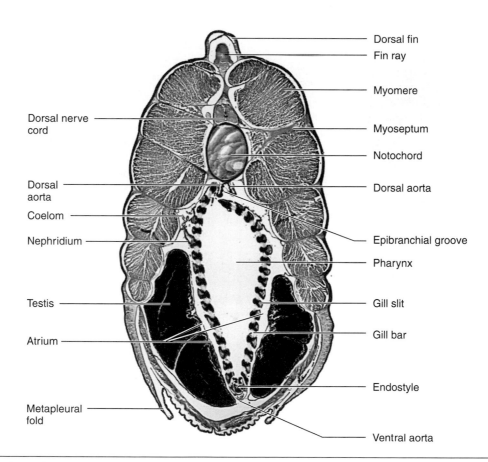

**FIGURE 15.6**  Lancelet, male, cross section through pharyngeal region.
Courtesy of Carolina Biological Supply Company, Burlington, NC.

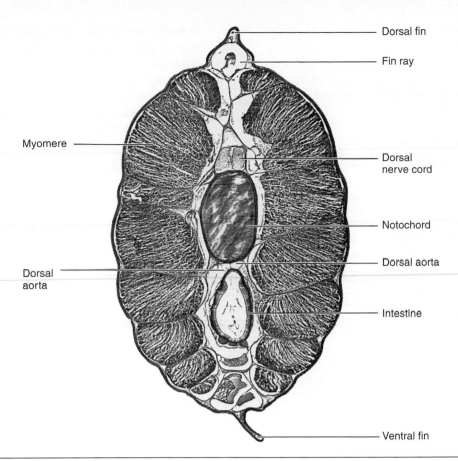

- Dorsal fin
- Fin ray
- Myomere
- Dorsal nerve cord
- Notochord
- Dorsal aorta
- Dorsal aorta
- Intestine
- Ventral fin

**FIGURE 15.7**  Lancelet, cross section through intestinal region.
Courtesy of Carolina Biological Supply Company, Burlington, NC.

## Demonstrations

1. Preserved lancelets
2. Specimens of solitary and colonial tunicates
3. Specimens of larvae and adult Urochordates

## Key Terms

**Endostyle**  a ciliated groove located on the ventral surface of the pharynx in tunicates and lancelets. Functions in the capture of food particles. Homologous with the thyroid gland of vertebrates.

**Gill slits**  paired openings in the lateral walls of the pharynx in urochordates and cephalochordates. Important in food capture and secondarily in gas exchange.

**Myomeres**  segmented muscle blocks arranged longitudinally along the dorsal portion of the lancelet body. Contraction of the myomeres, combined with the relatively stiff notochord, causes the lancelet body to flex, and produces effective lateral swimming and burrowing movements.

**Notochord**  a stiff supporting cartilaginous rod of mesodermal origin found in the tunicate tadpole larva and in adult cephalochordates. Absent in adult tunicates.

**Tadpole larva**  characteristic larval form of the tunicates with a notochord, dorsal tubular nerve cord, and paired gill slits. Superficially resembles an amphibian tadpole.

## Internet Resources

Visit the zoology website at http://www.mhhe.com/zoology to find live Internet links of each of the references listed below.

1. Animal Diversity Web, University of Michigan. Phylum Chordata. General characteristics of chordates, with the following links to urochordates and vertebrates.
   Links to classes:
   - Urochordata
   - Vertebrata
2. Introduction to the Cephalochordata. University of California at Berkeley, Museum of Paleontology. Images, photos, systematics, more information, and links.
3. Comparative Anatomy Chart. A lengthy table that compares various systems of the earthworm, the frog, the snake, the shark, the perch, the pigeon, and the pig.

## Critical Thinking Questions

1. Describe in detail the life cycle of the tunicate and explain why the larval form is believed to be of such morphological and evolutionary significance.

2. How does the adult tunicate vary in morphology from the larval form?

3. Compare the organization and morphology of a tunicate and a lancelet and explain their similarities and differences.

4. Discuss the advantages of filter feeding in a marine environment. Is this a common form of feeding? Identify the groups studied so far in which filter feeding has arisen.

5. Describe the main chordate features and explain the significance of these features.

6. Is the "tadpole" larva of tunicates really a tadpole? Why or why not? Do you think we should not use the word "tadpole" in this context? Support your reasoning.

## Suggested Readings

Barrington, E.J.W. 1965. *The Biology of Hemichordata and Protochordata*. Edinburgh: Oliver and Boyd.

Bone, Q. 1979. *The Origin of Chordates*. Oxford Biology Readers, No. 18. New York: Oxford University Press.

Stokes, M.D., and N.D. Holland. 1998. "The Lancelet." *American Scientist* 86:552–560.

# NOTES AND SKETCHES

# CHAPTER 16

## Shark Anatomy

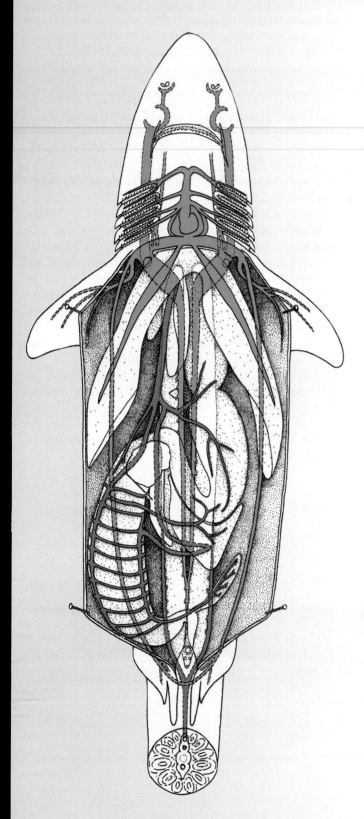

## OBJECTIVES

After completing the laboratory work in this chapter, you should be able to perform the following tasks:

1. Identify the principal external features of the dogfish shark.
2. Locate the pelvic and pectoral girdles of the shark and explain their function.
3. Locate and identify the parts of the digestive system of the shark and explain the function of each part.
4. Locate the parts of the male and female reproductive systems of the shark and give the function of each part.
5. Locate the principal arteries and veins of the shark and explain the pattern of blood circulation.
6. Demonstrate the parts of the heart and explain the function of each part.
7. Describe the hepatic and renal portal circulatory systems and trace their paths in a specimen.
8. Describe the pattern of branchial circulation in a shark, explain its importance, and point out the chief blood vessels involved.
9. Identify the main parts of the shark brain.
10. Locate the eleven pairs of cranial nerves on a specimen and list their names.

## The Dogfish Shark: *Squalus acanthias*

Sharks, skates, rays, and chimaeras are ancient fishes with a cartilaginous endoskeleton, biting jaws, paired appendages, and a tough, leathery skin that is usually covered with placoid scales. These fishes belong to the Superclass Elasmobranchiomorphii (Class Chondrichthyes) of the Subphylum Vertebrata and are commonly called elasmobranchs (because of their exposed gill openings) or cartilaginous fishes (because of the nature of their skeleton). Most of the members of this group are marine, although a few species have secondarily invaded fresh waters, as demonstrated by the freshwater sharks of Lake Nicaragua.

Paleontological studies of ancient sharks have revealed that this group evolved from ancestors with bony skeletons that inhabited fresh waters. Therefore, both the cartilaginous skeleton and the marine habitat of the elasmobranchs must be regarded as **specialized** (or secondarily derived) rather than as ancestral characteristics. Nonetheless, despite these specialized features, the elasmobranchs clearly illustrate many basic vertebrate features, and the shark has long been studied in zoology laboratories for this reason.

The spiny dogfish, *Squalus acanthias,* is a small shark (figure 16.1) common in shallow coastal waters on our Atlantic and Gulf coasts. A similar small shark from the Pacific coast, sometimes called *Squalus suckleyi* (which some authorities consider to belong to the same species as the Atlantic form), is also commonly used for laboratory study. Mature specimens usually range from 1–3 meters in length, although smaller, immature specimens are normally used for laboratory study. Formalin-preserved specimens with the arterial and/or venous systems injected with latex are most satisfactory for dissection.

## Materials List

### Preserved specimens
*Squalus acanthias,* mounted dissection
*Squalus,* pregnant female with embryos in uterus (demonstration)
*Squalus,* skeleton (demonstration)
Bony fish skeleton (demonstration)
Representative elasmobranchs (demonstration)

### Prepared microscope slides
Shark, placoid scales (demonstration)
Shark, retina (demonstration)

### Audiovisual materials
Anatomy of the Shark video

## External Anatomy

◆ Obtain a preserved dogfish shark and study the general form of the body; note the broad, flat **head,** the tapered **trunk,** and the laterally compressed **tail.** The surface of the body is covered with many tiny **placoid scales.** Rub your finger lightly over the surface of the skin and feel the sandpaperlike texture caused by the minute spines borne on each scale. *In which direction do the spines point? What is the significance of this orientation of the scales?*

Locate the paired **pectoral** (anterior) and **pelvic** (posterior) **fins.** In male sharks, the pelvic fins become enlarged and their inner borders are modified to form long, rodlike **claspers,** which aid in mating. Note also the two median **dorsal fins,** each with a sharp spine at its anterior edge, and the large **caudal** (tail) **fin.** The caudal fin consists of two lobes, a larger **dorsal lobe** and a smaller **ventral lobe.** This type of asymmetrical caudal fin with unequal dorsal and ventral lobes is called a **heterocercal tail.** Note how the end of the vertebral column curves upward into the dorsal lobe of the caudal fin.

On the head, find the ventral, curved, slitlike **mouth,** and anterior to the mouth locate the two **nostrils.** Inside the mouth note the numerous sharp-edged **teeth,** believed to be evolutionarily derived from placoid scales. The **eyes** are located on the sides of the head. Note the rudimentary **eyelids.**

Slightly behind and above the eyes, find the two **spiracles**—openings for water intake. The spiracles lead into the pharynx and represent a modified first pair of **gill slits.** Behind the mouth and near the ventral surface, find the five pairs of **external gill slits,** a fundamental chordate characteristic. Internally, the gill slits open into the **pharynx.**

Locate also the two light-colored **lateral lines** running posteriorly from the region of each spiracle to the tail on each side of the shark and the **cloaca opening** located ventrally between the two pelvic fins. The lateral lines of the shark represent a special kind of sensory system found only in fishes and in some larval amphibians. It appears to function in the perception of water movements, pressure changes, and current changes, and thus aids in orientation and locomotion.

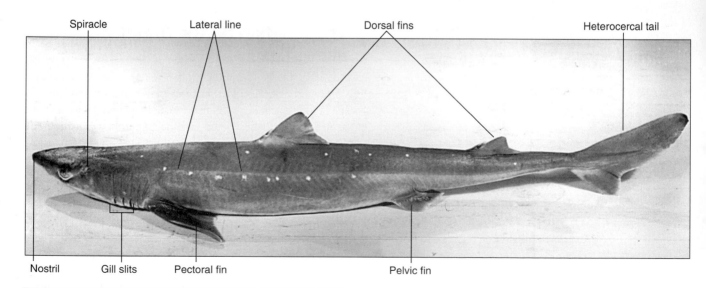

**FIGURE 16.1**  Shark, lateral view.
Courtesy of Carolina Biological Supply Company, Burlington, NC.

# Internal Anatomy

◆ Take care in your dissection to avoid damaging structures important for subsequent study. Do not tear, cut, or pierce parts until you are sure that you know what you are doing. Never cut and discard parts until you are sure of their identity and know that you will have no further use for them.

Be conservative in your cutting; often you will find that structures can be separated neatly and distinctly by teasing them free with a blunt instrument (a blunt probe, back of a scalpel blade, or even the handle of a scalpel) rather than by cutting. When cutting is necessary, make clean incisions with sharp instruments; dull scissors or a dull scalpel will tend to tear rather than to cut tissues and will lead to unsatisfactory results. Read the directions thoughtfully and follow them carefully *before* you begin your dissection.

## Coelom and Visceral Organs

◆ Place your specimen on its back and carefully locate with your fingers the cartilaginous **pectoral** and **pelvic girdles,** which support the pectoral and pelvic fins, respectively (figure 16.2). Also, using the same figure as a guide, estimate the relative thickness of the body wall in your specimen. Carefully make an incision through the body wall along the midventral line from the pectoral girdle backward through the pelvic girdle, cutting to one side of the cloaca and ending your incision at a point just posterior to it.

If the interior of the body cavity appears to be oily, rinse it out carefully with **cold** water. Take care not to disturb the position of the internal organs during the washing.

◆ Put the waste water from your washing in the container designated by your laboratory instructor for toxic liquid waste since it contains some of the chemicals used to preserve the sharks.

Now make two transverse incisions through the body wall, one to the rear of the pectoral fins and one anterior to the pelvic fins, each about 2 inches in length. Pin or tie back the flaps of tissue to expose the large coelomic cavity containing the visceral organs. Carefully study figures 16.2, 16.3, and 16.4, and identify the various structures visible within the coelom. Note the location, relative size, shape, color, and texture of each structure. During your subsequent study, try to relate each structure with its principal function or functions.

The coelom of the shark is divided into two portions, the **pericardial cavity,** found anterior to the pectoral girdle, and the **pleuroperitoneal cavity,** found posterior to the pectoral girdle. These two portions of the coelom are separated by a thin partition, the **transverse septum.**

The smooth lining tissue of the coelom is the **peritoneum,** which also covers the surface of the various organs suspended within the coelom. Dorsally, the peritoneum continues as a double epithelial membrane, the **dorsal mesentery** (figure 16.5), which supports the digestive tract within the coelom.

Locate the large esophagus anterior to the J-shaped **stomach.** The stomach is divided into a larger anterior **cardiac** region and a smaller posterior **pyloric** region (figure 16.6). The pyloric region lies beyond a sharp bend in the stomach and constitutes the lower portion of the J. The constriction between the pyloric region of the stomach and the small intestine is the **pylorus.** A sphincter muscle in this region controls the movement of food from the stomach into the **small intestine.**

The anterior segment of the small intestine is the **duodenum**—a short and narrow portion. The posterior, longer, segment of the small intestine is the **ileum.** Find the long **bile duct** extending from the liver to the duodenum. Within the ileum is the spiral valve, which aids the process of digestion in two ways: (1) it slows the passage of food through the ileum and gives more time for digestive enzymes to act, and (2) it increases the surface area for absorption of released nutrients.

Carefully make a longitudinal incision in the wall of the ileum with a scalpel to examine the spiral valve. Make a similar incision in the wall of the stomach to examine the inner lining of the stomach and the stomach contents. *Do you find any food material inside the stomach? What is the condition of the stomach contents? Can you identify any of the food materials? What inferences can you make from your observations about the diet of the shark and whether the shark had fed shortly before being captured and preserved?*

Posterior to the small intestine the digestive tract is continued by a short and narrow **colon,** which connects with a short **rectum.** Attached dorsally at the junction of the colon and rectum is a blind pocket called the **rectal gland** (figure 16.5), which plays an important role in maintaining the proper salt balance in the blood of sharks. The rectal gland secretes a fluid consisting mainly of a concentrated solution of sodium chloride. The rectum discharges into the cloaca, a common chamber into which the ducts of the digestive and urogenital systems also terminate. Technically, the opening from the rectum into the cloaca is the **anus,** and the opening from the cloaca to the exterior is the **cloacal opening** (also called the vent).

Also associated with the digestive system are two large digestive glands, the **liver** and the **pancreas.** The liver consists of three lobes, two long lobes on the **right** and **left** sides, and a shorter **median** lobe. Find the thin-walled, greenish **gallbladder** along the margin of the median lobe and the common bile duct leading from the anterior end of the gallbladder to the small intestine. The **pancreas** of the shark consists of two distinct parts—a round, flattened **ventral lobe** attached to the surface of the duodenum and a long, narrow **dorsal lobe** lying between the pyloric portion of the stomach and the duodenum.

Another large organ, the **spleen,** is also found attached to the stomach, although it is a part of the circulatory system rather than the digestive system. Locate the dark, triangular spleen closely applied around the outer curvature of the stomach.

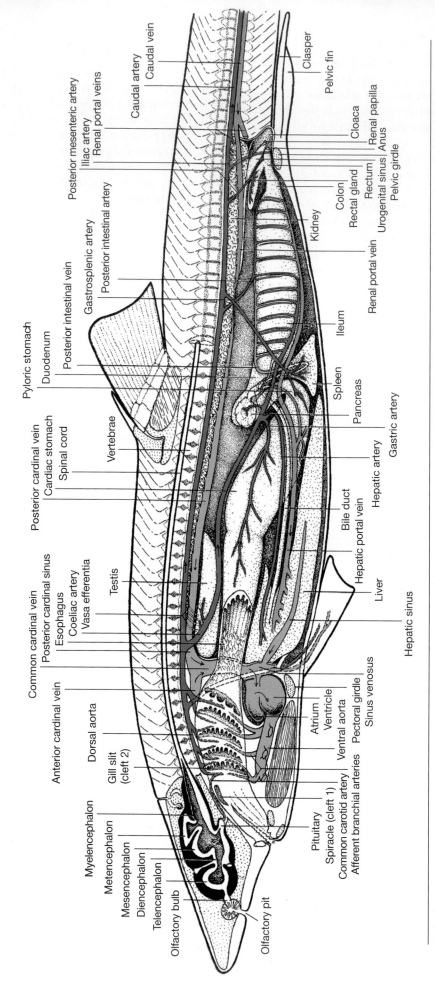

**FIGURE 16.2**  Shark, sagittal section. Oxygenated blood is red, deoxygenated blood is blue, and portal systems are yellow.

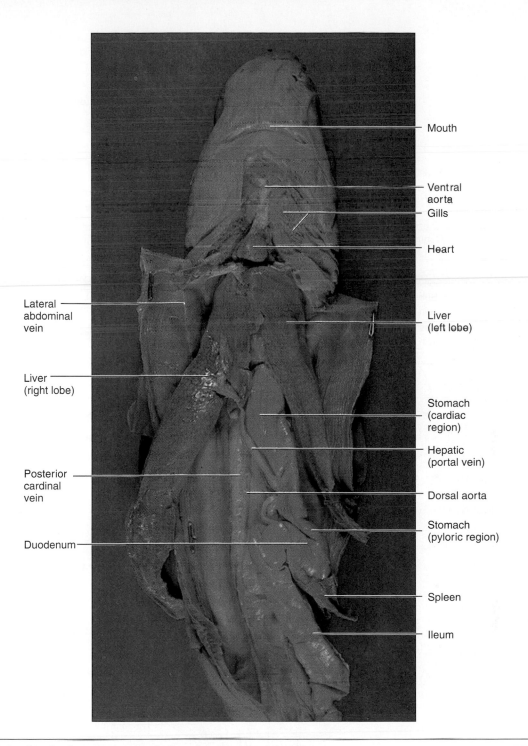

**FIGURE 16.3** Shark, dissected, ventral view showing internal organs.
Photograph by Ken Taylor.

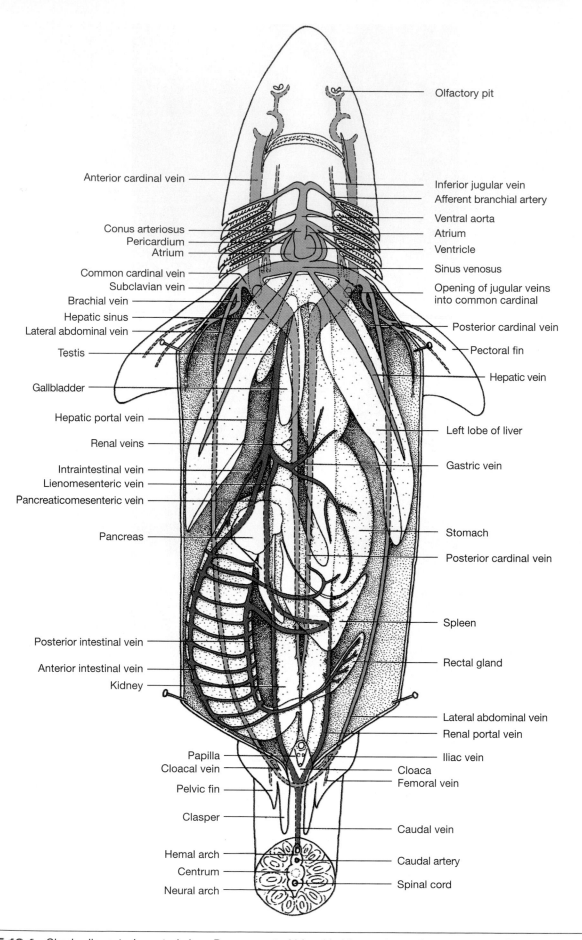

Olfactory pit

Anterior cardinal vein

Inferior jugular vein
Afferent branchial artery

Ventral aorta

Conus arteriosus
Pericardium
Atrium

Atrium
Ventricle

Sinus venosus

Common cardinal vein
Subclavian vein

Opening of jugular veins
into common cardinal

Brachial vein

Hepatic sinus

Posterior cardinal vein

Lateral abdominal vein

Testis

Pectoral fin

Hepatic vein

Gallbladder

Hepatic portal vein

Left lobe of liver

Renal veins

Intraintestinal vein

Gastric vein

Lienomesenteric vein

Pancreaticomesenteric vein

Pancreas

Stomach

Posterior cardinal vein

Spleen

Posterior intestinal vein

Rectal gland

Anterior intestinal vein

Kidney

Lateral abdominal vein

Renal portal vein

Papilla

Iliac vein

Cloacal vein

Cloaca

Pelvic fin

Femoral vein

Clasper

Caudal vein

Hemal arch

Caudal artery

Centrum

Spinal cord

Neural arch

**FIGURE 16.4**   Shark, dissected, ventral view. Deoxygenated blood is blue and portal systems are yellow.

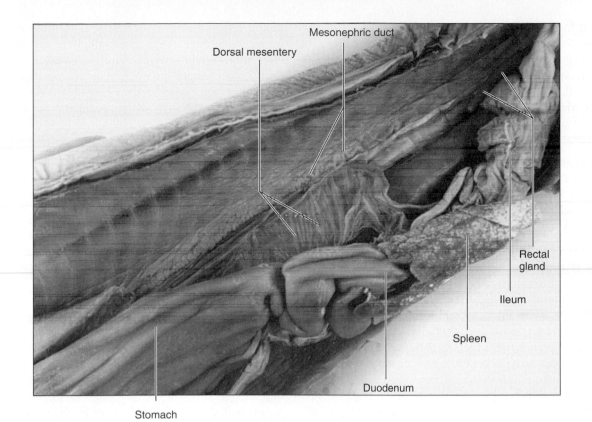

**FIGURE 16.5**  Shark, dissected, showing dorsal mesentery and abdominal organs of posterior region.
Photograph by Carol Majors.

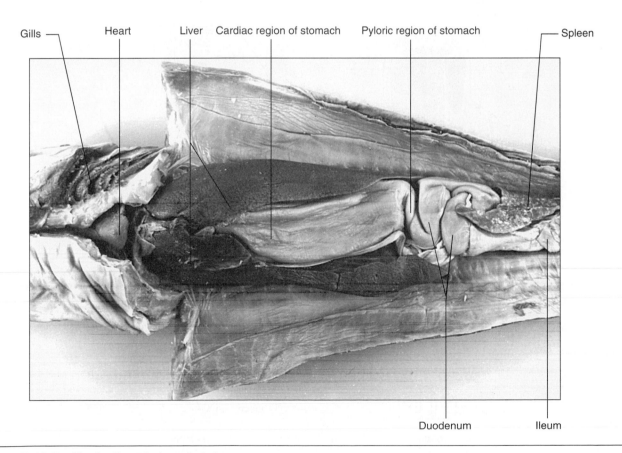

**FIGURE 16.6**  Shark, dissected, ventral view.
Photograph by Carol Majors.

## The Urogenital System

The reproductive system is poorly developed in the immature dogfish sharks usually provided for laboratory study. Therefore, it is advisable for you to supplement your observations of the reproductive system, particularly of the female shark, by viewing demonstrations of more mature specimens. Consult your instructor for information on the demonstrations available.

◆ Examine figure 16.7, which shows the urogenital systems of male and female sharks. **Do not remove** or **cut** any of the structures shown, but simply spread the abdominal organs apart to facilitate your observation. Note the long, flat **kidneys** closely applied to the body wall.

The kidneys of sharks, like those of all vertebrates, are actually **retroperitoneal;** that is, they are located **between** the parietal peritoneum and the body wall, and not suspended into the peritoneal cavity as are the other abdominal organs.

If you have a male specimen, observe the coiled **mesonephric ducts** as shown in figure 16.5 (also called the archinephric, pronephric, opisthonephric, or Wolffian ducts; sometimes incorrectly called ureters), which lie on the ventral surface of the kidneys and extend posteriorly to the cloaca. These ducts carry both urine from the kidneys and sperm from the testes. Locate the paired **testes**—a pair of elongated organs dorsal to the anterior end of the liver.

Leading from each of the testes and emptying into the mesonephric duct are several small efferent ductules (**vasa efferentia**). Within the cloaca, the right and left mesonephric ducts join and empty through a common **urogenital pore.** The urogenital pore is located at the tip of the **urogenital papilla,** a small, fleshy projection from the dorsal wall of the cloaca.

If you have a female specimen, locate the **ovaries**—a pair of oblong, lobed bodies situated near the dorsal body wall above the anterior portion of the liver. From the ovaries, a pair of slender **oviducts** extend posteriorly along the length of the body cavity. The two oviducts originate as a common duct with a single opening, the **ostium tubae,** located anterior to the liver and ventral to the esophagus. (Search carefully in this area; the ostium tubae is often difficult to find.) From the ostium tubae, the **oviducts** loop anteriorly and laterally over the anterior end of the liver and pass back posteriorly along the dorsal body wall.

Ripe eggs discharged from the ovaries enter the body cavity, pass through the ostium tubae (see arrows in figure 16.7), and enter the oviducts, where they may be fertilized. Most of the development of the embryos takes place in the **uterus,** an enlarged portion of each oviduct. The uteri open into the cloaca. Locate the enlarged **shell glands** (also called nidamental glands) attached to the oviducts. Note also the **mesonephric ducts** (you may require assistance from your

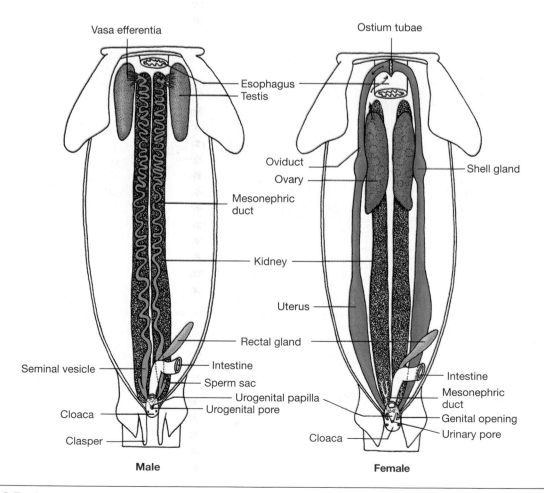

**FIGURE 16.7** Shark, urogenital system.

instructor since these ducts are sometimes difficult to find in immature specimens) of the female shark lying along the ventral surface of the kidney. Trace one of the mesonephric ducts posteriorly to its entrance in the cloaca. Inside the cloaca, the mesonephric ducts empty through a **urinary pore** located on a **urinary papilla.** *How do the male and female urogenital systems differ In this respect?* Locate a student in your class who has a specimen of the opposite sex and compare the male and female reproductive systems or (alternatively) study the demonstration materials provided by your instructor.

Dogfish sharks (*Squalus acanthias*) give birth to living young, unlike some other sharks, which are egg-laying **(oviparous).** The dogfish shark is **ovoviviparous** since more-or-less typical eggs are produced but are retained within the reproductive system of the female until hatching. Many of the mature female sharks used for laboratory study are pregnant because of the unusually long gestation period of this species, which ranges from 20 to 24 months. *Note:* Many sharks form a yolk sac placenta and give birth to living young without forming a shelled egg and thus are **viviparous.**

### The Circulatory System

◆ Before you attempt to make a detailed study of the vascular system, you should study the general plan of the circulatory system and the direction of blood flow from the heart through the arteries to the capillaries of various organs and back to the heart by the veins (figure 16.8).

In certain cases, blood from the capillary beds in the tissues does not return directly to the heart but passes through an extra capillary bed (network) as well as an entire venous system en route back to the heart. The veins connecting the two beds of capillaries are called **portal veins.** Two **portal systems** exist in the shark, the **hepatic portal system** and the **renal portal system.** All portal systems begin and end in capillary beds. Study figures 16.2 and 16.8 for this preliminary

survey and note the following principal structures: (1) the **heart,** which pumps blood to the gills where it is oxygenated and then flows to the dorsal aorta for distribution to various parts of the body; (2) the **hepatic portal system,** which returns blood chiefly from the digestive system to the liver and thence to the heart via the hepatic vein and sinus venosus; (3) the **renal portal system,** which returns blood from the posterior portion of the body to the kidneys, from which it goes to the heart via the postcardinal sinuses and the sinus venosus; and (4) the **anterior cardinal veins,** which collect blood from the head.

◆ From your study of the circulatory system, you should learn the following: (1) the names of the principal parts of the system, (2) the direction of blood flow in each part, (3) the organs served by the major blood vessels (both arteries and veins), and (4) the gains and losses from the blood as it flows through the capillaries of the various organs, particularly in regard to oxygen, carbon dioxide, nutrients, and nitrogenous wastes.

*The Arteries.* The arteries of the shark can be conveniently studied in three groups: (1) the **visceral arteries,** consisting principally of the dorsal aorta and its branches (figures 16.8 and 16.9); (2) the **afferent branchial arteries** and their branches, which carry blood to the gills; and (3) the **efferent branchial arteries,** which carry blood away from the gills and connect with the dorsal aorta.

### Visceral Arteries:

◆ Spread apart the organs and carefully separate the blood vessels from the mesenteries as necessary to locate the arteries. Study the principal arteries in your specimen, using figure 16.2 as a guide. Locate first the large median **dorsal aorta,** visible through the peritoneal lining of the dorsal wall of the coelom.

Posteriorly, the dorsal aorta is continued as the caudal artery, best seen in cross sections of the tail (figure 16.4).

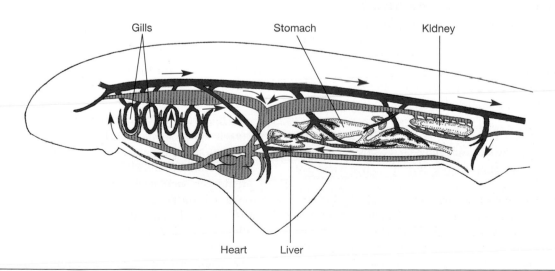

**FIGURE 16.8** Shark, diagram of circulatory system. Oxygenated blood is red, deoxygenated blood is blue, and portal systems are yellow.

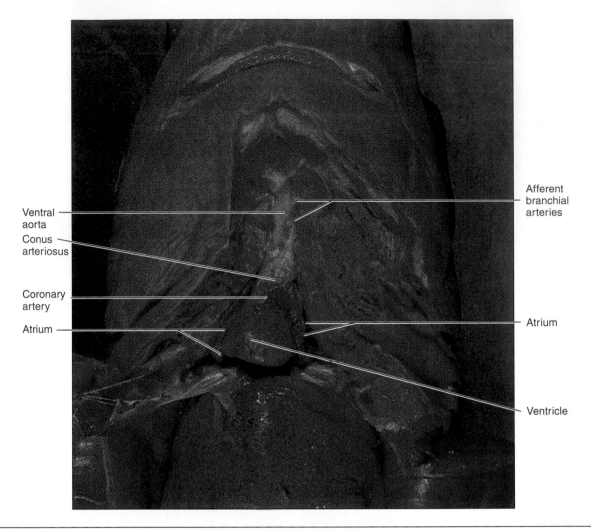

Ventral aorta

Conus arteriosus

Coronary artery

Atrium

Afferent branchial arteries

Atrium

Ventricle

**FIGURE 16.9** Shark, heart and afferent branchial arteries, ventral view.
Photograph by Ken Taylor.

Within the pleuroperitoneal cavity, the dorsal aorta gives off several arteries, some of which are paired and some of which are unpaired. The following is a list of the principal branches of the dorsal aorta that can be found within this cavity (anterior to posterior):

**Coeliac artery**—arises just posterior to the transverse septum and gives off branches to the gonads, esophagus, stomach, liver, and pancreas.

**Posterior intestinal artery**—arises from the aorta dorsal and posterior to the stomach and near the posterior end of the mesentery and supplies one side of the ileum and part of the spiral valve.

**Gastrosplenic (lienogastric) artery**—arises just behind the posterior intestinal (in some specimens the two arteries arise from the aorta as a single vessel and then split) and supplies blood to the spleen, stomach, and a portion of the pancreas.

**Posterior mesenteric artery**—arises from the aorta near the anterior end of the rectal gland, which it supplies.

**Iliac arteries (2)**—a pair of large arteries arising just anterior to the cloaca. Supply the pelvic fin and the posterior portion of the body wall.

**The Heart, Afferent Branchial Arteries, and Gills:**

◆ Slice off the skin and muscles from the ventral surface of the head posterior to the mouth. Carefully cut away the muscles just anterior to the pectoral girdle until you reach the membrane enclosing the pericardial cavity. Cut through the membrane to expose the heart within its cavity, and carefully cut through the pectoral girdle and remove a section of the pectoral girdle (about 3 mm wide) to further expose the heart. Consult figures 16.9, 16.10, and 16.11, and identify the parts of the heart and the surrounding blood vessels. Note that the shark has a **two-chambered heart** with a thick-walled, muscular **ventricle** and a thin-walled **atrium** (auricle). Identify also the **sinus venosus**—a flattened, thin-walled sac closely applied to the posterior surface of the ventricle

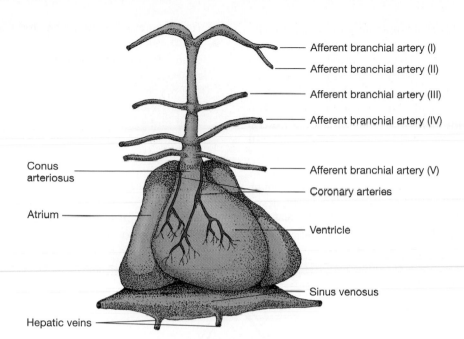

**FIGURE 16.10**   Shark heart, ventral view. (After Wischnitzer.)

Afferent branchial artery (I)
Afferent branchial artery (II)
Afferent branchial artery (III)
Afferent branchial artery (IV)
Conus arteriosus
Afferent branchial artery (V)
Coronary arteries
Atrium
Ventricle
Sinus venosus
Hepatic veins

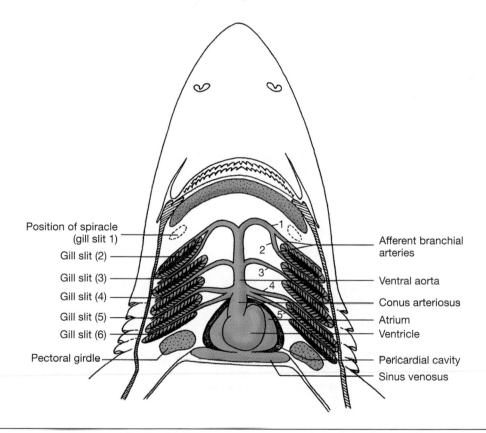

**FIGURE 16.11**   Shark, afferent branchial arteries and gills.

Position of spiracle (gill slit 1)
Gill slit (2)
Gill slit (3)
Gill slit (4)
Gill slit (5)
Gill slit (6)
Pectoral girdle
Afferent branchial arteries
Ventral aorta
Conus arteriosus
Atrium
Ventricle
Pericardial cavity
Sinus venosus

and lying between the ventricle and the transverse septum. Lift up the posterior end of the heart to facilitate your viewing of the sinus venosus. Note that the transverse septum separates the **pericardial cavity** from the **pleuroperitoneal cavity,** thus dividing the **coelom** into two distinct parts. Locate the muscular **conus arteriosus** extending anteriorly from the ventricle.

The shark is commonly said to have a two-chambered heart; however, the conus arteriosus and the sinus venosus also have cardiac muscle. Technically, therefore, the shark can be said to have a four-chambered heart!

Blood enters the heart through the sinus venosus, passes into the atrium, then on to the muscular ventricle, and is pumped out through the conus arteriosus into the ventral aorta. The ventral aorta extends anteriorly from the heart and gives rise to five pairs of **afferent branchial arteries.** Trace the ventral aorta forward, and on one side locate the afferent branchial arteries that carry blood to the gills for oxygenation. Cut away the lower portion of one of the gills and note the **cartilaginous bars** within the **gill arches:** these support the numerous **gill filaments.** Find also the **gill rakers,** cartilage-supported, fingerlike projections from the gill arches that guard the internal gill slits and prevent large food particles from entering the **gill chamber.** Count the number of gills and gill slits on your specimen. The **spiracle,** although nonfunctional as a gill in the shark, is customarily still designated as the first gill slit (figures 16.2 and 16.12). Note the many small branches of the afferent branchial arteries, which carry blood into the gill filaments where **gas exchange** occurs.

**Efferent Branchial, Subclavian, and Carotid Arteries:**

◆ Cut through the angle of the jaw on the left side of the head and continue the incision posteriorly through the pharynx, esophagus, and the ventral body wall to a point just beyond the pectoral girdle. Make a similar cut from the angle of the jaw on the right side through the pectoral girdle. Fold back the lower jaw and the parts immediately behind it to expose the roof of the mouth and pharynx. With your forceps and scalpel, carefully remove the membrane from the roof of the mouth and pharynx to expose the **efferent branchial arteries.**

Study figure 16.12 and continue your dissection until you have exposed all of the efferent branchial arteries and other major blood vessels in this region. Identify in your specimen

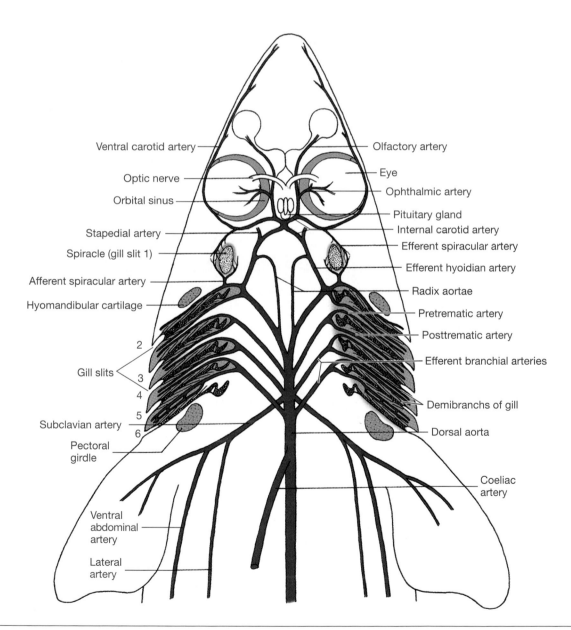

**FIGURE 16.12** Shark. Efferent branchial arteries, carotid arteries, and subclavian arteries are red, and orbital sinus is blue.

each of the blood vessels shown in figure 16.12. Note that each efferent branchial artery is formed by the union of two smaller arteries, a **pretrematic artery** and a **posttrematic artery,** which collect blood from the demibranchs ("half-gills") on the anterior and posterior sides of each gill slit. The pre- and posttrematic arteries are joined at both the dorsal and ventral ends of the gill slit, thus forming an **arterial collecting loop** (figure 16.12), which encircles the spiracle.

Find also the **dorsal aorta** and the paired **subclavian arteries** running toward the two lateral pectoral fins. Arising from each subclavian artery are two smaller arteries that carry blood posteriorly. These branches are the **ventral abdominal arteries** (larger and more lateral in location) and the **lateral arteries** (smaller and more medial in location).

Between the gills and the eyes, several other important arteries can also be found, including the **hyoidean epibranchial arteries** (or **efferent hyoidean arteries**), which arise from the arterial collecting loops around the first gill slits on each side and extend diagonally across the roof of the mouth. The hyoidean epibranchials form two branches, the **temporal arteries** and the **internal carotid arteries.**

*The Veins.*   The venous circulation (figures 16.2 and 16.8) of the shark includes three systems: (1) the **hepatic portal system,** which carries blood from the digestive tract to the liver; (2) the **renal portal system,** which carries blood from the tail region to the kidneys; and (3) the **systemic veins,** which collect blood from the other tissues and organs of the body. All three venous systems eventually empty into the two large common cardinal veins and, finally, into the sinus venosus.

**Hepatic Portal System:**
Study figures 16.3, 16.4, and 16.8 and identify on your specimen the **hepatic portal vein** extending anterior to the stomach and intestine toward the liver. Note that the hepatic portal vein runs along the bile duct and enters the liver. Inside the liver, it branches several times and terminates in several capillary beds.

Posteriorly, the hepatic portal vein is formed by the junction of three branches: (1) the **gastric vein** from the stomach (left branch), (2) the **lienomesenteric vein** with branches from the intestine and spleen (central branch), and (3) the **pancreaticomesenteric vein** with branches from the intestine, stomach, and pancreas (right branch).

Some of the food materials absorbed from the stomach and intestine are removed and then processed and stored in the liver before the blood is returned to the heart and circulated to the rest of the body.

**Renal Portal System:**
The caudal vein leads from the tail region anteriorly into the trunk. Near the level of the cloacal opening, the caudal vein divides into two **renal portal veins,** which pass lateral to the kidneys. The renal portal veins give off many small **afferent renal veins,** which empty into the kidneys (figure 16.4).

**Systemic Veins:**
The systemic veins collect blood from the various organs and return it to the heart. The systemic veins are more difficult to study than the arteries because most of them lack definite walls and appear as more or less open tissue spaces or sinuses. The **posterior cardinal veins** run along the sides of the dorsal aorta and carry blood forward to the large **posterior cardinal sinus.** The posterior cardinal veins collect blood from the kidneys via many small **efferent renal veins,** which drain the renal sinuses within the kidneys. In addition to the efferent renal veins, the posterior cardinal veins also collect blood from many small, segmentally arranged **parietal veins,** which collect blood from the muscles of the body wall.

Other important systemic veins include the two large **hepatic veins,** which drain the liver and carry blood to the sinus venosus, the **lateral abdominal veins,** and the **brachial veins** (from the pectoral fins), which unite to form the short **subclavian veins.** On each side, the subclavian vein joins with the **posterior cardinal sinus** and the **anterior cardinal vein** to form the short **common cardinal vein** (duct of Cuvier), which empties into the **sinus venosus.** The openings of the anterior cardinal veins can be found by slitting open one of the common cardinal veins and using your probe to locate the aperture of the anterior cardinal vein.

### Nervous System and Sense Organs
The nervous system of the shark, like that of other vertebrates, consists of two principal components: (1) the **central nervous system,** including the brain and the spinal cord, and (2) the **peripheral nervous system,** including the various nerves that connect the brain and spinal cord with various other parts of the body. The nerve fibers of the peripheral nervous system can be divided anatomically into those that innervate the outer portions of the body, such as the skin and voluntary muscles (the **somatic division**), and those that innervate the internal organs or viscera (the **visceral division**). Both the somatic and visceral divisions of the peripheral nervous system contain numerous **afferent** (sensory) and **efferent** (motor) **nerve fibers.** The **autonomic nervous system** is a specialized and very important part of the visceral division of the peripheral nervous system in higher vertebrates, although it is poorly developed in the shark. In higher vertebrates, the autonomic nervous system plays a vital role in the coordination of the smooth muscles and glands of the body; thus, it is intimately involved in such processes as digestion, regulation of heart rate, dilation and constriction of blood vessels, and regulation of blood glucose levels.

The basic organization of the vertebrate nervous system can therefore be outlined in the following manner:

Central nervous system
    Brain
    Spinal cord
Peripheral nervous system
    Somatic division
        Afferent neurons (sensory)
        Efferent neurons (motor)
    Visceral division
        Autonomic nervous system
        (plus a few motor neurons to the branchial
           musculature)

Your study of the nervous system of the shark will be limited to a brief survey of the brain and cranial nerves since these parts will suffice to introduce you to the pattern of organization of the vertebrate nervous system. Some further aspects of the peripheral nervous system of vertebrate animals will be studied in some of the later exercises in this manual.

The brain and cranial nerves of the shark are illustrated in figures 16.13 and 16.14. Study these figures carefully before undertaking the dissection of the nervous system. The successful dissection and study of the nervous system requires patience, care, and attention to detail. Haste and carelessness most often lead to disappointing results and confusion.

◆ After you have carefully studied figures 16.13 and 16.14, take the shark head saved from your earlier dissection and carefully remove the skin from the dorsal side. When all of the skin is removed from the dorsal surface, slice away thin sections of the cartilaginous skull (chondrocranium) until the dorsal and lateral surfaces of the brain are exposed. Be careful not to cut or tear away the delicate nerves that pass through the several small openings (**foramina**) of the skull.

On either side of the skull, near its posterior margin, are the two fused **otic capsules,** which enclose the **semicircular canals** of the inner ear. Carefully shave away the cartilage from the otic capsule and locate the semicircular canals, which are important organs of equilibrium (figure 16.13).

During the early stages of embryonic development, the brain of vertebrates becomes divided first into **three distinct lobes.** These three lobes are identified as the **forebrain** (prosencephalon), the **midbrain** (mesencephalon), and the **hindbrain** (rhombencephalon). This early embryonic differentiation of the three primary brain divisions (which together form the **brain stem**) reflects the evolutionary history of the vertebrate brain, since the brain of the earliest verte-

brates also consisted of three divisions, each associated closely with one of three major sense organs: the nose, the eye, and the ear (including the lateral line).

The brain of adult sharks and other recent vertebrates, however, consists of five major divisions rather than three. Two of the three primary brain divisions (the forebrain and hindbrain) divide again during later stages of embryonic development. The brain of **adult sharks** therefore consists of **five major divisions:** (1) the **telencephalon,** (2) the **diencephalon,** (3) the **mesencephalon,** (4) the **metencephalon,** and (5) the **myelencephalon.** Locate each of these five divisions of the shark brain on your specimen. These divisions and their relationships are summarized in table 16.1.

*Survey of the Brain*
1. The anteriormost **telencephalon** is represented by the **olfactory bulbs,** the **olfactory tracts,** and the **olfactory lobes.** Just behind the two olfactory bulbs are the less prominent **cerebral hemispheres.** They can be identified as slight swellings behind the olfactory lobes and are separated from the olfactory lobes by a shallow groove.
2. The **diencephalon** lies behind the telencephalon and appears on the dorsal surface of the brain as a narrow depressed area just anterior to the optic lobes. The diencephalon bears the **epiphysis** or pineal gland—a slender stalk extending anteriorly up through an opening in the roof of the skull. The pineal or "third eye" has long been an enigmatic organ of vertebrates; its function in the shark is not well understood. Ventrally the diencephalon bears the **infundibulum** and the **hypophysis,** which together comprise the **pituitary gland** of the shark. These structures will be studied later.
3. Behind the diencephalon are the two large lateral swellings, the **optic lobes**—important brain centers

## TABLE 16.1

### Summary of the Organization of the Shark Brain

| Embryonic Divisions | Adult Divisions | Component Structures | |
|---|---|---|---|
| Prosencephalon (forebrain) | Telencephalon | Olfactory bulbs<br>Olfactory tract<br>Olfactory lobes<br>Cerebral hemispheres | |
| | Diencephalon | Epiphysis (pineal gland)<br>Infundibulum<br>Hypophysis | Pituitary gland |
| Mesencephalon (midbrain) | Mesencephalon | Optic lobes | |
| Rhombencephalon (hindbrain) | Metencephalon | Cerebellum | |
| | Myelencephalon | Medulla | |

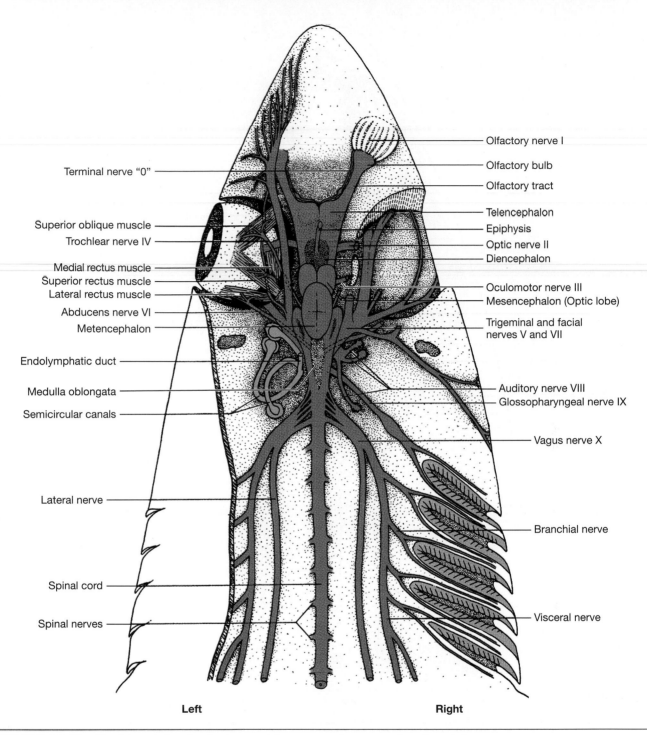

Terminal nerve "0"

Superior oblique muscle
Trochlear nerve IV

Medial rectus muscle
Superior rectus muscle
Lateral rectus muscle
Abducens nerve VI
Metencephalon

Endolymphatic duct

Medulla oblongata

Semicircular canals

Lateral nerve

Spinal cord

Spinal nerves

Olfactory nerve I
Olfactory bulb
Olfactory tract

Telencephalon
Epiphysis
Optic nerve II
Diencephalon

Oculomotor nerve III
Mesencephalon (Optic lobe)

Trigeminal and facial
nerves V and VII

Auditory nerve VIII
Glossopharyngeal nerve IX

Vagus nerve X

Branchial nerve

Visceral nerve

**Left**                    **Right**

**FIGURE 16.13**   Shark head showing brain, cranial nerves, and eye muscles, dorsal view.

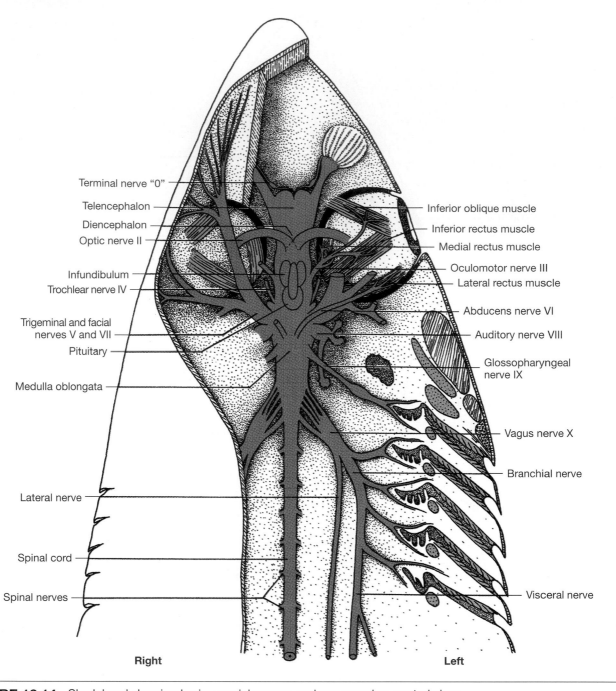

Terminal nerve "0"
Telencephalon
Diencephalon
Optic nerve II
Infundibulum
Trochlear nerve IV
Trigeminal and facial nerves V and VII
Pituitary
Medulla oblongata
Lateral nerve
Spinal cord
Spinal nerves

Inferior oblique muscle
Inferior rectus muscle
Medial rectus muscle
Oculomotor nerve III
Lateral rectus muscle
Abducens nerve VI
Auditory nerve VIII
Glossopharyngeal nerve IX
Vagus nerve X
Branchial nerve
Visceral nerve

**Right**

**Left**

**FIGURE 16.14** Shark head showing brain, cranial nerves, and eye muscles, ventral view.

that play an essential role in vision; here the optic nerves terminate. The optic lobes represent the **mesencephalon.**

4. The **metencephalon** consists mainly of the **cerebellum,** a single large median lobe lying behind the paired optic lobes.

5. The **myelencephalon** consists principally of two parts—the triangular **medulla oblongata,** the most posterior portion of the brain, which tapers into the connecting spinal cord, and the two lateral **auricular lobes.** The latter extend forward from the medulla

and can be seen adjacent to the posterior portion of the cerebellum. The auricular lobes serve as important centers of equilibrium.

*Cranial Nerves*

Sharks have eleven pairs of **cranial nerves.** Each of these nerves attaches to a specific portion of the brain and connects that part of the brain with some specific organ or organs of the body. Some of the cranial nerves consist wholly of **sensory neurons** (carry impulses to the brain), and other cranial nerves are entirely made up of **motor neurons** (carry impulses from the brain), and about half of them contain mixtures of sensory and motor elements.

Cranial nerves were first studied seriously in humans and were given names and numbers (I to XII) based on their location and function in humans. (It is customary to use Roman numerals to designate the cranial nerves.) Subsequent study of other vertebrates, however, has revealed that the human pattern of cranial nerves does not hold for all vertebrates. In sharks, for example, only 10 of the cranial nerves correspond to those of humans (nerves I through X). The eleventh cranial nerve in the shark is a small anterior nerve, the terminal nerve, not present in humans. It is designated as nerve 0.

Students often use mnemonic devices to help them remember the names of cranial nerves and other anatomical parts. In the case of the cranial nerves of the shark, one such saying is, "There, on old Olympus' top, the army favorites are great victories." See if you can make up another such saying in which the words start with the same letters as the 11 cranial nerves of the shark.

Locate each of the cranial nerves on your specimen with the aid of figures 16.13 and 16.14 and the list provided below.

**Nerve 0. Terminal nerve**—A delicate sensory nerve arising from the median surface of the olfactory lobe and extending into the nasal region. (It is numbered "0" because it was discovered after the numbering system for cranial nerves became established.)

**Nerve I. Olfactory nerve**—Carries sensory impulses from the olfactory epithelium within the olfactory sac to the olfactory bulb.

**Nerve II. Optic nerve**—Arises in the retina of the eye and runs to the ventral surface of the diencephalon where it crosses the brain and carries impulses to the optic lobe on the opposite side of the brain. The prominent crossing of the optic nerves on the ventral surface of the diencephalon is called the optic chiasma (to be studied later).

**Nerve III. Oculomotor nerve**—Originates in the mesencephalon and divides into four branches, which carry motor impulses to the muscles of the eye.

**Nerve IV. Trochlear nerve**—Originates from the dorsal surface of the mesencephalon and passes through the chondrocranium to carry impulses to the superior oblique muscle of the eye.

**Nerve V. Trigeminal nerve**—A mixed nerve (with both motor and sensory fibers), which arises from the anterior part of the medulla and enters the orbit. It is the largest of the cranial nerves in the shark and divides into four main branches: (1) the superficial ophthalmic nerve to the skin of the head, (2) the deep ophthalmic nerve to the skin of the snout, (3) the infraorbital nerve to the region of the mouth and ventral surface of the snout, and (4) the mandibular nerve to the jaw muscles and skin of the lower jaw.

**Nerve VI. Abducens nerve**—A motor nerve arising from the ventral surface of the medulla and carrying impulses to the external rectus muscle of the eye.

**Nerve VII. Facial nerve**—A mixed nerve with both motor and sensory fibers arising with the trigeminal nerve from the medulla. It is made up of three main branches: (1) a branch from the **superficial ophthalmic nerve** described previously, (2) the **buccal nerve** with sensory fibers from the mouth region, and (3) the **hyomandibular nerve** with sensory branches from the tongue, lateral line, and lower jaw.

**Nerve VIII. Auditory nerve**—Arises from the anterior end of the medulla and innervates the inner ear.

**Nerve IX. Glossopharyngeal nerve**—A mixed nerve arising from the medulla with branches to the first functional gill slit and to the roof of the mouth.

**Nerve X. Vagus nerve**—A large mixed nerve, which arises from several roots on the medulla. It divides to form three main branches: (1) the **lateral line trunk** to the lateral line, (2) the **branchial trunk** to the gills (except the first), and (3) the **visceral trunk** to the heart and abdominal organs.

**Ventral Surface of the Brain:**

◆ After you have completed your study of the dorsal parts of the brain and the cranial nerves, study figure 16.14 and carefully dissect away the anterior portion of the roof of the mouth and the cartilage underlying the brain. Study the ventral surface of the brain and locate the **optic chiasma** where the large optic nerves cross, the two lobes of the **infundibulum,** the **hypophysis,** and the origin of the **sixth cranial nerve** (abducens) described previously.

## Demonstrations

1. Dogfish skeleton, preserved or plastic mount
2. Microscope slide of placoid scale
3. Pregnant female shark showing embryos in uterus
4. Preserved mount of dissected triply injected shark
5. Preserved mount of shark brain
6. Preserved representatives of Superclass Elasmobranchiomorphii: skates, rays, chimaeras

## Key Terms

**Hepatic portal system**  portion of the venous circulation in sharks and higher vertebrates that collects blood from the stomach and intestine, and returns it to the liver, where many food materials absorbed from the gut are removed and processed for storage.

**Heterocercal tail**   type of tail found in sharks in which the caudal vertebra are deflected upward into the dorsal lobe. The tail is asymmetrical with the dorsal lobe larger and longer than the ventral lobe.

**Lateral line**   type of sense organ found in sharks and many other fishes and amphibians. Consists of a series of sensory cells usually found along the two sides of the body. Enables the animals to detect water currents, temperature changes, and electrical currents.

**Ostium tubae**   opening of the oviduct in the abdominal cavity of the shark and other vertebrates.

**Renal portal system**   part of the venous circulation in sharks (also other fishes, amphibians, reptiles, and birds; absent in mammals), which returns blood from the tail, hind limbs, and posterior portion of the body to the kidneys.

## Internet Resources

Visit the zoology website at http://www.mhhe.com/zoology to find live Internet links of each of the references listed below.

1. Dissection Table. A lengthy table that compares various systems of the earthworm, the frog, the snake, the shark, the perch, the pigeon, and the pig.

2. Dissection of the Shark. Terrific photos of external and internal organs (loads slowly); some pictures are labeled, others are not. Separate pictures of various aspects of external anatomy, individual pictures of various internal organs or organ systems.

3. Introduction to the Chondrichthyes. University of California at Berkeley, Museum of Paleontology. Images, photos, systematics, more information, and links.

## Critical Thinking Questions

1. Although the elasmobranchiomorphs are an ancient group of fishes, systematists regard them as having specialized rather than "primitive" characteristics. Why is this so? Do you agree?

2. Discuss the advances (if any) of the elasmobranchiomorphs over the arthropods. Can you say with certainty that one group is more "advanced" than another? How would you measure the comparative success of the various phyla studied so far?

3. Discuss the significance of an entirely closed circulatory system coupled with a heart that beats regularly (rather than irregularly as those found in most invertebrate groups).

4. What is the significance of dividing the coelom into a pericardial cavity and pleuroperitoneal cavity? Is this a sort of internal vertebrate "tagmatization" because it compartmentalizes distinct functions such as the gas exchange and digestive systems?

5. What is the physiological significance of the two portal systems (the hepatic portal system and the renal portal system)? What do several capillary beds offer that a single capillary bed cannot?

6. Describe and discuss the tremendous advances made by the evolution of the shark brain and nervous system over those of invertebrates.

7. How does the lateral line function? Do you suppose it would be easier to approach a shark by swimming downward toward its dorsal surface or by approaching it from the sides? Explain your reasoning.

## Suggested Readings

Ashley, L.M., and R.B. Chaisson. 1988. *Laboratory Anatomy of the Shark,* 5th ed. Dubuque, IA: Wm. C. Brown Company Publishers.

Gans, C. and T.S. Parsons. 1988. *A Photographic Atlas of Shark Anatomy: The Gross Morphology of Squalus acanthias.* Chicago: University of Chicago Press.

Hamlett, W.C. (ed.) 1999. *Sharks, Skates, and Rays: The Biology of Elasmobranch Fishes.* Baltimore, MD: Johns Hopkins University Press.

# CHAPTER 17

## Perch Anatomy

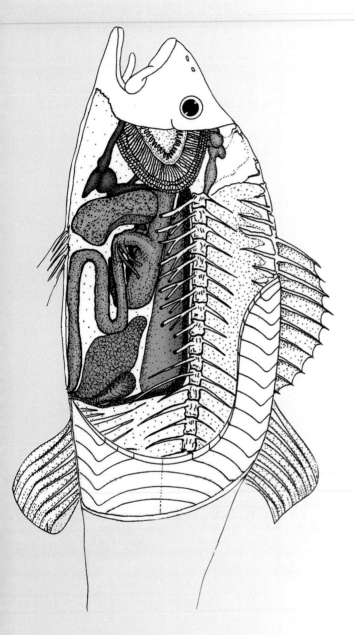

### OBJECTIVES

After completing the laboratory work in this chapter, you should be able to perform the following tasks:

1. Locate and name the main external features of a perch.
2. Locate the axial skeleton, appendicular skeleton, and visceral skeleton on a prepared perch skeleton and explain the function of each. Describe and point out the main parts of a trunk vertebra.
3. Describe and locate the main divisions of the perch musculature.
4. Locate the gills, show the main parts of a gill, and explain the function of each part.
5. Identify the parts of the digestive system.
6. Describe the heart of a perch and locate its main parts.
7. Describe the basic pattern of circulation in a perch.
8. List the five major divisions of the perch brain and demonstrate their location on a specimen.
9. Discuss the principal morphological similarities and differences between the perch and the shark.

## The Yellow Perch: *Perca flavescens*

The yellow perch, *Perca flavescens,* is a common bony fish found in lakes and streams throughout most of the United States; it is native to the central United States and southern Canada, and is widely stocked elsewhere. A similar species, *Perca fluviatilis,* the European perch, is common in Europe. Although less commonly studied in zoology and comparative anatomy laboratories than the shark, the perch is more typical of modern bony fishes.

The yellow perch is a member of the Class Osteichthyes, the bony fishes, the largest class of living vertebrates. Over 20,000 species of bony fishes have been described from the lakes, streams, rivers, and oceans of the world. Among the principal distinguishing features of the Osteichthyes are a **bony skeleton,** a **terminal mouth, dermal scales,** a **homocercal tail, paired nostrils,** and ears with **three semicircular canals.**

Mature specimens range from 6 to about 12 inches in length. Preserved or freshly killed specimens may be dissected, but preserved, latex-injected specimens are best suited for study of the circulatory system.

You should also study a goldfish or other small bony fish in an aquarium to observe swimming movements and other aspects of fish behavior.

## Materials List

**Living specimens**
  Small perch or goldfish
**Preserved specimens**
  Perch
  Perch, double- or triple-injected to show details of the circulatory system
**Prepared microscope slides**
  Ctenoid scale, whole mount
  Fish gill, cross section
  Fish skin, cross section to show origin of scales
  Freshwater fish, longitudinal section of head to show gills
**Plastic mounts**
  Fish heart
  Fish skeleton
  Perch skull
  Perch, dissected
**Audiovisual materials**
  Anatomy of the Perch video
**Miscellaneous**
  Aquarium
  Model of dissected perch

## External Anatomy

◆ Obtain a preserved perch and study the principal features of its external anatomy (figure 17.1). Note the streamlined **fusiform** (spindle-shaped) **body,** which is thickest about one-third of the distance from the mouth to the tail and tapers in both directions. Pick up the fish and look directly at the mouth from the front; observe the ovoid cross section of the fish. *How would this shape facilitate movement of the fish through the water?* Numerous mucous glands in the skin further aid in reducing resistance during movement through the water.

Identify the three regions of the body: the anterior **head,** which extends to the rear of the bony operculum covering the gills; the **trunk,** extending from the operculum to the anus; and the **tail,** extending from the anus and posteriorly.

On the head, find the two sets of double **nostrils,** two large **eyes** (no eyelids), and the large **mouth** equipped with **teeth.** *Where are the teeth found?* A bony **operculum** covers the gills on each side of the head; under each operculum are four **gills.** The operculum is attached at the front and on the dorsal side, but is open behind and ventrally, which provides for the release of water.

On the ventral surface, just anterior to the tail, locate the **anus** and the **urogenital opening(s).** Female perch have a single urogenital opening anterior to the anus; males have a separate genital pore and also a urinary pore located on a small urinary papilla.

Observe the several fins attached to the body: four unpaired **median fins** (two dorsal fins, one anal fin, and one caudal fin) and two sets of **paired fins** (two pectoral fins and two pelvic fins). All fins are membranous extensions of the skin supported by numerous bony **fin rays.** The fins aid in stabilizing the fish and in directing its movements through the water.

On each side of the fish, extending from the operculum behind the eye to the base of the tail, is a **lateral line.** The lateral line is a specialized sensory organ system that detects vibrations and current directions in the water. It appears to aid fishes in orientation, in avoiding obstacles in the water, and in escaping predators.

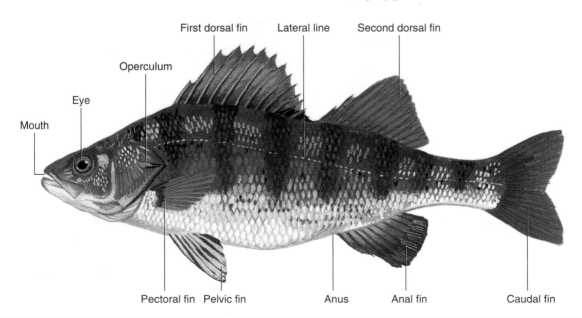

**FIGURE 17.1** Perch, external anatomy, lateral view.

Painting from *Fisherman's Guide: Fishes of the Southeastern United States,* by Charles S. Manooch III and Duane Raver, Jr. North Carolina State Museum of Natural History, Raleigh, NC, 1984. Used by permission.

The exterior surface of the perch is covered by a tough **skin** that contains many **mucous glands.** The skin produces the **scales,** which protect the surface of the body and are arranged in a well-ordered pattern of longitudinal and diagonal rows. Note how the posterior portion of each scale overlaps the anterior portion of the next scale. Each scale is formed in an epidermal pocket and extends posteriorly from the pocket. The scales grow continuously during the life of the fish and are not regenerated if lost. Seasonal differences in the growth of the fish are revealed by the deposition of new material around the margin of the scales; thus, a scientist can determine the age of a fish by microscopic examination of the "rings" in its scales. This is an important technique frequently used in research on the biology of fishes. Similar annual depositions are found in certain bones, such as the **otoliths** in the ear.

◆ Remove a scale from your specimen, make a wet mount on a microscope slide, and observe it under low power. Note the numerous concentric ridges **(annuli)** on the scale and the many fine **teeth** on the posterior portion of the scale. This type of scale is called a **ctenoid** ("comb") **scale** because of the presence of these teeth.

Draw a ctenoid scale in figure 17.2.

## Internal Anatomy

### Skeletal System

The scales, fin rays, and some of the bones of the skull of the perch represent elements of a **dermal exoskeleton,** but the chief supporting structure of the body consists of a **bony endoskeleton.**

◆ Observe a prepared skeleton of a perch or other bony fish on demonstration. Locate the **axial skeleton** consisting of the skull, the vertebral column, the ribs, and the medial fins. The **appendicular skeleton** is made up of the pectoral girdle, the pectoral fins, and a small pelvic girdle that supports the pelvic fins.

The **vertebral column** is made up of many individual **vertebrae.** The trunk vertebrae have a large cylindrical **centrum** with a dorsal **neural arch** through which the dorsal nerve cord passes, and a single **neural spine.** Lateral processes on each side of the trunk vertebrae articulate with the ribs.

◆ Draw an anterior view of a trunk vertebra in figure 17.3. Label each part.

The caudal vertebrae also have a ventral **haemal arch,** through which the caudal artery passes, and a supporting **haemal spine.**

There is also a **visceral skeleton,** first formed of cartilage and later replaced by bone, which supports the gills. There are **seven paired visceral arches** that correspond to similar structures in the skeleton of the shark. The upper part of the first arch connects to the skull; the second arch (hyoid arch) supports the tongue; the four gill arches each support a gill; and the last arch has no gill.

We shall observe the gill arches later when we study the gills.

### Muscular System

The muscular system of the perch is relatively simple compared to that of terrestrial vertebrates. Most of the body musculature consists of **segmental muscles** (myotomes) (figure 17.4). Contractions of these myotomes result in flexing of the body, which aids in swimming. Adjacent myotomes are separated by a **myoseptum** of connective tissue. The myotomes are also separated into dorsal and ventral portions by a **transverse septum.** The muscle segments dorsal to the transverse septum are called **epaxial muscles,** and the

**FIGURE 17.2** Drawing of a ctenoid scale.

**FIGURE 17.3** Drawing of a trunk vertebra.

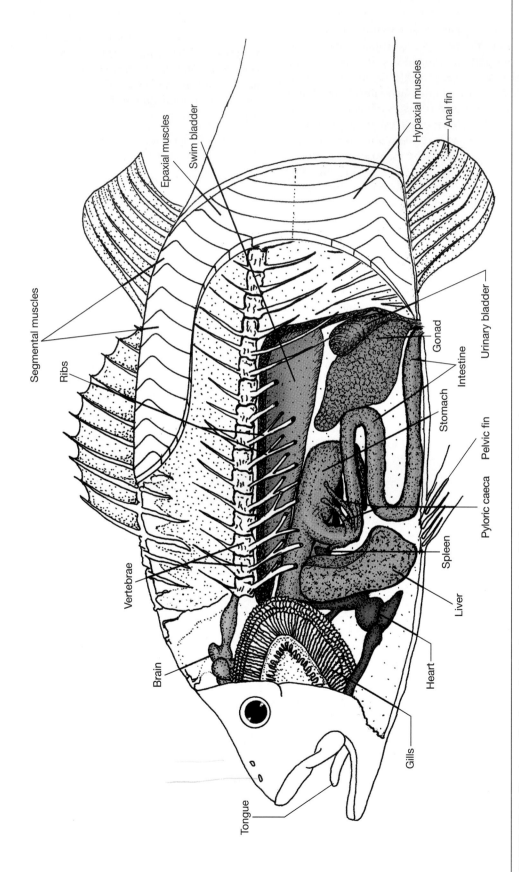

**FIGURE 17.4**  Perch, partly dissected to show muscles and internal organs, lateral view.

muscle segments ventral to the transverse septum are called **hypaxial muscles.**

More specialized muscles in the head region serve to move the lateral fins, mouthparts, jaws, gill opercula, gill arches, and associated parts.

### Respiratory System

◆ Carefully cut away the bony operculum from one side of the perch to expose the **gills** (figure 17.5). Locate the four gills within the gill chamber. Observe the numerous fingerlike **gill filaments** extending posteriorly from each gill (figure 17.6). The large surface area of these filaments facilitates gas exchange within the **capillary beds** of each filament.

Remove one gill and locate the bony **gill arch,** which supports the gill and the hard, fingerlike projections, the **gill rakers,** which protect the gills and prevent passage of coarse material through the gills. Each filament consists of many thin **lamellae,** which contain the capillaries and provide a large surface area for gas exchange.

### Coelom and Visceral Organs

The **coelom** of the perch consists of a large **peritoneal cavity** and a small **pericardial cavity.** The peritoneal cavity contains the stomach, liver, and other digestive organs, swim bladder, and other visceral organs. The pericardial cavity is located anterior to the peritoneal cavity and encloses the heart. The location of the principal internal organs is illustrated in figure 17.4.

◆ To study the internal organs, make a longitudinal cut along the ventral abdominal wall with your scalpel.

Start your incision just anterior to the anus and carefully cut anteriorly to the level of the pelvic girdle. Take care not to cut too deeply or you may damage internal organs to be studied later.

◆ After you have completed the midventral incision, make a second incision from the posterior end of the first incision and cut dorsally to the level of the lateral line. Make a similar incision from the anterior end of the midventral incision. Raise this portion of body wall to locate the visceral organs in the peritoneal cavity. The outer lining of the cavity is the parietal ("wall") peritoneum. Cut through the peritoneum if you have not already done so in opening up the body wall and observe the large **liver.** Beneath the liver, locate the short **esophagus,** the **stomach,** and the **small intestine.** Extending from the anterior part of the small intestine are three short sacs—the **pyloric caeca.** Posteriorly, the small intestine empties into the **large intestine,** which terminates at the anus.

Dorsal to the digestive tract, find the large **swim bladder** or air bladder. (It is often deflated in preserved specimens.) The swim bladder is a hollow, gas- or air-filled sac that serves as a buoyancy organ. Altering the volume of gas within the swim bladder assists the perch in compensating for the differences in the specific gravity between its body and that of the surrounding water while swimming at various depths.

Above the swim bladder and underneath the vertebrae are two long, dark **kidneys.** As in the shark, the kidneys are retroperitoneal. Other organs in the peritoneal cavity include the **spleen,** an elongate organ lying along the posterior

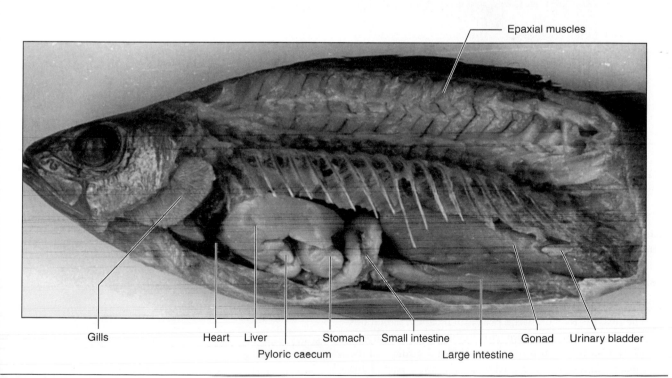

Epaxial muscles

Gills     Heart   Liver     Stomach    Small intestine      Gonad   Urinary bladder

Pyloric caecum      Large intestine

**FIGURE 17.5**   Perch, dissected to show visceral organs, lateral view.
Photograph by Carol Majors.

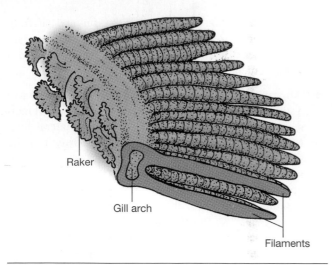

**FIGURE 17.6** Perch, portion of a gill.

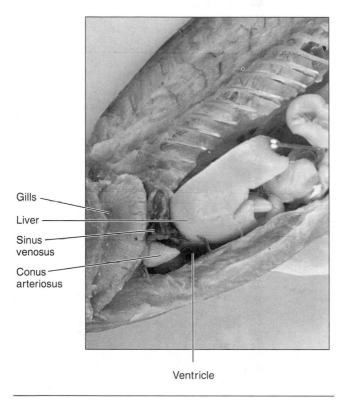

**FIGURE 17.7** Perch heart, dissected.
Photograph by Carol Majors.

Like other fishes, the perch has a **two-chambered heart** (figure 17.8). Find the thin-walled **atrium** and the thick-walled, muscular **ventricle**. Blood passes from the **sinus venosus** to the **atrium** and from the atrium to the muscular **ventricle**. Contraction of the ventricle forces the blood into the short **conus arteriosus** and out through the short **ventral aorta**. From the ventral aorta, the blood passes to the gills via four different pairs of **branchial arteries**.

The **afferent branchial arteries** lead to extensive capillary beds in the lamellae of the gills, where the blood is oxygenated (figure 17.9). The oxygenated blood is collected by the **efferent branchial arteries**, which empty into the **dorsal aorta**. From the dorsal aorta, arteries carry oxygenated blood to the organs and tissue of the head, trunk, and caudal regions. Some of the principal blood vessels are shown in figure 17.10.

The venous system of the perch consists of two main divisions: (1) the **hepatic portal system,** and (2) the **systemic veins.** The perch and other modern bony fishes lack a well-developed renal portal system as seen in the shark. The hepatic portal system consists of veins that collect blood from the stomach, intestine, and other visceral organs, and carry it to capillary beds in the liver. From the liver, the blood is collected by the hepatic vein and is carried via the posterior cardinals to the sinus venosus and the heart for recirculation.

The principal systemic veins include a pair of large **anterior cardinal veins** that collect blood returning from the head region and a pair of **posterior cardinals** that collect blood from the posterior regions, including the liver.

### Urogenital System

The **kidneys** are two long, slender organs lying dorsal to the swim bladder and ventral to the vertebral column. They filter nitrogenous wastes from the blood and empty posteriorly

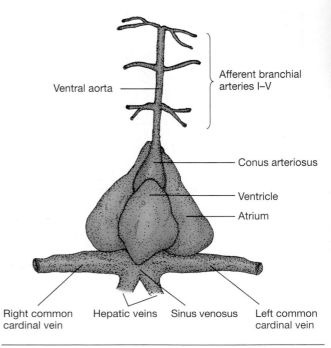

**FIGURE 17.8** Perch heart and associated blood vessels, ventral view.

surface of the stomach; the **pancreas,** on the ventral surface of the intestine (often difficult to find); the **gonads,** posterior to the stomach and dorsal to the intestine; and the **urinary bladder,** found posterior to the gonads.

### Circulatory System

◆ The **heart** is located in the pericardial cavity, which lies ventral to the gills and anterior to the pelvic fins (figure 17.7). Carefully cut through the pectoral girdle and the muscles anterior to the girdle and cut away part of the lateral body wall to expose the heart and major blood vessels.

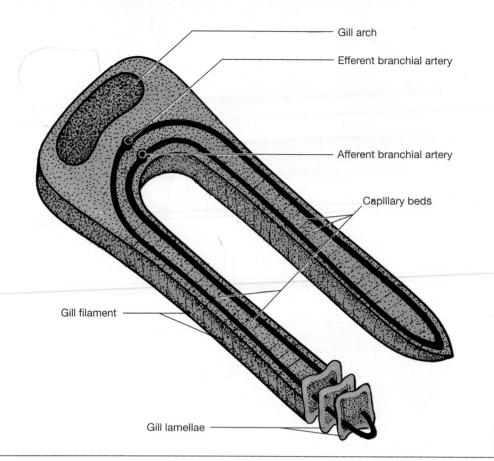

**FIGURE 17.9**  Perch gill, pattern of circulation within filaments.

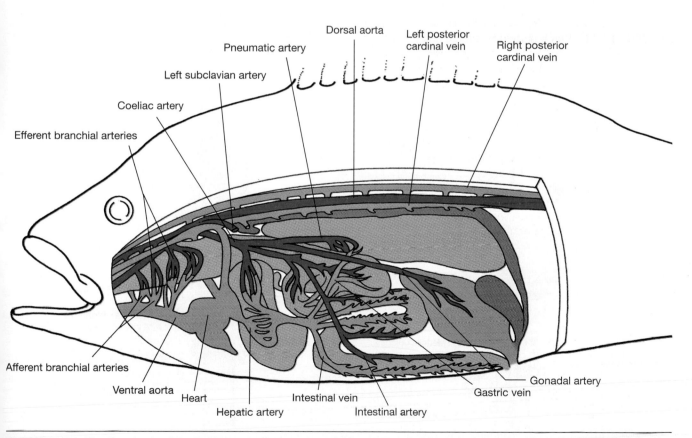

**FIGURE 17.10**  Perch circulatory system, major arteries and veins. Oxygenated blood is red, and deoxygenated blood is medium blue.

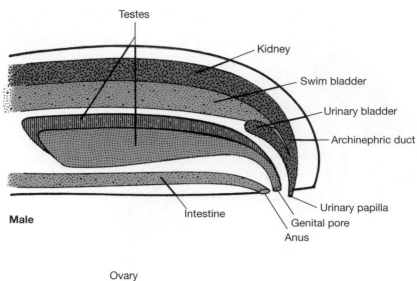

**Male**

- Testes
- Kidney
- Swim bladder
- Urinary bladder
- Archinephric duct
- Urinary papilla
- Genital pore
- Anus
- Intestine

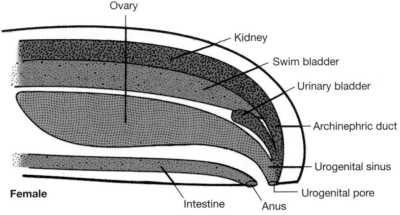

**Female**

- Ovary
- Kidney
- Swim bladder
- Urinary bladder
- Archinephric duct
- Urogenital sinus
- Urogenital pore
- Anus
- Intestine

**FIGURE 17.11** Perch urogenital system, diagrammatic.

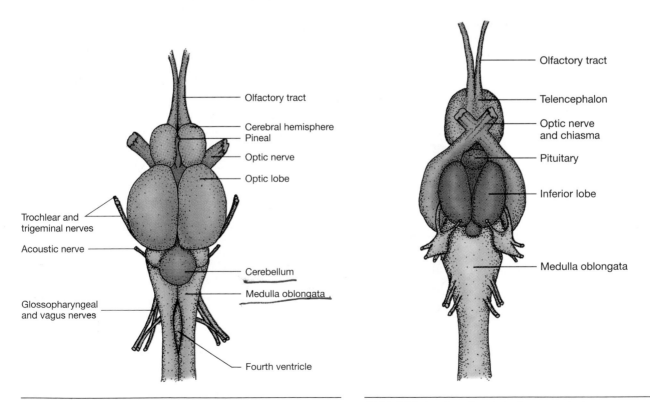

Dorsal view labels:
- Olfactory tract
- Cerebral hemisphere
- Pineal
- Optic nerve
- Optic lobe
- Trochlear and trigeminal nerves
- Acoustic nerve
- Glossopharyngeal and vagus nerves
- Cerebellum
- Medulla oblongata
- Fourth ventricle

Ventral view labels:
- Olfactory tract
- Telencephalon
- Optic nerve and chiasma
- Pituitary
- Inferior lobe
- Medulla oblongata

**FIGURE 17.12** Perch brain, dorsal view.

**FIGURE 17.13** Perch brain, ventral view.

**TABLE 17.1**

Components of the Perch Brain

| Division | Chief Structures |
|---|---|
| Telencephalon | Olfactory lobes, cerebral hemispheres |
| Diencephalon | Thalamus, hypothalamus, pineal, pituitary |
| Mesencephalon | Optic lobes |
| Metencephalon | Cerebellum |
| Myelencephalon | Medulla oblongata |

**TABLE 17.2**

Cranial Nerves of the Perch

| Nerve | Function |
|---|---|
| I. Olfactory | Olfaction (smell) |
| II. Optic | Vision |
| III. Oculomotor | Eye movements |
| IV. Trochlear | Superior oblique muscle of eye |
| V. Trigeminal | Jaw muscles, touch |
| VI. Abducens | Lateral rectus muscle of eye |
| VII. Facial | Taste, lateral line, skin of head |
| VIII. Acoustic | Inner ear and lateral line |
| IX. Glossopharyngeal | Gill muscles and lateral line |
| X. Vagus | Gills, heart, anterior part of digestive tract, lateral line |

through the **archinephric** (or Wolffian) **ducts,** which lead to the **urinary bladder.** From the bladder, urine passes into the **urogenital sinus** and out through the **urogenital pore** (figure 17.11). In the male, the urinary pore and the genital openings are separate. In the female, there is a common urogenital pore through which both systems empty.

The reproductive system of the **male** includes two long-lobed **testes** lying posterior to the stomach and ventral to the swim bladder. Two **vasa deferentia** carry the sperm to a common **genital sinus,** which opens via the **genital pore.**

In the **female** is a single, fused **ovary,** a large, saclike structure that releases the eggs through a short **oviduct** to the **urogenital pore.**

### Nervous System

Like the nervous system of the shark, that of the perch consists of two main divisions: the **central nervous system** (brain and spinal cord) and the **peripheral nervous system** (nerves connecting the brain and spinal cord with other parts of the body).

We shall confine our brief study of the nervous system of the perch largely to the **brain** (figures 17.12 and 17.13). The brain of the adult perch consists of five major divisions: (1) the **telencephalon,** (2) the **diencephalon,** (3) the **mesencephalon,** (4) the **metencephalon,** and (5) the **myelencephalon.** The specific parts of the brain making up these five divisions are summarized in table 17.1.

◆ To expose the brain for study, you must remove the skin from the dorsal surface of the skull behind the eyes (figure 17.14). Carefully shave away the bony roof of the skull above the brain, taking care not to damage the delicate tissues beneath. You will find the brain enclosed in a gelatinous mass that must be removed to expose the brain. Also around the brain is a pigmented membrane. Carefully remove the membrane and identify the principal structures of the brain with the aid of figures 17.12 and 17.13.

Associated with the brain of the perch and other bony fishes are **ten pairs of cranial nerves** (table 17.2). Nerve 0, the terminal nerve of the shark, is absent in bony fishes.

The spinal cord leads posteriorly from the brain to the tail and passes through the neural arches of the vertebrae. One pair of **spinal nerves** arises from the spinal cord in each segment.

### Demonstrations

1. Microscope slide of ctenoid scale
2. Microscope slide of fish skin showing origin of scale
3. Mounted or plastic-embedded fish skeleton
4. Model of dissected perch
5. Fish heart, plastic mount
6. Microscope slide of fish gill, sectioned through filaments
7. Living fish in aquarium

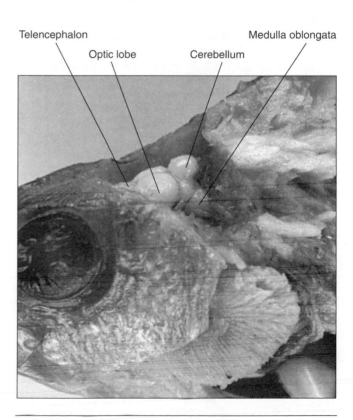

**FIGURE 17.14** Perch brain, dissected.
Photograph by Carol Majors.

## Key Terms

**Appendicular skeleton**  portion of the endoskeleton that supports the appendages. Consists of the anterior pectoral girdle and the posterior pelvic girdle.

**Archinephric duct**  a primitive type of kidney duct that collects urine from a series of segmentally arranged nephrons; also called a Wolffian duct.

**Axial skeleton**  portion of the endoskeleton that supports the longitudinal axis of the body. Consists of the skull, the vertebral column, the ribs, and the medial fins in bony fishes.

**Ctenoid scale**  type of dermal scale made of a thin sheet of bonelike material, circular or oval in shape, and bearing spines ("teeth") on the posterior margin.

**Homocercal tail**  type of tail with upper and lower parts of the tail symmetrical and with the vertebral column ending at the center of the base; found in most bony fishes.

**Swim bladder**  a gas- or air-filled sac found in the abdominal cavity of most bony fishes; serves mainly as a hydrostatic organ in modern fishes.

**Visceral skeleton**  portion of the endoskeleton that supports the gills. Consists of seven paired gill arches.

## Internet Resources

Visit the zoology website at http://www.mhhe.com/zoology to find live Internet links of each of the references listed below.
1. Animal Diversity Web, University of Michigan.
    Actinopterygii.
2. Dissection Table. A lengthy table that compares various systems of the earthworm, the frog, the snake, the shark, the perch, the pigeon, and the pig.
3. Dissection of the Perch. Several pictures of internal organs.
4. Dissection of the Blue Mackerel. This fish is different from the perch in many ways, but the photographs in this web site clearly illustrate internal organs. This may require a bit of time if you have a slow browser.

## Critical Thinking Questions

1. Discuss the components and functions of the axial skeleton, the appendicular skeleton, and the visceral skeleton. What advantages (if any) does the rigid endoskeleton have over a cartilaginous endoskeleton of the shark?

2. Describe the basic morphological similarities and differences between the shark and the perch.

3. Speculate on the advantages and disadvantages of the heterocercal tail of the shark versus the homocercal tail of the perch. Think about maneuverability, thrust, steering ability, etc.

4. Describe in some detail the operations of the gas exchange system. When water warms, it loses its free oxygen. How would this affect the gas exchange system of a bony fish? Could the fish literally "drown" if the water temperature got too high? Explain.

## Suggested Readings

Bond, C.E. 1996. *Biology of Fishes*. Philadelphia: Saunders College Publishers.

Cailliet, G.M., M.S. Love, and A.W. Eberling. 1986. *Fishes: A Field and Laboratory Manual on Their Structure, Identification, and Natural History*. Belmont, CA: Wadsworth.

Chaisson, R.B., and W.J. Radke. 1991. *Laboratory Anatomy of the Perch*. 4th ed. Dubuque, IA: Wm. C. Brown Publishers.

Eddy, S., Underhill, J.C., Bamrick, J., Crawley, E.T., and W.G. Jacques. 1978. *How to Know the Freshwater Fishes*. 3d ed. Dubuque, IA: Wm. C. Brown Publishers.

# C H A P T E R 18

## Frog Anatomy

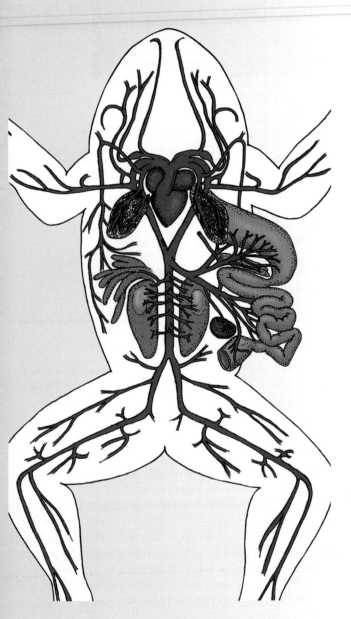

## OBJECTIVES

After completing the laboratory work in this chapter, you should be able to perform the following tasks:

1. Identify the main structures within the oral cavity of the frog and explain the function of each.
2. Describe the external features (secondary sexual characteristics) that distinguish mature male and female frogs.
3. Describe the major divisions of the frog skeleton and explain the chief function(s) of each division.
4. Describe the principal parts of the digestive system of the frog and explain their functions.
5. Describe the structure of the frog heart and identify its major parts.
6. Describe the respiratory system of the frog and identify the main structures involved in gas exchange.
7. Describe the basic pattern of blood circulation in the frog and identify the principal blood vessels involved.
8. Identify the important parts of the urogenital system of both male and female frogs and explain their functions.
9. Identify the five main regions of the frog brain and the components of each. Briefly explain the main function of each part.
10. Describe the stages of frog metamorphosis.

## *Rana pipiens* or *Rana catesbeiana*

Frogs are the most commonly studied representatives of the Class Amphibia, Subphylum Vertebrata, Phylum Chordata (figure 18.1). Although there is no "typical vertebrate" any more than there is a "typical person," a study of frog anatomy does serve to illustrate effectively the basic body organization of a vertebrate animal. The descriptions in this chapter are based primarily on *Rana pipiens* (the grass or leopard frog), but *Rana catesbeiana* (the bullfrog) can also be used for these studies.

It is important to remember that the amphibians are transitional animals that typically live a portion of their lives in the water and another portion on land. Thus, they exhibit a peculiar mixture of characteristics—some representing adaptations for terrestrial life and some representing adaptations for life in the water. Mating nearly always occurs in the water, since the eggs lack the protective outer coverings that

Eyes

External naris

Tympanic membrane

**FIGURE 18.1** Leopard frog, *Rana pipiens*.
Photograph courtesy of Carolina Biological Supply Company, Burlington, NC.

permit the terrestrial existence of birds and reptiles. Amphibian eggs hatch into **tadpoles,** an immature, larval form with gills and a muscular tail for swimming. Later the tadpoles metamorphose into four-legged adults, which are semiterrestrial and become sexually mature. Some amphibians, however, spend all of their lives in the water, and a very few spend all their lives on land, having evolved special mechanisms to protect their eggs from desiccation.

The skin of adult amphibians is smooth and usually moist. Generally the adults live in wet or moist environments, since they are very susceptible to water loss through the skin. Toads are rather exceptional amphibians that have developed a tough, horny skin that reduces water loss and thus allows them to spend more time on land.

Since the amphibians represent an evolutionary transition between the fishes and the terrestrial animals, they also exhibit numerous morphological advances over the fishes. The skull is broad, flat, and lighter in weight; they have jointed tetrapod limbs rather than fins for locomotion; adult amphibians often develop lungs for breathing; a three-chambered heart is formed; and the circulatory system exhibits two distinct circuits, a **systemic division** to supply the body organs and a **pulmonary division** to carry blood to and from the lungs.

Frogs exhibit most of the typical amphibian features, but they also show some features peculiar to their own mode of life. The salamanders more nearly represent the typical features of the Amphibia. Some of the specialized features of the adult frog include the following: (1) the absence of a tail, (2) the loss of certain skull bones, (3) the posterior attachment of the tongue, (4) the absence of ribs, (5) the lack of a distinct neck, and (6) the powerful and highly developed hindlimbs.

## Materials List

**Living specimens**
  *Rana pipiens* (or *R. catesbeiana*)
**Preserved specimens**
  *Rana pipiens* (or *R. catesbeiana*)
  Frog skeleton
  Representative vertebrate skeletons
**Audiovisual materials**
  Anatomy of the Frog video

## External Anatomy and Behavior

◆ Study a living frog and note the smooth, moist, and pliable skin. Observe the pattern of coloration of the dorsal and ventral surfaces of the body.

*How does the coloration differ on the two surfaces?* Compare the color pattern of your specimen with those of others in the laboratory. *How much variation in color patterns do you observe?* The specific pattern of spots has been shown to be genetically determined.

Note the broad, flat head with the large mouth, the nostrils or **external nares,** two conspicuous **eyes,** and the circular **tympanic membranes** located behind the eyes. Bordering the eye is a fleshy lower eyelid and a less-prominent upper eyelid. A third transparent inner eyelid, the **nictitating membrane,** helps keep the eye moist while the frog is on land and also helps protect the eye under water from abrasion.

◆ Observe demonstrations of living frogs in an aquarium and in a terrarium. Study a frog in the aquarium and observe its swimming behavior.

*How does the frog propel itself through the water? Which limbs are most prominent in swimming? What adaptations of the limbs are most important for swimming? What is the posture of the frog while it is floating?* Compare the behavior of frogs in an aquarium filled with water at room temperature and the behavior of frogs in an aquarium filled with cold water. *Which frogs are more active? How can you explain the difference in their activity? Which frogs remain submerged for longer periods of time: those in the cold aquarium or those in the warm (room temperature) aquarium?* Frogs are **ectothermic** or poikilothermic, which means that their body temperature fluctuates with the thermal environment around them. *How does this fact help you to explain the difference in behavior between the frogs in the cold aquarium and those in the warm aquarium?*

◆ Observe also the behavior of frogs in a terrarium.

When exposed to the air, frogs breathe by drawing air through the external nares into the oral cavity. Later, air is forced into the lungs by closing the internal nares and raising the floor of the mouth. Some gas exchange with the blood also occurs across the moist epithelium of the oral cavity and through the skin covering the exterior of the body.

◆ Compare the locomotion of frogs in the water with frogs in a terrarium or on a laboratory table. *How are the limbs used in locomotion on land?*

To observe the feeding behavior of a frog, place some living house flies, fruit flies, or other small insects in a small aquarium with a frog. *How does the frog react to the food organisms? How are the insects captured?*

## Skeletal System

The skeleton of the frog (figures 18.2 and 18.3) consists chiefly of bone and cartilage. It supports the various parts of the body, protects delicate organs such as the brain and spinal cord, and provides surfaces for attachment of the skeletal muscles.

◆ Study a prepared skeleton of a bullfrog (plastic-embedded specimens are excellent for this purpose) and compare the general organization of the frog skeleton to that of other vertebrates on demonstration.

The skeleton of **vertebrates** consists of the **somatic skeleton** (skeleton of the body wall and appendages) and the **visceral skeleton** (skeleton of the pharyngeal wall—prominent in fishes as support for the gills and as a part of the jaws, but much reduced in higher vertebrates). In the frog, the visceral skeleton is represented principally by the **hyoid apparatus,** a small bone and cartilage structure that helps support the floor of the mouth, the base of the tongue, and parts of the jaws and larynx.

The major components of the frog skeleton and some of their relationships are shown in table 18.1.

The **skull** consists of three main parts (figure 18.2): (1) the narrow braincase or **cranium;** (2) the paired **sensory capsules** of the ears, nose, and the large eye sockets, or orbits, for the eyes; and (3) the **visceral skeleton**

## TABLE 18.1

### Components of the Frog Skeleton

Visceral Skeleton
    Hyoid apparatus
Somatic Skeleton
    Axial skeleton
        Vertebral column
        Sternum
        Skull
    Appendicular skeleton
        Pectoral girdle
        Forelimbs
            Humerus
            Radio-ulna
            Carpals
            Metacarpals
            Phalanges
        Pelvic girdle
        Hindlimbs
            Femur
            Tibiofibula
            Tarsals
            Metatarsals
            Phalanges

(consisting of parts of the jaws, hyoid apparatus, and the laryngeal cartilages).

◆ Study the figures and locate the various bones of the skull on your specimen.

The **vertebral column** is made up of 10 vertebrae. The first vertebra, the **atlas,** articulates with the base of the skull (figure 18.2). It has no transverse processes and is the only **cervical** (neck) **vertebra** in the frog. The next seven vertebrae are the **abdominal vertebrae** (figure 18.3). Caudal to the abdominal vertebrae is a large **sacrum** with two strong transverse processes that join with the **ileum.** The last vertebra is the long **urostyle** (figure 18.3).

◆ Study the parts of an abdominal vertebra with the aid of figure 18.3.

## Muscular System (Optional)

Effective study of the muscular system of the frog takes time, patience, and good dissection technique. There are some 200 muscles in the frog, so a complete study of its musculature would take a good deal of time. In this exercise, we shall concentrate only on a few muscles to illustrate the organization of the muscular system and to learn something about the relationship between muscles, bones, and other parts of the body.

Some of the principal muscles of the frog are illustrated in figures 18.4 and 18.5. The skeletal muscles of the adult are well-adapted for swimming and for locomotion on land. Note the large, highly developed muscles associated with the hindlimbs.

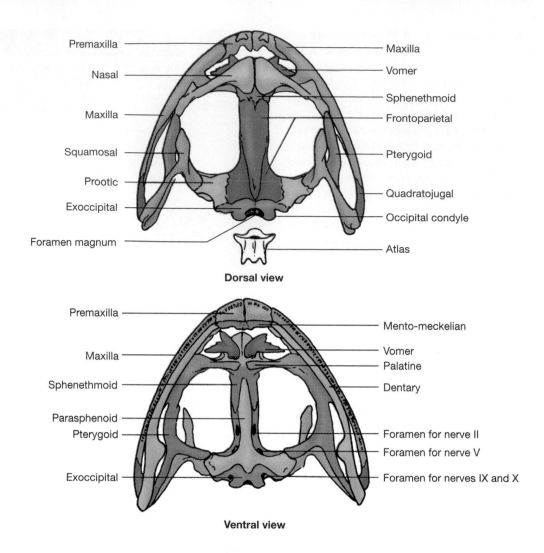

Premaxilla — — Maxilla
Nasal — — Vomer
— Sphenethmoid
Maxilla — — Frontoparietal
Squamosal — — Pterygoid
Prootic — — Quadratojugal
Exoccipital — — Occipital condyle
Foramen magnum — — Atlas

**Dorsal view**

Premaxilla — — Mento-meckelian
— Vomer
Maxilla — — Palatine
Sphenethmoid — — Dentary
Parasphenoid —
Pterygoid — — Foramen for nerve II
— Foramen for nerve V
Exoccipital — — Foramen for nerves IX and X

**Ventral view**

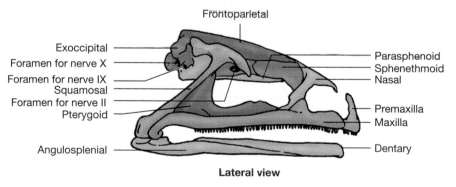

Frontoparietal

Exoccipital — — Parasphenoid
Foramen for nerve X — — Sphenethmoid
Foramen for nerve IX — — Nasal
Squamosal —
Foramen for nerve II —
Pterygoid — — Premaxilla
— Maxilla
Angulosplenial — — Dentary

**Lateral view**

**FIGURE 18.2** Frog, skull.

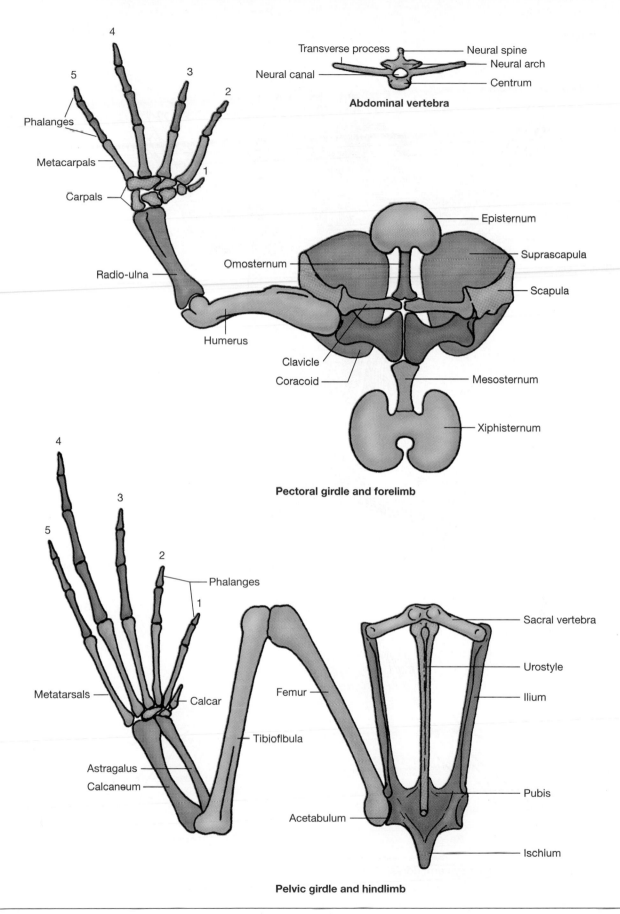

4
5
Phalanges
3
2
Metacarpals
Carpals
1

Transverse process — Neural spine
Neural arch
Neural canal — Centrum

**Abdominal vertebra**

Radio-ulna

Omosternum

Episternum
Suprascapula
Scapula

Humerus

Clavicle
Coracoid

Mesosternum

Xiphisternum

**Pectoral girdle and forelimb**

4
3
5
2
Phalanges
1
Metatarsals — Calcar

Femur

Sacral vertebra
Urostyle
Ilium

Tibloflbula

Astragalus
Calcaneum

Acetabulum

Pubis

Ischium

**Pelvic girdle and hindlimb**

**FIGURE 18.3**  Frog, skeleton.

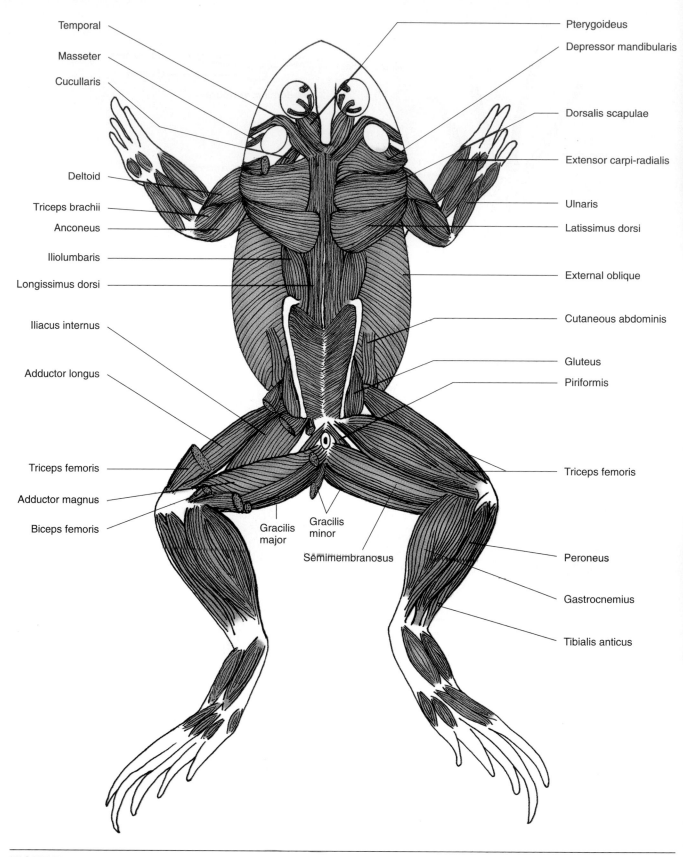

Temporal

Masseter

Cucullaris

Deltoid

Triceps brachii

Anconeus

Iliolumbaris

Longissimus dorsi

Iliacus internus

Adductor longus

Triceps femoris

Adductor magnus

Biceps femoris

Pterygoideus

Depressor mandibularis

Dorsalis scapulae

Extensor carpi-radialis

Ulnaris

Latissimus dorsi

External oblique

Cutaneous abdominis

Gluteus

Piriformis

Triceps femoris

Peroneus

Gastrocnemius

Tibialis anticus

Gracilis
major

Gracilis
minor

Semimembranosus

**FIGURE 18.4**   Frog, muscular system, dorsal view.

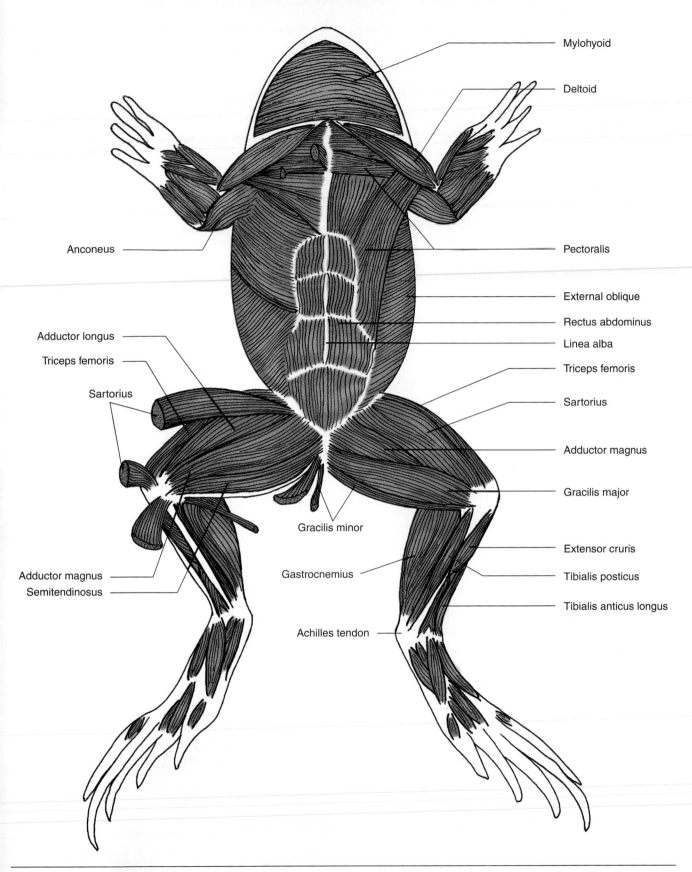

Mylohyoid

Deltoid

Anconeus

Pectoralls

External oblique

Rectus abdominus

Linea alba

Triceps femoris

Sartorius

Adductor magnus

Gracilis major

Extensor cruris

Tibialis posticus

Tibialis anticus longus

Adductor longus

Triceps femoris

Sartorius

Gracilis minor

Adductor magnus

Semitendinosus

Gastrocnemius

Achilles tendon

**FIGURE 18.5**   Frog, muscular system, ventral view.

A skeletal muscle typically consists of a fixed end, called the **origin,** and a movable end, called the **insertion.** The fleshy middle portion is called the **belly.** Most of the skeletal muscles taper at their ends into tough white cords of connective tissue, the **tendons,** which serve to attach them to bones or to other muscles.

The movement or effect caused by a muscle is its **action.** Thus, each skeletal muscle has a characteristic origin, insertion, and action. Most muscles are arranged in pairs or groups that are **antagonistic:** they have opposing actions.

Muscles are classified according to their actions. Some general classes of skeletal muscles are as follows: **extensors**—muscles that straighten or extend a part; **flexors**—muscles that bend one part toward another part; **adductors**—muscles that pull a part back toward the axis of the body; and **abductors**—muscles that draw a part away from the axis of the body.

◆ For a detailed study of the muscles of the frog, you must first remove the skin from a specimen. Specimens especially preserved in alcohol are best for this purpose although formalin-preserved specimens may also be used.

⚠ The chemicals used for preservation can be irritating to the skin and eyes. You should wear plastic or rubber gloves and protective eye gear when dissecting preserved specimens. See the section on laboratory safety following the preface of this book for further information about the properties of chemicals used for preservation and for other safety precautions. If you follow a few simple safety rules, work in the biology lab can be safe and rewarding.

## Skinning the Frog

1. Place the frog ventral surface up in a dissecting pan. Carefully cut through the skin along the midventral line from the anterior tip of the jaw to the region of the cloaca. Pull the skin up and away from the underlying muscles to avoid damage to the muscles.
2. Make two transverse cuts through the skin around the body anterior to the forelimbs and anterior to the hindlimbs.
3. Make additional cuts around the opening of the cloaca, and around the eyes and tympanic membranes on each side of the head.
4. Carefully loosen the skin from the underlying muscles with a **blunt** (not sharp!) **instrument.** Observe the thin white sheets of connective tissue between the skin and the muscles. Note the large areas where the skin is not attached to the muscles. These areas are **subcutaneous lymph sacs** where this colorless fluid collects in a living frog.
5. Gently peel off the skin to expose the muscles of the abdomen, head, and appendages. Loosen the skin at the base of each limb and carefully pull from the base toward the distal end of the appendage, turning the skin inside out.

After you have finished skinning the frog, compare your specimen with figures 18.4 and 18.5. Identify first the principal muscles of the abdomen. Carefully free and separate the muscles with a blunt instrument as you work. Follow the direction of the muscle fibers as you gently separate one muscle from another.

### Muscles of the Hindlimb

A study of the muscles of the well-developed hindlimbs provides an excellent introduction to the principles of muscle anatomy. Table 18.2 lists the principal muscles of the hindlimb. Start with the muscles of the dorsal surface of the thigh (upper leg) and identify the five principal muscles on this surface. Observe how these muscles provide control of the various movements of the leg and foot.

Continue your study with the muscles of the ventral surface of the thigh and then identify the muscles of the shank (lower leg). *How many pairs of antagonistic muscles can you identify from the hindlimb?*

### Oral Cavity

◆ Obtain a preserved or anesthetized frog from your instructor to study the anatomy of the **oral cavity.**

Frogs can be anesthetized by immersion in a solution of tricaine (ethyl m-aminobenzoate methanesulfonate) or urethane. You and your instructor should make sure that your frog is properly anesthetized before you proceed.

Place the frog on its back and make a small cut with your scissors at each corner of the mouth so the jaws can be opened widely. Rinse out the oral cavity with cold water if there is an accumulation of mucus. Consult figure 18.6 and locate the following structures in the oral cavity of your specimen: the **maxillary teeth** on the margin of the upper jaw, the **vomerine teeth** on the palate (these teeth are used for holding food rather than for chewing since the food is swallowed whole); the **internal nares** (probe through them from the **external nares**); and the openings into the **eustachian tubes.** Observe also the **tongue** and its attachment.

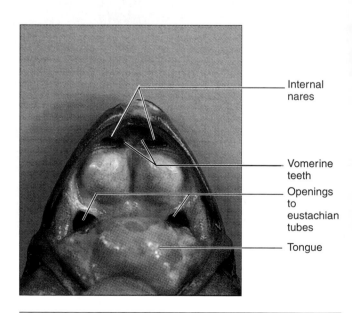

**FIGURE 18.6** Frog, oral cavity, ventral view.
Courtesy of Carolina Biological Supply Company, Burlington, NC.

# TABLE 18.2

## Chief Muscles of the Thigh and Shank

| Muscle | Location | Origin | Insertion | Action |
|---|---|---|---|---|
| **Dorsal Muscles of the Thigh** | | | | |
| Triceps femoris (three parts) | Lateral surface of thigh | Ilium, acetabulum | Tibiofibula | Flexes thigh and extends shank |
|   Rectus anterior femoris | | Ilium | Fascia attached to tibiofibula | Flexes thigh and extends shank |
|   Vastus internus | | Acetabulum | Tibiofibula | Flexes thigh and extends shank |
|   Vastus externus | | Ilium | Vastus internus | Flexes thigh and extends shank |
| Gluteus | Medial and anterior to triceps femoris | Ilium | Femur | Rotates thigh |
| Semimembranosus | Medial to triceps femoris | Ischium and pubis | Tibiofibula | Extends thigh and flexes shank |
| Biceps femoris (iliofibularis) | Between triceps femoris and semimembranosus | Ilium | Tibiofibula and femur | Extends and adducts thigh; flexes shank |
| Piriformis | Small muscle near cloacal opening; between biceps femoris and semimembranosus | Urostyle | Femur | Extends and rotates thigh |
| **Ventral Muscles of the Thigh** | | | | |
| Sartorius | Large, flat midventral muscle | Pubis | Tibiofibula | Flexes thigh and shank |
| Gracilis major (rectus internus major) | Medial to sartorius | Pubis | Tibiofibula | Extends thigh and flexes shank |
| Gracilis minor (rectus internus minor) | Medial to gracilis major | Pubis | Tibiofibula | Extends thigh and flexes shank |
| Adductor magnus | Medial and partly beneath sartorius | Ischium and pubis | Femur | Adducts thigh |
| Adductor longus | Beneath sartorius | Pubis | Femur | Adducts thigh |
| Semitendinosus | Under and between gracilis minor and adductor magnus | Ischium | Tibiofibula | Extends and adducts thigh, flexes knee |
| **Muscles of the Shank** | | | | |
| Gastrocnemius | Medial surface of shank | Femur | Achilles tendon | Flexes shank and foot |
| Peroneus | Lateral to gastrocnemius | Femur | Distal end of tibiofibula | Extends shank and foot |
| Tibialis anticus longus (tibialis anterior longus) | Anterior to tibiofibula | Femur | Tarsal bones | Lifts foot and flexes ankle |
| Tibialis posticus (tibialis posterior) | Posterior to tibiofibula | Tibiofibula | Tarsal bones | Flexes foot |
| Extensor cruris | Anterior to and partly beneath tibialis anticus longus | Femur | Tibiofibula | Extends shank |

◆ *What advantage does this peculiar attachment of the tongue have for the frog?*

Posterior to the oral cavity (behind the tongue) is the **pharynx.** At the rear of the pharynx is the opening into the esophagus; on its ventral surface find the **glottis**—a slitlike opening into the **larynx.** A pair of short **bronchi** connect the larynx to the two lungs.

The males of many species of frog have openings into two **vocal sacs** located at the rear of the mouth just anterior to the openings to the eustachian tubes. Inflation of the vocal sacs serves to amplify the croaking sounds. Vocal sacs are absent in female frogs.

## Internal Anatomy

◆ Before continuing with your study of internal anatomy of the frog, you should review the general instructions for dissection provided in Chapter 8.

If you are dissecting an **anesthetized** specimen, take care to minimize the cutting of blood vessels to prevent the blood from obscuring your view of the internal organs. Keep your specimen covered with cold water during dissection. Fasten the frog, ventral side up, to the bottom of the dissection pan with pins through the tip of the jaw and all the limbs.

If you have a **preserved** specimen for dissection, it may be saturated with the preservation fluid or it may simply be moist and come in a plastic bag. In either case, the tissues should be moist and pliable.

◆ Study the general location of internal organs of the frog in figures 18.7 and 18.8 to make certain that you understand the approximate location of the internal organs before you begin your dissection. Proceed carefully in your dissection to avoid unnecessary damage to body parts that you may need to study later.

Lift the skin from the body with your forceps and make a **longitudinal cut** slightly to one side of the midventral line and forward from the pelvis to the tip of the lower jaw.

Observe the subcutaneous lymph spaces and the attachment of the skin to the body as you free the skin from the ventral body surface. Make short transverse cuts in the skin at the anterior and posterior ends of the trunk and pin back the flaps of skin on both sides.

Consult figure 18.5 and locate the following three large muscles before continuing with your dissection: the **pectoralis,** the **rectus abdominis,** and the **external oblique.** Note also the whitish **linea alba** (the midventral connective tissue joining the lateral muscles) along the midventral line. The ventral abdominal vein lies beneath the muscular body wall along the midventral line.

Now lift the muscles of the abdomen with your forceps and make a **longitudinal incision** through the body wall with your scissors a little to one side of the linea alba. Cut through the bony **sternum,** which supports the forelimbs (figure 18.3), and up to the posterior end of the lower jaw,

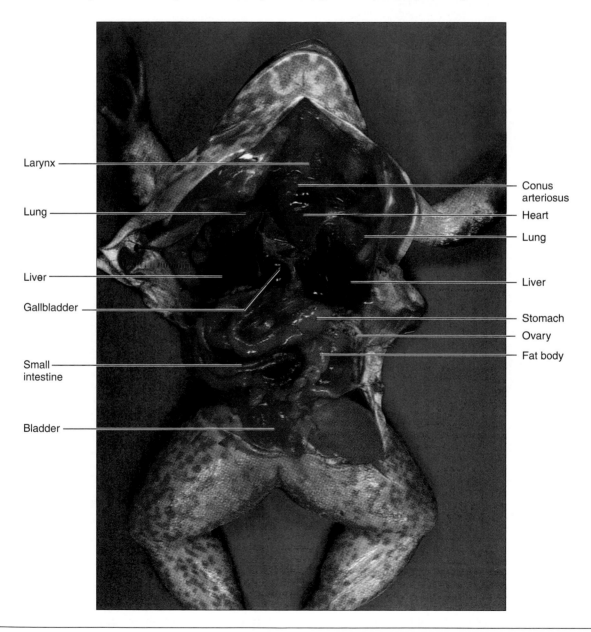

**FIGURE 18.7** Frog, internal organs, female, ventral view.
Photograph by Ken Taylor.

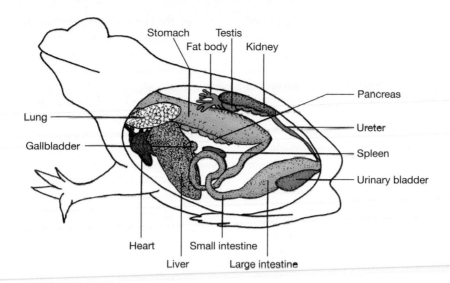

Stomach  Testis
Fat body  Kidney

Lung

Gallbladder

Pancreas

Ureter

Spleen

Urinary bladder

Heart    Small intestine

Liver    Large intestine

**FIGURE 18.8**   Frog, internal organs, lateral view.

taking care not to cut the ventral abdominal vein. Avoid damage to the internal organs within the coelom. Trace the ventral abdominal vein to the liver and make two short transverse incisions through the body wall just in front of the liver so the muscles can be pinned back to expose the internal organs. Remove a 10–12 mm section from the middle portion of the sternum and pin the forelimbs back to provide access to the internal organs in this region as shown in figure 18.7.

### Survey of Internal Organs

Now that you have completed the preliminary phase of your dissection, make a brief survey of the internal anatomy, referring to the figures cited in the following description. Note the position, size, shape, color, and texture of each organ. Consider also the relationship of each organ to other organs and attempt to determine the function(s) of each organ as you work. If you have a female specimen, and the ovaries and oviducts are filled with eggs and greatly enlarged, consult figure 18.14 and carefully remove the ovary and oviduct from one side of the body to facilitate your study of the other parts.

The location of the principal internal organs is illustrated in figures 18.7 and 18.8. The large, three-lobed **liver,** located just posterior to the pectoral girdle, is one of the most prominent internal organs. The liver is dark in color in both preserved and living specimens. Adjacent to the liver on the right side of the frog (viewed from the ventral surface as in a dissecting pan) is the curved, tubular **stomach.**

The **heart** is enclosed in a membranous sac, the **pericardial cavity,** which lies anterior to the liver and partly beneath the pectoral girdle. You will open the pericardial cavity later to study the heart and its relationship with its attached blood vessels. The two **lungs** are located on the sides of the visceral cavity posterior and lateral to the heart and dorsal to the liver. In a living or freshly killed frog, the lungs will be semitransparent, pink, and inflated. In preserved specimens, they will often be deflated.

If your specimen is a mature female, much of the visceral cavity will be filled with many black and white **eggs** enclosed in the large, membranous **ovaries.**

After you have located these major interior organs, you should proceed with a more detailed study of specific organ systems as directed by your laboratory instructor and as described in the following sections.

### Digestive System

Trace the pathway of the digestive system starting with the mouth and the **oral cavity.** Behind the **tongue,** locate the **pharynx** and the opening into the **esophagus.** The esophagus is a short, cylindrical tube that passes food material to the **stomach,** where the food is stored temporarily and digestion begins. The muscular stomach also provides a thorough mixing of the food mass by its contractions. At the posterior end of the stomach, find the **pyloric valve,** a sphincter that controls the release of the stomach contents into the small intestine.

The anterior segment of the small intestine is the **duodenum,** which receives secretions from the liver and pancreas through the common bile duct. Behind the duodenum is the convoluted **ileum,** the posterior section of the small intestine where digestion is completed and where most absorption of nutrients into the bloodstream occurs. The ileum empties into the **large intestine,** where most water as well as certain vitamins and ions are absorbed. The large intestine is also where the undigested residue is temporarily stored as fecal material. Posteriorly the large intestine narrows and empties into the **cloaca.** The cloaca is a common chamber that collects materials from the digestive, excretory, and reproductive systems prior to their discharge through the cloacal opening.

### Respiratory System

Adult frogs accomplish the necessary exchange of gases by using three different parts of their body: the **skin,** the **lungs,** and the **lining** of the **oral cavity** and **pharynx.** Locate the following parts of the respiratory system of the frog, which supplement the gas exchange accomplished through the skin.

Find the **external nares** located on the anterior part of the head (figure 18.1) and the **internal nares** on the roof of the oral cavity (figure 18.6). Air enters via the nares and passes into the **nasal cavity,** which connects the internal and external nares on each side. From the **pharynx,** posterior to the oral cavity, air passes through the **glottis** into the **larynx,** a cartilaginous tube that divides posteriorly to form two **bronchial tubes** or **bronchi.** The two bronchial tubes connect with the two **lungs.** The lungs are located dorsal to the heart and liver, one on each side of the body (figure 18.7). Inside each lung are thousands of tiny **air sacs,** each surrounded by numerous capillaries. In these sacs, the exchange of gases with the blood takes place.

## Circulatory System

Amphibians have a **three-chambered heart** with two **atria** and a **ventricle** (figure 18.7). In the adult frog, part of the venous blood moves from the heart to the lungs and then returns to the heart before it is pumped out to the various organs of the body. There are two major divisions of the circulatory system: (1) the **pulmocutaneous circulation** and (2) the **systemic circulation.** Preserved frogs that have been injected with latex are most satisfactory for detailed study of the circulatory system. Injected specimens may be singly injected (arteries only) or doubly injected (both arteries and veins injected with latex). Arteries usually are filled with red latex and veins are filled with blue latex, except for the pulmonary artery and vein.

*Heart.* Carefully remove the enveloping membrane, the **pericardium,** from around the heart. Consult figures 18.7 and 18.9, and identify the thin-walled right and left **atria,** the thick-walled **ventricle,** and the **conus arteriosus,** which divides into two branches, the right and left **truncus arteriosus.** Lift the ventricle and find the dark-colored, thin-walled, triangular **sinus venosus** on its dorsal side (figure 18.12), the **posterior vena cava** entering it at the posterior end, and the right and left **anterior vena cavae** entering it at the anterior end.

*Arterial System.* Study the arterial system with the aid of figures 18.9, 18.10, and 18.11. Note the left and right branches of the **truncus arteriosus,** each giving rise to three large arteries called aortic arches.

1. The **common carotid artery.** This divides into
   a. the **external carotid** (lingual artery) leading to the ventral part of the head and to the tongue;
   b. the **internal carotid** to the dorsal part of the head and brain. (Note the carotid gland.)
2. The **systemic arch.** The two systemic arches encircle the heart to pass around the pharynx and unite to form the large **dorsal aorta.** Note the arteries leading from each systemic arch:
   a. the **short occipitovertebral artery** dividing into the **occipital artery** leading to the skull and the **vertebral artery** leading to the vertebral column;
   b. the **subclavian artery,** giving off branches to the shoulder region and extending into the arm where it becomes the **brachial artery;**
   c. the **dorsal aorta;**

d. the **coeliacomesenteric,** which divides into the **coeliac** and the **mesenteric;**
e. the **right** and **left gastric arteries** to the stomach, the **pancreatic artery** to the pancreas, and the **hepatic artery** to the liver (these latter arteries are all branches of the coeliac);
f. the **mesenteric artery** to the intestine, with branches to the spleen (the **splenic artery**) and to the rectum and large intestine **(posterior mesenteric artery);**
g. the **urogenital arteries** to the kidneys, gonads, and fat bodies (**renal artery** to kidneys and **genital artery** to gonads). (The urogenital arteries vary considerably in number and arrangement in different individuals. Sometimes they do not inject well because of their small size and may be difficult to find. *How many are there in your specimen?);* the **lumbar arteries** to the dorsal body wall (several pairs of small arteries arising from the dorsal surface of the aorta—not shown in the figure, also small and often hard to find);
h. the **common iliac arteries,** divisions of the dorsal aorta;
i. the **epigastric artery** to the bladder and body wall of that region;
j. the **femoral artery** in the thigh;
k. the **sciatic artery,** a continuation of the external iliac artery furnishing branches to most of the muscles of the leg; and
l. the **peroneal** and **tibial arteries,** divisions of the sciatic to lower parts of the leg as shown in figure 18.11.
3. The **pulmocutaneous artery.** This divides into
   a. the **pulmonary artery** going to the lung and
   b. the **cutaneous artery** going to the skin. Note that this artery and its branches carry **deoxygenated** blood.

*Venous System.* Veins are blood vessels that carry blood **toward** the heart. Study the venous system of your specimen with the aid of figure 18.12 and locate the following veins.

1. The two (right and left) **anterior venae cavae** into which enter: (a) the **external jugular vein** with its branches, the **lingual vein** from the tongue and floor of the mouth and **maxillary vein** or mandibular from the jaw; (b) the **innominate vein** into which flows the **internal jugular vein** from the deeper parts of the head and the **subscapular vein** from the shoulder; and (c) the **subclavian vein** formed by the union of the **musculocutaneous vein** from the muscles and skin of the side and back, and the **brachial vein** from the forelimb.
2. The **posterior vena cava** into which enter: (a) the **hepatic veins** from the liver; (b) the **renal veins** from the kidneys; and (c) the **genital veins** from the gonads.
3. The **hepatic portal system** consists of: (a) the **abdominal vein,** which enters the liver and is formed

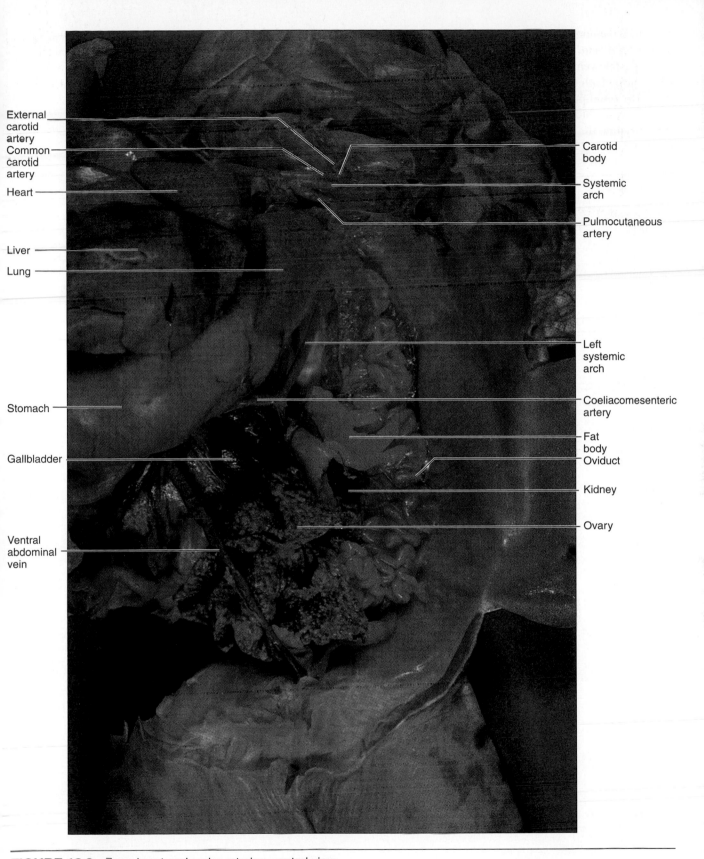

External carotid artery

Common carotid artery

Heart

Liver

Lung

Stomach

Gallbladder

Ventral abdominal vein

Carotid body

Systemic arch

Pulmocutaneous artery

Left systemic arch

Coeliacomesenteric artery

Fat body

Oviduct

Kidney

Ovary

**FIGURE 18.9** Frog, heart and major arteries, ventral view.
Photograph by Ken Taylor.

by the union of the two **pelvic veins;** (b) the **hepatic portal vein,** which carries blood from the stomach (**gastric vein**), intestine (**mesenteric vein**), and the spleen (**splenic vein**).

4. The **renal portal system.** The **renal portal veins** carry the blood to the kidneys. They receive the blood from (a) the **dorsolumbar vein;** (b) a branch of the

**femoral vein** from the hindlimb; and (c) the **sciatic vein** from the thigh. Note that the blood from the femoral vein may go to the kidney by way of the renal portal vein or to the liver by way of the pelvic and abdominal veins.

5. The **pulmonary veins.** These veins carry **oxygenated** blood from the lungs and unite to enter the left auricle.

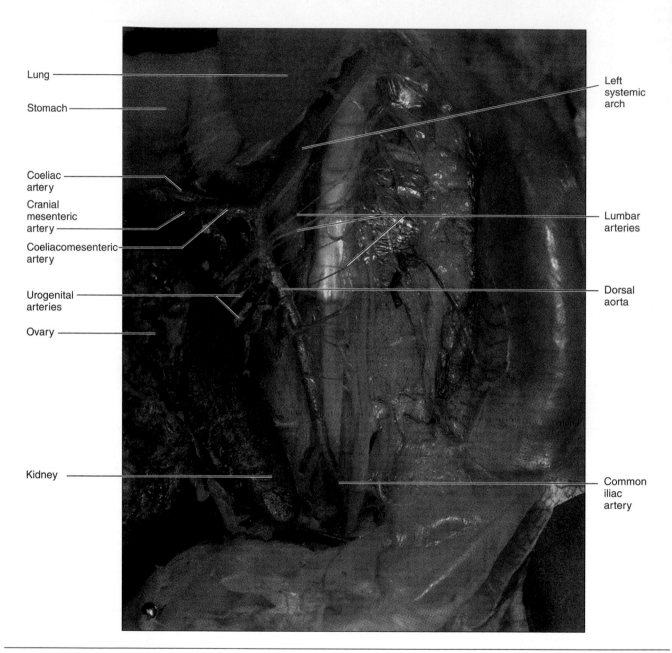

**FIGURE 18.10**  Frog, dorsal aorta and abdominal arteries, ventral view.
Photograph by Ken Taylor.

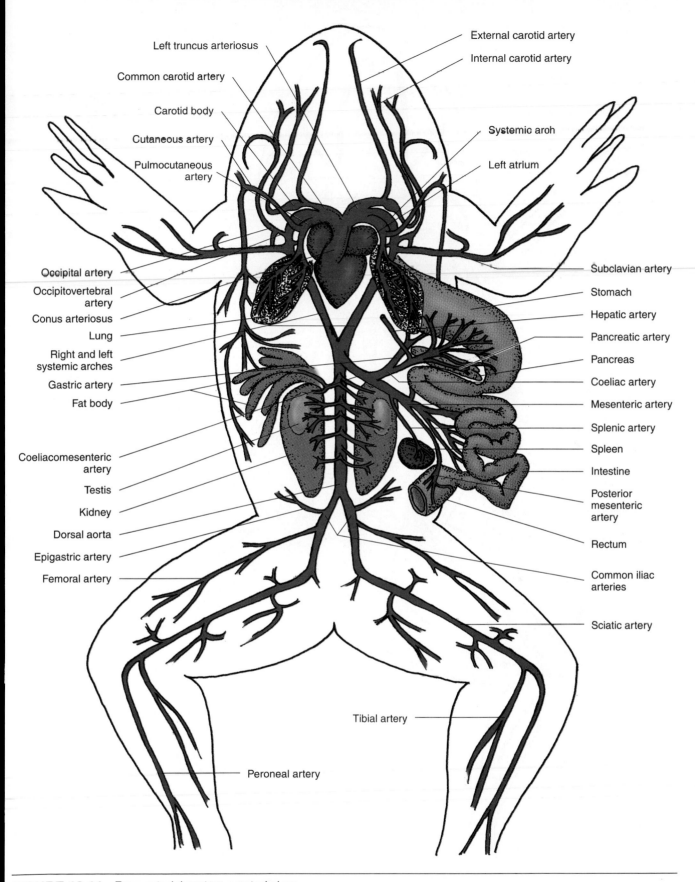

Left truncus arteriosus

Common carotid artery

Carotid body

Cutaneous artery

Pulmocutaneous
artery

External carotid artery

Internal carotid artery

Systemic arch

Left atrium

Occipital artery

Occipitovertebral
artery

Conus arteriosus

Lung

Right and left
systemic arches

Gastric artery

Fat body

Coeliacomesenteric
artery

Testis

Kidney

Dorsal aorta

Epigastric artery

Femoral artery

Subclavian artery

Stomach

Hepatic artery

Pancreatic artery

Pancreas

Coeliac artery

Mesenteric artery

Splenic artery

Spleen

Intestine

Posterior
mesenteric
artery

Rectum

Common iliac
arteries

Sciatic artery

Tibial artery

Peroneal artery

**FIGURE 18.11**  Frog, arterial system, ventral view.

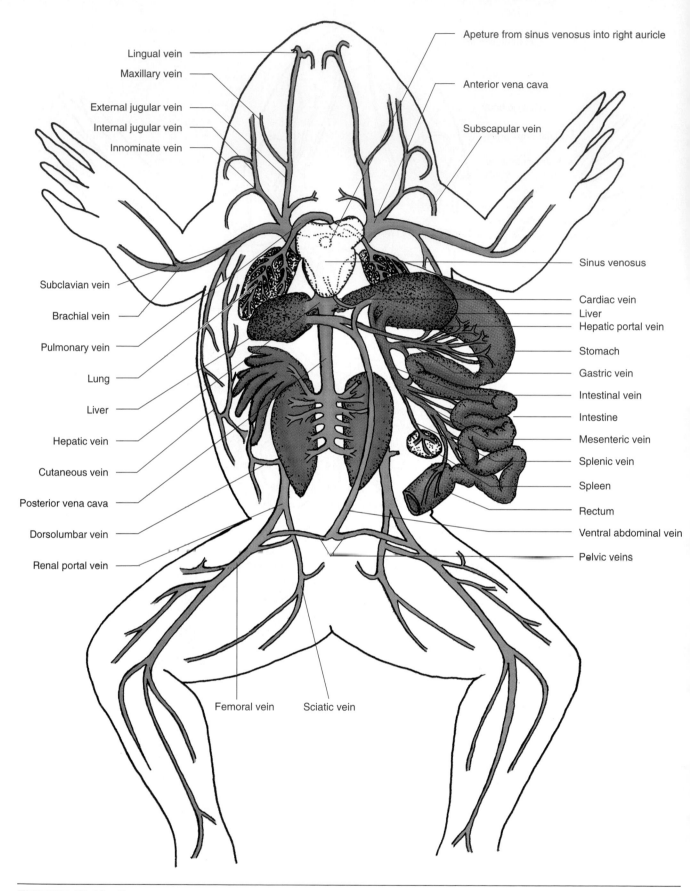

Lingual vein

Maxillary vein

External jugular vein

Internal jugular vein

Innominate vein

Subclavian vein

Brachial vein

Pulmonary vein

Lung

Liver

Hepatic vein

Cutaneous vein

Posterior vena cava

Dorsolumbar vein

Renal portal vein

Apeture from sinus venosus into right auricle

Anterior vena cava

Subscapular vein

Sinus venosus

Cardiac vein

Liver

Hepatic portal vein

Stomach

Gastric vein

Intestinal vein

Intestine

Mesenteric vein

Splenic vein

Spleen

Rectum

Ventral abdominal vein

Pelvic veins

Femoral vein     Sciatic vein

**FIGURE 18.12** Frog, venous system, ventral view. Veins are blue, and hepatic portal system is yellow.

## The Urogenital System

The excretory and reproductive organs are closely associated, and together comprise the **urogenital system.** If your specimen is a female with large ovaries concealing the other visceral organs, consult figures 18.13, 18.14, and 18.15, and carefully remove one of the ovaries. The excretory structures are similar in the two sexes of the frog. With the help of figure 18.13, again note the two **kidneys,** each with the adrenal gland on its ventral surface; the **ureter** leading from the posterior border of the kidney to the cloaca; the **urinary bladder,** which empties into the cloaca on the ventral side; and the **fat bodies.**

The kidneys are the major excretory organs of the frog. They are responsible for the removal of most of the wastes from the body, although some wastes are also lost through the skin. The kidneys remove most of the nitrogenous wastes from the blood, which are excreted in soluble form as **urea** and **ammonia.** These important organs also play a major role in **homeostasis,** the maintenance of a constant internal environment by removing excess ions and other substances from the blood and by conserving those substances of limited availability.

Specimens should be distributed in the class so that both male and female specimens are available at each table. After you have completed the study of the urogenital system on your own specimen, find a specimen of the opposite sex and make a comparative study so that you will be familiar with the organization of the urogenital system in both the male and female.

In a female specimen (figures 18.13 and 18.14), locate on one side the **ovary** attached dorsally and suspended into the coelom by a **mesentery;** the coiled **oviduct** with a funnel-shaped opening, the **ostium,** at its anterior end; and a **uterus,** an enlargement near the cloaca. Cut open the cloaca and use your probe to find the openings of the oviducts, the ureters, and the urinary bladder.

Near the time of their maturation, the eggs are released from the ovaries into the coelom and pass through the ostia into the oviducts. As the eggs pass down the oviduct, they are covered with several layers of jellylike material secreted by glands in the walls of the oviducts. The eggs collect in the uteri before they are discharged through the cloaca and into the water. Fertilization is external: the male mounts the female and discharges sperm on the eggs as they are discharged. This mating posture is called **amplexus.**

In a male specimen (figures 18.13 and 18.15), locate the two **testes,** each suspended from the dorsal abdominal wall by a mesentery, and the several **vasa efferentia,** the small ducts that carry sperm from the testes to the kidneys. The sperm are carried from the kidneys to the cloaca via the **ureters,** which serve as genital ducts in the male. Male leopard frogs also frequently have vestigial oviducts (mesonephric ducts) alongside the kidneys.

## Nervous System

The nervous system of vertebrates consists of three main parts: (1) the **central nervous system,** which includes the brain and the spinal cord; (2) the **peripheral nervous system,** which includes the nerves extending from the central nervous system; and (3) the **autonomic nervous system,** a specialized portion of the peripheral nervous system that regulates the glands of the body and the visceral organs.

◆ Successful study of the nervous system of the frog requires careful dissection. Do not rush the dissection, because you may damage important structures necessary for your study. Carefully remove the skin from the dorsal surface of the head between the eyes and along the vertebral column. Clear away the muscles and connective tissue underlying the skin to expose the skull and vertebral column. With your scalpel, carefully shave thin sections of bone from the skull until you expose the brain. Then carefully pick away additional small pieces of bone with your forceps until the brain is completely exposed. Continue the same procedure to expose the vertebral column and the spinal cord, but leave the exposed nervous system in place.

***The Brain.*** Consult figures 18.16 and 18.17 and identify the five main regions of the brain: the **telencephalon,** the **diencephalon,** the **mesencephalon,** the **metencephalon,** and the **myelencephalon.** The anterior-most telencephalon bears four distinct lobes: two **olfactory lobes** and, posterior to them, two **cerebral hemispheres.** Extending anteriorly from the olfactory lobes are the **olfactory nerves,** which carry impulses from the olfactory epithelium to the brain.

Posterior to the telencephalon is the diencephalon, a diamond-shaped depressed area directly behind and somewhat between the posterior portion of the cerebral hemispheres. The **pineal gland** (epiphysis) is a small, inconspicuous body attached to the dorsal wall of the diencephalon, which probably will be removed in your dissection of the skull. The stalk by which it was attached, however, may still be visible. The lateral walls of the diencephalon constitute the **thalamus.** Below the diencephalon lies the **pituitary gland,** which you will study later. The principal function of the telencephalon and diencephalon is the integration of olfactory signals. Chemical perception is of great importance to the frog, both for feeding and defense.

The mesencephalon, immediately posterior to the diencephalon, bears two large **optic lobes** that serve to integrate nerve impulses from the eyes. The optic lobes in the frog also provide perhaps the most important overall coordination of sensory information, a function carried out mainly by the cerebral hemispheres in higher vertebrates. Frogs will attack an insect seen as potential food long before they can smell it.

Posterior to the mesencephalon find the metencephalon, which is represented by the **cerebellum,** a narrow transverse portion of the brain lying immediately posterior to the optic lobes.

The most posterior part of the brain is the myelencephalon, consisting of the elongated **medulla oblongata,** which tapers gradually into the **spinal cord.** The anterior portion of the medulla oblongata has a thin roof that is frequently removed during dissection. The cerebellum and

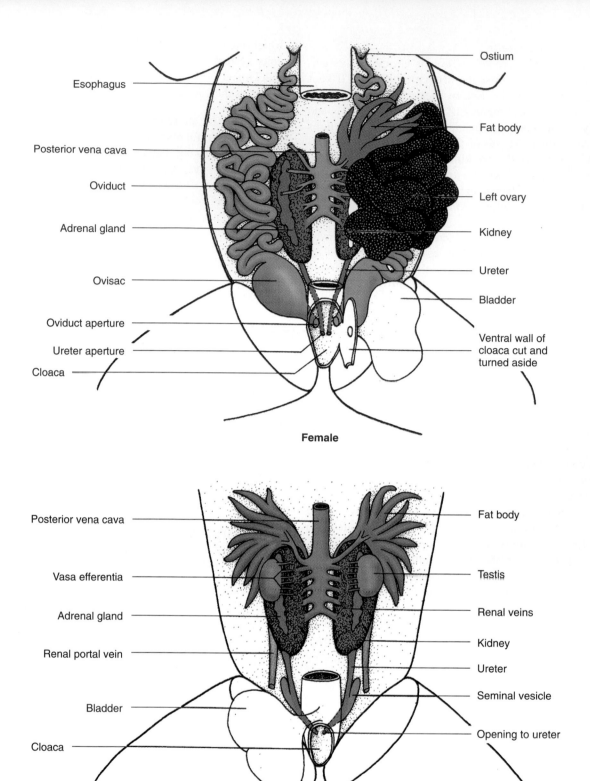

Esophagus

Posterior vena cava

Oviduct

Adrenal gland

Ovisac

Oviduct aperture

Ureter aperture

Cloaca

Ostium

Fat body

Left ovary

Kidney

Ureter

Bladder

Ventral wall of
cloaca cut and
turned aside

**Female**

Posterior vena cava

Vasa efferentia

Adrenal gland

Renal portal vein

Bladder

Cloaca

Fat body

Testis

Renal veins

Kidney

Ureter

Seminal vesicle

Opening to ureter

**Male**

**FIGURE 18.13** Frog, urogenital system, male and female, ventral view. Systemic veins are blue, and hepatic portal veins are yellow.

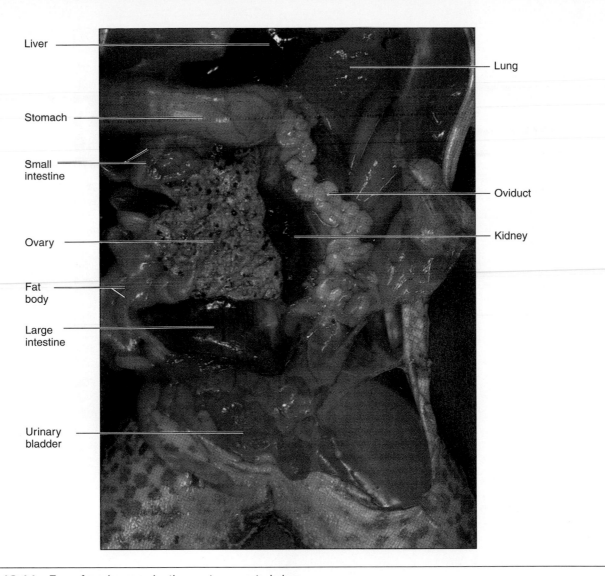

Liver

Stomach

Small
intestine

Ovary

Fat
body

Large
intestine

Urinary
bladder

Lung

Oviduct

Kidney

**FIGURE 18.14**  Frog, female reproductive system, ventral view.
Photograph by Ken Taylor.

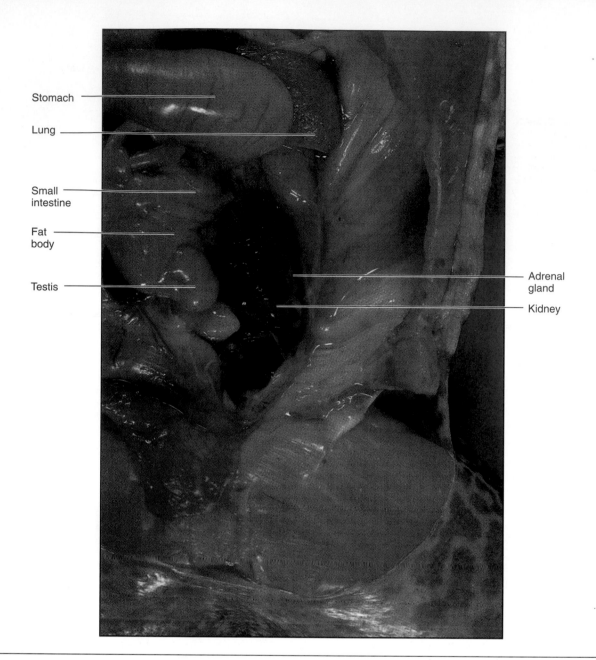

Stomach

Lung

Small
intestine

Fat
body

Testis

Adrenal
gland

Kidney

**FIGURE 18.15**   Frog, male reproductive system, ventral view.
Photograph by Ken Taylor.

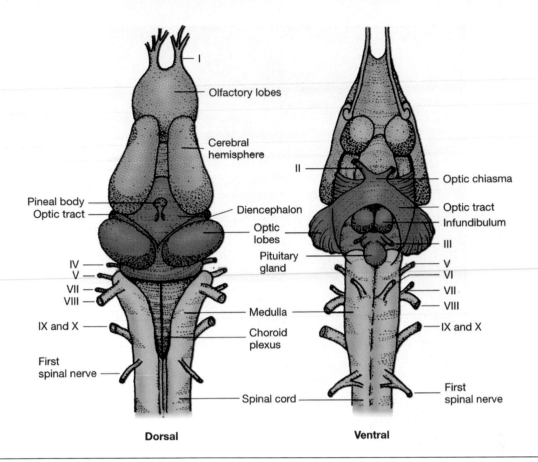

**FIGURE 18.16**   Frog, brain and cranial nerves.

medulla receive and integrate signals from the ears and from the skeletal muscles. They serve to monitor the general state of activity in the body and also coordinate gas exchange, feeding, and muscular activity.

◆ When you have completed your study of the dorsal side of the brain, **carefully** detach the anterior portion of the brain and lift it gently from the floor of the skull. Continue detaching the brain and spinal cord from its ventral connections and gradually free the entire nervous system. Place the isolated nervous system ventral-side-up in your dissecting pan and study the structures visible from this aspect.

Note the crossed optic nerves, called the **optic chiasma,** on the ventral side of the diencephalon. Observe how the nerves from the left eye carry their impulses across the brain to the right optic lobe. The optic nerves are the **second pair** of **cranial nerves** extending from the brain to various parts of the body. The olfactory nerves extending anterior to the olfactory lobes are the **first pair** of **cranial nerves.**

*Cranial Nerves.*   There are 10 pairs of cranial nerves in the frog, extending from the brain to various parts of the body. (Fishes and amphibians generally have 10 distinct pairs of cranial nerves; reptiles, birds, and mammals have 12 pairs.) Refer to figure 18.16 and locate the following cranial nerves.

**Nerve I**          Olfactory, extends from the naris to the olfactory lobe.

**Nerve II**         Optic, extends from the eye to the ventral side of the diencephalon. Note the optic chiasma (figure 18.16).

**Nerve III**        Oculomotor, extends from the ventral side of the mesencephalon to the muscles of the eye.

**Nerve IV**         Trochlear, extends from the dorsal side of the mesencephalon to the superior oblique muscle of the eye.

**Nerve V**          Trigeminal, closely associated with Nerve VII, the facial, from the side of the medulla to the skin of the face, muscles of the jaw, and tongue.

**Nerve VI**         Abducens, extends from the ventral surface of the medulla to the external rectus muscle of the eye.

**Nerve VII**        Facial, previously mentioned in connection with Nerve V.

**Nerve VIII**       Auditory (statoacoustic), extends from the side of the medulla to the ear.

**Nerve IX**         Glossopharyngeal, extends from the side of the medulla to the muscles and membranes of the tongue and pharynx.

**Nerve X**          Vagus, closely associated with Nerve IX, extends from the side of the medulla to the heart, lung, and digestive organs.

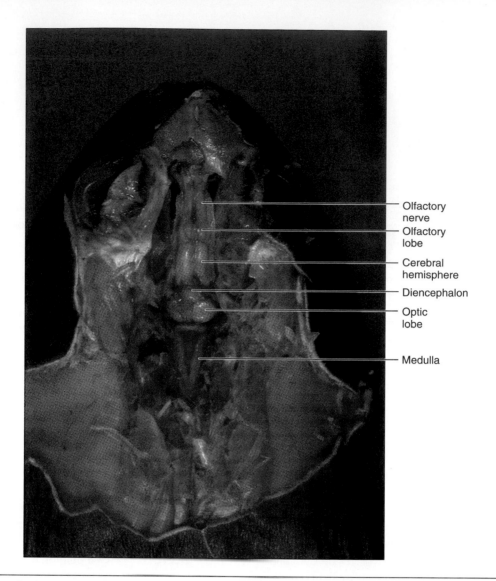

Olfactory
nerve

Olfactory
lobe

Cerebral
hemisphere

Diencephalon

Optic
lobe

Medulla

**FIGURE 18.17**  Frog, brain, dorsal view.
Photograph by Ken Taylor.

Next, observe the ventral outgrowth from the diencephalon, the **infundibulum,** which corresponds to the posterior lobe of the pituitary gland in birds and mammals. Located posterior and proximal to the infundibulum is a dorsal outgrowth originating from the roof of the mouth, the **anterior lobe of the pituitary** (also called the hypophysis). Together these two lobes constitute the **pituitary gland,** an important gland of internal secretion (**endocrine gland).** Note that the adult pituitary gland develops from two different sources—a ventral outgrowth from the diencephalon (the infundibulum) and a dorsal outgrowth from the oral cavity (the hypophysis). The hypophysis frequently adheres to the floor of the skull and may be detached during your removal of the brain. If you do not find it on the isolated nervous system, search on the floor of the skull in the appropriate location.

*Spinal Cord and Spinal Nerves.*    Observe now the spinal cord. Note that the terminal portion of the spinal cord in the adult frog is threadlike and is called the **filum terminale.** Ten pairs of **spinal nerves** are attached to the spinal cord

(figure 18.18). Each nerve has two roots: the **dorsal root** and the **ventral root.** Be careful not to disturb the sympathetic nerves while tracing the spinal nerves with which they unite. Note the large size of the **second spinal nerve.** Find the **brachial plexus** and the **sciatic plexus** where several of the spinal nerves leading to the forelimbs and hindlimbs extend from the spinal cord separately, then unite, and later rebranch. Trace the large **sciatic nerve** to the sciatic plexus and follow its principal branches into the hindlimb.

*Autonomic Nervous System.*    Examine figure 18.19, which shows the autonomic system in solid black. Note the two **sympathetic nerve trunks,** which originate anteriorly as the Gasserian ganglia and extend backward along the systemic arches and dorsal aorta. Find the **sympathetic ganglia;** the **sympathetic nerves** connecting with the spinal nerves; the **cardiac plexus;** the **solar plexus;** the large peripheral **splanchnic nerves** with branches to the stomach, intestines, and other organs; and the delicate **peripheral nerves,** which pass to the gonads, kidneys, spleen, adrenal glands, and other organs.

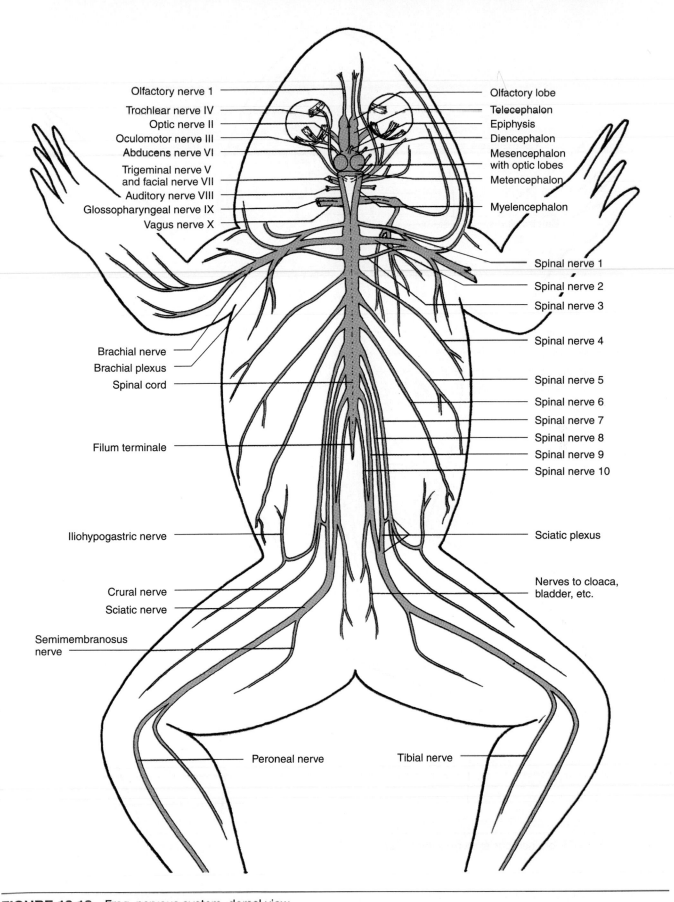

Olfactory nerve 1
Trochlear nerve IV
Optic nerve II
Oculomotor nerve III
Abducens nerve VI
Trigeminal nerve V and facial nerve VII
Auditory nerve VIII
Glossopharyngeal nerve IX
Vagus nerve X

Olfactory lobe
Telecephalon
Epiphysis
Diencephalon
Mesencephalon with optic lobes
Metencephalon
Myelencephalon

Spinal nerve 1
Spinal nerve 2
Spinal nerve 3
Spinal nerve 4

Brachial nerve
Brachial plexus
Spinal cord

Spinal nerve 5
Spinal nerve 6
Spinal nerve 7
Spinal nerve 8
Spinal nerve 9
Spinal nerve 10

Filum terminale

Iliohypogastric nerve

Sciatic plexus

Nerves to cloaca, bladder, etc.

Crural nerve
Sciatic nerve

Semimembranosus nerve

Peroneal nerve

Tibial nerve

**FIGURE 18.18**   Frog, nervous system, dorsal view.

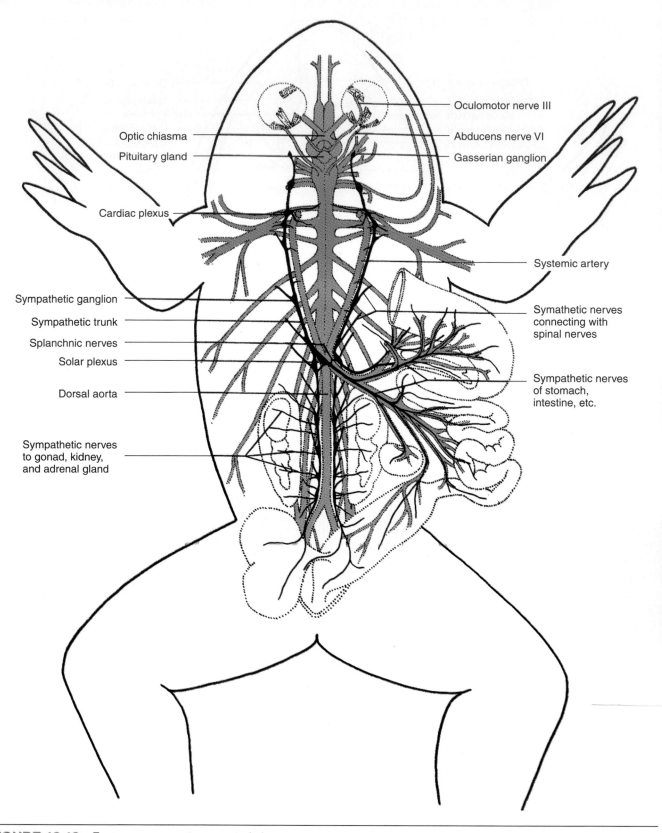

Optic chiasma

Pituitary gland

Cardiac plexus

Oculomotor nerve III

Abducens nerve VI

Gasserian ganglion

Systemic artery

Sympathetic ganglion

Sympathetic trunk

Splanchnic nerves

Solar plexus

Dorsal aorta

Sympathetic nerves
to gonad, kidney,
and adrenal gland

Symathetic nerves
connecting with
spinal nerves

Sympathetic nerves
of stomach,
intestine, etc.

**FIGURE 18.19**   Frog, nervous system, ventral view.

## Demonstrations

1. Bullfrogs—injected and dissected to show the various systems
2. Various vertebrate skeletons for comparison with the bullfrog
3. Beating of the frog's heart and circulation of the blood through capillaries of the web of the frog's foot

## Key Terms

**Abductor muscle**   a skeletal muscle that pulls an appendage or body part away from the central axis of the body.

**Adductor muscle**   a skeletal muscle that pulls an appendage or body part toward the central axis of the body.

**Autonomic nervous system**   division of the vertebrate nervous system that controls involuntary functions, such as breathing or heart rate. Consists of branches from several cranial and spinal nerves leading to various internal organs.

**Central nervous system**   portion of the vertebrate nervous system consisting of the brain and the spinal cord.

**Cranium**   portion of the skull that encloses the brain; the braincase.

**Ectothermal** (poikilothermal)   refers to an organism whose body temperature varies with that of the thermal environment.

**Extensor muscle**   a skeletal muscle that extends or straightens an appendage.

**Flexor muscle**   a skeletal muscle that bends an appendage.

**Homeostasis**   maintenance of a steady internal physiological state.

**Peripheral nervous system**   portion of the nervous system outside the central nervous system made up of the nerves connecting the organs and tissues with the central nervous system.

**Tadpole**   larval form (typical of amphibians) with an ovoid body and a thin muscular tail.

**Urogenital system**   combined organ system of vertebrates specialized for excretion and reproduction.

**Vertebra**   unit of the spinal (vertebral) column or "backbone" of the frog and other vertebrates.

**Visceral skeleton**   skeletal parts associated with the support of the gills in fishes and other primitive vertebrates; vestiges remain in the frog and higher vertebrates mainly in the hyoid apparatus that helps support the tongue and trachea.

## Internet Resources

Visit the zoology website at http://www.mhhe.com/zoology to find live Internet links of each of the references listed below.

1. Interactive Frog Dissection. So far, this is my favorite site, as it shows very clear pictures and animations of the entire process from putting the frog in the pan to identifying internal organs. This site is designed to prepare high school students for frog dissection. It is an interesting review for beginning zoology students or an introduction to frog anatomy for zoology students who did not dissect a frog in high school. The video may not work with all systems or platforms.

2. Stanford University X-rays of the Frog. A few X-rays of the frog; the most useful one is the labeled picture of the skeleton of the frog.

3. The Virtual Frog Project. This was designed at the Lawrence Berkeley National Laboratory using X-ray CT imaging, MRI imaging and other visualization techniques to make clickable three-dimensional images of the frog and its internal organs.

4. Comparison of the Common Frog Skeleton to the Human Skeleton. Text review of the differences and similarities between the skeletons.

5. Frog Dissection. One unlabeled photo of internal organs.

6. Introduction to the Amphibia. This University of California at Berkeley, Museum of Paleontology site provides links to information on the fossil record, life history, and systematics of the amphibians.

## Critical Thinking Questions

1. Describe the metamorphosis of a frog. What changes occur in the major organ systems? Speculate on the evolutionary history of amphibians.

2. What special adaptations evolved to protect amphibian eggs from desiccation?

3. Is it fair to say that amphibians—for all their terrestrial aptitudes—are still tied to a watery environment (whether that water be within the mouth of a parent frog, a pond, or a foamy nest excreted by a parent)?

4. Are frogs endotherms or ectotherms? What limitations (if any) does this thermoregulatory strategy place on them?

5. Discuss the significance of antagonistic pairs of muscles operating against the bones of an endoskeletal system.

6. What advantages are there for the rear attachment of a frog's tongue?

7. Discuss the circulatory system of the frog and the apparent advances over those of the shark and bony fish. Are these changes possibly the result of a terrestrial existence? Explain.

## Suggested Readings

Blaustein, A.R., and D.B. Wake. 1995. "The Puzzle of Declining Amphibian Populations." *Scientific American* 272:52–57.

Chaisson, R.B., and R.A. Underhill. 1994. *Laboratory Anatomy of the Frog and Toad,* 6th ed. Dubuque, IA: Wm. C. Brown Communications, Inc.

Cloudsley-Thompson, J.L. 1999. *The Diversity of Amphibians and Reptiles: An Introduction.* New York: Springer Verlag.

Conant, R., and J.T. Collins. 1991. *A Field Guide to Reptiles and Amphibians,* 3d ed. The Peterson Field Guide Series. Boston: Houghton Mifflan Company.

Duellman, W.E., and L. Traub. 1994. *Biology of Amphibians.* Baltimore: Johns Hopkins University Press.

Stebbins, R.C., and N.W. Cohen. 1995. *A Natural History of Amphibians.* Princeton: Princeton University Press.

# NOTES AND SKETCHES

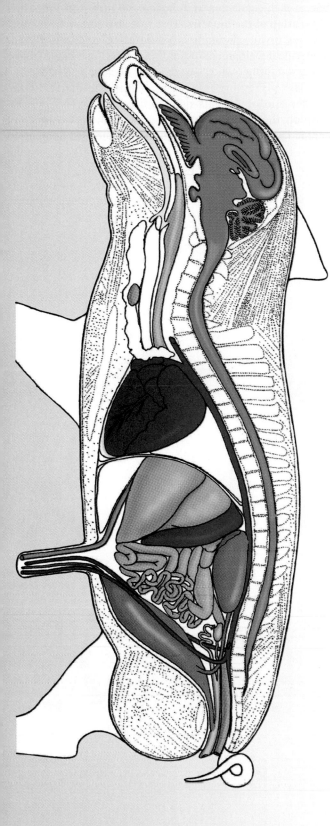

# CHAPTER 19

## Fetal Pig Anatomy

### OBJECTIVES

After completing the laboratory work in this chapter, you should be able to perform the following tasks:

1. Locate and identify the principal external features of a fetal pig.
2. Identify the chief components of the axial and appendicular skeletons of the fetal pig.
3. Locate five major skeletal muscles of the fetal pig and explain their origins, insertions, and actions. Distinguish between extension, flexion, adduction, and abduction.
4. Demonstrate on a specimen the divisions of the coelom and the five principal mesenteries that support the abdominal organs.
5. Locate and identify the organs of the digestive system of the fetal pig.
6. Find and explain the function of the parts of the urogenital system of the fetal pig. Demonstrate the differences between male and female specimens.
7. Trace the main circulatory pathway on a specimen. Explain the main changes in circulation that occur at birth.
8. Point out the similarities and differences among the hearts of a fish, a frog, and a fetal pig or other mammal. Discuss the significance of the differences.
9. Locate the main arteries and veins on a dissected specimen.
10. Identify the five divisions of the mammalian brain on a pig or sheep brain, and point out the brain parts that make up these divisions.

## The Fetal Pig: *Sus scrofa*

The fetal pig is often chosen for study as a representative mammal because it contains the organs and organ systems typical of most mammals. The principal distinguishing characteristics of the Order Mammalia include the following: a body surface covered with hair; an integument (skin) with several types of glands; a skull with two occipital condyles; seven cervical (neck) vertebrae; teeth borne on bony jaws; movable eyelids and fleshy external ears (pinnae); a four-chambered heart; a persistent left aorta; and a muscular diaphragm separating the thoracic and abdominal cavities. Mammals also are "warm-blooded" or **endothermic,** which means they produce heat to maintain a constant and elevated body temperature. The young develop within the uterus of the female, have a placental attachment for fetal nourishment, and are enveloped by special fetal membranes (amnion, chorion, and allantois). Milk to nourish the young after birth is produced by mammary glands. Other common representatives of this group include moles, bats, whales, mice, deer, monkeys, horses, cattle, and humans.

The fetal pig serves well as an example of mammalian anatomy because of its convenient size, availability, and relatively low cost. Fetal pigs are removed from the uteri of pregnant sows sold to meat packers. The gestation period in swine is 16 to 17 weeks; full-term fetal pigs measure about 30 cm (12 in) in length and weigh 1–1.5 kg (2–3 lbs). At 14 weeks, pig embryos average about 23 cm (9 in) in length. The number of pigs in a litter averages about seven to 12, although occasionally the litter may go as high as 18.

Many of the anatomical features of the fetal pig are typical of mammals, although some special features related to its embryonic condition will also be apparent during your study. Among the typical features of mammals are the **four**

**appendages, mammary glands, lungs, four-chambered heart, persistent left aortic arch, muscular diaphragm** that divides the coelomic cavity (into anterior pericardial and pleural cavities, and a posterior abdominal cavity), and endothermy (**homeothermy**). Specialized embryonic features include the **rudimentary development of the reproductive system**, the **low degree of ossification** in many bones of the skeleton, and the **soft texture and indistinct separation of the muscles.** There are also certain interesting features of the circulatory system associated with the intrauterine development of most mammals. The latter will be discussed further when we study the circulatory system.

The pig embryo develops within the uterus of the mother and obtains its nourishment and oxygen supply through the umbilical cord, which attaches to the placenta. The **placenta** is an important structural adaptation found in most mammals. It is made up partly of uterine (maternal) tissues and partly of embryonic tissues and provides an important channel for the supply of nutrients and oxygen and for the removal of metabolic wastes from the embryo.

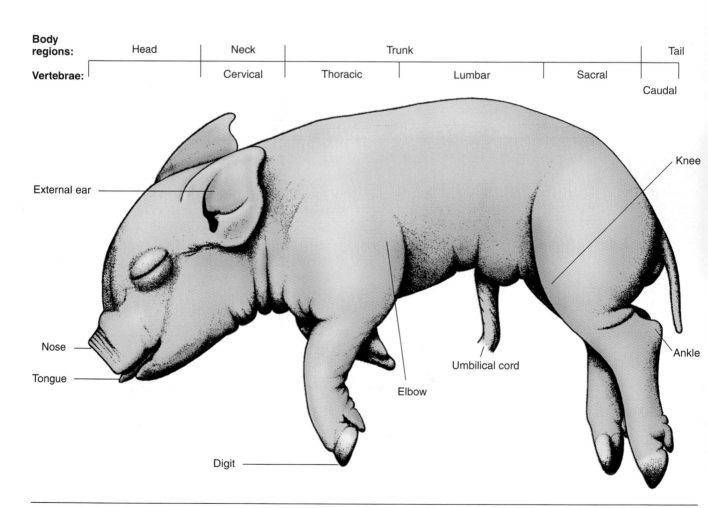

**FIGURE 19.1**   Fetal pig, external features.

**Living specimens**
Mammalian spermatozoa (demonstration)
**Preserved specimens**
Fetal pig
Cat skeleton (demonstration)
Mammalian placenta (demonstration)
Mammalian lung (demonstration)
Fetal pig, injected and dissected to illustrate circulatory system (demonstration)
Mammalian heart, dissected (demonstration)
Sheep brain (demonstration)
Fetal pig, dissected to show cranial and spinal nerves (demonstration)
**Prepared microscope slides**
Pig testis, cross section (demonstration)
Mammalian spermatozoa (demonstration)
Mammalian lung, cross section (demonstration)
Mammalian ovary, cross section (demonstration)
Mammalian spinal cord, cross section (demonstration)
**Audiovisual materials**
Anatomy of the Fetal Pig video
**Other**
Mounted cat skeleton

## External Anatomy

◆ Obtain a preserved fetal pig from your instructor and first study the general organization of the mammalian body and the principal features of its external anatomy (figure 19.1). Fresh hearts and brains can often be obtained from the meat department of local food stores.

Note that the body of the fetal pig is divided into three major regions—the **head, neck,** and **trunk.** At the posterior end of the trunk is the **tail.** The trunk consists of two principal subdivisions, the anterior **thorax** and the posterior **abdomen,** separated internally by the muscular diaphragm. Observe the two pairs of appendages, one pair of **forelimbs** and one pair of **hindlimbs.** Note that the limbs of the pig are directed ventrally rather than laterally as in the case of the frog studied earlier. This difference in orientation of the appendages represents an important advance of higher vertebrates and is closely linked with the evolution of the rapid and efficient locomotion characteristic of most mammals.

### Head

Principal external features of the head include the large **mouth,** the **eyes,** and the **ears.** Note the elongated **snout** with a pair of nostrils at the anterior end. On the snout, find the stiff sensory hairs or **vibrissae.** Observe also the strong lower jaws below the mouth and the hard, rounded dorsal cap between the ears. This bony "cap" represents a portion of the braincase or **cranium,** which protects the soft brain.

### Neck

Connecting the head with the trunk is a short, stout **neck.** The neck is equipped with powerful dorsal muscles. Squeeze the dorsal portion of the neck with your fingers to feel these muscles. *What is the significance of these muscles in the characteristic rooting of pigs?* Locate these neck muscles in figure 19.6.

### Thorax

The thoracic region comprises the anterior half of the trunk and includes the **pectoral girdle; the forelimbs;** and the **rib cage,** which encloses the lungs. Observe carefully the forelimbs and the feet. Note that the feet consist of **four digits.** The pentadactyl ("five-fingered") structure of the primitive vertebrate limb has been secondarily lost. The third and fourth digits are large and make up the two halves of the "cloven hoof" of the pig. The second and fifth digits are present but are reduced in size and are directed posteriorly. Feel the ribs and sternum in the thoracic region. These important skeletal parts provide protection for the lungs, heart, and major blood vessels within the pleural cavity.

### Abdomen

The abdominal region lies posterior to the thorax and contains the **peritoneal cavity** within which the abdominal organs are suspended. Observe the ventral **umbilical cord.** The point of attachment of the umbilical cord to the abdominal wall of the fetus is called the **umbilicus;** after birth, the scar (on the young pig's abdomen), marking the former site of the umbilicus, becomes the **navel.** Observe the severed end of the umbilical cord and note that there are **four large tubes** or channels within the tube. These tubes include an **umbilical vein,** two **umbilical arteries,** and an **allantoic duct.** The former three channels provide important routes for the exchange of food, oxygen, and other material between the mother and embryo; the latter represents an important channel for the disposal of metabolic waste products produced by the embryo.

At the posterior end of the abdomen are the **pelvic girdle** and the **hindlimbs.** Note the basic similarity in the organization of the forelimbs and hindlimbs. Find the several pairs of nipples or **mammae** on the ventral abdominal wall. Also locate the anus at the base of the tail.

You can determine the sex of your specimen by observing certain external features in the abdominal region. Male specimens exhibit a **urogenital opening** just posterior to the umbilical cord and two scrotal sacs at the posterior end of the body, just below the anus. In older specimens, the **testes** descend into these scrotal sacs from the peritoneal cavity. Female specimens exhibit a urogenital opening, the **vulva,** just ventral to the anus. A urogenital papilla protrudes from the vulval area.

### Tail

The **tail** is short, flexible, and usually curled. Its skeletal axis is a continuation of the vertebral column.

## Skeletal System

The general organization of the mammalian body can best be understood after study of the skeleton, the underlying supporting framework for the body. Unfortunately, the skeleton of the fetal pig is not very suitable for such study because in fetal pigs the skeleton is incompletely ossified and, consequently, some of the bones are not distinct. Therefore, you should refer to a mounted skeleton of a cat or similar vertebrate on demonstration for the purpose of general orientation. Figure 19.2 is provided to assist you in identifying various parts of the cat skeleton; compare it with figure 19.3, which illustrates the skeleton of a late-term pig embryo.

The skeleton of the pig and other mammals is made up of two divisions: the **axial skeleton** (skeleton of the body axis including the skull) and the **appendicular skeleton.** The appendicular skeleton provides articulation for the four limbs with the axial skeleton. The skull is a complex structure consisting of many fused bones that enclose and protect the brain and other organs of the head. The cat skull is more convenient to study than the skull of the pig and is more typical of the adult condition. Both skulls are made up of the same bones, although their shapes vary slightly. Figure 19.4 will help you identify the principal bones of the skull.

The principal bones of the cat skeleton are listed in the following outline. The vertebral column of the pig and other mammals exhibit some differences in the specific numbers of the vertebrae in some regions.

### Components of the Skeleton

A. **Axial skeleton**
   1. Skull
   2. Mandible (lower jaw)
   3. Hyoid apparatus (several small bones that support the tongue and larynx; remnant of the visceral skeleton of fishes and amphibians)
   4. Vertebral column
      Cervical vertebrae (7)
      Thoracic vertebrae (13)
      Lumbar vertebrae (7)
      Sacral vertebrae (3)
      Caudal vertebrae (20–23)
   5. Ribs
   6. Sternum

B. **Appendicular skeleton**
   1. Pectoral girdle
      Scapula
   2. Forelimb
      Humerus
      Radius
      Ulna
      Carpals
      Metacarpals
      Phalanges
   3. Pelvic girdle
      Ilium ⎫
      Ischium ⎬ fused to form the innominate
      Pubis ⎭
   4. Hindlimb
      Femur
      Patella
      Tibia
      Fibula

Tarsals (The calcaneus, or heel bone, is the largest of the tarsals.)
Metatarsals
Phalanges

## Muscular System

The muscles of the fetal pig are soft, incompletely developed, and easily torn. Special care is therefore necessary if you are to be successful in your study of the muscular system.

A skeletal muscle typically consists of a fixed end, called the **origin,** and a movable end, called the **insertion.** The fleshy middle portion of the muscle is called the **belly.** The whitish connective tissue covering the muscles is the **deep fascia,** and skeletal muscles are connected to cartilage, bone, ligaments, or skin by this fascia, by tendons, or by aponeuroses (singular: aponeurosis)—thin flat sheets of tough connective tissue.

The movement or effect caused by a muscle is its **action;** thus, each skeletal muscle has a characteristic origin, insertion, and action. Most muscles are arranged in pairs or groups that are **antagonistic:** they have opposite actions. Several actions of muscles have special names. Some common actions are the following:

| | |
|---|---|
| **Adduction** | moving the distal end of a bone closer to the ventral median line of the body |
| **Abduction** | moving the distal end of a bone farther away from the ventral median line of the body |
| **Flexion** | bending a limb at a joint |
| **Extension** | straightening a limb |

◆ The pig must be skinned before you can study its muscles. Make a midventral incision as indicated in figure 19.5. Extend the cut forward to the lower jaw and backward to the hindlimbs. **Take care to cut only through the skin; do not cut into the underlying musculature.** Next make an incision through the skin on the medial (inner) surface of the right forelimb to the hoof. Make a similar incision on the right hindlimb. Remove the skin from the right side of the body, and from the right forelimb and hindlimb.

After you have removed the skin, you will observe a thin whitish layer of connective tissue. This is the **superficial fascia,** which must be removed to expose the underlying muscles. Carefully remove the superficial fascia and study first the muscles of the dorsal side (figure 19.6). Then study the muscles of the ventral surface with the aid of figure 19.7. Tables 19.1 and 19.2 give the origins, insertions, and actions of the most prominent muscles of the pig.

## General Internal Anatomy

Place your specimen ventral-side-up in a dissecting pan and tie a stout cord to one of the forelimbs. Pass the cord under the pan and attach the cord to the other leg in order to secure the specimen in place, to spread the legs apart, and to provide access to the ventral body surface. Repeat the tying

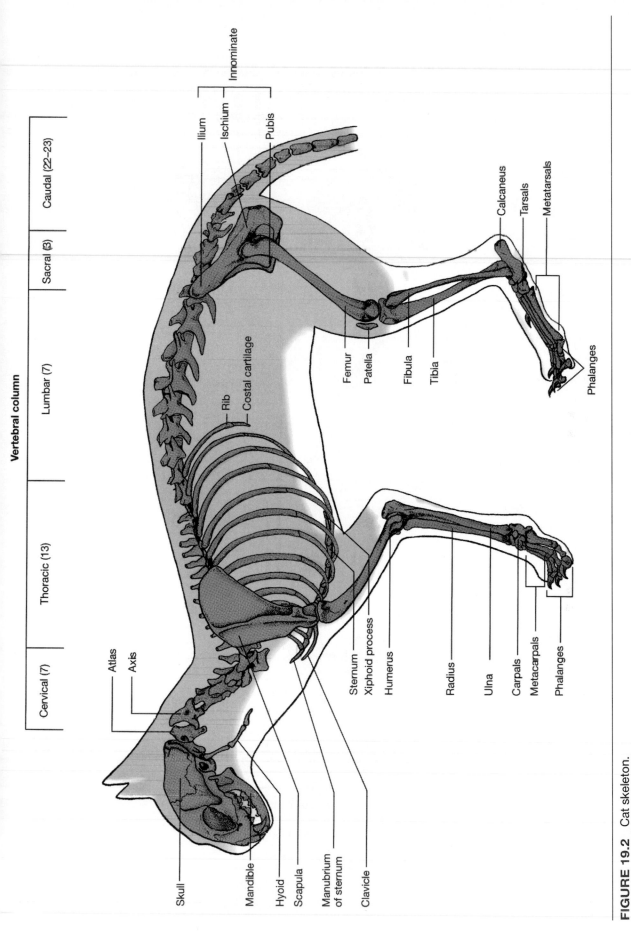

**FIGURE 19.2** Cat skeleton.

From Kendall/Hunt Publishing Company Biology Plate Series Part I, *Zoology*. Copyright © 1975 by Kendall/Hunt Publishing Company. Reprinted with permission.

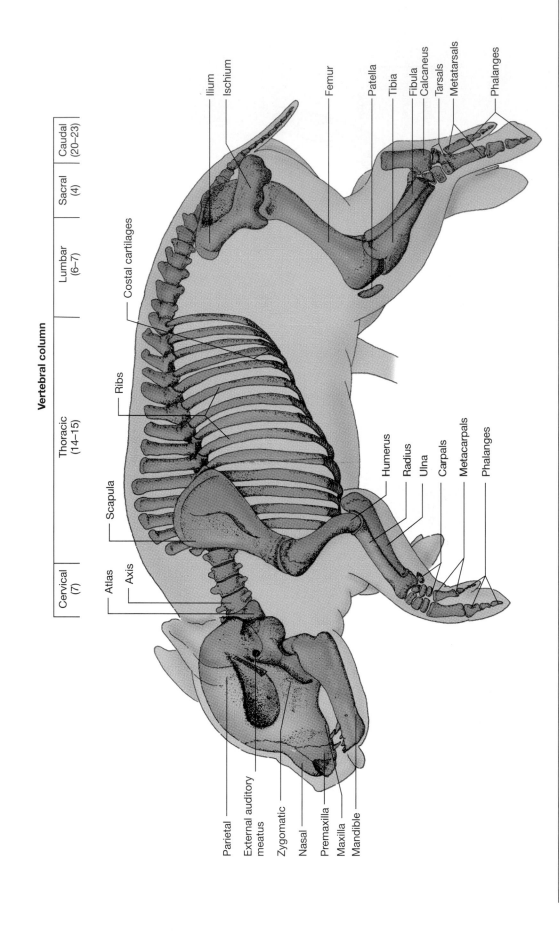

**FIGURE 19.3** Fetal pig skeleton.

From Kendall/Hunt Publishing Company Biology Plate Series Part I, *Zoology*. Copyright © 1975 by Kendall/Hunt Publishing Company. Reprinted with permission.

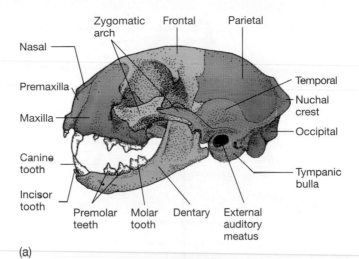

(a)

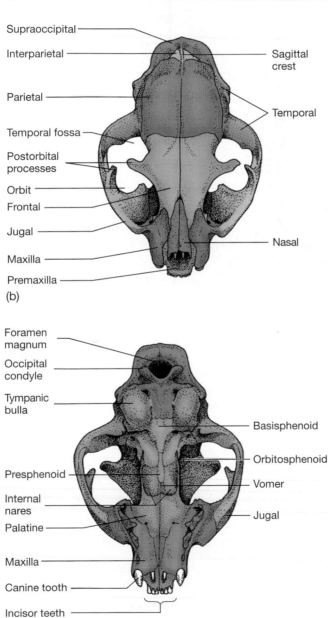

(b)

(c)

**FIGURE 19.4** Cat skull, (a) lateral view, (b) dorsal view, (c) ventral view.

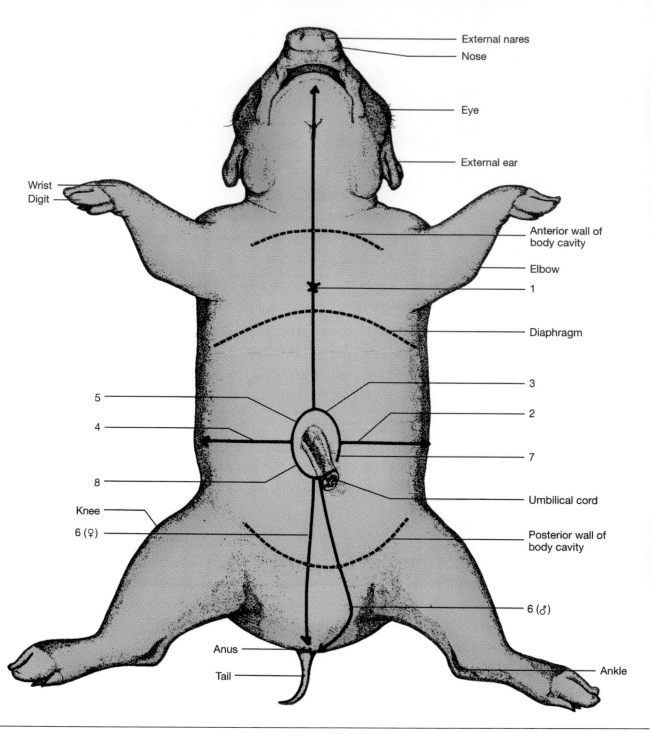

External nares
Nose
Eye
External ear
Wrist
Digit
Anterior wall of body cavity
Elbow
1
Diaphragm
3
5
2
4
7
8
Umbilical cord
Knee
6 (♀)
Posterior wall of body cavity
6 (♂)
Anus
Tail
Ankle

**FIGURE 19.5**   Fetal pig, ventral view with lines to indicate incisions for dissection. Numbers show sequence of recommended incisions.

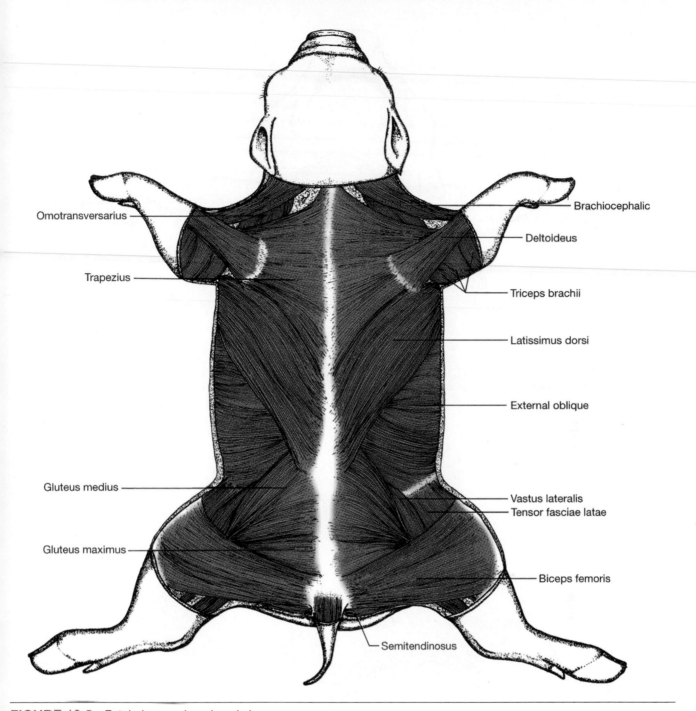

Omotransversarius

Brachiocephalic

Deltoideus

Trapezius

Triceps brachii

Latissimus dorsi

External oblique

Gluteus medius

Vastus lateralis

Tensor fasciae latae

Gluteus maximus

Biceps femoris

Semitendinosus

**FIGURE 19.6**   Fetal pig muscles, dorsal view.

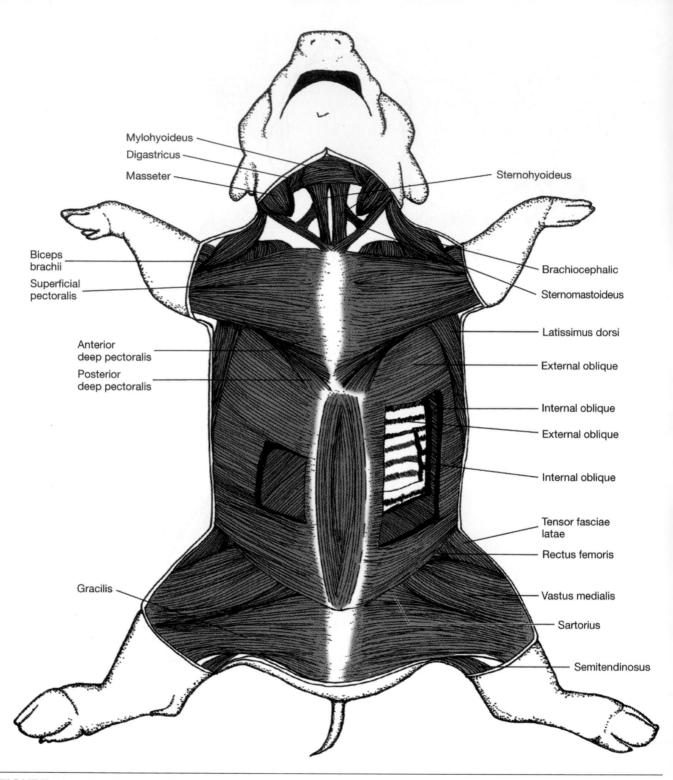

**FIGURE 19.7**   Fetal pig muscles, ventral view.

## TABLE 19.1

Muscles of the Dorsal Region

| Muscle | Origin | Insertion | Action |
|---|---|---|---|
| **Back and Side** | | | |
| Latissimus dorsi | Lumbar and thoracic vertebrae | Humerus | Draws humerus upward and backward |
| Trapezius | Skull; cervical and thoracic vertebrae | Scapula | Elevates shoulder |
| External oblique | Last 9 or 10 ribs, lumbodorsal fascia | Linea alba, ilium | Flexes trunk, constricts abdomen |
| **Forelimb** | | | |
| Splenius | Thoracic vertebrae | Skull, cervical vertebrae | Elevates head and neck |
| Deltoideus | Scapula | Humerus | Raises humerus |
| Brachiocephalic | Skull | Humerus, shoulder muscles | Raises head |
| Triceps brachii | Scapula, humerus | Ulna | Extends forearm |
| **Pelvic Region and Hindlimb** | | | |
| Gluteus medius | Longissimus dorsi muscle, ilium, sacroiliac | Femur | Abducts thigh |
| Vastus lateralis | Femur | Patella and tibia | Extends shank |
| Tensor fasciae latae | Ilium | Patella and tibia | Flexes hip joint, extends knee joint |
| Gluteus maximus | Sacral and caudal vertebrae | Fascia lata | Abducts thigh |
| Biceps femoris | Ischium, sacrum | Patella, leg, thigh | Abducts and extends limb |
| Semitendinosus | Caudal vertebrae, ischium | Tibia, calcaneus | Extends hip, flexes knee joint |

process with the hindlimbs. Study figure 19.8 to obtain a general idea of the thickness of the body wall and the approximate location of the internal organs before starting your dissection.

◆ Begin your dissection by making an incision through the skin and muscles of the ventral body wall overlying the sternum. Locate the sternum between the forelimbs and make a longitudinal incision as shown in figure 19.5. Continue the incision anteriorly to the level of the lower jaw and posteriorly to a point just anterior to the umbilical cord. Use a sharp scalpel for your dissection and cut cleanly through the skin and muscles. Take care not to damage any underlying organs.

Make a second incision from the region of the umbilical cord toward the side of the body and then connect the first and second incisions by cutting around the umbilicus to leave intact a portion of the ventral body wall immediately surrounding the umbilical cord. The fourth incision should be made laterally from the umbilical region toward the side of the pig. Continue your incisions in the sequence indicated in figure 19.5 to provide good access to the internal organs of the pig.

## Neck Region

◆ Remove the skin from the ventral surface of the neck, the lower jaw, and the left side of the head up to the base of the ear. Also, cut through the muscles on the ventral surface of the neck to expose the thymus gland. Take care to avoid damage to the several large blood vessels in the neck.

The large, spongy tissue of the **thymus gland** covers the ventral surface of the larynx and trachea. Identify the large cartilaginous **larynx** and the smaller **trachea** that extends posteriorly and splits into two **bronchi** that connect with the two lungs (figure 19.8). Locate the cartilaginous rings of the trachea that ensure a continuous passage for air to pass to and from the lungs. Just behind the larynx, locate the small bilobed **thyroid gland** on the surface of the trachea just posterior to the larynx. The thyroid gland is darker in color than the thymus and the trachea. Just ventral to the ear, find the large, light-colored **parotid gland** (figure 19.21). Ventral and slightly anterior to the parotid gland, find the smaller **submaxillary gland**. Also, locate the flat, narrow **sublingual gland** anterior to the submaxillary gland. The sublingual gland lies deep to the mylohyoid and geniohyoid muscles and surrounds the anterior portion of the submaxillary duct. The ducts from both the submaxillary and sublingual glands open into the floor of the mouth cavity.

## The Coelom and Its Divisions

The coelom of the pig is divided into two major regions by a muscular partition, the **diaphragm** (figure 19.9). Anterior to the diaphragm is the **thoracic cavity,** which is subdivided into **two pleural cavities** containing the lungs, and the **pericardial cavity** (mediastinum), which contains the heart and the large blood vessels connected with the heart. The thin epithelial lining of the pleural cavity is the **pleura;** that of the pericardial cavity is the **pericardium.** Posterior to the diaphragm is the large **peritoneal** (abdominal) **cavity** within which the abdominal organs are suspended. Each of the

# TABLE 19.2

## Muscles of the Ventral Region

| Muscle | Origin | Insertion | Action |
|---|---|---|---|
| **Head and Neck Region** | | | |
| Mylohyoideus | Mandibles | Hyoid bone | Raises floor of mouth, tongue, and hyoid bone |
| Digastricus | Mastoid process of skull | Mandible | Depresses mandible |
| Masseter | Zygomatic arch of skull | Mandible | Elevates jaw, closes mouth |
| Sternohyoideus | Sternum | Hyoid bone | Retracts and depresses hyoid and base of tongue |
| **Forelimb and Pectoral Girdle** | | | |
| Brachiocephalic | Nuchal crest, mastoid process of skull | Humerus and shoulder | Inclines or extends head, draws forelimb forward |
| Biceps brachii | Scapula | Radius, ulna | Flexes forearm |
| Superficial pectoralis | Sternum | Humerus | Adducts humerus |
| Anterior deep pectoralis | Sternum | Scapula, supraspinatus muscle | Adducts and retracts forelimb |
| Posterior deep pectoralis | Sternum, 4th to 9th ribs | Humerus | Retracts and adducts forelimb |
| **Side and Belly** | | | |
| Latissimus dorsi | Lumbar and thoracic vertebrae | Humerus | Draws humerus upward and backward |
| External oblique | Last 9 or 10 ribs | Linea alba, ilium | Constricts abdomen, flexes trunk |
| Internal oblique | Lumbodorsal fascia | Linea alba | Constricts abdomen, flexes trunk |
| External intercostal | Ribs | Adjacent rib | Pulls ribs together |
| Internal intercostal | Ribs | Adjacent rib | Pulls ribs together |
| Rectus abdominis | Pubic symphysis | Sternum | Constricts abdomen, bends vertebral column |
| **Pelvic Girdle and Hindlimb** | | | |
| Tensor fasciae latae | Ilium | Patella and tibia | Flexes hip joint, extends knee joint |
| Rectus femoris | Ilium | Patella and tibia | Extends shank |
| Vastus medialis | Femur | Patella and tibia | Extends shank |
| Semitendinosus | Ischium | Tibia | Extends hip, flexes limb |
| Sartorius | Iliac fascia and tendon of psoas minor muscle | Patella and tibia | Abducts hindlimb, flexes hip joint |
| Gracilis | Pubis | Patella and tibia | Adducts hindlimb |

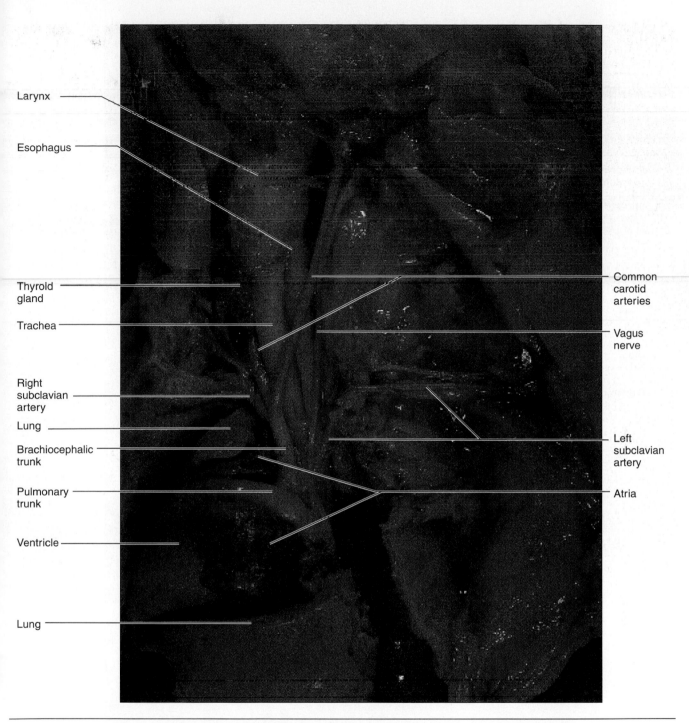

**FIGURE 19.8** Fetal pig, dissected, neck and thoracic region, ventrolateral view.
Photograph by Ken Taylor.

Larynx

Esophagus

Thyroid gland

Trachea

Right subclavian artery

Lung

Brachiocephalic trunk

Pulmonary trunk

Ventricle

Lung

Common carotid arteries

Vagus nerve

Left subclavian artery

Atria

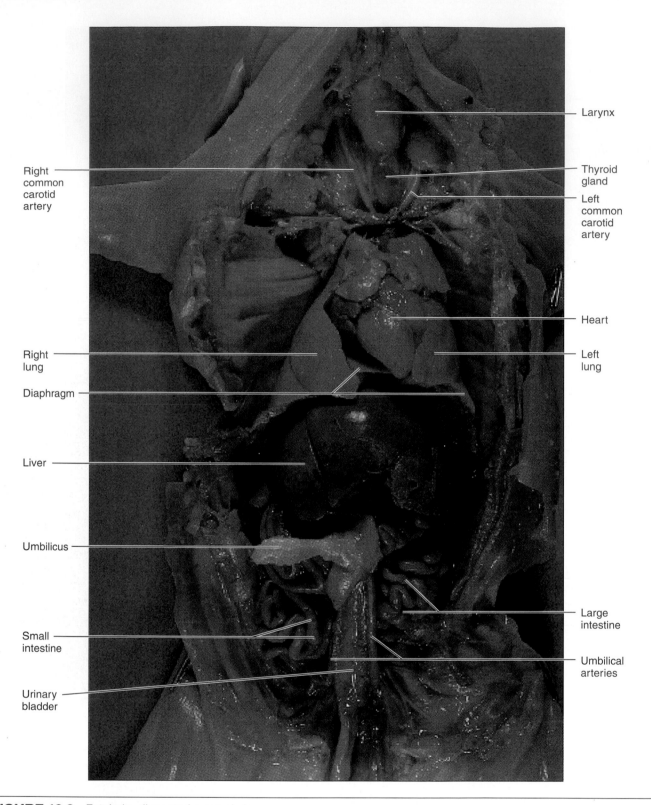

**FIGURE 19.9**  Fetal pig, dissected, ventral view.
Photograph by Ken Taylor.

abdominal organs is enclosed in a thin layer of mesodermal epithelium (**visceral peritoneum**); a similar layer also covers the inner surface of the body wall (**parietal peritoneum**). The double sheets of peritoneum that support the abdominal organs are called **mesenteries.**

Several mesenteries that support specific organs can be identified: (1) the **falciform ligament,** which connects the liver to the ventral body wall; (2) the **coronary ligament,** which attaches the anterior surface of the liver to the diaphragm; (3) the **greater omentum,** which attaches the greater curvature of the stomach to the colon and most of the small intestine; (4) the **lesser omentum,** which attaches the lesser curvature of the stomach to the liver, and connects the anterior portion of the duodenum; and (5) the **mesentery proper,** which suspends the small intestine from the dorsal midline. Other smaller mesenteries support the colon, the rectum, and the gonads from the dorsal body wall.

### Thoracic Cavity
Within the thoracic cavity, identify the **diaphragm,** the two **pleural cavities** enclosing the lungs, and the **pericardial cavity** around the heart.

The heart of the pig, like that of all mammals, consists of four chambers: **two atria** (auricles) and **two ventricles.**

◆ *How does the heart of mammals compare in this regard with that of the frog and the shark?*

Locate the four lobes of the right lung. *How many lobes of the left lung can you locate?* Observe the demonstration materials showing the internal structure of the mammalian lung.

### Peritoneal (Abdominal) Cavity
Within the peritoneal cavity (figures 19.9, 19.10, 19.11, and 19.12) identify the four-lobed **liver,** the **gallbladder,** the common **bile duct,** and the **stomach.** (In addition to the four main lobes of the liver, there is also a small caudate lobe attached to the posterior portion of the right lateral lobe.) Find the **esophagus** where it connects to the stomach.

The upper portion of the stomach, into which the esophagus empties, is called the **cardiac region.** It empties into the large **fundus,** seen externally as a large bulge on the left side of the stomach. The lower part of the stomach, which empties into the small intestine, is the **pyloric region.** The constriction between the stomach and small intestine is the **pylorus.** Passage of material from the stomach into the small intestine is regulated by a specialized circular muscle, the **pyloric sphincter.**

Locate the elongated, dark-red **spleen** along the greater curvature of the stomach and the light-colored granular **pancreas** found posterior to the stomach in the curve of the duodenum. Also observe the loosely coiled **small intestine** on the right and the tightly coiled **large intestine** on the left. Near the junction of the small intestine with the large intestine, locate a blind sac—the **caecum.** At the junction of the small and large intestines is the **ileocolic valve,** which regulates passage of materials into the large intestine. Posteriorly, the large intestine is modified to form the **rectum.**

Find the chief mesenteries that support the abdominal organs: **falciform ligament, coronary ligament, greater omentum, lesser omentum,** and **mesentery proper.**

Locate also the **urinary bladder,** which extends dorsally and posteriorly from the umbilicus. On either side of the bladder find the two **umbilical arteries** (figure 19.9).

After you have completed your general survey of internal anatomy, sever the small intestine about 3 cm posterior to the pylorus and cut the mesentery close to the intestine. Remove the small and large intestines from the body cavity after you have cut through the rectum about 5 cm from its posterior end. Observe the large blood vessels that supply the intestines (figures 19.16 and 19.20) as you remove this portion of the digestive tract. Examine the interior of the large and small intestines and of the caecum. *How do their interior walls differ in structure?* Notice also the numerous, light-colored lymph nodes found between the two layers of the mesentery. *Why do you think lymph nodes would be found in this area?*

## The Urogenital System
The excretory and reproductive (genital) systems are best studied together because they are closely related in development and because both systems use common ducts.

### The Excretory (Urinary) System
Refer to figures 19.13 and 19.14, and locate the kidneys on your specimen. A **ureter** leads from each kidney to the large **urinary bladder** (an enlargement of the embryonic allantois). Find the **allantoic duct,** which leads from the bladder through the umbilical cord. Note also the **renal arteries** (figure 19.17), which enter the kidneys and the large renal veins (figure 19.21) through which blood leaves the kidneys. Locate also the two large umbilical arteries (figure 19.9), which enter the umbilical cord. Along the medial border of each kidney, find the long, narrow, whitish **adrenal glands.** The adrenals are important endocrine glands.

◆ The specimens provided for each laboratory section should consist of both male and female pigs. Study the genital (reproductive) system of your specimen according to the following directions and then study a specimen of the opposite sex on a specimen assigned to one of your classmates.

### The Female Genital System
◆ Study figures 19.10, 19.12, and 19.13, and locate the principal female reproductive organs. Then cut through the skin, muscles, and pelvic bone along the midventral line of your specimen and pull the limbs wide apart to locate these parts on your specimen. Find the two small, bean-shaped **ovaries** near the posterior end of the peritoneal cavity.

Attached to the dorsal surface of each ovary is a small convoluted duct, the **fallopian tube** (oviduct). Locate the wide, ciliated funnel, called the **infundibulum,** at the terminus of each fallopian tube. The opening into the end of the fallopian tube is the **abdominal ostium.** Trace the fallopian

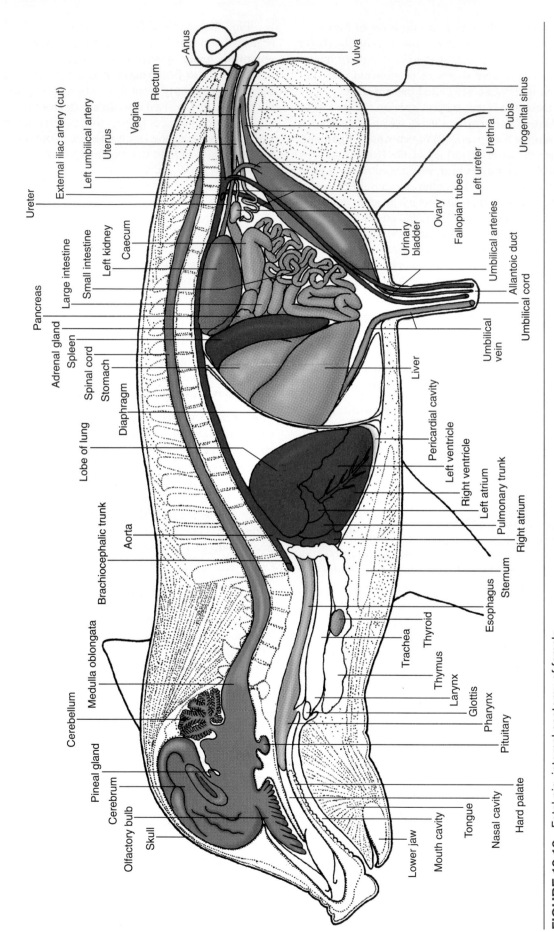

**FIGURE 19.10** Fetal pig, internal anatomy of female.

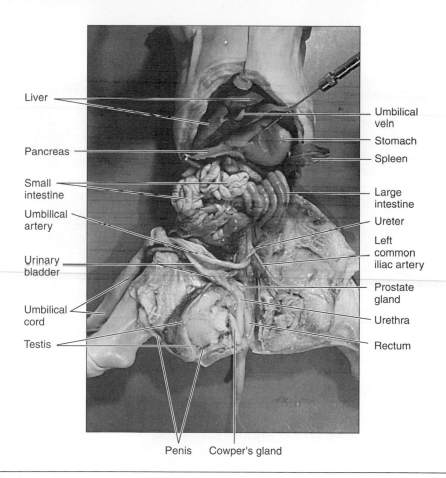

Liver

Pancreas

Small
intestine

Umbilical
artery

Urinary
bladder

Umbilical
cord

Testis

Umbilical
veln

Stomach

Spleen

Large
intestine

Ureter

Left
common
iliac artery

Prostate
gland

Urethra

Rectum

Penis     Cowper's gland

**FIGURE 19.11**   Fetal pig, abdominal organs, male.
Photograph by John Vercoe.

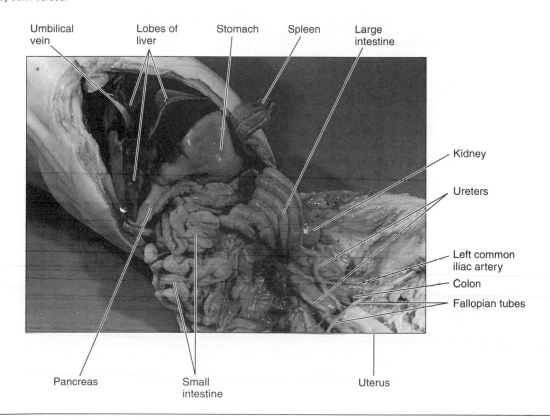

Umbilical
vein

Lobes of
liver

Stomach

Spleen

Large
intestine

Kidney

Ureters

Left common
iliac artery

Colon

Fallopian tubes

Pancreas

Small
intestine

Uterus

**FIGURE 19.12**   Fetal pig, abdominal organs, female.
Photograph by John Vercoe.

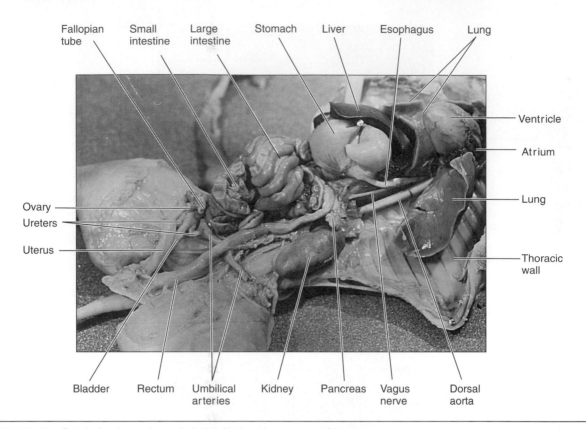

Fallopian tube · Small intestine · Large intestine · Stomach · Liver · Esophagus · Lung

Ventricle

Atrium

Lung

Thoracic wall

Ovary
Ureters
Uterus

Bladder · Rectum · Umbilical arteries · Kidney · Pancreas · Vagus nerve · Dorsal aorta

**FIGURE 19.13**   Fetal pig, thoracic and abdominal region organs, female.
Photograph by John Vercoe.

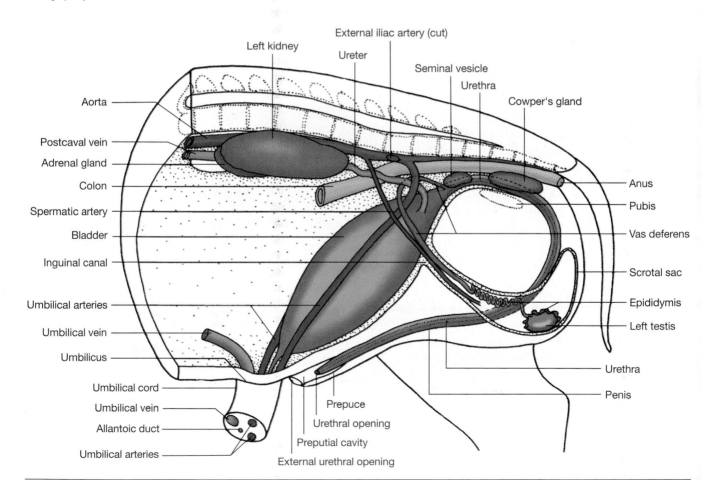

Left kidney

External iliac artery (cut)

Ureter

Seminal vesicle

Urethra

Cowper's gland

Aorta

Postcaval vein

Adrenal gland

Colon

Spermatic artery

Bladder

Inguinal canal

Umbilical arteries

Umbilical vein

Umbilicus

Umbilical cord

Umbilical vein

Allantoic duct

Umbilical arteries

Anus

Pubis

Vas deferens

Scrotal sac

Epididymis

Left testis

Urethra

Penis

Prepuce

Urethral opening

Preputial cavity

External urethral opening

**FIGURE 19.14**   Fetal pig, urogenital system, male. Arteries are red, and veins are blue.

tubes to the larger **horns of the uterus** (where embryonic development takes place), which connect with the small median **body of the uterus.** The uterus continues posteriorly as a thick, muscular tube, the **vagina.** The vagina and the urethra both empty into a common chamber, the **urogenital sinus** (vestibule). Carefully cut open the urogenital sinus by making an incision along one side, and use a blunt probe to locate the openings of the urethra and the vagina. Also, search on the ventral floor of the urogenital sinus for the **clitoris,** a small, rounded papilla. *Do not* confuse the clitoris (homologous with the penis of the male) with the larger externally obvious **genital papilla.** The exterior opening of the urogenital sinus is the **vulva,** a slitlike opening located immediately ventral to the anus. The genital papilla is an external ventral projection of the vulva.

Study also the demonstrations of microscopic sections of a mammalian ovary to see germ cells in various stages of development and of pig embryos at various stages.

### *The Male Genital System*

◆ Study figures 19.11, 19.14, and 19.15, and note the relative position of the principal male reproductive organs. Cut through the scrotal sac and down through the pubis about 3 mm to one side of the midventral line, pull the hindlimbs carefully apart, and identify the principal male organs.

First locate the two **testes** (singular: testis). In larger and older specimens, the testes will be located within two scrotal sacs (collectively referred to as the scrotum), but in younger specimens the testes may not yet have descended from the peritoneal (abdominal) cavity where they initially grow and differentiate. As the male pig grows larger, the developing testes descend from the peritoneal cavity through the **inguinal canal** into the scrotal sacs. In younger specimens, the testes may be found anywhere along this descending path.

Now locate the **epididymis,** a mass of coiled tubules lying along one side of each testis. The epididymis connects with the **vas deferens,** which passes through the inguinal canal along the **spermatic artery** and **vein,** and crosses over the **ureter** to enter the **urethra** (figures 19.11 and 19.14).

Locate the **penis,** a long, thin cylinder lying beneath the skin just posterior to the umbilical cord. Cut through the skin overlying the penis to free the penis and trace it posteriorly to its junction with the urethra. Dorsal to the urethra, near the entrance of the **vasa deferentia,** find the two small **seminal vesicles.** Adjacent to the seminal vesicles is the **prostate gland** (poorly developed in younger specimens). Also note the two **Cowper's glands** (bulbourethral glands) lying alongside the urethra near its junction with the penis.

The tissue overlying the penis on the ventral abdominal wall is called the **prepuce,** and the cavity enclosed within the prepuce is the **preputial cavity.** Opening from the preputial cavity to the exterior is the **external urethral orifice.**

Observe the demonstration of the pig testis showing **germ cells** in various stages of maturation. See also the demonstrations of mammalian **spermatozoa.**

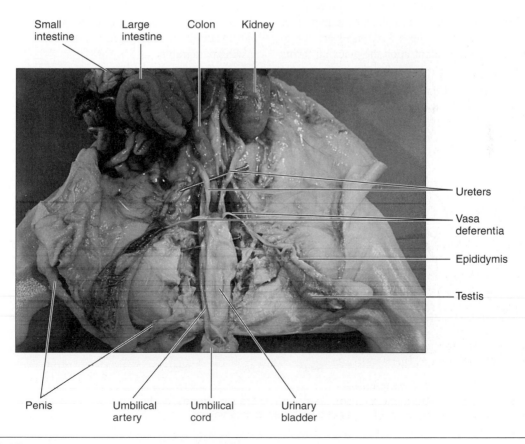

**FIGURE 19.15** Fetal pig, urogenital system, dissected, male.
Photograph by John Vercoe.

# Circulatory System

To assure your understanding of circulation in mammals, you should review the basic organization of the mammalian circulatory system prior to undertaking a dissection and study of the circulatory system of the pig. Consult figure 19.16 for the general pattern of mammalian circulation. Among these essential features are the following:

1. A four-chambered heart, which provides an efficient separation of the systemic and pulmonary divisions of the circulatory system.
2. Oxygenated blood from the lungs is carried to the heart by the **pulmonary veins,** which empty into the left atrium. From the left atrium, the blood passes to the left ventricle, which pumps it out through the aorta and its branches to all parts of the body **except the lungs.**
3. Blood from the organs of the body (except the lungs) is returned to the right atrium by the large precaval and postcaval veins. From the right atrium, it goes to the right ventricle and then back via the **pulmonary arteries** to the lungs to be oxygenated.
4. Mammals have a **hepatic portal system,** but do not have a **renal portal system,** characteristic of many lower vertebrates such as the frog and the shark. The latter has been lost during the evolution of the mammals.

## Fetal Circulation

As noted previously, the circulatory system of the fetal pig differs in some important respects from the typical pattern of adult circulation in a mammal. These differences are due to the fact that the fetus is dependent upon the placenta for its supply of oxygen and nutrients, and for the removal of carbon dioxide and nitrogenous wastes. At birth, specific and important changes occur in the pattern of circulation of the pig and other placental mammals, and a change to the adult pattern of circulation is effected.

Consult your textbook for further details on the pattern of fetal circulation in mammals and the important circulatory changes at birth. Also, observe the demonstration materials in the laboratory, which further illustrate the nature of the placenta and its relationship to the fetus in mammals.

## Heart and Arterial System

Observe the heart and identify its four chambers, the muscular ventricles and the two thin-walled atria. Blood from the systemic veins enters the vena cava and then flows into the right atrium, followed by the right ventricle. From the right ventricle, the blood is pumped through the pulmonary arteries to the lungs. From the lungs, the blood returns to the left atrium through the pulmonary veins. From the left atrium, the blood passes to the left ventricle, where it is pumped into the aorta and then to the various organs of the body. We will study the internal structure of the heart later.

On the top (anterior surface) of the heart, locate the arch of the **aorta,** the most dorsal structure, and the more ventral **pulmonary trunk,** which exits the right ventricle and splits into the left and right pulmonary arteries. Note also that the upper part of the pulmonary trunk (where the left and right pulmonary arteries have their origin) connects with the **ductus arteriosus,** a temporary connection between the pulmonary trunk and the aorta that allows the blood to bypass the lungs in the fetal pig prior to birth. Study figure 19.17 and locate the major systemic arteries described in the following paragraphs.

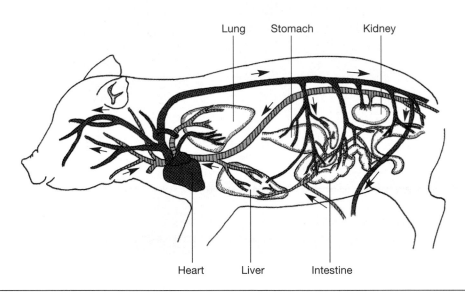

**FIGURE 19.16** Adult pig, pattern of circulation. The heart, systemic arteries, and pulmonary veins are red, pulmonary arteries and systemic veins are blue, and the hepatic portal system is yellow.

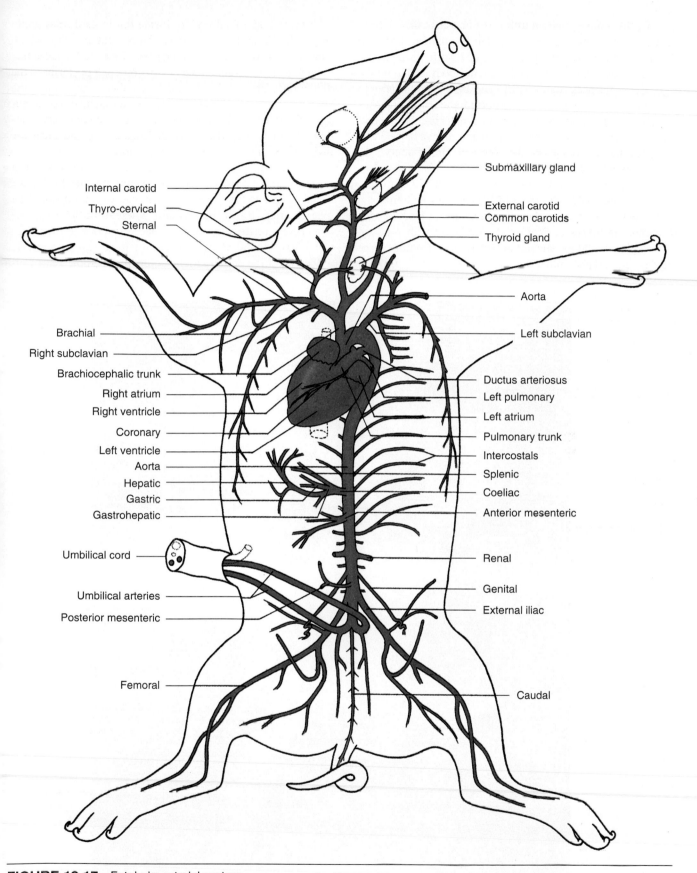

Internal carotid

Thyro-cervical

Sternal

Submaxillary gland

External carotid
Common carotids

Thyroid gland

Aorta

Brachial

Right subclavian

Brachiocephalic trunk

Right atrium

Right ventricle

Coronary

Left ventricle

Aorta

Hepatic

Gastric

Gastrohepatic

Left subclavian

Ductus arteriosus

Left pulmonary

Left atrium

Pulmonary trunk

Intercostals

Splenic

Coeliac

Anterior mesenteric

Umbilical cord

Umbilical arteries

Posterior mesenteric

Renal

Genital

External iliac

Femoral

Caudal

**FIGURE 19.17**  Fetal pig, arterial system.

The **brachiocephalic trunk** is the first large branch from the arch of the aorta and leads toward the head; it branches into the **right subclavian artery,** the external branch of which supplies the right forelimb, the right anterior body wall, and the right side of the neck (figures 19.17 and 19.18). Just anterior to its origin from the aorta, the **brachiocephalic trunk** branches into the left and right **common carotid arteries,** the external branch of which supplies the skull, brain, tongue, cheek, and face. The internal branch of the common carotid artery supplies the brain and the skull. The left subclavian artery arises independently from the aorta and just to the left of the brachiocephalic trunk. Locate the series of arteries branching from the right subclavian artery: the large **brachial artery** leads into the forelimb, the **thyro-cervical artery** leads to the thyroid and parotid glands, and the **sternal artery** leads to the muscles of the thoracic and abdominal walls.

In the abdominal region, locate the large **dorsal aorta** (figure 19.19) and the numerous **intercostal arteries,** which supply the muscles between the ribs. Trace the large **coeliac artery** from its origin on the dorsal aorta just posterior to the diaphragm. Three branches of the coeliac artery supply the stomach (**gastric artery**); the spleen, stomach, and pancreas (**splenic artery**); and the stomach, liver, pancreas, and duodenum (**gastrohepatic artery**). Locate also the **anterior mesenteric artery** leading from the dorsal aorta to the pancreas, small intestine, and large intestine. It originates just posterior to the coeliac artery. Two **renal arteries** lead to the kidneys, and two **genital arteries** lead to the gonads. The ovarian artery travels horizontally whereas the spermatic artery travels posteriorly to the gonad. Posterior to the origin of the two genital arteries, locate the **posterior mesenteric artery** leading ventrally to the colon and rectum.

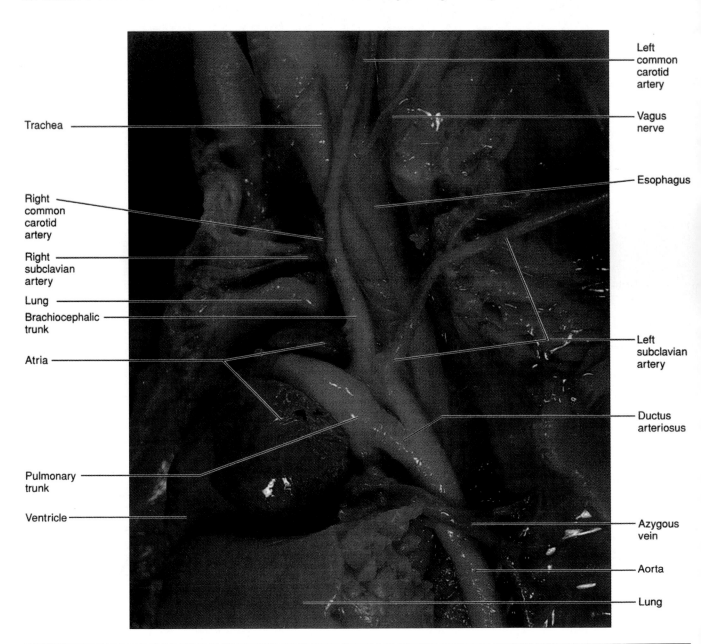

**FIGURE 19.18** Fetal pig, dissection showing heart and major arteries attached.
Photograph by Ken Taylor.

Near the posterior end of the abdominal cavity, the dorsal aorta divides into two large **common iliac arteries,** which supply the hindlimbs (figures 19.20 and 19.21). Posterior to the origin of the two external iliac arteries, locate the two large **umbilical arteries** leading into the umbilical cord and the single **caudal artery** continuing into the tail region. The umbilical arteries are fetal vessels that carry blood to the embryonic portion of the placenta. These arteries degenerate after birth, and only those branches of the umbilical artery supplying the bladder persist in the adult pig. The umbilical arteries also branch into the internal and external iliac arteries.

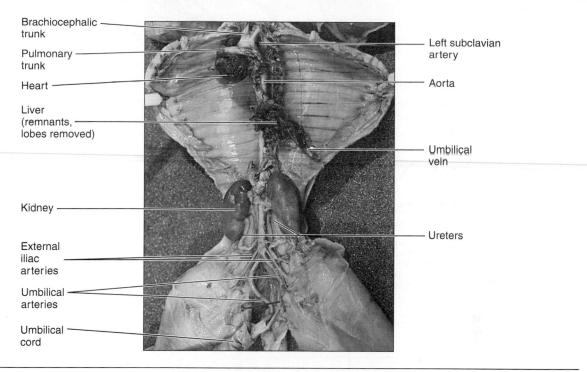

Brachiocephalic trunk
Pulmonary trunk
Heart
Liver (remnants, lobes removed)
Kidney
External iliac arteries
Umbilical arteries
Umbilical cord
Left subclavian artery
Aorta
Umbilical vein
Ureters

**FIGURE 19.19** Fetal pig, principal blood vessels of the thoracic and abdominal areas.
Photograph by John Vercoe.

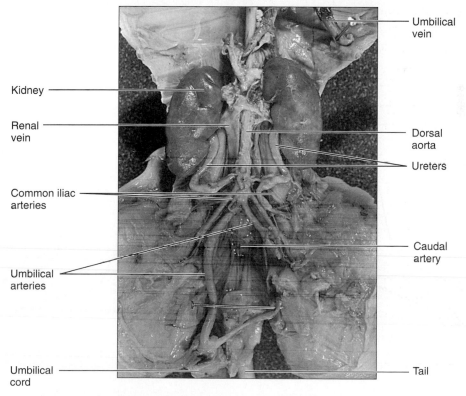

Kidney
Renal vein
Common iliac arteries
Umbilical arteries
Umbilical cord
Umbilical vein
Dorsal aorta
Ureters
Caudal artery
Tail

**FIGURE 19.20** Fetal pig, principal blood vessels of the abdominal and pelvic areas.
Photograph by John Vercoe.

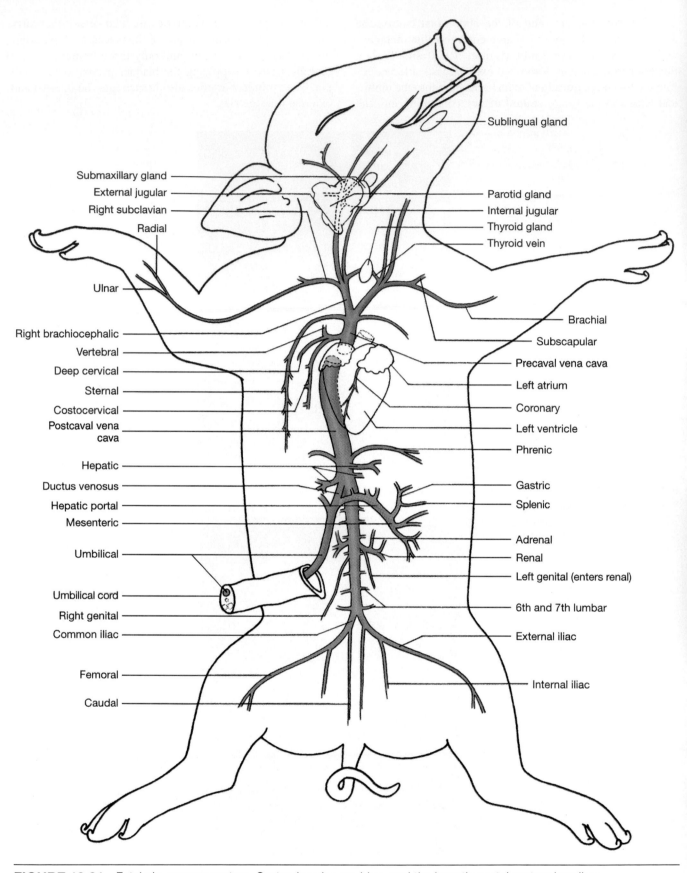

**FIGURE 19.21**  Fetal pig, venous system. Systemic veins are blue, and the hepatic portal system is yellow.

Sublingual gland

Submaxillary gland

External jugular

Right subclavian

Radial

Ulnar

Right brachiocephalic

Vertebral

Deep cervical

Sternal

Costocervical

Postcaval vena
cava

Hepatic

Ductus venosus

Hepatic portal

Mesenteric

Umbilical

Umbilical cord

Right genital

Common iliac

Femoral

Caudal

Parotid gland

Internal jugular

Thyroid gland

Thyroid vein

Brachial

Subscapular

Precaval vena cava

Left atrium

Coronary

Left ventricle

Phrenic

Gastric

Splenic

Adrenal

Renal

Left genital (enters renal)

6th and 7th lumbar

External iliac

Internal iliac

Trace each of the arteries listed from its origin along the route to the chief organs that it supplies with blood and observe its main branches. When you complete your study of the arterial system, you should be able to identify each of the main arteries and to describe the chief organs supplied by each.

After you have completed your study of the systemic arteries, you should next study the pulmonary circulation. Locate the **pulmonary trunk** leading from the right ventricle (figure 19.17) and trace it to the point where it branches into the **right** and **left pulmonary arteries.** These arteries carry the blood to the lungs and are relatively small in the fetus, but they become much larger at birth when the lungs become functional. Note also the short duct connecting the pulmonary trunk with the aorta. This duct is called the **ductus arteriosus** (figure 19.18): It is a special fetal structure. The duct ceases to function after birth and gradually degenerates to form a fibrous cord, the **ligamentum arteriosum.** Remember, in the adult pig, all of the blood in the pulmonary circulation passes to the lungs for oxygenation.

Another special feature of the fetal circulation is also related to the mixing of blood from the pulmonary and systemic divisions. Within the heart, blood passes from the right atrium into the left atrium through a temporary opening, the **foramen ovale,** located in the median wall of the heart. Blood returning to the heart through the postcaval veins is shunted through this opening and directly back into systemic circulation, thus bypassing the lungs. The foramen ovale becomes sealed at birth, and all of the blood from the right atrium thereafter goes to the lungs via the right ventricle, pulmonary trunk, and the remainder of the pathway. The former site of the foramen ovale appears as an oval depression in the medial wall of the heart of the adult pig and is called the **fossa ovalis.**

### Venous System

Study figures 19.16 and 19.21 and identify the major veins of the pig, including the **precaval vena cava,** the **right** and **left jugular trunks,** the **subclavian veins,** the **brachial veins,** the **external** and **internal jugular veins,** the **postcaval veins,** and the **hepatic veins.** Locate also the **hepatic portal vein** with its three main branches: the **gastric vein** carrying blood from the stomach, the **splenic vein** carrying blood from the spleen, and the **mesenteric vein** carrying blood chiefly from the small intestine. Find the **renal veins** leading from the kidneys, and from the hindlimbs find the **common iliac veins** into which blood flows from the **internal** and **external iliac veins.**

Locate the **umbilical vein** in the umbilical cord once again and trace it forward. Note the various branches that carry blood to the **hepatic portal vein** and into the liver. Note also that part of the blood from the **umbilical vein** is shunted directly to the **postcaval vena cava** through the short **ductus venosus** located in the liver. Shortly after birth, the umbilical vein and the ductus venosus cease to function and begin to degenerate. Like the umbilical arteries, the ductus arteriosus, and the foramen ovale, the umbilical vein and the ductus venosus are functional only in the fetus.

◆ Compare the blood of the umbilical vein with that of the umbilical arteries in regard to the content of (1) oxygen, (2) nitrogenous wastes, (3) carbon dioxide, and (4) nutrients.

Observe the several **pulmonary veins** arising from the lungs and uniting to form two main **pulmonary veins,** which enter the left atrium of the heart.

### Heart Anatomy

◆ You should supplement your study of the heart with a demonstration specimen of a dissected adult pig or cow heart. The structure of the four-chambered heart is similar in all mammals (figure 19.22). On the dissected heart, find the four chambers (two atria, two ventricles), and the valves between the atria and ventricles. Review the connection of the heart with the major blood vessels and the pattern of circulation in the pig with the aid of figures 19.16, 19.17, and 19.21.

After you have completed the review of adult heart structure, remove the heart of your fetal pig by severing the large blood vessels (about 1–2 cm from the heart) and by freeing the heart and the blood vessels from the surrounding tissues. Make a clean cut across the heart to expose the interior of the two ventricles. Wash out the clots of blood with cold water and note the interior structure of the heart.

Insert the tips of your scissors into the cavity within the right ventricle and cut through the ventricular wall to expose the cavity within the right atrium. Note the opening, or orifice, between the right ventricle and right atrium guarded by the **tricuspid valve.** Make a similar incision on the left side of the heart and observe the orifice between the two left chambers of the heart guarded by the **bicuspid valve.** *How does the bicuspid valve differ in structure from the tricuspid valve?* Next locate the opening from the left ventricle into the aorta guarded by the **semilunar valves.** Note that these semilunar valves consist of three pouchlike structures.

Review the general plan of circulation in the fetal pig and note especially those features of **fetal circulation** that distinguish it from the pattern of **adult circulation** (table 19.3).

| TABLE 19.3 |
|---|
| Summary of Changes in Fetal Circulation Occurring at Birth |

**Arterial System**
 1. Ductus arteriosus closes
 2. Umbilical arteries degenerate
**Venous System**
 1. Ductus venosus closes
 2. Umbilical vein degenerates
**Heart**
 1. Foramen ovale closes

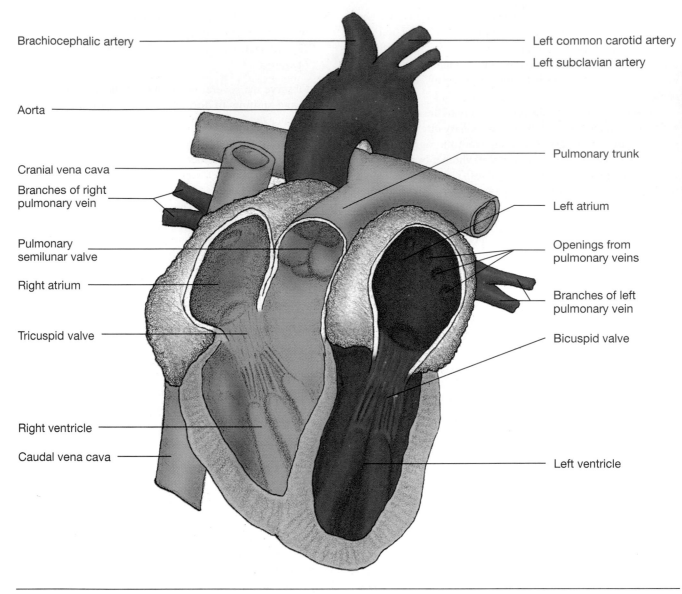

Brachiocephalic artery

Aorta

Cranial vena cava

Branches of right
pulmonary vein

Pulmonary
semilunar valve

Right atrium

Tricuspid valve

Right ventricle

Caudal vena cava

Left common carotid artery

Left subclavian artery

Pulmonary trunk

Left atrium

Openings from
pulmonary veins

Branches of left
pulmonary vein

Bicuspid valve

Left ventricle

**FIGURE 19.22**    Mammalian heart, dissected, anterior view. Schematic based on human and rat models.

## Nervous System

Review figure 19.23, and observe the location and relative
position of the principal components of the central nervous
system of the fetal pig. Note the large anterior **brain** enclosed in the bony **braincase** of the skull and the **spinal
cord** within the **neural canal** of the vertebral column.

◆   Cut away the skin, muscles, and dorsal half of the skull
to expose the brain. Remove the covering tissues carefully to avoid damage to the brain.

Note the three membranes or **meninges** surrounding the
brain. The tough outer layer that adheres to the skull is the
**dura mater;** underneath is the delicate **arachnoid layer,**
and the thin layer that dips into the crevices called **sulci** (singular: sulcus) of the brain is the **pia mater.** The same three
meninges also cover the spinal cord but are more difficult to
observe than on the surface of the brain.

Cut away the left side of the skull and remove the dura
mater. Locate and study the five main regions of the pig
brain listed in table 19.4 (figure 19.24).

It is often more convenient and satisfactory to study the
anatomy of the mammalian brain with a specimen that has
been preserved and prepared for this purpose. The brain of

### TABLE 19.4

#### Principal Regions of the Mammalian Brain

| Region | Principal Structure(s) |
|---|---|
| Telencephalon | Cerebral hemispheres |
| Diencephalon | Pituitary gland |
| Mesencephalon | Corpora quadrigemina |
| Metencephalon | Cerebellum |
| Myelencephalon | Medulla oblongata |

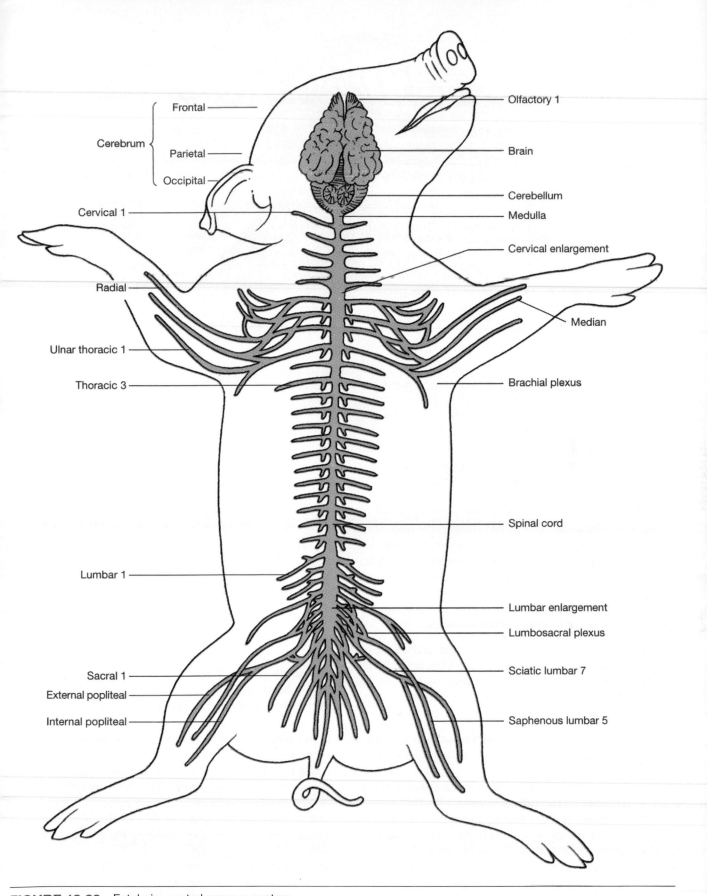

Frontal
Cerebrum
Parietal
Occipital
Cervical 1
Radial
Ulnar thoracic 1
Thoracic 3
Lumbar 1
Sacral 1
External popliteal
Internal popliteal

Olfactory 1
Brain
Cerebellum
Medulla
Cervical enlargement
Median
Brachial plexus
Spinal cord
Lumbar enlargement
Lumbosacral plexus
Sciatic lumbar 7
Saphenous lumbar 5

**FIGURE 19.23**   Fetal pig, central nervous system.

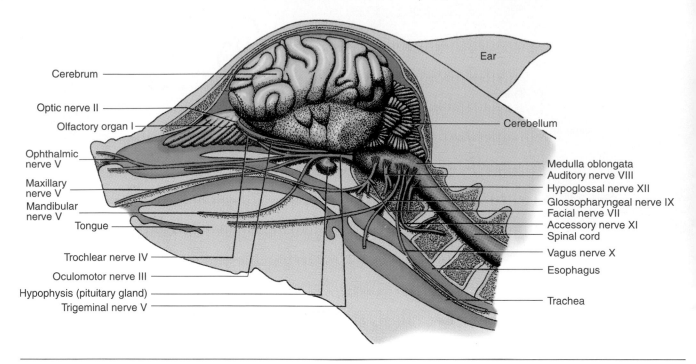

Cerebrum

Optic nerve II

Olfactory organ I

Ophthalmic nerve V

Maxillary nerve V

Mandibular nerve V

Tongue

Trochlear nerve IV

Oculomotor nerve III

Hypophysis (pituitary gland)

Trigeminal nerve V

Ear

Cerebellum

Medulla oblongata
Auditory nerve VIII
Hypoglossal nerve XII
Glossopharyngeal nerve IX
Facial nerve VII
Accessory nerve XI
Spinal cord

Vagus nerve X

Esophagus

Trachea

**FIGURE 19.24**   Fetal pig, lateral view of head with brain and cranial nerves.

the sheep and the cat are most commonly used. The principal parts of the sheep brain are illustrated in figure 19.25. Figure 19.26 shows the cranial nerves on the ventral surface of a dissected cat brain.

You should also review your previous studies of the brains of the shark and the frog, and compare the brain of the pig with those of these two "lower" vertebrates. *What are the principal differences in brain structure among these animals? Which parts of the brain are more highly developed in the pig? Which structures are less well-developed or absent in the pig? Can you relate these structural differences to differences in behavior of the three species?*

◆ Dissect away the skin, connective tissue, and muscles to expose the anterior portion of the vertebral column. Next remove the neural arches of several of the cervical (neck) vertebrae to observe the spinal cord within the neural canal.

Note in figure 19.23 the **enlargements** in the spinal cord in the **brachial** and in the **lumbosacral regions** and the **nerve plexes** (networks) associated with these enlargements. Locate the **cervical** and **lumbosacral nerve plexes** on your specimen by carefully dissecting away the tissues in these two regions of the vertebral column. After you have removed the muscles and connective tissues from the brachial region of the vertebral column, you will find an interconnected network of tough, whitish nerves that connect with the spinal cord. This is the **brachial plexus;** it is made up of branches from several of the spinal nerves of the pig.

There are 33 pairs of spinal nerves in the pig; other mammals may have more or fewer spinal nerves. The cat, for example, has 38 pairs of spinal nerves. The 33 pairs of

spinal nerves in the pig include eight pairs in the cervical region, 14 pairs in the thoracic region, seven pairs in the lumbar region, and four pairs in the sacral region.

◆ Remove the skin, connective tissue, and muscles in the lumbosacral region to locate the **lumbosacral plexus.**

◆ Use your scalpel or scissors to cut out a section of the spinal cord about 3 cm in length from the cervical region. Sever also the nerves attached to the spinal cord in this region, leaving enough nerve fiber to study the attachment of the nerves to the spinal cord.

Observe that each spinal nerve is formed by the union of **two roots,** one **dorsal** and one **ventral.** The former carries principally **sensory fibers,** and the latter carries principally **motor fibers.** Also observe the microscopic demonstration illustrating a cross section of a mammalian spinal cord.

The pig has 12 pairs of cranial nerves, the same number found in cats (figure 19.26) and humans. Nerves I–X correspond with those of the frog and the shark studied in previous exercises. Posterior to these are Nerve XI, the **spinal accessory nerve,** and Nerve XII, the **hypoglossal nerve** (figure 19.24). The spinal accessory nerve (XI) arises from several roots on the lateral surface of the spinal cord and medulla, and is made up of several motor and sensory fibers connected with the shoulder muscles. The hypoglossal nerve arises from several roots on the ventral surface of the medulla, and contains sensory and motor fibers connected with the tongue.

A helpful mnemonic device used by many students to help them remember the names of the cranial nerves of the pig and other mammals is:

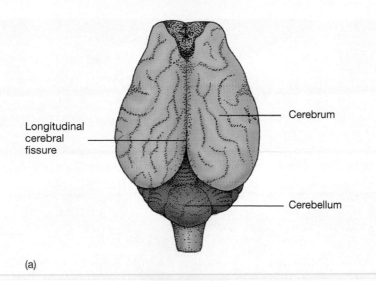

(a)

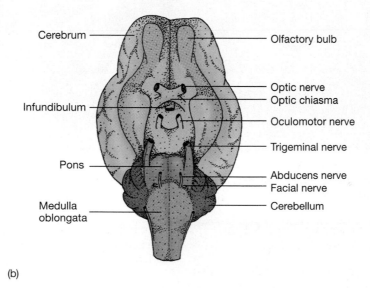

(b)

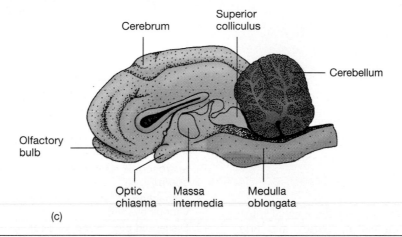

(c)

**FIGURE 19.25** Sheep brain: (a) dorsal view, (b) ventral view, and (c) sagittal section.

Fetal Pig Anatomy     **323**

Olfactory bulb
Optic nerve II
Trigeminal nerve V
Trochlear nerve IV
Accessory nerve XI

Oculomotor nerve III
Acoustic nerve VIII
Facial nerve VII
Vagus nerve X
Glossopharyngeal nerve IX
Hypoglossal nerve XII
Spinal nerve

**FIGURE 19.26**   Cat brain, dissected, ventral view showing cranial nerves.
Courtesy of Carolina Biological Supply Company, Burlington, NC.

## On Old Olympus' Towering Top, A Finn And German Viewed A Hop.

Note that the first letter of each word corresponds to the first letter in the names of the twelve cranial nerves of mammals:

   I. **O**lfactory
  II. **O**ptic
 III. **O**culomotor
 IV. **T**rochlear
  V. **T**rigeminal
 VI. **A**bducens
VII. **F**acial
VIII. **A**uditory
 IX. **G**lossopharyngeal
  X. **V**agus
 XI. **A**ccessory
XII. **H**ypoglossal

◆ Carefully free the brain, starting from the severed spinal cord in the cervical region and proceeding anteriorly, cutting nerves close to the skull on each side and gently freeing the brain from the skull. As you work, identify as many of the cranial nerves as possible, using figures 19.24, 19.25, and 19.26 as guides.

Find the **pituitary gland** and **optic chiasma** on the ventral surface, and with a *sharp* scalpel carefully bisect the brain from front to rear into equal right and left halves. Locate and study the various brain structures on your specimen with the aid of figures 19.24 and 19.25.

## Demonstrations

1. Placentae of pig and other mammals
2. Skeleton of fetal pig
3. Mounted cat skeleton
4. Fresh or dried specimens of mammalian lung
5. Microscope slides of mammalian lung tissue
6. Microscope slide showing germ cells in mammalian ovary
7. Microscope slides showing mammalian testis tissue and developing spermatozoa
8. Microscopic demonstration of stained mammalian sperm
9. Dissection of fetal pigs to illustrate injected arterial and venous systems
10. Dissected pig, sheep, or beef hearts to illustrate interior chambers and valves
11. Preserved sheep brains, whole and sections
12. Dissected fetal pigs to show cranial and spinal nerves
13. Microscope slide to show cross section of mammalian spinal cord

# Key Terms

**Appendicular skeleton**  portion of the skeleton of the pig and other vertebrates that includes the bones of the appendages and the girdles, which provide articulation of the limbs with the axial skeleton.

**Axial skeleton**  portion of the skeleton that provides support for the main axis of the body, including the skull, the vertebral column, and so on.

**Mesentery**  a double sheet of mesodermal epithelium that serves to support various internal organs in the coelom of vertebrates.

**Pericardial cavity**  portion of the coelom that encloses the heart; lined by the pericardium.

**Peritoneal cavity**  (Abdominal cavity) portion of the coelom lying posterior to the diaphragm; contains the abdominal organs.

**Placenta**  structure formed by a combination of maternal and embryonic tissues that attaches the developing embryo to the inside wall of the mammalian uterus. It also provides channels for the passage of nutrients and gases to the embryo and removal of wastes from the embryo without providing direct connection between the two separate circulatory systems. The placenta also produces certain reproductive hormones.

**Pleural cavity**  portion of the coelom that encloses a lung; lined by the pleura.

**Thoracic cavity**  portion of the coelom lying anterior to the diaphragm; subdivided in the pig and other mammals into the medial pericardial cavity and two lateral pleural cavities.

**Umbilical cord**  the cord that attaches the fetus to the placenta of the mother; contains the umbilical arteries, the umbilical vein, and the allantoic duct.

# Internet Resources

Visit the zoology website at http://www.mhhe.com/zoology to find live Internet links of each of the references listed below.

1. Animal Diversity Web, University of Michigan. Suidae (Pigs and Hogs). Information on the morphology of the pig.
2. Fetal Pig Dissection Humor. Some awful, but humerous, jokes.
3. Fetal Pig Dissection Guide, Hints and Tips. Five links to a good amount of material that may improve your techniques, as well as an explanation of the reasons why we dissect fetal pigs, found on the home page of the Fetal Pig Dissection Guide, by J.S. Miller, Goshen College.
4. Dissection Table. A lengthy table that compares various systems of the earthworm, the frog, the snake, the shark, the perch, the pigeon, and the pig.
5. The Virtual Pig Dissection. You can click on different systems to see the diagrams of the dissection, all with labels. This is the best pig dissection site I have found. It really is a good dissection guide.

6. University of Minnesota Pig Dissection Page. This page links to various pages on different organ systems, as well as a discussion group. Very high resolution photos, but loads slowly.
7. Fetal Pig Dissection. Color photographs of internal organs with labels.
8. A must-see site: Fetal Pig Dissection on the WWW. Done by instructors at Lakeview High School. Thirty-seven photographs in order of the steps of the dissection.

# Critical Thinking Questions

1. Discuss the advantages of mammalian circulation (if any) over those of amphibians and fish. Again, does greater complexity always mean "better?" Is a four-chambered heart better than a three-chambered heart, or is a three-chambered heart better than a two-chambered heart? Explain your reasoning.

2. Discuss the changes that occur in circulation in the pig at birth. Specifically, the fetal pig emerges from an almost marine environment in the uterus to a terrestrial environment (in a matter of minutes). What would happen if any of these changes did not occur?

3. What is the advantage (if any) of a placenta and live birth, rather than a marsupium or pouch and an undeveloped fetus as in marsupial mammals?

4. Compare the vertebral and appendicular skeletons of the fetal pig with those of the amphibian and fish. What are the similarities? What are the differences? Are there any remnants of the visceral skeleton of the fish identifiable in the amphibians and mammals?

5. Discuss the evolutionary advances of homeothermy (constant endothermy) versus ectothermy. Which system is metabolically "more expensive." Why? Support your reasoning.

# Suggested Readings

Chaisson, R.B., Odlaug, T.O., and W.J. Radke. 1997. *Laboratory Anatomy of the Fetal Pig*, 11th ed. Dubuque, IA: Times Mirror Higher Education Group, Inc..

Gilbert, S.G., 1996. *Pictorial Anatomy of the Fetal Pig*. Seattle: University of Washington Press.

Walker, W.F., Jr. 1974. *Dissection of the Fetal Pig*. San Francisco: W.H. Freeman Company.

# NOTES AND SKETCHES

# CHAPTER 20
## Rat Anatomy

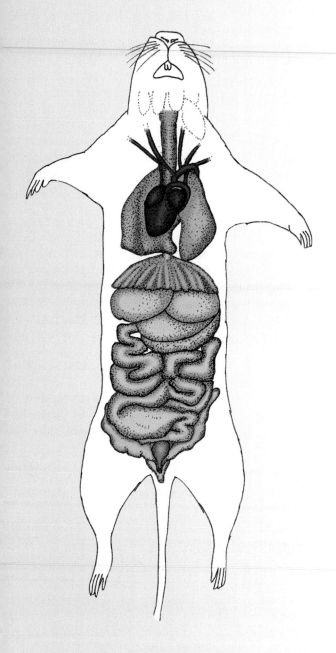

## OBJECTIVES

After completing the laboratory work in this chapter, you should be able to perform the following tasks:

1. Locate and identify the principal external features of the rat. Explain which external features distinguish it as a mammal.
2. Identify the main elements of the axial and appendicular skeletons of the rat.
3. Locate five major skeletal muscles of the rat and identify their origins, insertions, and actions. Distinguish between extension, flexion, adduction, and abduction.
4. Locate and identify the principal organs of the digestive system of the rat.
5. Explain the general pattern of circulation in the rat and other mammals, and discuss the significance of the four-chambered heart.
6. Describe and identify the principal organs of the urogenital system of the rat. Demonstrate differences between male and female rats.
7. Identify the main parts of the brain of the rat or sheep, and discuss the basic organization of the mammalian nervous system.

## Introduction

The white laboratory rat, an albino form of the Norway rat *Rattus norvegicus,* is often studied as a representative mammal because it exhibits most of the organs and organ systems typical of mammals. Among the distinguishing features of mammals (Class Mammalia) are a body surface covered with hair, an integument (skin) with mammary and other types of glands, a skull with two occipital condyles, seven cervical (neck) vertebrae, teeth borne on bony jaws, movable eyelids and fleshy external ears (pinnae), a four-chambered heart, a persistent left aorta, and a muscular diaphragm separating the thoracic and abdominal cavities. Mammals also are "warm-blooded" or **endothermic** (or homeothermic). *Most mammals* maintain a constant and elevated body temperature under normal conditions and rely on internally produced heat from metabolism to maintain their high body temperature. The young develop within the uterus of the female, have a placental attachment for nourishment, and are enveloped by special fetal membranes (amnion, chorion, and allantois). Milk to nourish the young after birth is produced by mammary glands. Other common representatives of this group include moles, bats, whales, mice, deer, monkeys, horses, cattle, and humans.

## Classification

Subphylum Vertebrata
Class Mammalia
Order Rodentia
Suborder Myomorpha
Family Muridae
Genus and species *Rattus norvegicus*

## Materials List

**Preserved specimens**
White rat, double- or triple-injected
Sheep brain
Beef heart
**Plastic or liquid mounts**
Rat dissection
**Prepared microscope slides**
Rat testis, cross section
Rat kidney, median section
Rat heart, longitudinal section
**Other**
Mounted rat skeleton
Mounted cat skeleton

## External Anatomy

◆ Select a preserved specimen and place it in a dissecting pan. Observe the two principal external features that distinguish the rat as a mammal: the **hair** covering most of the body and the paired **mammary glands.** Locate the paired nipples or **teats** on the ventral surface of the trunk between the forelimbs and hindlimbs. *How many pairs are present? Are they present on both male and female rats?*

The body of the rat is divided into an anterior **head** connected to a cylindrical **trunk** by a short, thick **neck.** A long **tail** extends posteriorly from the trunk. Interiorly, the trunk is divided by a muscular **diaphragm** into an anterior **thorax** and a posterior **abdomen.** The diaphragm is another distinguishing feature of mammals; it is a muscular sheet that divides and separates the coelom into the paired **anterior pleural** (and pericardial) **cavities** and the **posterior abdominal cavity.** The diaphragm is lacking in birds, reptiles, and other lower vertebrates.

◆ Observe the long tail. *How does its exterior surface differ from that of the remainder of the body?*

Study the cone-shaped head, which has an elongate (prognathous) face. On the head, find the two **eyes** and two **ears.** The eyes have upper and lower eyelids and a reduced third eyelid or **nictitating membrane,** which can be found on the medial portion of the eye opening beneath the two outer lids.

Note that each ear has an external fold of tissue called the **pinna,** which aids in directing sound waves to the opening of the ear, the **external auditory meatus.** Other important sense organs located on the head are the **vibrissae**—long sensory hairs that provide the rat with a very effective sense of touch. The vibrissae can tell a rat in an instant if a hole is large enough for him to crawl into. Note that most of the vibrissae are attached to the upper lip. *Where else do you find vibrissae?*

Observe the well-developed upper and lower **lips** surrounding the mouth. In the center of the upper lip, find the **philtrum,** a groove or cleft separating the lip into right and left halves. Locate the two **external nares** above the upper lip. Inside the mouth find the two long, sharp **incisor teeth** that are characteristic of rodents. These two incisors grow continuously, and the rat wears them down by its gnawing habits. Molars or grinding teeth are found farther back in the jaw and will be studied later.

Observe one of the **forelimbs** and note that the limb consists of an upper portion, a lower portion, and a handlike portion much like the structure of your own arm. Most of the bones in the rat forelimb are homologous with those of a human. Embryological and anatomical studies have shown that these bones have the same origin in the rat and in the human.

Locate the horny **claws** at the tip of each digit and the walking pads in the palm area. These pads are less well-developed in the rat than they are in the cat and catlike animals because rats have evolved a tendency to walk on the digits rather than on the entire palm or sole.

Near the base of the tail locate the **anus.** You can determine the sex of your specimen by studying adjacent urogenital structures. Male rats have a **scrotum** ventral to the anus, which holds the two **testes** during reproductive season. (At other times, the testes are suspended inside the abdominal cavity, and the scrotum is an empty sac.) Anterior to the scrotum is the **penis** with the opening of the male urogenital system at the end of the penis. The penis is usually withdrawn into a sheath of skin, the **prepuce.**

If you have a female rat, find the **vaginal opening** ventral to the anus and the separate opening of the **urethra** in front of the **clitoris,** which is located ventral to the vagina.

## Skeletal System

The skeleton of all vertebrates exhibits a common basic design, and an examination of the skeletons of a frog, a rat, a cat, and a human reveals many fundamental similarities. Since the bones of the rat are small, biology and zoology students often use a cat skeleton to learn more about the organization of the mammalian skeleton.

◆ The skeleton of the rat is illustrated in figure 20.1. Study a mounted cat skeleton and locate the major parts of the skeleton with the aid of figure 19.2 and compare it with the illustration of the cat skeleton.

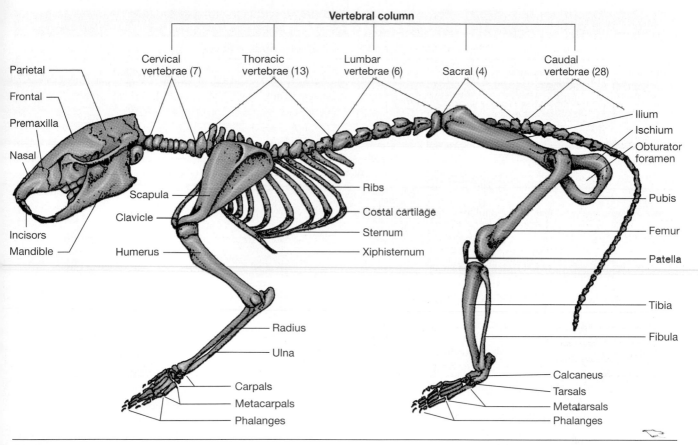

**FIGURE 20.1** Rat skeleton.

From Kendall/Hunt Publishing Company Biology Plate Series Part 1, *Zoology*. Copyright © 1975 by Kendall/Hunt Publishing Company. Reprinted with permission.

## Muscular System (Optional)

A thorough study of the muscular system of the rat would take more time than is available in most introductory zoology courses. In this exercise, we will limit our study to a few of the major muscles of the ventral surface of the thorax and shoulders to provide an example of the organization of the muscular system.

◆ Before you start your dissection, you should review the information in Chapter 19 about the structure and action of skeletal muscles and be sure that you understand the meaning of such terms as **origin, insertion, action, antagonistic muscles, extension, flexion, abduction,** and **adduction.**

### Skinning

◆ Prior to the study of the muscles, you must first skin the appropriate portion of the body. Most preserved and injected rats have an incision on the midventral surface of the neck where they were injected. Start with this opening and free the skin from the underlying muscle and continue the incision posteriorly, stopping just anterior to the anus as shown in figure 20.2. Extend the incision anteriorly to the mouth and then cut around the side of the mouth.

◆ Next, make two incisions in the chest region between the forelimbs at right angles to the first incision. Extend these lateral incisions to the wrist of each forearm, as

indicated in figure 20.2. Carefully separate the skin from the underlying tissue with a blunt probe to expose the superficial (surface) muscles of the trunk and shoulder regions. You may later wish to make similar lateral incisions on the surface of the hindlimbs to study them in more detail.

◆ Locate and study the representative muscles listed on table 20.1 with the aid of figure 20.3. Carefully separate each muscle from adjacent ones with a blunt probe and identify its origin (fixed end), insertion (movable end), and its action (what it does or what it moves).

## Internal Anatomy

### Mouth and Pharynx

We will begin our study of internal organs with the mouth and pharynx.

◆ Cut the muscles, skin, and bones at the corner of the mouth on each side with a pair of scissors or bone cutters. Push down the lower jaw and examine the interior of the **oral cavity.** Locate the muscular **tongue** attached at the rear of the oral cavity.

The anterior part of the tongue is attached ventrally with a thin sheet of tissue, the **frenulum.** Rats have two types of teeth, **incisors** (cutting teeth) and **molars** (grinding

# TABLE 20.1

## Major Muscles of the Ventral Surface and Shoulders

| Muscle | Location | Origin | Insertion | Action |
|---|---|---|---|---|
| Rectus abdominis | Paired muscles adjacent to midventral line of connective tissue (linea alba) | Pubis | First rib, sternum | Supports abdominal organs; flexes vertebral column |
| External oblique | Flat sheet covering ventral surface of trunk | Ribs and spinal fascia | Ilium, pubis, linea alba | Supports abdominal organs |
| Internal oblique | Flat sheet lying under external oblique | Spinal fascia, ilium | Cartilages of false ribs, linea alba | Supports abdominal organs |
| Pectoralis major | Large triangular muscle lying midventral to shoulder girdle | Sternum | Humerus | Adducts forelimb |
| Pectoralis minor | Posterior and beneath pectoralis major | Sternum | Humerus | Adducts forelimb |
| Biceps brachii | Anterior ventral surface of humerus | Scapula | Radius | Flexes lower part of forelimb |
| Triceps brachii | Posteriodorsal surface of humerus | Humerus | Ulna | Extends lower part of forelimb |
| Sternomastoideus | Side of neck anterior to pectoralis major | Sternum | Mastoid process of skull | Turns head sideways or back of head down |
| Sternohyoideus | Medial to sternomastoid | Occipital bone of skull | Hyoid bone | Elevates hyoid |

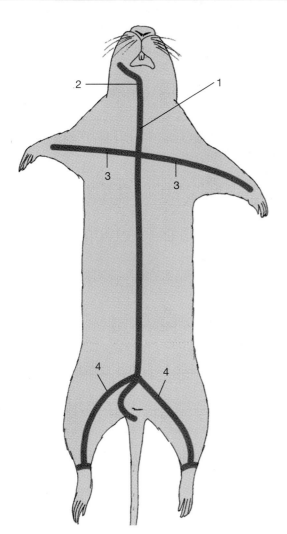

**FIGURE 20.2**   Rat, ventral view, incisions for skinning

teeth). *How many of each type do you find?* Three pairs of **salivary glands** empty into the oral cavity and provide lubrication for the food mass. These glands are the **parotids** (most lateral), the **mandibular glands** (most posterior), and the **sublingual glands** (anterior and lateral to the mandibular glands).

Find the **hard palate** covering the anterior part of the roof of the oral cavity and the **soft palate** covering the posterior portion. The **pharynx** is located at the rear of the oral cavity and consists of three portions: the **oropharynx,** which is below the soft palate; the **nasopharynx,** which receives air from the external nares lies above the soft palate; and the **laryngopharynx,** which is located at the rear and connects with the esophagus.

On the floor of the laryngopharynx find the **glottis,** a slitlike opening into the trachea, which is covered by a small flap of tissue, the **epiglottis.** Opening into the nasopharynx are the **internal nares** and the **eustachian tubes;** the latter connect with the middle ears. The nasopharynx leads to the more posterior laryngopharynx.

See page 305 for a description of the mammalian coelom.

## Other Internal Organs

◆ To study other features of internal anatomy, you must cut through the ventral muscles to expose the organs within the coelomic cavity. Make an incision through the muscles just to the right of the linea alba (ventral midline). Start just anterior to the anus and continue anteriorly to the neck region. Cut through the pectoral girdle and pull apart the ribs in the chest region to expose the heart and lungs. Make two transverse cuts through the skin and superficial muscles on each side just behind the pectoral girdle and another pair of cuts just in front

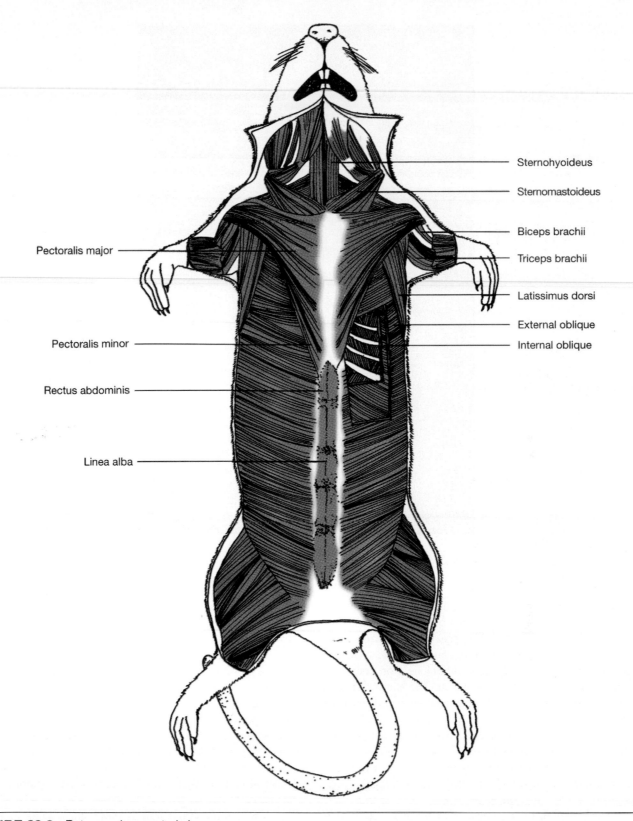

Pectoralis major

Pectoralis minor

Rectus abdominis

Linea alba

Sternohyoideus

Sternomastoideus

Biceps brachii

Triceps brachii

Latissimus dorsi

External oblique

Internal oblique

**FIGURE 20.3** Rat, muscles, ventral view.

From Kendall/Hunt Publishing Company Biology Plate Series Part 1, *Zoology*. Copyright © 1975 by Kendall/Hunt Publishing Company. Reprinted with permission.

of the pelvic girdle (figure 20.4). Take care not to damage the organs inside the abdominal cavity beneath the superficial muscle layers. Free the diaphragm from the body wall and pin aside the skin and body wall to expose the abdominal organs as in figure 20.4.

## Digestive System

The digestive system consists of the mouth, oral cavity, pharynx, esophagus, stomach, small intestine, large intestine, and rectum, and terminates at the anus. Several parts of the digestive system are shown in figures 20.4 and 20.5. The

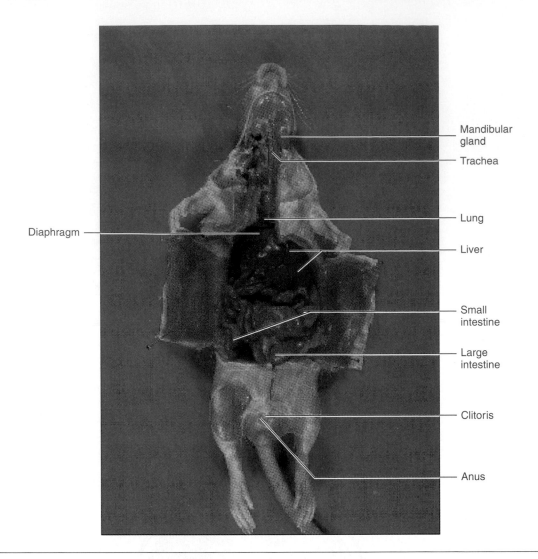

Mandibular gland

Trachea

Lung

Liver

Small intestine

Large intestine

Clitoris

Anus

Diaphragm

**FIGURE 20.4** Rat, dissected, internal organs, female, ventral view.
Photograph by Ken Taylor.

**esophagus** is a cylindrical tube leading from the pharynx to the stomach. It enters the **stomach** on the **lesser curvature** (medial concave surface). The **greater curvature** of the stomach refers to the larger convex lateral surface. Trace the path of the esophagus (dorsal to the trachea) on your specimen to its entrance into the stomach. The stomach lies beneath the liver on the left side of the abdomen and consists of three portions: a **forestomach** (fundus), which serves mainly for temporary storage; a glandular **middle portion,** which secretes mucus, hydrochloric acid, and pepsin (a protease); and a smaller posterior **pyloric region** with a muscular **pyloric valve,** which controls the passage of food material into the small intestine.

The small **intestine** consists of three specialized regions: the **duodenum,** the **jejunum,** and the **ileum.** The anterior portion of the small intestine is the duodenum. It is 25–30 cm in length and receives ducts from the liver and pancreas. Most digestion continues here. Distal to the duodenum is the shorter jejunum. The remaining portion of the small intestine posterior to the jejunum is the ileum. The exterior surface of the ileum often feels lumpy because of the presence of numerous lymph nodes embedded in its walls. These lymph nodes are involved in the absorption of lipids from the intestine.

Locate the large **liver,** which has four lobes. Bile ducts from each of the lobes join to form a single duct carrying bile to the duodenum. Note that rats typically lack or have a reduced gallbladder. Anterior to the liver, locate the flat, muscular **diaphragm** (figure 20.5), which partitions the coelom into the thoracic and abdominal cavities. The **pancreas** is not a discrete organ in the rat, but consists of scattered patches of tan or pinkish glandular tissue embedded in the **mesentery** (the thin sheet of connective tissue) between the duodenum and the stomach.

The ileum empties into the large intestine, or colon. At the junction of the ileum and colon, find the **caecum,** a blind sac that extends posteriorly from the **ileocolic valve** between the ileum and colon. From this junction, the colon extends anteriorly as the **ascending colon.** The **transverse colon** extends across the abdominal cavity adjacent to the diaphragm, and, in the region of the stomach, it loops posteriorly as the **descending colon.**

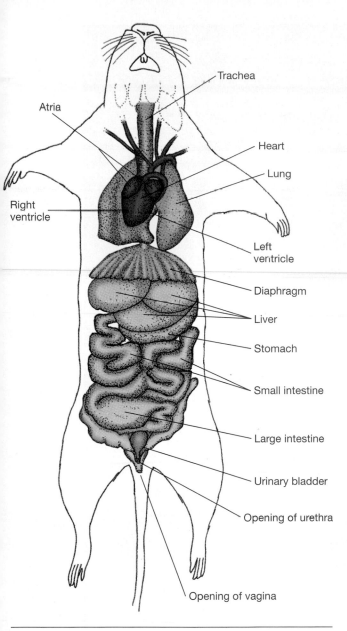

Atria

Trachea

Heart

Lung

Right ventricle

Left ventricle

Diaphragm

Liver

Stomach

Small intestine

Large intestine

Urinary bladder

Opening of urethra

Opening of vagina

**FIGURE 20.5** Rat, internal organs, female.

The caecum and colon are important in the resorption of ions and water from the gut contents. Also, numerous mucous glands are found in the epithelium of the colon, which facilitate movement of the contents. The terminal portion of the colon is the muscular **rectum,** which leads to the **anus.**

## Respiratory System

During respiration, air enters the external nares and passes through the nasal passages to the nasopharynx and downward through the glottis into the **larynx.** The air is warmed and filtered as it passes through the nasal passages. The presence of vocal cords in the lateral walls of the larynx allows rats to make audible sounds, possibly for communication.

From the larynx, air passes to the **trachea,** a hollow tube supported by a series of incomplete cartilaginous rings (figure 20.6). Trace the path of the trachea from the larynx to the thoracic cavity where it branches into **two primary bronchi.** The primary bronchi enter the lungs, branch further into many **branchioles,** and terminate in many highly vascularized **alveoli** where the gas exchange takes place.

The exchange of air in the lungs of the rat and other mammals is brought about by changes in the volume of the thoracic cavity, which contains the lungs. During normal breathing, **inspiration,** the process of taking in air, results from the contraction of the dome-shaped diaphragm and of the external intercostal muscles located between adjacent ribs. These contractions enlarge the thoracic cavity, lower the pressure in the cavity below the external atmospheric pressure, and allow outside air to enter the lungs. **Expiration,** the expulsion of air from the lungs, occurs when the diaphragm and the external intercostal muscles relax, and the volume of the thoracic cavity decreases.

Mammals are therefore **negative pressure breathers** since air enters the lungs because of the negative pressure created inside the thoracic cavity by muscular contractions. In contrast, frogs and other amphibians are **positive pressure breathers** because in these animals air is forced into the lungs by swallowing-like movements produced by the muscles of the mouth cavity and throat, thus increasing the pressure within the lungs above that of external atmospheric pressure.

## Circulatory System

The circulatory system of the rat exhibits the same basic pattern characteristic of all mammals, including humans. There are two major features: (1) a **four-chambered heart,** and (2) two separate divisions for circulation of the blood—a **systemic circuit** and a **pulmonary circuit.**

The systemic division carries oxygenated blood from the heart through a branching system of arteries to most organs and tissues of the body, collects deoxygenated blood from these organs via a system of capillaries and veins, and brings the blood back to the heart. The pulmonary division carries deoxygenated blood from the heart to the lungs for oxygenation and returns the oxygenated blood to the heart.

This pattern of circulation shown by mammals represents a major advance over the patterns exhibited by "lower" vertebrates such as the shark or the bony fish, which have a two-chambered heart and a single circulatory circuit, and the frog, with its three-chambered heart and an incomplete double circuit, which allows some mixing of oxygenated and deoxygenated blood in the heart.

In mammals, deoxygenated blood from the body (except the lungs) returns to the right atrium by the large cranial vena cava and caudal vena cava (figure 20.7). From the right atrium, blood goes to the right ventricle and is pumped via the pulmonary arteries to the lungs to be oxygenated.

Oxygenated blood from the lungs passes through the pulmonary veins to the left atrium of the heart. From the left atrium, the blood goes to the left ventricle and is pumped through the aorta and its branches to all parts of the body except the lungs.

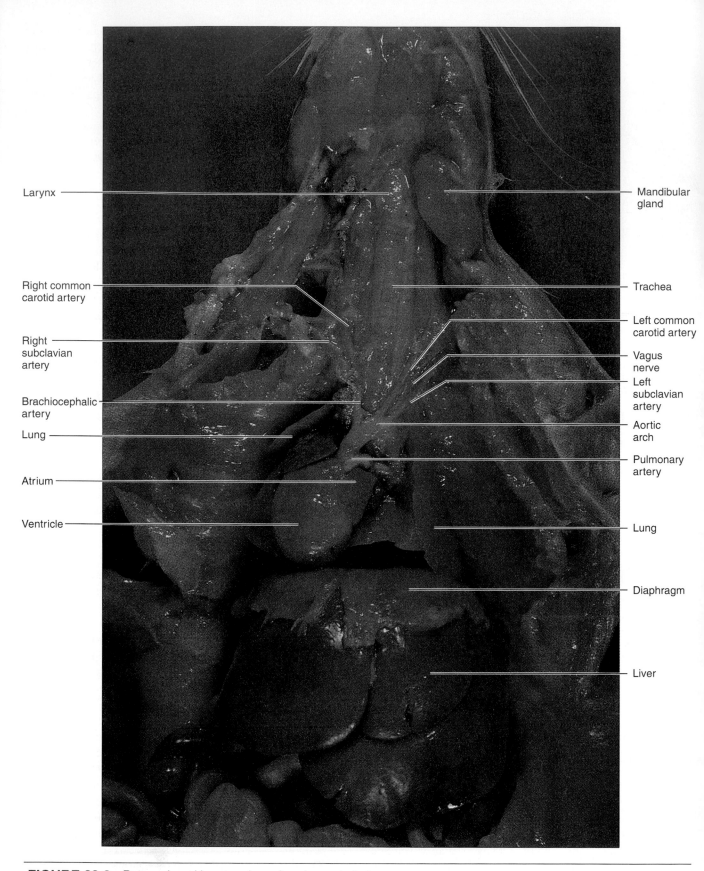

Larynx

Right common carotid artery

Right subclavian artery

Brachiocephalic artery

Lung

Atrium

Ventricle

Mandibular gland

Trachea

Left common carotid artery

Vagus nerve

Left subclavian artery

Aortic arch

Pulmonary artery

Lung

Diaphragm

Liver

**FIGURE 20.6** Rat, neck and heart regions, female, ventral view.
Photograph by Ken Taylor.

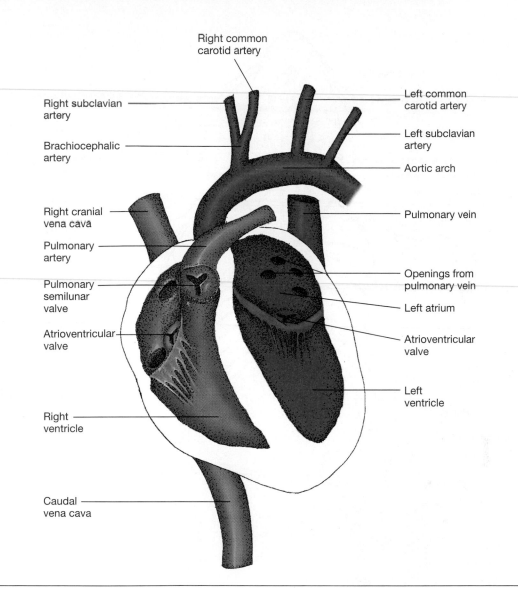

Right common
carotid artery

Right subclavian
artery

Brachiocephalic
artery

Right cranial
vena cava

Pulmonary
artery

Pulmonary
semilunar
valve

Atrioventricular
valve

Right
ventricle

Caudal
vena cava

Left common
carotid artery

Left subclavian
artery

Aortic arch

Pulmonary vein

Openings from
pulmonary vein

Left atrium

Atrioventricular
valve

Left
ventricle

**FIGURE 20.7**  Rat, heart, longitudinal section.

The best specimens for study of the anatomy of the circulatory system are preserved rats that have been injected with colored latex. Specimens may be single-, double-, or triple-injected. Single-injected specimens have red latex injected into the arterial system. In double-injected specimens, the arteries are red and the veins are blue; triple-injected specimens are similar to double-injected specimens, but also have yellow latex injected into the hepatic portal system.

### Heart and Arterial Circulation

◆ Carefully remove the pericardial membranes surrounding the heart. Take care not to rupture the blood vessels attached to the heart. Study figure 20.7 and locate the two small, saclike **atria** lying on the surface of the two muscular **ventricles.** Note that the ventricles are unequal in size; *which ventricle is the larger?* Examine a microscope slide of heart tissue to observe the difference between the thick, muscular ventricles and the thin-walled atria.

Extending anteriorly from the heart, locate the large **aorta,** which exits from the left ventricle, curves to the left, and passes dorsally to the heart (figure 20.6). This large curvature forms the **aortic arch,** which gives off five branches. The first two branches of the aortic arch are two **coronary arteries,** which supply the muscles of the heart. They arise within the heart and are often difficult to locate. The other three branches of the aortic arch are more prominent and are easier to locate. These branches are the **brachiocephalic artery,** the **left common carotid artery,** and the **left subclavian artery** (figures 20.6, 20.7, and 20.8).

The brachiocephalic artery extends forward and divides to form the **right common carotid artery,** which supplies the head and brain, and the **right subclavian artery,** which supplies the right forelimb and shoulder. The **left common carotid artery** branches from the aortic arch to the left of the brachiocephalic and passes anteriorly along the left side of the trachea. It gives rise to several branches that supply blood to the left side of the neck and head. The third branch

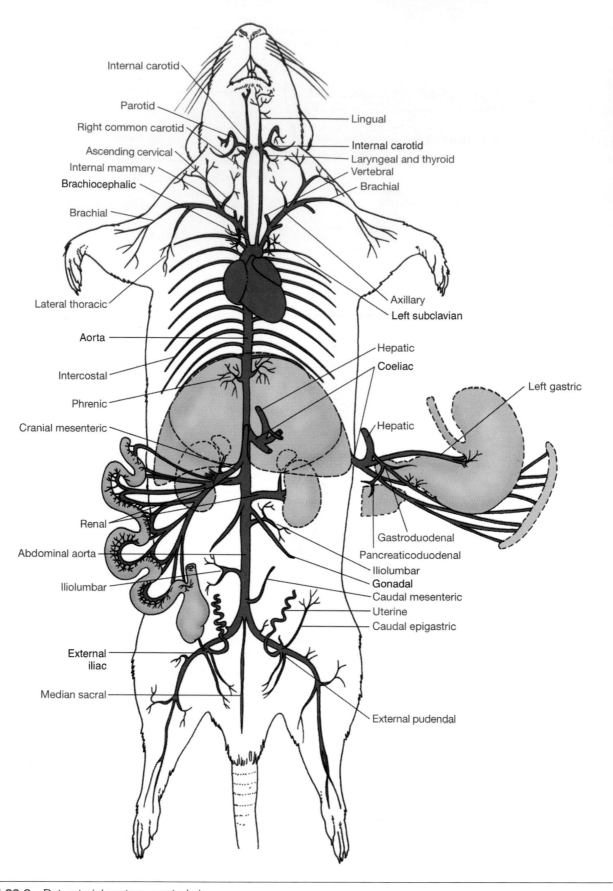

**FIGURE 20.8**   Rat, arterial system, ventral view.

from the aortic arch is the **left subclavian artery,** which supplies the left forearm, the chest, and the neck through several branches.

Posterior to the heart, the aortic arch continues as the **dorsal aorta,** which runs along the dorsal wall of the coelom. Along its course from the arch to the posterior end, the dorsal aorta gives off several major branches. *How many of these branches can you locate on your specimen?*

## Branches of the Aorta Posterior to the Heart

**Phrenic arteries**—carry blood to the diaphragm

**Lumbales arteries**—supply back muscles and adrenal glands

**Coeliac artery**—unpaired artery supplying the spleen, liver, stomach, pancreas, and duodenum

**Cranial mesenteric artery**—unpaired artery to the mesentery and small intestine; connects with caudal mesenteric

**Renal arteries**—supply the kidneys

**Spermatic arteries** (male)—supply the testes **or**

**Uterine arteries** (female)—supply the ovaries and uterus

**Iliolumbar arteries**—supply dorsal wall muscles

**Lumbar arteries**—supply back muscles

**Caudal mesenteric artery**—unpaired artery to colon and rectum; connects with cranial mesenteric

**Common iliac arteries**—branch several times to carry blood to the hindlimbs, scrotum, and penis (male) or uterus and vagina (female)

**Median sacral artery**—supplies blood to the tail

## Venous Circulation

The pattern of venous circulation consists of four distinct parts: (1) pulmonary circulation from the lungs, (2) a system that provides for the collection of blood from the head and anterior parts of the body, (3) a system that provides for the collection of blood from the abdomen and posterior part of the body, and (4) the hepatic portal system. We will consider each of these portions of the venous circulation separately.

Veins are thin-walled vessels that carry blood back to the heart after the blood has traversed capillary beds in the tissues. Veins are filled with blue latex in injected specimens, but because of their thin walls some veins burst from injection pressure, and often smaller veins may be hard to find because they are not well-injected.

*Pulmonary Circulation.* Deoxygenated blood leaves the right ventricle of the heart via the large pulmonary trunk, which divides into the left and right pulmonary arteries just dorsal to the heart. After the release of carbon dioxide and the uptake of oxygen in the capillaries surrounding the alveoli in the lungs, blood is returned to the heart via the right and left pulmonary veins.

*Systemic Veins.* System blood returns to the heart from the head, neck, forelimbs, and thoracic regions via a pair of large **cranial vena cava veins,** which empty into the right atrium (figure 20.9). The best method for studying the venous system is to start at the right atrium and trace the veins away from the heart. This allows you to start with the larger veins that are usually the best injected and therefore easier to find.

Locate the left and right cranial venae cavae, which empty into the right atrium. Each of these cranial vena cava veins is formed at the level of the first rib by the junction of three veins: the **internal jugular,** the **external jugular,** and the **subclavian.** The internal jugular receives blood from the trachea, larynx, and neck. The external jugular receives blood from the shoulder and the head. The subclavian receives blood from the forearm and part of the shoulder. Locate these veins on your specimen and observe how they come together to form the cranial vena cava.

The **caudal vena cava** is a single vein extending from the posterior end of the abdominal cavity to the heart. Locate the attachment of this vein to the heart.

Several other vessels join the posterior vena cava in the abdominal cavity. Locate the following:

**Hepatic veins**—collect blood from the liver. *How many do you find in your specimen?*

**Phrenic veins**—enter the caudal vena cava where it passes through the liver

**Renal veins**—collect blood from the kidneys (also from adrenal glands and left gonad)

**Right gonadal vein**—collects blood from the right gonad

**Iliolumbar veins**—carry blood from the dorsal lumbar region

**Common iliac veins**—large paired veins that join to form the caudal vena cava

Each common iliac vein collects blood from several smaller veins from the hindlimbs and posterior abdomen, including the pelvic veins from the pelvic region, the caudal veins from the tail, and the femoral veins from the hindlimbs.

*Hepatic Portal System.* An elaborate venous system in the rat collects blood from parts of the digestive tract and carries this nutrient-rich blood to the liver. This is the hepatic portal system. (See Chapter 19 for a discussion of portal systems.) In the liver, most of these nutrients are removed for storage or chemically changed before they are released into general circulation. The principal vessel is the hepatic portal vein, which carries blood from the intestines to the liver. Find the hepatic portal vein in the mesenteries posterior to the liver. If your specimen is triple-injected, this vessel will be filled with yellow latex. If your specimen is double-injected, the hepatic portal vein will be filled with reddish-brown coagulated blood. The hepatic portal collects blood from several smaller veins leading from the small intestine, large

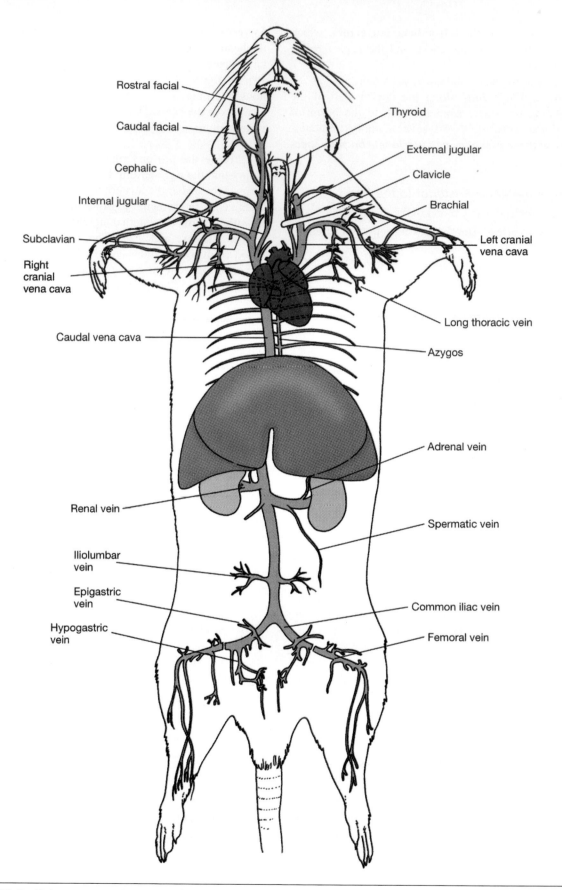

**FIGURE 20.9**  Rat, venous system, ventral view.

intestine, stomach, pancreas, and spleen. From the liver, blood passes through the hepatic vein to the caudal vena cava and back to the heart, as we have seen.

### Spleen and Thymus

Two accessory organs that have important roles in the circulatory system are the **spleen** and the **thymus.** The spleen is a dark, elongated structure attached to the mesentery along the greater curvature (larger convex surface) of the stomach. The spleen serves as a reservoir for the storage of red blood cells and is important in the immune system. It functions also as part of the lymphatic system in filtering out and neutralizing infectious agents.

The thymus in adult rats consists of a small mass of tissue on the ventral surface of the trachea anterior to the heart and near the branching of the two bronchi from the trachea.

The thymus is most well-developed in young rats where it serves to produce white blood cells. It decreases in size as rats age. The thymus is made up largely of lymphatic tissue and functions in immunity and other forms of defense against infectious agents.

## Urogenital System

The excretory and reproductive organs of the rat are closely related both embryologically and anatomically and are commonly considered together as components of a single system, the **urogenital system.** The chief components of this system involved with excretion are the **kidneys,** the **ureters,** the **urinary bladder,** and the **urethra.** The paired kidneys are closely held to the dorsal wall by a layer of peritoneum that separates them from the abdominal cavity. Attached to the surface of each kidney is a thin sheet of reddish tissue, the adrenal gland. The adrenal glands are endocrine organs that produce several important hormones. The kidneys serve to concentrate nitrogenous wastes of metabolism and produce urine. Urea is actually produced in the liver and is transported to the kidneys. The concentration of wastes and the production of a concentrated urine is an important water conservation adaptation of rats and many other terrestrial animals.

◆ Cut through the peritoneum covering one kidney, remove the fat deposits around the kidney, and make a longitudinal section of the kidney with a razor blade or scalpel. Observe that the kidney tissue consists of an outer **cortex** and an inner **medulla** area, which is darker in color.

The inner concave surface of the kidney where the ureter and blood vessels attach is the **hilus.** Identify the **renal artery** and **renal vein** adjacent to the large **ureter,** which carries the urine to the **urinary bladder.** If your specimen is a female, trace the **urethra** from the urinary bladder to its external opening adjacent to the clitoris. If you have a male specimen, trace the urethra to the penis.

### Male Reproductive Organs

The male reproductive system is illustrated in figures 20.10 and 20.11. Sperm are produced in the paired **testes.** During breeding season, the testes are located in the **scrotum,** a

large sac located ventral to the anus. At other times, the testes are usually retracted through the inguinal canal into the posterior part of the abdominal cavity. Locate the testes in the scrotum or abdominal cavity, and carefully cut through the connective tissue surrounding one testis. Find the **epididymis,** a coiled tube attached to the surface of the testis where mature sperm are stored (figure 20.12). A smaller tubule, the **vas deferens,** carries the sperm from the epididymis to the **urethra.** Follow the vas deferens from one testis and observe that it joins the vas deferens from the other testis where they empty together into the urethra. The urethra carries the sperm from the two vasa deferentia to the penis, from which it is deposited in the vagina of the female during copulation. The sperm are suspended in the seminal fluid or **semen** made up of the secretions of several glands associated with the urethra and penis.

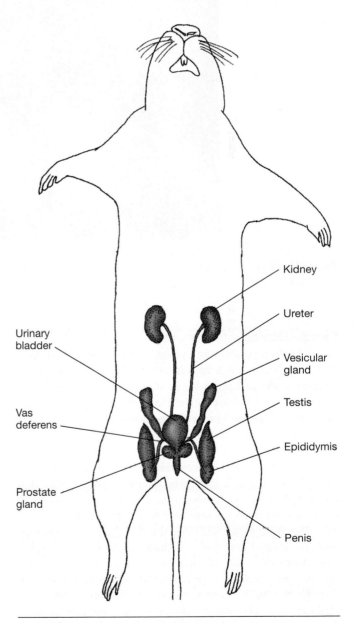

**FIGURE 20.10**  Rat, urogenital system, male.

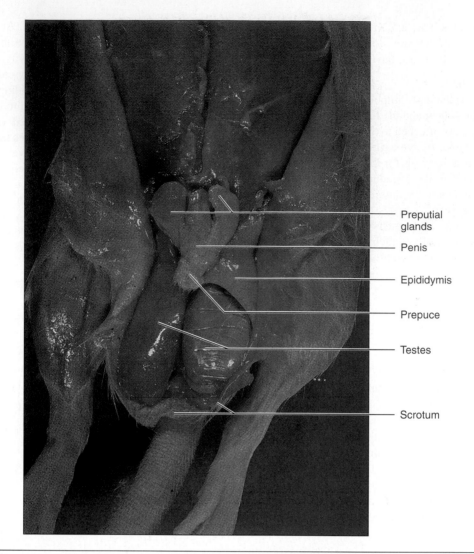

- Preputial glands
- Penis
- Epididymis
- Prepuce
- Testes
- Scrotum

**FIGURE 20.11**   Rat, urogenital system, male, scrotum dissected to show testes.
Photograph by Ken Taylor.

### Female Reproductive Organs

The female reproductive system is illustrated in figures 20.5 and 20.13. Eggs or **ova** are produced in the paired **ovaries** found just posterior to the kidneys. The ovary is a small mass of follicles embedded in mesentery rather than in a compact organ. The **oviducts** in the rat are small, short tubules that lead from the funnel-shaped ostia where ova are received from the ovaries. The oviducts connect with the two branches of the bicornuate **uterus.** The uterus consists of two large branches or **horns of the uterus** that open separately into the **vagina.** These **uterine horns** in the horns of the uterus represent a significant morphological adaptation that facilitates the simultaneous development of several embryos and multiple births. The vagina leads from the body of the uterus to the exterior and receives sperm during copulation. Mature ova descending through the oviduct are fertilized in the uteri. The fertilized eggs are subsequently implanted into the uterine wall. The **gestation period** (time of uterine development before birth) of the laboratory rat is about 21–22 days. A litter averages between 6–8 young.

## Nervous System

The nervous system of mammals consists of three major divisions: (1) the central nervous system, consisting of the brain and spinal cord; (2) the peripheral nervous system, made up of the myelinated sensory and motor nerves extending from the central nervous system; and (3) the autonomic nervous system, a system of visceral nonmyelinated nerves that regulate the glands and visceral organs. In this exercise, we will confine our study mainly to the brain and the central nervous system (figure 20.14).

### Brain

The brain of the rat is small and is more difficult to study than that of larger mammals. We will therefore supplement our study of the rat brain with the larger but similar brain of the sheep.

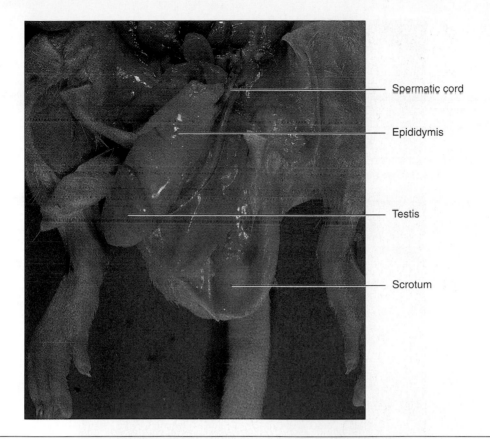

Spermatic cord

Epididymis

Testis

Scrotum

**FIGURE 20.12** Rat, urogenital system, male, testes dissected to show internal details.
Photograph by Ken Taylor.

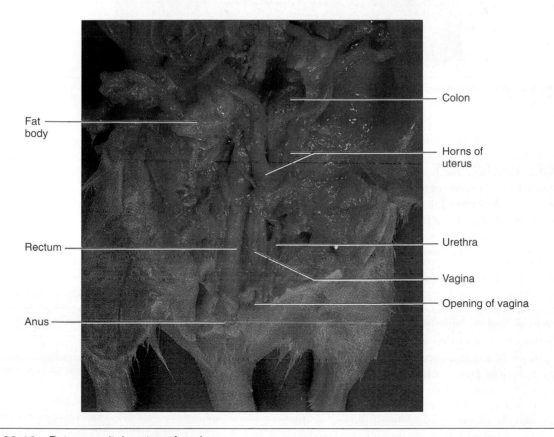

Fat body

Rectum

Anus

Colon

Horns of uterus

Urethra

Vagina

Opening of vagina

**FIGURE 20.13** Rat urogenital system, female.
Photograph by Ken Taylor.

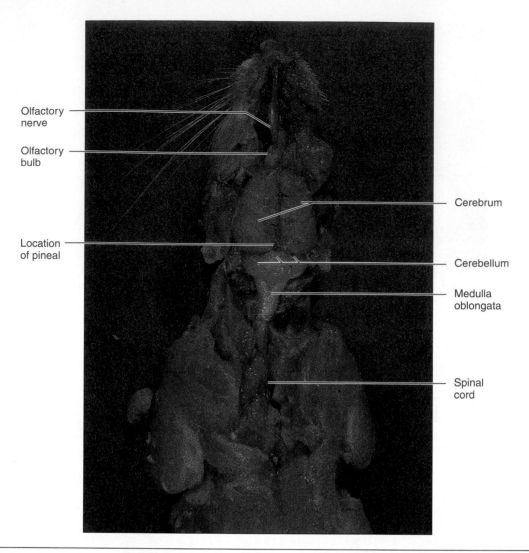

Olfactory nerve

Olfactory bulb

Location of pineal

Cerebrum

Cerebellum

Medulla oblongata

Spinal cord

**FIGURE 20.14** Rat, brain, dorsal view.
Photograph by Ken Taylor.

◆ Remove the skin and hair from the top of the head between the eyes and ears, and cut away the muscles at the back of the head near the occipital region. Use bone shears to cut away a portion of the bone at the back of the skull to expose the brain. Take care not to cut into the brain. When it is exposed, use forceps to pick away small pieces of bone to further study the brain (figure 20.14).

Observe that the brain is covered by several tough outer layers of connective tissue, the meninges. The brain of the rat and other mammals consists of five principal regions that can be traced to its embryological development. These anatomical regions and the principal structures in the adult brain representing these divisions are listed in table 20.2.

Locate first the two large anterior **cerebral hemispheres** separated by a median groove or **fissure.** Anterior to the cerebral hemispheres find the two smaller **olfactory bulbs** to which the olfactory nerves connect (figure 20.15). At the posterior margin of the cerebral hemispheres in the cerebral fissure is the **pineal body.** Posterior to the cerebral hemispheres on the dorsal surface find the **cerebellum,**

consisting of three lobes—a right and left hemisphere and the vermis—each with a folded surface. The cerebellum is the principal center responsible for coordination of body movements.

The **medulla oblongata** is the most posterior part of the brain and connects with the spinal cord. The medulla regulates such vital processes as respiration, heart rate, blood pressure, and hormonal secretion.

## TABLE 20.2
### Principal Regions of the Mammalian Brain

| Region | Principal Structures |
| --- | --- |
| Telencephalon | Olfactory bulbs, cerebral hemispheres |
| Diencephalon | Pituitary gland |
| Mesencephalon | Corpora quadrigemina |
| Metencephalon | Cerebellum, pons, part of medulla oblongata |
| Myelencephalon | Part of medulla oblongata |

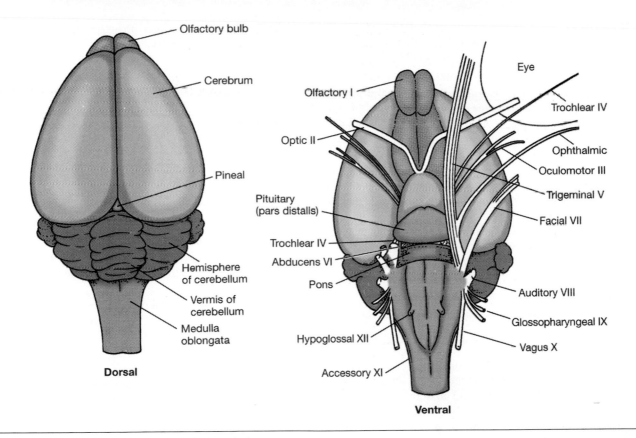

**FIGURE 20.15**    Rat, brain, dorsal and ventral views.

◆ After you have studied these structures on the dorsal surface of the brain, carefully detach the brain from the braincase. Try to retain a portion of the several pairs of cranial nerves attached to the ventral surface of the brain so that you can locate where these nerves arise from the brain.

On the ventral surface of the brain, locate the anterior **olfactory bulbs** with two large olfactory tracts leading posteriorly to the cerebrum, and the **optic chiasma** where the two large **optic nerves** cross. Posterior to the optic chiasma is the **pituitary gland,** which may have been left on the floor of the braincase when the brain was removed from the cranium. Behind the pituitary, a transverse tract of nerve fibers, the **pons,** connects the left and right hemispheres of the cerebellum.

Twelve pairs of **cranial nerves** arise from the ventral surface of the brain (figure 20.15). We have already noted olfactory tracts and the optic nerves, which are the first and second pairs of cranial nerves. The twelve pairs of cranial nerves and their locations are listed in table 20.3.

### Sheep Brain

For further information on the anatomy of the mammalian brain, study a preserved sheep brain as described in Chapter 19, "Fetal Pig Anatomy" (figure 19.25). Identify the principal structures of the sheep brain and compare them with the rat brain. **What differences do you find?**

### TABLE 20.3

#### Cranial Nerves of the Rat

| No. Name | Location | Function(s) |
|---|---|---|
| I. Olfactory | Olfactory bulbs | Smell |
| II. Optic | Diencephalon | Vision |
| III. Oculomotor | Ventral surface of mesencephalon | Eye movements |
| IV. Trochlear | Dorsal surface of mesencephalon | Eye movements |
| V. Trigeminal | Pons | Innervates skin, vibrissae, jaw muscles, tongue, and teeth |
| VI. Abducens | Medulla oblongata | Eye movements |
| VII. Facial | Medulla oblongata | Innervates jaw muscles |
| VIII. Auditory | Medulla oblongata | Hearing |
| IX. Glossopharyngeal | Medulla oblongata | Innervates pharynx and tongue |
| X. Vagus | Medulla oblongata | Innervates larynx, heart, lungs, diaphragm, and stomach |
| XI. Accessory | Medulla oblongata | Innervates neck muscles and pharyngeal organs |
| XII. Hypoglossal | Medulla oblongata | Tongue movements |

## Spinal Cord

The spinal cord extends posteriorly from the medulla oblongata and is covered by meninges. The spinal cord passes through the vertebrae in the vertebral column, and several pairs of **spinal nerves** extend from the spinal cord through openings between the vertebrae. The spinal nerves are usually classified in groups according to the region where they arise from the spinal cord: **cervical, thoracic, lumbar, sacral,** and **caudal.**

### Demonstrations

1. Liquid or plastic mount of dissected rat
2. Microscope slide with longitudinal section of a rat heart
3. Microscope slide with cross section of rat testis
4. Microscope slide with cross section of rat ovary
5. Microscope slide with median section of rat kidney
6. Microscope slide with rat sperm

## Key Terms

**Diaphragm**  muscular sheet located anterior to the liver that separates the thoracic and abdominal cavities in mammals.

**Hepatic portal system**  portion of the venous circulation in many vertebrates that collects blood from capillary beds in the stomach and intestine, and returns it to the liver; here the blood passes through a second capillary bed where nutrients are removed prior to returning the blood to the heart.

**Mesentery**  double sheet of mesodermal tissue that supports various internal organs in the coelom of vertebrates.

**Pulmonary circulation**  portion of mammalian circulatory system that carries blood from the heart to the lungs and returns it to the heart.

**Systemic circulation**  portion of the mammalian circulatory system that carries blood from the heart to and from all organs of the body except the lungs.

**Vibrissae**  long sensory hairs on the face of the rat.

## Internet Resources

Visit the zoology website at http://www.mhhe.com/zoology to find live Internet links of each of the references listed below.

1. Animal Diversity Web, University of Michigan. Raccoon skeleton labeled (very similar to a cat). Very useful for comparative anatomy.
2. Animal Diversity Web, University of Michigan. Rodentia. A description of rodents, with links to other sites that further describe morphological characteristics.

3. Clickable Human Skeleton. This interactive site features the anatomy of the skeletal system. This may not work on all systems or platforms.
4. Complete Muscle Tables of the Human Body. This site contains comprehensive information on human muscles.

## Critical Thinking Questions

1. What are the features of the rat that distinguish it as a mammal? Are these the same features present in the pig?

2. Discuss the basic organization of the mammalian nervous system. What components affect intelligence? How would you measure the relative difference in intelligence between the rat and the pig?

3. Explain how the fetal membranes (amnion, chorion, and allantois) function. In the evolutionary sense, why was the development of these membranes so important?

4. Discuss the difference between negative pressure breathers and positive pressure breathers. Does one method have advantages over another in a terrestrial environment?

5. Outline the urogenital systems of both male and female rats. Point out homologous organs wherever you are able (e.g., testis—ovaries).

## Suggested Readings

Chaisson, R.B. 1994. *Laboratory Anatomy of the White Rat*, 5th ed. Dubuque, IA: The McGraw-Hill Companies, Inc.

Feldhammer, G.A., Drickamer, L.C., Vessey, S.H., and J.F. Merritt. 1999. *Mammalogy: Adaptation, Diversity, and Ecology*. Dubuque, IA: The McGraw-Hill Companies, Inc.

Vaughn, T.A., Czaplewski, N., and J. Ryan. 1998. *Mammalogy*. Philadelphia: Saunders College Publishing.

Walker, W.F., Jr. 1998. *Anatomy and Dissection of the Rat*. 3d ed. San Francisco: W.H. Freeman Company.

# NOTES AND SKETCHES

# NOTES AND SKETCHES

# CREDITS

## *PHOTOS*

**Chapter One Opener:** Courtesy of Wolfe Sales Corporation and Carolina Biological Supply Company, Burlington, NC.

**Chapter Eleven Opener:** Courtesy of Carolina Biological Supply Company, Burlington, NC.

## *ILLUSTRATIONS*

### Chapter Two

**Chapter Two Opener:** From Stuart Ira Fox, *A Laboratory Guide to Human Physiology,* 6th ed. Copyright © 1993 Wm. C. Brown Communications, Inc. Used by permission of McGraw-Hill Publishers Inc., New York, NY. All Rights Reserved. **2.3:** From Stuart Ira Fox, *A Laboratory Guide to Human Physiology,* 6th ed. Copyright © 1993 Wm. C. Brown Communications, Inc. Used by permission of McGraw-Hill Publishers Inc., New York, NY. All Rights Reserved. **2.11:** From Kent M. Van De Graaff and Stuart Ira Fox, *Concepts of Human Anatomy and Physiology,* 4th ed. Copyright © 1995 Wm. C. Brown Communications, Inc. Used by permission of McGraw-Hill Publishers Inc., New York, NY. All Rights Reserved. **2.12:** From Kent M. Van De Graaff and Stuart Ira Fox, *Concepts of Human Anatomy and Physiology,* 4th ed. Copyright © 1995 Wm. C. Brown Communications, Inc. Used by permission of McGraw-Hill Publishers Inc., New York, NY. All Rights Reserved. **2.29:** From Kent M. Van De Graaff and Stuart Ira Fox, *Concepts of Human Anatomy and Physiology,* 4th ed. Copyright © 1995 Wm. C. Brown Communications, Inc. Used by permission of McGraw-Hill Publishers Inc., New York, NY. All Rights Reserved.

### Chapter Three

**Chapter Three Opener:** From Sylvia S. Mader, *Human Biology,* 4th ed. Copyright ©1995 Wm. C. Brown Communications, Inc. Used by permission of McGraw-Hill Publishers Inc., New York, NY. All Rights Reserved. **3.1:** From Stuart Ira Fox, *Human Physiology,* 4th ed. Copyright © 1993 Wm. C. Brown Communications, Inc. Used by permission of McGraw-Hill Publishers Inc., New York, NY. All Rights Reserved. **3.2:** From Sylvia S. Mader, *Human Biology,* 4th ed. Copyright © 1995 Wm. C. Brown Communications, Inc. Used by permission of McGraw-Hill Publishers Inc., New York, NY. All Rights Reserved. **3.5:** From Sylvia S. Mader, *Human Biology,* 4th ed. Copyright © 1995 Wm. C. Brown Communications, Inc. Used by permission of McGraw-Hill Publishers Inc., New York, NY. All Rights Reserved. **3.9a:** From Sylvia S. Mader, *Human Biology,* 4th ed. Copyright © 1995 Wm. C. Brown Communications, Inc. Used by permission of McGraw-Hill Publishers Inc., New York, NY. All Rights Reserved. **3.9b:** From Sylvia S. Mader, *Human Biology,* 4th ed. Copyright © 1995 Wm. C. Brown Communications, Inc. Used by permission of McGraw-Hill Publishers Inc., New York, NY. All Rights Reserved.

### Chapter Four

**4.9a:** From Sylvia S. Mader, *Inquiry into Life,* 7th ed. Copyright © 1994 Wm. C. Brown Communications, Inc. Used by permission of McGraw-Hill Publishers Inc., New York, NY. All Rights Reserved.

### Chapter Five

**Chapter Five Opener:** From Ricki Lewis, *Life.* Copyright © 1992 Wm. C. Brown Communications, Inc. Used by permission of McGraw-Hill Publishers Inc., New York, NY. All Rights Reserved. **5.1:** From Sylvia S. Mader, *Biology,* 4th ed. Copyright © 1993 Wm. C. Brown Communications, Inc. Used by permission of McGraw-Hill Publishers Inc., New York, NY. All Rights Reserved. **5.8:** From Ricki Lewis, *Life.* Copyright © 1992 Wm. C. Brown Communications, Inc. Used by

permission of McGraw-Hill Publishers Inc., New York, NY. All Rights Reserved. **5.28:** From Eldon D. Enger, et al., *Concepts in Biology,* 7th ed. Copyright © 1994 Wm. C. Brown Communications, Inc. Used by permission of McGraw-Hill Publishers Inc., New York, NY. All Rights Reserved.

### Chapter Eight

**8.1:** From Sylvia S. Mader, *Biology,* 4th ed. Copyright © 1993 Wm. C. Brown Communications, Inc. Used by permission of McGraw-Hill Publishers Inc., New York, NY. All Rights Reserved.

### Chapter Fourteen

**Chapter Fourteen Opener, and 14.4:** From Sylvia S. Mader, *Inquiry into Life,* 7th ed. Copyright © 1994 Wm. C. Brown Communications, Inc. Used by permission of McGraw-Hill Publishers Inc., New York, NY. All Rights Reserved.

# INDEX

## A

Abdomen
  of cockroach, 209, 210
  of crayfish, 201, 202
  of fetal pig, 297, 308, 309, 310, 311
  of grasshopper, 215, 216
  of horseshoe crab, 192, 197
  of rat, 332–33
Abdominal arteries
  of crayfish, 204, 206
  of shark, 252, 253
Abdominal extensor muscles, 204, 206
Abdominal flexor muscles, 204, 206
Abdominal ganglion, 214
Abdominal veins
  of frog, 280, 281, 282, 284
  of shark, 245, 246, 253
Abdominal vertebrae, 271, 273
Abducens nerve
  of frog, 289, 291, 292
  of rat, 343
  of shark, 255, 256, 257
  of sheep, 323
Acanthocephala, 146, 147
Accessory glands, 212
Accessory nerve
  of cat, 324
  of fetal pig, 322
  of rat, 343
Achilles tendon, 275
Acoelomates, 122, 123, 128
Acontia, 116–17
Acoustic nerve
  of cat, 324
  of yellow perch, 266
Acrania, 233–34
Actin, 28
*Actinosphaerium,* 77
Adaptive radiation, 159
Adductor longus muscle, 274, 275, 277
Adductor magnus muscle, 274, 275, 277
Adductor muscles, of mussel, 162, 163, 165
Adhesive pads, 111, 112
Adrenal gland
  of fetal pig, 309, 310, 312
  of frog, 286, 288
Aedeagus, 212
African sleeping sickness, 83
Agnatha, 234
Air sacs
  of cockroach, 210
  of frog, 280
Albumen gland, 168
Alcynaria, 117–18

Allantoic duct, 297, 309, 310, 312
Allantois, 60, 64, 65, 67, 68, 69
Alligator, 234
Ambulacral groove, 225, 226, 228
Ambulacral regions, 228
Amitosis, 87
Amnion, 60, 62, 63
Amniotic egg, 59–60
Amoeba, 74–77
*Amoeba proteus,* 74–77
Amoeboid cells, 100
Amphibia, 234. *See also Rana pipiens*
Amplexus, 55, 285
Ampulla, 227, 228
Amylase, 210
Anal tubercle, 196, 197
Anaphase
  of meiosis, 42–43, 44
  of mitosis, 38, 39
Anatomical planes, 124
Anconeus muscle, 274, 275
*Ancylostoma duodenale,* 151–52
*Anguillula aceti,* 153–54
Animal pole, 51–52, 55, 57
Animal safety, xiv, xvi–xvii
Animal-vegetal axis, 55
Annelida, 175–87
  classification of, 176
  Hirudinea, 184–86
  Oligochaeta, 180–84
  Polychaeta, 176–80
  segmentation in, 123, 176
Annuli, 185
Antennae
  of cockroach, 209
  of crayfish, 201, 202, 203, 206
  of grasshopper, 214, 215
  of *Peripatus,* 219
  of *Philodina,* 155
  of water flea, 198, 199–200
Antennary arteries, 204, 206
Antennules, 201, 202, 203, 206
Anthozoa, 106, 116–18
Anus
  of *Anguillula aceti,* 154
  of *Ascaris,* 147, 148
  of *Caenorhabditis elegans,* 153
  of cockroach, 210
  of crayfish, 201, 205, 206
  of earthworm, 180, 182
  of fetal pig, 310, 312
  of freshwater mussel, 164
  of horseshoe crab, 193
  of leeches, 186
  of mussel, 164

Anus—*Cont.*
  of *Philodina,* 155, 156
  of rat, 328, 332, 333, 341
  of rotifers, 156
  of shark, 243, 244
  of squid, 171
  of water flea, 199
  of yellow perch, 260, 266
Aorta
  embryonic, 62
  of fetal pig, 310, 312, 314, 315, 316,
    317, 320
  of frog, 280, 282, 283, 292
  of lancelets, 237, 238, 239
  of mussel, 165
  of rat, 336, 337
  of shark, 244, 245, 246, 249, 250, 251, 252
  of yellow perch, 264, 265
Aortic arches
  of chick embryo, 62, 63, 64, 66, 67, 68
  of earthworm, 183
  of rat, 334, 335
Apicomplexa, 74, 89–92
Aplacophora, 161
Apodeme, 196
Apopyles, 100
Appendages
  of crayfish, 202, 203
  of fetal pig, 296, 297
  of horseshoe crab, 192
Appendicular skeleton, 298, 299
Arachnida, 191
Arachnoid layer, 320
*Arbacia punctulata,* 229
*Arcella,* 76–77
Archenteron, 54, 57
Archinephric ducts
  of shark, 247
  of yellow perch, 266, 267
*Arenicola,* 180
*Argiope,* 193–98
Aristotle's Lantern, 228
Arms
  of octopus, 171
  of squid, 169, 170, 171
  of starfish, 225, 227
Arolium, 215, 216
Arrector pili muscle, 21
Arterial collecting loop, 252, 253
Arterial system
  of crayfish, 204
  of fetal pig, 314–19
  of frog, 280, 281, 282, 283
  of rat, 335–37
  of shark, 249–50, 252–53
Arthrobranchs, of crayfish, 201
Arthropoda, 123, 190–217
  Chelicerata, 192–95
  classification of, 190–91
  Crustacea, 195, 198–208
  exoskeleton of, 190
  Insecta, 191, 208–17
  major features of, 190
  metamorphosis in, 217
  Onychophora, 217, 219
  segmentation in, 123
*Ascaris lumbricoides,* 147–50
Asconoid sponge, 98, 99
Asexual reproduction, 130–31

*Asplanchna,* 158
Aster, 38
Aster rays, 37
*Asterias,* 225–28
Asteroidea, 225–28
Asterozoa, 224–25
Asymmetry, 122
Atlas
  of fetal pig, 300
  of frog, 271
Atriopore, 237
Atrioventricular valve, 335
Atrium
  embryonic, 62
  of fetal pig, 307, 309, 310, 312, 315, 320
  of frog, 280, 283
  of lancelets, 237, 238
  of rat, 333, 334
  of shark, 244, 250, 251
  of yellow perch, 264
Auditory nerve
  of fetal pig, 322
  of frog, 289, 291
  of rat, 343
  of shark, 255, 256, 257
*Aurelia,* 115, 116
Auricular lobes, 256
Autonomic nervous system
  of frog, 285, 290
  of shark, 253
Aves, 234
Axial skeleton, 298, 299
Axis, 300
Axon, 28, 30, 31
Azygous vein, 337, 338

# B

Bailer, 204
Barnacles, 191
Basal disc
  of *Hydra,* 107
  of *Metridium,* 116
Basisphenoid, 301
Basket stars, 224
Basophils, 22
Beaks, of squid, 169, 170
Biceps brachii muscle
  of fetal pig, 306
  of rat, 330, 331
Biceps femoris muscle
  of fetal pig, 304
  of frog, 274, 277
Bicuspid valve, 319, 320
Bilateral cleavage, 52
Bilateral symmetry, 122
Bile duct
  of fetal pig, 309
  of shark, 243, 244
Bipinnaria larvae, of starfish, 228
Birds, 234
Bivalvia, 161, 162–67
Bivium, 225
Bladder
  of fetal pig, 308, 310, 311, 312, 313
  of frog, 278, 279, 285, 286, 287
  of rat, 333, 339
  of yellow perch, 262, 263, 264,
    266, 267

Blastocoel, 54
Blastoderm, 68
Blastopore, 54, 57
Blastostyle, 112, 113
Blood, 21–24
Blood sinuses, 167
Bone, 26
Bony fishes, 234, 259–67
Book gills, 192, 193
Book lung, 196
Book lung cover, 197
Box jellyfish, 106
Brachial artery
    of fetal pig, 315, 316
    of frog, 280
Brachial nerve, 291
Brachial plexus
    of fetal pig, 321, 322
    of frog, 290, 291
Brachial vein
    of fetal pig, 318, 319
    of frog, 280, 284
    of shark, 246
Brachiocephalic artery
    of fetal pig, 320
    of rat, 335
Brachiocephalic muscle, 303, 305, 306
Brachiocephalic trunk, 307, 310, 315, 316, 317
Brain
    of cat, 323
    of Clonorchis sinensis, 133
    of crayfish, 206, 207
    of earthworm, 182, 183, 184
    of fetal pig, 320, 321, 322
    of free-living flatworm, 129
    of frog, 285, 289, 290
    of liver fluke, 133, 134
    of rat, 342–43
    of shark, 254–56
    of sheep, 323, 344
    of squid, 169
    of water flea, 199, 200
    of yellow perch, 262, 266, 267
Brain stem, 254
Branchial arteries
    of lancelets, 237
    of shark, 244, 246, 249, 250, 251, 252
    of yellow perch, 264, 265
Branchial nerve, 255, 256
Branchiopoda, 191, 195, 198–201
Branchiostegites, of crayfish, 204
Branchiostoma, 236–39
Bristleworms (Oligochaeta), 176
Brittle stars, 225
Bronchial tubes, 280
Brood chamber, 200
Buccal cavity
    of earthworm, 181, 182
    of free-living flatworm, 131
Buccal nerve, 257
Buds, of Hydra, 107

C

Caecum
    of cockroach, 210, 211
    of fetal pig, 309, 310
    of lancelets, 236
    of liver fluke, 133, 134

of rat, 333
of squid, 170, 171
of starfish, 226, 227
of water flea, 199
of yellow perch, 262, 263
Caenorhabditis elegans, 152–53
Calcaneus
    of fetal pig, 300
    of rat, 329
Calcarea, 97, 98
Calcareous sponges, 98
Calciferous glands, 182, 183
Calcispongiae, 98
Canaliculi, 26
Carapace
    of crayfish, 201
    of horseshoe crab, 192
    of spider, 194
    of water flea, 198–99
Cardiac chamber, 205
Cardiac muscle, 28, 29
Cardiac plexus, 290, 292
Cardiac vein, 284
Cardinal sinus, 253
Cardinal veins
    of shark, 244, 245, 246, 249, 253
    of yellow perch, 264, 265
Carosafe, xiv, xv
Carotid artery
    of fetal pig, 307, 308, 315, 316, 320
    of frog, 280, 281, 283
    of rat, 334, 335
    of shark, 244, 252, 253
Carotid body, 283
Carpals
    of fetal pig, 300
    of rat, 329
Carpi-radialis muscle, 274
Cartilage, 25, 26
Cat
    brain of, 323
    skeletal system of, 298, 299
    skull of, 301
Cat tapeworm, 141, 142, 143
Caudal artery
    of fetal pig, 315, 316, 317
    of shark, 244, 246
Caudal fold, 63
Caudal vein, 244, 246
Celiac artery, 336, 337
Celiacomesenteric artery, 280
Cell, 16, 17
Cell cycle, 36, 38, 40. See also Meiosis; Mitosis
Cell membrane, 16, 75
Cell theory, 15–16
Centipedes, 191
Central canal, 238
Central disc, 225
Centriole, 17, 18, 38
Centrolecithal eggs, 51
Centromere, 38, 44
Centrosome, 38
Centrum
    of shark, 246
    of yellow perch, 261
Cephalic ganglia, 169
Cephalization, 123
Cephalochordata, 234, 236–39
Cephalopoda, 160, 161, 168–71

Cephalothorax, 201
*Ceratium,* 81, 82
Cercariae
    of liver fluke, 133, 135
    of sheep liver fluke, 136
    of trematodes, 135
Cercus, 210, 213
Cerebellum
    of fetal pig, 310
    of frog, 285
    of rat, 342, 343
    of shark, 256
    of sheep, 323
    of yellow perch, 266
Cerebral ganglion, 133, 134
Cerebral hemisphere
    of frog, 285, 290
    of shark, 254
Cerebral vesicle, 237
Cerebropleural ganglia, 167
Cerebrum
    of fetal pig, 310, 322
    of rat, 342
    of sheep, 323
Cervical groove
    of crayfish, 201
    of spider, 196
Cervical nerve plexus, 321, 322
Cervical region, of cockroach, 209
Cervical vertebra, 271
Cestoda, 128, 138–41
Chagas disease, 83
Chelae, 201
Chelate legs, 192, 193
Chelicerae
    of horseshoe crab, 192, 193
    of spider, 194
Chelicerata, 190–91, 192–95
Chelipeds, 201
Chemoreceptors
    of crayfish, 207
    of water flea, 199
Chiasmata, 44
Chick embryo, 59–69
    24-hour, 60–61
    48-hour, 62–65
    72-hour, 65–67
    96-hour, 68–69
Chilaria, 192, 193
Chilopoda, 191
Chitin, 190
Chitinous bristles, 205
Chitinous ring, 169
Chitons, 160, 161
Chlorogogue cells
    of earthworm, 182, 183, 184
    of sandworm, 178
Chloroplasts
    of *Euglena,* 78
    of *Volvox,* 79, 80
Choanocytes, 96, 100, 102
Choanoflagellates, 92, 96–97
Chondrichthyes, 234
Chordata, 123, 233–39
    Cephalochordata, 236–39
    classification of, 233–34
    Urochordata, 234–36
Chorioallantoic membrane, 60, 64–65, 67

Chorion, 60
Chromatophores, 169–70
Chromosomes
    diploid, 42
    haploid, 42
    homologous, 42
    in meiosis, 44
    in mitosis, 37–38
Cilia
    vs. flagella, 87
    of *Metridium,* 116, 117
    of mussel, 164
    of *Paramecium caudatum,* 85, 86–87
Ciliated grooves
    of *Aurelia,* 115
    of *Metridium,* 116
Ciliophora, 74, 85–89
Cingulum, 156
Ciplopoda, 191
Circular canals
    of *Aurelia,* 115
    of *Gonionemus,* 111, 112
    of starfish, 227, 228
Circulatory system
    of Bivalvia, 165, 166
    of *Branchiostoma,* 237
    of cockroach, 210
    of crayfish, 204
    of earthworm, 183
    of fetal pig, 314–19
    of frog, 280–84
    of lancelets, 237
    of mussel, 166
    of *Peripatus,* 219
    of rat, 333–39
    of sea squirts, 236
    of shark, 244, 246, 249–53
    of water flea, 200
    of yellow perch, 264, 265
Circumesophageal connectives, 212
Circumpharyngeal connectives, 183
Cirripedia, 191
Cirrus
    of *Euplotes,* 88
    of lancelets, 236, 237
    of sandworm, 178
    of tapeworm, 143
Clams, 161
Clamworm, 176–80
Claspers, of shark, 242, 244, 246
Clavicle, 329
Claws
    of grasshopper, 215, 216
    of rat, 328
Cleavage, embryonic, 49, 50–52
Clitellum, 180
Clitoris
    of fetal pig, 313
    of rat, 328, 332
Cloaca, 67
    of *Anguillula aceti,* 154
    of *Ascaris,* 149
    embryonic, 67
    of frog, 279, 286
    of *Philodina,* 155, 156
    of shark, 242, 243, 244, 246
Cloacal vein, 246
*Clonorchis (Opisthorchis) sinensis,* 133–35

Clypeus
  of grasshopper, 215
  of spider, 194
Cnidaria, 105–18
  Anthozoa, 116–18
  *Aurelia,* 115, 116
  body forms of, 106
  classification of, 106
  corals, 117–18
  *Gonionemus,* 110–12
  *Hydra,* 106–10
  Hydrozoa, 106–14
  *Metridium,* 116–17
  *Obelia,* 112–13
  *Physalia,* 113–14
  Scyphozoan, 115–16
Cnidocytes, 106, 108, 109
Coagulation, 24
Coccidiosis, 90
Cockroach, 208–12
Coelenterates. *See* Cnidaria
Coelenteron, 106
Coeliac artery
  of fetal pig, 315, 316
  of frog, 282, 283
  of shark, 244, 250, 252
  of yellow perch, 265
Coeliacomesenteric artery, 281, 282, 283
Coelom, 122–23
  of *Branchiostoma,* 238
  of earthworm, 181, 182, 184, 185
  embryonic, 54, 57
  of fetal pig, 305, 308, 309
  of Hirudinea, 186
  of lancelets, 238
  of leeches, 186
  of Mollusca, 160
  of *Nereis,* 179
  of sandworm, 178, 179
  of shark, 243–47
  of starfish, 226, 227
  of yellow perch, 263–64
Coelomate, 122–23
Coelomocytes, 228
Coenosarc, 112, 113
Collar, of squid, 169
Colon
  of cockroach, 210
  of fetal pig, 311, 312, 313
  of rat, 333, 341
  of shark, 243, 244
Colonial theory, of multicellular animal
  evolution, 92–93
Colulus, 196, 197
Columnar epithelium, 18, 19, 20
Compound microscope, 1–8
  focusing for, 5
  illumination for, 4–5
  magnification for, 3–4, 5, 7
  object size measurement with, 4, 7–8
  parts of, 2–3
  precautions with, 6–7
  procedure for, 6
  resolving power of, 4
Conduction, nerve, 30
Conjugation
  of *Euplotes,* 88
  of *Paramecium caudatum,* 87

Connective tissue, 21–26
  loose, 24, 25
Contractile progeins, 87
Contractile vacuoles
  of amoeba, 75, 76
  of *Euglena,* 78
  of *Paramecium caudatum,* 85, 86
  of *Volvox,* 79
Conus arteriosus
  of frog, 278, 280, 283
  of shark, 246, 250, 251
  of yellow perch, 264
Convergent evolution, 169
Copulatory organ, of spider, 194, 196
Copulatory spicules
  of *Anguillula aceti,* 154
  of *Ascaris,* 148, 149
Corals, 106, 117–18
Cornea, of crayfish, 207–8
Corona, of *Philodina,* 155, 156
Coronary artery
  of rat, 335
  of shark, 250
Coronary ligament, 309
Costal cartilages, 300
Cowper's gland, 311, 312, 313
Cranial nerves
  of fetal pig, 322, 324
  of frog, 289–90
  of rat, 343
  of shark, 255, 256–57
  of yellow perch, 267
Cranial vena cava veins, 337, 338
Craniata, 234
Cranium
  of fetal pig, 297
  of frog, 271
Crayfish, 201–8
Crest, of *Physalia,* 114
Crinoidea, 224
Crinozoa, 224
Crop
  of cockroach, 210, 211
  of earthworm, 182
  of leeches, 186
  of snail, 167, 168
Crossing over, 44
Crural nerve, 291
Crustacea, 191, 195, 198–208
Crystalline style, 164
Ctenoid scales, 261
Cuboidal epithelium, 18, 19–20
Cubozoa, 106
Cucullaris muscle, 274
*Cucumaria frondosa,* 230
Cutaneous abdominis muscle, 274
Cutaneous artery, 280, 283
Cutaneous vein, 284
Cuticle
  of *Ascaris,* 150
  of *Caenorhabditis elegans,* 152
  of earthworm, 181
  of grasshopper, 214
  of sandworm, 179
  of spider, 194
Cutting plates, of hookworm, 151
Cuttlefish, 161
Cyclomorphosis, 199

Cytokinesis, 36
Cytopharynx, 86
Cytoplasm, 16
Cytoplasmic bridges, 79, 80
Cytopyge, 86
Cytoskeleton, 18
Cytostome
 of *Paramecium caudatum,* 85, 86
 of *Peranema,* 79

# D

*Daphnia,* 195, 198–201
Dart sac, 168
Definitive host
 of liver fluke, 133
 of sheep liver fluke, 137
Deltoid muscle, 274, 275
Deltoideus muscle, 303, 305
Demospongiae, 97, 98
Dendrite, 28
Dendrites, 30, 31
Depressor mandibularis muscle, 274
Dermal branchiae
 of sea urchin, 228
 of starfish, 226, 228
Dermis, 21
 of lancelets, 238
Determinate development, 52
Determination, embryonic, 50, 52
Development
 of chick, 59–69
 embryonic, 49–69. *See also* Embryo
 of frog, 54–59
 of starfish, 52–54
Diaphragm
 of fetal pig, 305, 308, 309, 310
 of rat, 328, 332, 333, 334
*Didinium,* 88
Diencephalon, 62, 63, 64, 68, 69
 of fetal pig, 320
 of frog, 285, 290, 291
 of rat, 342
 of shark, 244, 254, 255, 256
 of yellow perch, 267
Differentiation, embryonic, 50
*Difflugia,* 76–77
Digastricus muscle, 306
Digestive cells, of *Hydra,* 109
Digestive gland
 of crayfish, 202, 206, 207
 of mussel, 164
 of snail, 167, 168
Digestive system
 of *Amoeba,* 76
 of Bivalvia, 164
 of *Caenorhabditis elegans,* 152–53
 of cockroach, 210–12
 of crayfish, 205
 of earthworm, 181–82
 of *Fasciola hepatica,* 137
 of free-living flatworm, 132–33
 of frog, 279
 of grasshopper, 216, 217
 of *Helix,* 167
 of Hirudinea, 186
 in *Hydra,* 109
 of liver fluke, 133
 of mussel, 164

 in *Paramecium,* 86
 in *Peranema,* 79
 of *Peripatus,* 219
 of rat, 330–33
 of shark, 243
 of snail, 167
 of spider, 196
 of starfish, 226
 of water flea, 200
Dilator muscles, 181
Dinoflagellates, 81, 82
Dioecious reproduction, 80, 110, 149
*Diopatra,* 179–80
Diploblastic, definition of, 105
Diploblastic tissue
 construction, 122
*Dipylidium caninum,* 138–41
Discoidal cleavage, 52
Dissection, 123–25
Dog tapeworm, 138–41
Dogfish shark, 241–58
Dorsalis scapulae muscle, 274
Dorsolumbar vein, 282, 284
*Drosophila,* 217, 218
Ductus arteriosus, 314, 315, 319
Ductus deferens, 168
Ductus venosus, 318, 319
*Dugesia,* 129–33
Duodenum
 of frog, 279
 of rat, 332
 of shark, 243, 244, 245, 247
Dura mater, of fetal pig, 320

# E

Ears, of fetal pig, 296, 297
Earthworm, 180–84
*Echinarachnius parma,* 229
Echinodermata, 123, 223–30
 Asteroidea, 225–28
 classification of, 224–25
 Echinoidea, 228–29
 Holothuroidea, 229–30
Echinoidea, 224, 228–29
Echinozoa, 224
Ectoderm, 54, 57
Ectoplasm, 75
Eggs, 50, 51
 amniotic, 60
*Eimeria,* 90
Ejaculatory duct
 of *Ascaris,* 149
 of cockroach, 212, 213
 of sheep liver fluke, 136, 138
Elasmobranchiomorphii, 234, 241–58
Elastic fibers, 24
Electron microscope, 10–11
Embryo, 49–69
 of chick, 59–69
 cleavage of, 50–52
 determinate development of, 52
 of frog, 54–59
 indeterminate development of, 52
 of starfish, 52–54
Endoderm, 54, 57
Endolymphatic duct, 255
Endoplasm, 75
Endoplasmic reticulum, 17, 18

Endoskeleton
  of corals, 118
  of echinoderms, 223
  of sea urchin, 228
  of yellow perch, 261
Endosome, of *Euglena,* 78
Endostyle
  of lancelets, 238
  of sea squirts, 236
*Entamoeba histolytica,* 77
*Enterobius vermicularis,* 152
Eosinophils, 22, 23
Epaxial muscles, 261, 262, 263
*Ephydatla,* 101
Ephyra larvae, of *Aurelia,* 115
Epicranium, 215
Epidermis, 21, 105
  of free-living flatworm, 129, 131
  of *Hydra,* 107, 109
  of lancelets, 238
Epididymis
  of fetal pig, 312, 313
  of rat, 339, 340, 341
Epigastric artery, 280, 283
Epigastric furrow, 196, 197
Epiglottis, 330
Epigynum, 196, 197
*Epiphanes senta,* 158
Epipharyngeal groove, 238
Epiphysis
  of frog, 285, 291
  of shark, 254, 255
Epiproct, 215
Epitheliomuscular cells, 109
Epithelium, 17–21
  columnar, 18, 19, 20
  cuboidal, 18, 19–20
  simple, 18–20
  stratified, 20–21
Erythrocytes, 22, 23
  *Plasmodium* life cycle in, 90–92
Esophagus, 205
  of cockroach, 211
  of earthworm, 182
  of fetal pig, 307, 309, 310, 312
  of frog, 279, 286
  of liver fluke, 133, 134
  of mussel, 164
  of *Philodina,* 156
  of rat, 330–31
  of shark, 243, 244
  of sheep liver fluke, 137
  of snail, 167
  of water flea, 200
  of yellow perch, 263
Ethylene glycol, xiv, xv
*Euplotes,* 88
Eustachian tube
  of frog, 276
  of rat, 330
Evolution, convergent, 169
Excretory canals
  of *Ascaris,* 150
  of tapeworm, 138, 140, 143
Excretory pores
  of crayfish, 207
  of earthworm, 181
Excretory system
  of Bivalvia, 166

  of *Caenorhabditis elegans,* 152
  of crayfish, 205
  of earthworm, 183
  of fetal pig, 309
  of free-living flatworm, 129, 132
  of liver fluke, 134
  of starfish, 228
  of water flea, 200
Exoskeleton
  of cockroach, 209
  of grasshopper, 214
  of spider, 194
  of yellow perch, 261
Extensor carpi-radialis muscle, 274
Extensor cruris muscle, 275, 277
Extensor muscles, 205
External auditory meatus
  of cat, 301
  of fetal pig, 300
  of rat, 328
Extraembryonic membranes, 60
Eyes
  of cockroach, 209
  of crayfish, 201, 202, 206–7
  embryonic, 63, 66, 67, 69
  of fetal pig, 296, 297
  of frog, 270
  of grasshopper, 214, 215
  of horseshoe crab, 192
  of octopus, 171
  of *Peripatus,* 219
  of rat, 328
  of shark, 242, 252, 255, 256
  of spider, 194, 198
  of squid, 169
  of water flea, 199, 200
  of yellow perch, 260
Eyespot
  of lancelets, 236, 237
  of starfish, 226, 227, 228

## F

Facial nerve
  of cat, 324
  of fetal pig, 322
  of frog, 289, 291
  of rat, 343
  of shark, 255, 256, 257
  of sheep, 323
Falciform ligament, 309
Fallopian tube, 309, 311, 312
Fan worm, 179
Fang, of spider, 194, 195, 197
*Fasciola hepatica,* 135–38
Fat body
  of cockroach, 210
  of frog, 278, 279, 281, 283, 285, 286, 287, 288
  of rat, 341
Feather-duster worm, 179
Feeding
  by Bivalvia, 162, 164
  by free-living flatworm, 132–33
  by horseshoe crab, 192–93
  by *Hydra,* 110
  by lancelets, 237
  by *Loligo,* 169
  by *Metridium,* 116
  by *Nereis,* 177–78

Feeding—*Cont.*
  by *Paramecium caudatum,* 86
  by sea squirts, 236
  by spider, 195
  by starfish, 226
Femoral artery
  of fetal pig, 315
  of frog, 280, 283
Femoral vein
  of frog, 282, 284
  of shark, 246
Femur
  of fetal pig, 300
  of rat, 329
Fertilization, 40, 49
Fertilization membrane, 55
Fetal pig, 295–325
  skinning of, 298, 302
Fibula
  of fetal pig, 300
  of rat, 329
Filum terminale, 290, 291
Fin(s)
  of lancelets, 237, 238, 239
  of shark, 242, 244, 246
  of squid, 169
  of yellow perch, 260, 262
Fin rays
  of lancelets, 237, 238, 239
  of yellow perch, 260
Fission
  of *Metridium,* 117
  of *Paramecium caudatum,* 87
Flagellated chambers, of leucon-type
  sponge, 101
Flagellates, 78–85
Flagellum
  of *Ceratium,* 82
  vs. cilia, 87
  of *Euglena,* 78
  of *Peranema,* 78–79
  of snail, 168
  of *Trypanosoma,* 84
  of *Volvox,* 79, 80
Flame bulbs, of free-living flatworm, 130, 132
Flatworms, 123, 127–43
  free-living, 128, 129–33
  parasitic, 128, 133–38
Flexion, embryonic, 62, 64, 65
Flexor muscles, 205
Float, of *Physalia,* 114
Flukes, 128, 133–38
Focusing, of compound microscope, 5
Food groove, of mussel, 162
Food vacuoles
  of amoeba, 75, 76
  of *Paramecium caudatum,* 85, 86
  of *Peranema,* 79
Foot
  of fetal pig, 296, 297
  of mussel, 162, 163, 164, 165
  of snail, 167
Foramen magnum, 301
Foramen ovale, 319
Foraminiferans, 77
Forelimbs, of fetal pig, 297
Forewings, of grasshopper, 215, 216
Formaldehyde, xiv, xv

Frenulum, 330
Frog, 234, 269–94
  embryo of, 54–59
  skinning of, 276
Frontal bone
  of cat, 301
  of rat, 329
Funnel, of squid, 169, 170, 171
Funnel retractor muscles, 170, 171

# G

Gallbladder
  of fetal pig, 309
  of frog, 278, 279, 281
  of shark, 243, 246
Gametes, 40, 50
Gametogenesis, 42, 44, 49
Ganglion
  abdominal, 214
  cephalic, 169
  cerebral, 133, 134
  cerebropleural, 167
  of free-living flatworm, 129
  segmental, 207
  of snail, 167
  subesophageal, 212
  supraesophageal, 207, 212, 214
  suprapharyngeal, 183
Garden centipede, 191
Garden snail, 167–68
Garden spider, 193–98
Gas exchange
  in shark, 252
  in starfish, 228
  in yellow perch, 263
Gasserian ganglion, 292
Gastric artery
  of fetal pig, 315, 316
  of frog, 280, 283
  of shark, 244
Gastric filaments, of *Aurelia,* 115
Gastric mill, 205
Gastric muscles, 204
Gastric pouches, 115
Gastric vein
  of fetal pig, 318, 319
  of frog, 282, 284
  of shark, 246, 253
  of yellow perch, 265
Gastrocnemius muscle, 274, 275, 277
Gastrodermal cells
  of *Ascaris,* 150
  of sandworm, 179
Gastrodermis, 105
  of free-living flatworm, 129, 131
  of *Hydra,* 107, 109
  of sandworm, 178
Gastrohepatic artery, 315, 316
Gastroicha, 147
Gastroliths, 205
Gastropoda, 160, 161
Gastrosplenic artery, 244
Gastrotricha, 146
Gastrovascular cavity
  of free-living flatworm, 129, 130, 131
  of *Hydra,* 107, 109
  of *Metridium,* 116, 117

Gastrulation, 50, 54, 57
Gemmules, 101, 103
Genital artery
    of fetal pig, 315, 316
    of frog, 280
Genital operculum, 192, 193
Genital papilla, 313
Genital pore
    of *Anguillula aceti,* 154
    of *Ascaris,* 148, 149
    of tapeworm, 138, 140, 141, 143
    of yellow perch, 266, 267
Genital sinus, 267
Genital vein, 280
Germ cells, 42, 313
Giant fibers, 184
*Giardia,* 85
Gill(s)
    of crayfish, 201, 202, 204
    of mussel, 162, 163, 164, 165
    of shark, 245, 247, 249
    of squid, 170, 171
    of yellow perch, 260, 262, 263
Gill arches, 64, 65
    of shark, 252
    of yellow perch, 263, 264, 265
Gill bars, 236, 237, 238
Gill chamber, 252
Gill clefts, 64, 65
Gill filaments
    of shark, 252
    of yellow perch, 263, 264, 265
Gill rakers
    of shark, 252
    of yellow perch, 263, 264
Gill slits
    of lancelets, 236, 238
    of sea squirts, 234
    of shark, 242, 244, 251, 252
Gizzard, of earthworm, 182
Gland cells
    of free-living flatworm, 129,
        131, 133
    of *Hydra,* 109
Glass sponges, 98
Glochidia, 166
Glossopharyngeal nerve
    of cat, 324
    of fetal pig, 322
    of frog, 289, 291
    of rat, 343
    of shark, 255, 256, 257
    of yellow perch, 266
Glottis
    of fetal pig, 310
    of frog, 277, 280
    of rat, 330
Gluteus muscle
    of fetal pig, 303, 304, 305
    of frog, 274, 277
Gnathobases, 192, 193
Golgi apparatus, 17, 18
Gonadal artery, 265
Gonadal vein, 337, 338
Gonangia, 112, 113
Gonidia, 80
*Gonionemus,* 110–12
Gonopods, 201

Gonopores
    of *Obelia,* 112
    of spider, 196
Gonotheca, 112, 113
*Gonyaulax,* 81
Gordiacea, 147
Gracilis muscle, 274, 275, 277
    of fetal pig, 306
    of frog, 274, 275, 277
Grasshopper, 212, 214–17
Green glands, 205, 206
Ground substance, of connective tissue, 24
Growth, embryonic, 50
Gullet
    of *Euglena,* 78
    of *Metridium,* 116
*Gymnodinium,* 81

# H

Haemal arch, 261
Haemal spine, 261
Hagfish, 234
Hair bulb, 21
Hanging drop, 75
Haversian canal, 26
Head
    of cockroach, 209
    of fetal pig, 296, 297
    of *Nereis,* 177
Headfold, of chick embryo, 60
Heart
    of chick embryo, 62, 63, 64, 65, 67, 68, 69
    of cockroach, 210
    of crayfish, 204, 206
    of earthworm, 182
    of fetal pig, 308, 314, 315, 316, 317, 319, 320
    of frog, 278, 279, 280, 281
    of grasshopper, 216, 217
    mammalian, 320
    of mussel, 163, 164, 165, 166
    of rat, 333–37
    of sea squirts, 236
    of shark, 245, 246, 247, 249, 250–52
    of snail, 167
    of spider, 196
    of squid, 170, 171
    of water flea, 199, 200
    of yellow perch, 262, 264, 265, 267
Heart urchins, 224
Hectocotyl, 169
*Helix,* 167–68
Hemal arch, 246
Hemocoel
    of cockroach, 210
    of grasshopper, 216, 217
    of *Peripatus,* 219
    of water flea, 200
Hemocyanin, 165, 204
Hemoglobin, 22
Hepatic artery
    of crayfish, 204, 206
    of fetal pig, 315
    of frog, 280, 283
    of shark, 244
    of yellow perch, 265
Hepatic portal system
    of fetal pig, 314

Hepatic portal system—*Cont.*
  of frog, 280, 282
  of rat, 337, 339
  of shark, 249, 253
  of yellow perch, 264
Hepatic portal vein
  of fetal pig, 318, 319
  of frog, 282, 284
  of lancelets, 237
  of shark, 244, 245, 246
Hepatic sinus, 246
Hepatic vein
  of fetal pig, 318, 319
  of frog, 280, 284
  of lancelets, 237, 238
  of rat, 337, 338
  of shark, 246, 251, 253
Hepatopancreas, 205
Hermaphroditism
  of *Caenorhabditis elegans,* 153
  of *Scypha,* 100
Heterocercal tail, 242
*Heterodera schactii,* 155
Hexactinellida, 97, 98
Hindlimb, of frog, 273, 276
Hindlimb bud, 69
Hindwings, of grasshopper, 215, 216
Hinge, of mussel, 163
Hinge ligament, 162
Hirudinea, 176, 184–86
Holoblastic cleavage, 51, 57
Holothuroidea, 224, 229–30
Homology, 202
Hook, of tapeworm, 138
Hookworms, 151–52
Horny sponges, 98
Horsehair worms, 147
Horseshoe crab, 191, 192–93
Humerus
  of fetal pig, 300
  of rat, 329
Hyaline cartilage, 25
Hyalospongiae, 98
*Hydra,* 106–10
Hydranths, 112, 113
Hydrocaulus, 112
*Hydroides,* 179
Hydroids, 106, 112
Hydromedusa, 110–12
Hydrotheca, of *Obelia,* 112, 113
Hydrozoa, 106–14
Hyoid apparatus, 271
Hyoidean artery, 252, 253
Hyomandibular cartilage, 252
Hyomandibular nerve, 257
Hyopharynx, 216
Hypaxial muscles, 262, 263
Hypodermal cells
  of *Caenorhabditis elegans,* 152
  of sandworm, 179
Hypodermis, 21
  of *Ascaris,* 150
  of crayfish, 204
  of grasshopper, 214
  of *Philodina,* 156
  of spider, 194
Hypoglossal nerve
  of cat, 324

  of fetal pig, 322
  of rat, 343
Hypopharynx, 209
Hypophysis, 254, 257
Hypostome
  of *Hydra,* 107, 108
  of *Obelia,* 113

# I

Ileocolic valve
  of fetal pig, 309
  of rat, 333
Ileum
  of cockroach, 210
  of frog, 271, 279
  of rat, 332
  of shark, 243, 244, 245, 247
Iliac artery
  of fetal pig, 310, 311, 312, 315, 316, 317
  of frog, 280, 282, 283
  of rat, 336, 337
  of shark, 244, 250
Iliac vein
  of fetal pig, 318, 319
  of rat, 337, 338
  of shark, 246
Iliacus internus muscle, 274
Iliohypogastric nerve, 291
Iliolumbar artery, 336, 337
Iliolumbar vein, 337, 338
Iliolumbaris muscle, 274
Ilium
  of fetal pig, 300
  of rat, 329
Illumination, for compound microscope, 4–5
Images, with compound microscope, 3
Imaginal disc, 217
Incurrent canals, 100
Indeterminate development, 52
Infundibulum
  of frog, 290
  of shark, 254, 256, 257
  of sheep, 323
Inguinal canal, 312, 313
Ink sac, 170, 171
Innominate artery, 334, 335
Innominate vein, 280, 284
Insecta, 191, 208–17
Instruments, for dissection, 124, 125
Intercalated discs, 28, 29
Intercostal artery, 315, 316
Intercostal muscles, of fetal pig, 306
Interference microscopy, 9–10
Interkinesis, 42, 44
Intermediate host
  of liver fluke, 133
  of sheep liver fluke, 136
Interparietal bone, 301
Intersegmental membrane, 210
Interstitial cells, 109
Intestinal artery
  of shark, 244, 250
  of yellow perch, 265
Intestinal vein
  of frog, 284
  of shark, 244, 246
  of yellow perch, 265

Intestine, 205
  of *Anguillula aceti*, 154
  of *Ascaris*, 148, 149
  of *Ascaris lumbricoides*, 147, 150
  of *Caenorhabditis elegans*, 153
  of crayfish, 206, 207
  of earthworm, 182
  of fetal pig, 308, 309, 311, 312, 313, 314
  of frog, 278, 279, 283, 284, 287, 288
  of grasshopper, 216, 217
  of leeches, 186
  of mussel, 163, 164
  of *Philodina*, 155, 156
  of rat, 332, 333
  of sandworm, 178, 179
  of sheep liver fluke, 137
  of snail, 168
  of yellow perch, 262, 263
Intraintestinal vein, 246
Ischium
  of fetal pig, 300
  of rat, 329
Isolecithal eggs, 51
Isopropanol, xiv, xv

# J

Jaw
  of fetal pig, 310
  of sandworm, 177, 178
Jejunum, 332
Jelly coats, 55
Jellyfish, 106
Jugular trunk, 318, 319
Jugular vein
  of fetal pig, 318, 319
  of frog, 280, 284
  of rat, 337, 338
  of shark, 246

# K

Karyotype, 37
Keratin, 20
Kidney
  of fetal pig, 310, 311, 312, 313, 314, 317
  of frog, 279, 281, 282, 285, 286, 287, 288
  of rat, 339
  of shark, 244, 246, 248, 249
  of yellow perch, 263, 264, 266
Kinorhyncha, 146, 147
Koehler illumination, 4–5

# L

Labial palps, 162, 163
Labium
  of cockroach, 209
  of grasshopper, 214, 215
  of spider, 196, 197
  of water flea, 200
Labrum
  of cockroach, 209
  of grasshopper, 214, 215
  of water flea, 200
Lacunae
  of bone, 26
  of cartilage matrix, 25, 26

Lamellae, of bone, 26
Lamprey, 234
Lancelets, 234, 236–39
Larum, 196
Laryngopharynx, 330
Larynx
  of fetal pig, 305, 307, 308, 310
  of frog, 277, 278, 280
  of rat, 333, 334
Lateral artery, 252, 253
Lateral line
  of shark, 242
  of yellow perch, 260
Lateral nerve, 255, 256
Latissimus dorsi muscle
  of fetal pig, 303, 305, 306
  of frog, 274
Leathery body wall, 229
Leeches (Hirudinea), 176, 184–86
Legs
  of crayfish, 201, 202, 203
  of grasshopper, 215, 216
  of spider, 193, 196, 197
*Leishmania*, 85
*Leptosynapta tennuis*, 230
Leuconoid sponges, 98, 101
*Leucosolenia*, 99
Leukocytes, 22–24
Lienogastric artery, 250
Lienomesenteric vein, 246, 253
Ligamentum arteriosum, 319
Limb bud, 66, 67
Limpets, 160, 161
*Limulus polyphemus*, 192–93
Linea alba, 278
Lines of growth, 162
Lingual vein, 280, 284
Lips
  of *Ascaris*, 147, 148, 149
  of rat, 328
Liver
  of fetal pig, 308, 309, 310, 311, 312, 314
  of frog, 278, 279, 281, 284, 287
  of rat, 332, 333, 334
  of shark, 243, 244, 245, 246, 249
  of yellow perch, 262, 263
Liver fluke, 133–38
Lizard, 234
*Loligo*, 168–71
Longissimus dorsi muscle, 274
Longitudinal fission, 117
Loricifera, 146, 147
Lug worm, 180
Lumbales artery, 336, 337
Lumbar artery
  of frog, 282
  of rat, 336, 337
Lumbosacral nerve plexus, 321, 322
*Lumbricus terrestris*, 180–84
Lungs
  of fetal pig, 305, 307, 308, 310, 312, 314
  of frog, 277, 278, 279, 280, 281, 282, 283, 284, 287, 288
  of rat, 332, 333, 334
  of snail, 167, 168
Lung slit, 196, 197
Lymphocytes, 22, 23
Lysosomes, 17, 18

# M

Macrogametocytes, 91, 92
Macronucleus, 85, 86
Macrophages, 24, 25
Madreporite, 225, 226, 227
Magnification, for compound microscope, 3–4, 5, 7
Malacostraca, 191, 201–8
Malaria, 90–92
Malpighian tubules
    of cockroach, 211, 212
    of grasshopper, 216, 217
Mammae, 297
Mammalia, 234. *See also Rattus norvegicus; Sus scrofa*
Mammary glands, 328
Mandible
    of cockroach, 209
    of crayfish, 202, 203
    of fetal pig, 300
    of grasshopper, 214, 215
    of rat, 329
    of water flea, 198, 200
Mandibular gland, 330, 332, 334
Mandibular muscles, 204–5
Mandibular nerve, 322
Mantle, of mussel, 162, 163
Mantle cavity, 162, 164
Manubrium, 111, 112
Marsupium, 164, 166
Masseter muscle
    of fetal pig, 306
    of frog, 274
Mastax, 155, 156
Mastigophora, 74, 78, 79–83
Maxilla
    of cat, 301
    of cockroach, 209
    of crayfish, 202, 203
    of fetal pig, 300
    of grasshopper, 214
    of spider, 194, 196, 197
    of water flea, 200
Maxillary gland, 199, 200
Maxillary nerve, 322
Maxillary vein, 280, 284
Maxilliped, 202, 203, 204
Median sacral artery, 336, 337
Medulla oblongata
    of fetal pig, 310, 322
    of frog, 285, 290
    of rat, 342, 343
    of shark, 255, 256
    of sheep, 323
    of yellow perch, 266
Medusa buds, 112, 113
Meiosis, 35–36, 40, 42–45
    vs. mitosis, 42, 45
    observation of, 44–45
    stages of, 42–43, 44
*Mellita quinquiesperforata,* 229
Membranelles, 88
Meninges, 320
Merostomata, 191
Merozoites, 90–91
Mesencephalon, 62, 63, 64, 68, 69
    of fetal pig, 320
    of frog, 285, 291
    of rat, 342

of shark, 244, 254, 255
of yellow perch, 267
Mesenchyme, 122
Mesenteric artery
    of fetal pig, 315, 316
    of frog, 280, 282, 283
    of rat, 336, 337
    of shark, 244, 250
Mesenteric vein
    of fetal pig, 318, 319
    of frog, 282, 284
Mesentery
    of fetal pig, 309
    of rat, 332
    of shark, 243, 247
Mesoderm, 57
Mesoglea, 105–6
    of *Gonionemus,* 112
    of *Hydra,* 107, 109
Mesohyl, 96
Mesolecithal eggs, 51
Mesonephric ducts, 248
Mesothorax, 215, 216
Metaboly, 78
Metacarpals
    of fetal pig, 300
    of rat, 329
Metacercariae
    of liver fluke, 133, 135
    of sheep liver fluke, 137
Metamorphosis, 217
Metaphase
    of meiosis, 42–43, 44
    of mitosis, 38, 39
Metapleural fold, 238
Metatarsals
    of fetal pig, 300
    of rat, 329
Metathorax, 215, 216
Metencephalon, 62, 63, 64, 68, 69
    of fetal pig, 320
    of frog, 285, 291
    of rat, 342
    of shark, 244, 254, 255
    of yellow perch, 267
Metric units, in microscopy, 4
*Metridium,* 116–17
*Microciona,* 103
Microfilaments, 17
Microgametocytes, 91, 92
Micrometer, ocular, 8
Micronucleus, 85, 86
Microscope
    compound, 1–8. *See also* Compound microscope
    electron, 10–11
    interference, 9–10
    phase contrast, 9
    stereoscopic, 8–9
Microtubules, 87
Millipedes, 191
Miracidium
    of liver fluke, 133, 135
    of sheep liver fluke, 136
Mite, 191
Mitochondria, 17, 18
Mitosis, 35, 36–40
    vs. meiosis, 42, 45

in *Paramecium*, 87
stages of, 38, 39
Mitotic apparatus, 36–38
Mitotic spindle, 38
*Molgula*, 234–36
Mollusca, 123, 159–73
  Bivalvia, 162–67
  Cephalopoda, 168–71
  classification of, 160–61
  Gastropoda, 167–68
Monocytes, 22, 23
Monoecious reproduction, 80, 110, 131
Monogenea, 128
Monophyletic, definition of, 93
Monoplacophora, 160, 161
Morphogenesis, 50
Morphology, 121–25
  body cavity in, 122–23
  body symmetry in, 122
  cephalization in, 123
  segmentation in, 123
  tissue construction grade in, 122
Mosaic development, 52
Mosquito, 90–91
Motor end plate, 29, 31
Motor neuron, 29, 256–57
Mouth, 205
  of *Ancylostoma*, 151
  of *Anguillula aceti*, 154
  of *Ascaris*, 147, 148, 149
  of *Ascaris lumbricoides*, 147, 148, 149
  of *Caenorhabditis elegans*, 153
  of crayfish, 201
  of earthworm, 181, 182
  of fetal pig, 296, 297, 310
  of free-living flatworm, 131
  of frog, 276–77, 279
  of *Gonionemus*, 112
  of hookworm, 151
  of horseshoe crab, 192, 193
  of *Hydra*, 107, 108
  of lancelets, 236, 237
  of leeches, 185, 186
  of liver fluke, 133, 134
  of *Metridium*, 116
  of mussel, 162, 164
  of octopus, 171
  of *Peripatus*, 219
  of *Philodina*, 155, 156
  of rat, 330
  of sea urchin, 228
  of shark, 242, 245
  of sheep liver fluke, 137, 138
  of snail, 167, 168
  of spider, 196
  of squid, 169, 170
  of starfish, 226
  of water flea, 199, 200
  of yellow perch, 260
Mouthparts
  of cockroach, 209
  of grasshopper, 214–15
Movement
  of amoeba, 76
  of *Euglena*, 78
  of frog, 271, 276
  of *Gonionemus*, 112
  of *Hydra*, 108
  of *Loligo*, 169

of mussel, 164
of *Paramecium caudatum*, 86–87
of *Peranema*, 79
by squid, 169
Mucous glands, 261
Multicellular animals, evolution of, 92–93
Muscle, 26–28
Muscle bands, 150
Muscular system
  of Bivalvia, 165
  of *Caenorhabditis elegans*, 152
  of crayfish, 204–5
  of earthworm, 184, 185
  of fetal pig, 298, 303–6
  of free-living flatworm, 129, 131, 133
  of frog, 271, 274–76, 277
  of lancelets, 237
  of mussel, 162, 164, 165
  of rat, 329–31
  of sandworm, 178, 179
  of sea squirts, 234, 235
  of starfish, 228
  of yellow perch, 261–63
Musculocutaneous vein, 280
Mussel, 161
  freshwater, 162–67
Mutualism, 83
Myelencephalon, 62, 63, 64, 68, 69
  of fetal pig, 320
  of frog, 285, 291
  of rat, 342
  of shark, 244, 254
  of yellow perch, 267
Myelin, 30
Myelin sheath, 31
Mylencephalon, 62
Mylohyoid muscle, 275
Mylohyoideus muscle, 306
Myofibrils, 27
Myomeres, 237, 238, 239
Myoseptum, 261
Myosin, 28

# N

Nacreous layer, of mussel, 162
Nares, 276, 279–80
Nasal bone
  of cat, 301
  of fetal pig, 300
  of rat, 329
Nasal cavity, 280, 310
Nasal pit, 65, 66
Nasal pits, 64, 65
Nasopharynx, 330
*Nautilus*, 161
Navel, 297
*Necator americanus*, 151–52
Neck, of fetal pig, 296, 297, 305, 307
Nematocysts, 106
  of *Gonionemus*, 112
  of *Hydra*, 108–9
  of *Obelia*, 113
Nematoda, 123, 147–55
  free-living, 152–55
  parasitic, 147–52
Nematomorpha, 146, 147
Nephridiopore, 185

Nephridium
  of earthworm, 182, 183, 184, 185
  of lancelets, 238
  of leeches, 185
  of mussel, 163, 164, 166
  of *Peripatus,* 219
  of sandworm, 179
  of snail, 168
  of squid, 171
Nephrostome, 182, 183
*Nereis virens,* 176–80
Nerve cord
  of *Ascaris,* 150
  of *Caenorhabditis elegans,* 153
  of cockroach, 212, 214
  of crayfish, 206, 207, 208
  of earthworm, 182, 183, 184
  of free-living flatworm, 130, 131
  of grasshopper, 216, 217
  of lancelets, 237, 238, 239
  of liver fluke, 134
  of sandworm, 178, 179
  of sea squirts, 234, 235
  of tapeworm, 138, 140
  of water flea, 200
Nerve net
  of Cnidaria, 106
  of *Hydra,* 109–10
Nerve ring
  of *Caenorhabditis elegans,* 153
  of starfish, 228
Nervous system
  of Bivalvia, 166–67
  of *Caenorhabditis elegans,* 153
  of cockroach, 213, 214
  of crayfish, 206
  of earthworm, 183
  of fetal pig, 320–24
  of free-living flatworm, 129
  of frog, 285, 289–92
  of *Helix,* 167
  of *Hydra,* 109–10
  of rat, 340–44
  of shark, 253–57
  of snail, 167
  of starfish, 228
  of water flea, 200
  of yellow perch, 266, 267
Neural arch
  of shark, 246
  of yellow perch, 261
Neural canal, 320
Neural fold, 61
Neural groove, 57, 61
Neural plate, 61
Neural spine, 261
Neural tube, 57, 60–61, 63
Neural vessels, 183
Neuron, 28–31, 109
Neuropodium, 177
Neurulation, 57
Neutrophils, 22, 23
Nictitating membrane
  of frog, 270
  of rat, 328
Nidamental glands, 171
Nipples, 297
Nodes of Ranvier, 30
Nomarski interference microscope, 9–10

Nonelastic fibers, 24
Nostrils
  of frog, 270
  of shark, 242
Notochord, 57, 61, 62
  of lancelets, 237, 238, 239
  of sea squirts, 234, 235
Notopodium, 177, 178
Nuchal crest, 301
Nuclear dimorphism, 86
Nuclear envelope, 17, 18
Nuclear pores, 17
Nucleolus, 17, 18
Nucleus, 16, 18
  of amoeba, 75, 76
  of *Euglena,* 78
  of *Paramecium caudatum,* 85, 86
  of *Peranema,* 79
  of *Plasmodium,* 90
  of *Volvox,* 79, 80

# *O*

*Obelia,* 112–13
Oblique muscles
  of fetal pig, 303, 304, 305, 306
  of frog, 274, 275, 278
  of rat, 330, 331
  of shark, 255, 256
Obturator foramen, 329
Occipital artery, 280, 283
Occipital bone, 301
Occipital condyle, 301
Occipitovertebral artery, 280, 283
Ocelli
  of grasshopper, 214, 215
  of sea squirts, 234
  of spider, 194, 198
Octocorals, 117–18
Octopus, 161, 171
Ocular micrometer, 8
Oculomotor nerve
  of cat, 324
  of fetal pig, 322
  of frog, 289, 291, 292
  of rat, 343
  of shark, 255, 256, 257
  of sheep, 323
Olfactory artery, 252
Olfactory bulb
  of cat, 324
  of fetal pig, 310
  of rat, 342, 343
  of shark, 244, 254, 255
  of sheep, 323
Olfactory lobe, 285, 290, 291
Olfactory nerve
  of frog, 285, 289, 290, 291
  of rat, 342, 343
  of shark, 255, 257
Olfactory pit, 246
Olfactory tract, 266
Oligochaeta, 176, 180–84
Omentum, 309
Ommatidia, 207
Onychophora, 217, 219
Oogenesis, 42, 44
Operculum, 260
Ophiuroidea, 225

Ophthalmic artery
  of crayfish, 204, 206
  of shark, 252
Ophthalmic nerve
  of fetal pig, 322
  of shark, 257
Opisthosoma, 193, 194, 197, 198
Optic chiasma
  of fetal pig, 324
  of frog, 289, 292
  of rat, 343
  of shark, 257
  of sheep, 323
Optic lobe
  of frog, 285, 290
  of shark, 254, 256
Optic nerve
  of cat, 324
  of fetal pig, 322
  of frog, 289, 291
  of rat, 343
  of shark, 252, 255, 256, 257
  of sheep, 323
  of yellow perch, 266
Oral arms, 115
Oral cavity, 276–77
Oral disc, 116, 117
Oral groove, 85, 86
Oral hood, 236
Oral lobes, 112
Oral spines, 226
Orbit, 301
Orbital sinus, 252
Orbitosphenoid, 301
Organ, 16–17
Organelles, 16, 17, 18
Organism, 17
Organogenesis, 50
Oropharynx, 330
Osmoregulation
  in Bivalvia, 166
  in free-living flatworm, 132
Ossicles, 226
Osteichthyes, 234, 259–67
Ostia, 117
Ostium tubae, 248
Otic capsules, 254
Otic vesicles, 62, 64, 65, 66, 67, 68, 69
Otoliths, 261
Ova, 50
Ovary, 205
  of Anguillula aceti, 154
  of Ascaris, 148, 149, 150
  of Ascaris lumbricoides, 149, 150
  of Caenorhabditis elegans, 153
  of cockroach, 212, 213
  of crayfish, 202, 206
  of earthworm, 182, 183
  of fetal pig, 310, 312
  of free-living flatworm, 130
  of frog, 278, 281, 282, 285, 286, 287
  of Hydra, 108, 110, 111
  of leeches, 185
  of liver fluke, 133, 134
  of mussel, 164
  of Peripatus, 219
  of Philodina, 156
  of rat, 340
  of shark, 248

  of sheep liver fluke, 136, 137
  of squid, 171
  of tapeworm, 140, 141, 143
  of water flea, 199, 200
  of yellow perch, 267
Oviducal gland, 171
Oviduct, 205
  of Anguillula aceti, 154
  of Ascaris, 148, 149, 150
  of Ascaris lumbricoides, 149, 150
  of Caenorhabditis elegans, 153
  of cockroach, 212, 213
  of crayfish, 206
  of fetal pig, 309
  of frog, 281, 285, 286, 287
  of rat, 340
  of shark, 248
  of sheep liver fluke, 136, 137
  of snail, 168
  of squid, 171
  of tapeworm, 138, 140, 143
  of water flea, 200
  of yellow perch, 267
Ovipositor, of grasshopper, 215
Ovisac, 286
Oysters, 161

## P

Palate
  of fetal pig, 310
  of rat, 330
Palatine bone, 301
Pallial line, 162
Palps
  of grasshopper, 215
  of sandworm, 177
Pancreas
  of fetal pig, 309, 310, 311, 312
  of frog, 279, 283
  of rat, 332
  of shark, 243, 244, 246
  of yellow perch, 264
Pancreatic artery, 280, 283
Pancreaticomesenteric vein, 246, 253
Papilla
  genital, 313
  sensory, 162
  urinary, 249
  urogenital, 248
Parabasal body, 84
Paramecium caudatum, 85–89
Paramylum grains, 78
Parapodium
  of Nereis, 177
  of sandworm, 177, 178
Parenchyma
  of flatworms, 128
  of free-living flatworm, 129
Parietal bone
  of cat, 301
  of fetal pig, 300
  of rat, 329
Parietal veins, 253
Parietal vessels, 183
Parotid gland
  of fetal pig, 305
  of rat, 330

Patella
    of fetal pig, 300
    of rat, 329
Paturon, 194, 197
Pauropoda, 191
Pax 6 gene, 169
Pectoral girdle
    of fetal pig, 297
    of shark, 243, 244, 251, 252
Pectoralis muscle
    of fetal pig, 306
    of frog, 275, 278
    of rat, 330, 331
Pedal disc, 117
Pedal ganglia, 167
Pedal laceration, 117
Pedicel, 194, 196
Pedicellariae
    of sea urchin, 228
    of starfish, 226
Pedipalps
    of horseshoe crab, 193
    of spider, 193, 194, 196
Pelecypoda, 160
Pellicle
    of *Euglena*, 78
    of *Paramecium caudatum*, 85
*Pelomyxa carolinensis,* 76
Pelvic girdle
    of frog, 273
    of shark, 243, 244
Pelvic vein, 282, 284
Pen, of squid, 169
Penis
    of fetal pig, 311, 312, 313
    of free-living flatworm, 130, 131
    of leeches, 185
    of rat, 328, 339, 340
    of sheep liver fluke, 136, 138
    of snail, 168
*Peranema,* 78–79
*Perca flavescens,* 259–67
Perch, 259–67
Pereiopods, 202, 204
Pericardial cavity
    of fetal pig, 305, 309, 310
    of mussel, 165
    of shark, 243, 251
    of yellow perch, 263
Pericardial sinus
    of crayfish, 204
    of mussel, 164
Pericardium
    of fetal pig, 305
    of frog, 280
    of mussel, 165
    of shark, 246
Periostracum, 162, 164
*Peripatus,* 217, 219
Peripheral nervous system
    of frog, 285, 290
    of shark, 253
    of yellow perch, 267
*Periplaneta americana,* 208–12
Perisarc, 112, 113
Peristome
    of sea urchin, 228
    of starfish, 225, 226
Peristomium, 177

Peritoneal cavity
    of fetal pig, 297, 305, 308, 309, 310, 311
    of yellow perch, 263
Peritoneum
    of earthworm, 184, 185
    of fetal pig, 309
    of sandworm, 179
    of shark, 243
    of starfish, 226
Peroneal artery, 280, 283
Peroneal nerve, 291
Peroneus muscle, 274, 277
Peroxisomes, 18
*Pfisteria piscida,* 81, 82
Phalanges
    of fetal pig, 300
    of rat, 329
Pharyngeal bulb
    of *Anguillula aceti,* 154
    of *Caenorhabditis elegans,* 153
Pharyngeal gill slits, 234
Pharynx
    of *Anguillula aceti,* 154
    of *Ascaris,* 148, 149
    of *Caenorhabditis elegans,* 153
    of earthworm, 181, 182
    of fetal pig, 310
    of free-living flatworm, 130, 131
    of frog, 277, 279, 280
    of leeches, 185, 186
    of liver fluke, 133, 134
    of *Philodina,* 155, 156
    of rat, 330
    of shark, 242
    of sheep liver fluke, 137
    of snail, 167
Phase contrast microscopy, 9
*Philodina,* 155–57
Philtrum, 328
Phototrophism, 78
Phrenic artery, 336, 337
Phrenic vein, 337, 338
*Physalia,* 113–14
Pia matter, of fetal pig, 320
Pig, fetal, 295–325
Pinacocytes, 96, 100
Pinacoderm, 96
Pineal body, 342
Pineal gland
    of fetal pig, 310
    of frog, 285
    of shark, 254
    of yellow perch, 266
Pinna, 328
Pinworm, 152
Piriformis muscle, 274, 277
Piston, 169
Pituitary gland
    of fetal pig, 310, 324
    of frog, 285, 290, 292
    of rat, 343
    of shark, 244, 252, 254, 256
    of yellow perch, 266
Placoid scales, 242
Planarians, 129–33
Planes, anatomical, 124
Planula larvae
    of *Aurelia,* 115
    of *Gonionemus,* 112

of *Metridium,* 117
of *Obelia,* 113
Plasma, 21, 204
Plasma cell, electromicrograph of, 10
Plasma membrane, 18
Plasmagel, 75, 76
Plasmasol, 75, 76
*Plasmodium,* 90–92
Platelets, 22, 24
Platyhelminthes, 123, 127–43
    Cestoda, 138–43
    classification of, 128
    Trematoda, 133–38
    Turbellaria, 129–33
Pleura, 305
Pleural cavity, 305, 309
Pleuroperitoneal cavity, 243, 251
Plume worm, 179–80
Pneumatic artery, 265
*Podophyra,* 89
Polar body, 42
Polian vesicles, 227, 228
Polychaeta, 176–80
Polymorphism, 112
Polyp
    of *Aurelia,* 115
    of corals, 117
    of *Gonionemus,* 112
    of *Obelia,* 112
Polyphyletic, definition of, 93
Polyplacophora, 160, 161
Pons, 323
Porifera, 95–103
    body organization of, 99–101
    classification of, 98
    regeneration of, 103
*Porpita,* 114
Portal veins, 249
Portuguese man-of-war, 113–14
Postabdomen, of water flea, 200
Postcaval vein, 312, 318, 319
Postorbital processes, 301
Posttrematic artery, 252, 253
Precaval vena cava, 318, 319
Premaxilla
    of cat, 301
    of fetal pig, 300
    of rat, 329
Prepuce
    of fetal pig, 312, 313
    of rat, 340
Preputial cavity, 313
Preputial glands, 340
Presphenoid, 301
Pretrematic artery, 252, 253
Priapulida, 146, 147
Primitive streak, 61, 62
Prismatic layer, of mussel, 162
*Procambarus,* 201–8
Proglottids, 138, 140, 141, 142, 143
Prophase
    of meiosis, 42–43, 44
    of mitosis, 38, 39
Prosencephalon, 63, 254
Prosoma, 194, 197, 198
Prosopyles, 100
Prostate gland
    of fetal pig, 311, 313
    of rat, 339

Prostomium
    of earthworm, 180, 182
    of sandworm, 177
*Proterospongia,* 83
Prothorax, 215, 216
Protozoa, 73–93
    *Amoeba proteus,* 74–77
    Apicomplexa, 89–92
    Ciliophora, 85–89
    classification of, 74
    *Euglena,* 78
    *Paramecium caudatum,* 85–89
    *Peranema,* 78–79
    *Plasmodium,* 90–92
    Sarcomastigophora, 74–84
    *Volvox,* 79–81
Protractor muscles, 163, 165
Proventriculus, 210, 211
Pseudocoelom, 122
Pseudocoelomates, 122, 123, 145–57. *See also*
    Roundworms
Pseudopodia, 75
Pterygoideus muscle, 274
Pubis
    of fetal pig, 310
    of rat, 329
Pulmocutaneous artery, 280, 281, 283
Pulmocutaneous circulation, 280
Pulmonary artery
    of fetal pig, 314, 315, 319
    of rat, 334, 335
Pulmonary trunk, 307, 310, 314, 315, 317,
    319, 320
    of rat, 337
Pulmonary vein
    of fetal pig, 314, 318, 319, 320
    of frog, 282, 284
    of rat, 335
Pycnogonida, 191
Pyloric chamber, 205, 207
Pyloric valve
    of cockroach, 210
    of frog, 279
Pylorus, 243

# R

Radial canals
    of *Aurelia,* 115
    of *Gonionemus,* 111, 112
    of *Scypha,* 100
    of starfish, 227, 228
Radial cleavage, 51–52
Radial nerve, 228
Radial symmetry, 106, 115, 122, 223
Radiating canals, 85, 86
Radiolarians, 77
Radius, 300
Radix aortae, 252
Radula, 167, 168
*Rana pipiens,* 269–94
    embryo of, 54–59
    skinning of, 276
Random sample, 40, 41
*Rattus norvegicus,* 327–44
    skinning of, 329
Ray, 234
Rectal gland, 243, 246, 247
Rectal pads, 212

Rectum
  of *Anguillula aceti,* 154
  of *Caenorhabditis elegans,* 153
  of cockroach, 210, 211
  of fetal pig, 309, 310, 311, 312
  of frog, 284
  of leeches, 186
  of mussel, 164
  of rat, 333, 341
  of shark, 243, 244
  of squid, 170, 171
  of water flea, 200
Rectus abdominis muscle
  of fetal pig, 306
  of frog, 275, 278
  of rat, 330, 331
Rectus muscles, of shark, 255, 256
Red tides, 81
Redia
  of liver fluke, 133, 135
  of sheep liver fluke, 136
Regeneration
  of *Hydra,* 110
  of sponges, 103
Renal artery
  of fetal pig, 309, 315, 316
  of frog, 280
  of rat, 336, 337, 339
Renal portal system
  of fetal pig, 314
  of frog, 282
  of shark, 249, 253
Renal portal vein
  of frog, 284, 286
  of shark, 244, 246, 253
Renal vein
  of fetal pig, 317, 318, 319
  of frog, 280, 286
  of rat, 337, 338, 339
  of shark, 246, 253
Renette gland
  of *Ascaris,* 150
  of *Ascaris lumbricoides,* 150
Reproduction
  of amoeba, 76
  of *Ancylostoma,* 151
  of *Anguillula aceti,* 154
  of *Ascaris,* 150
  of *Ascaris lumbricoides,* 148, 149, 150
  asexual, 130–31
  of *Aurelia,* 115
  of Bivalvia, 166
  of *Caenorhabditis elegans,* 153
  of *Clonorchis sinensis,* 133
  of cockroach, 212, 213
  of crayfish, 205
  dioecious, 80, 110, 149
  of earthworm, 183
  of *Fasciola hepatica,* 136–38
  of fetal pig, 309–13
  of free-living flatworm, 129, 131–32
  of frog, 285, 287, 288
  of *Gonionemus,* 112
  of *Helix,* 167–68
  of hookworm, 151–52
  of *Hydra,* 110, 111
  of *Loligo,* 170–71
  of *Metridium,* 117
  monoecious, 80, 110, 131

  of *Nereis,* 178
  of *Obelia,* 112–13
  of *Paramecium,* 87
  of *Paramecium caudatum,* 87
  of *Peranema,* 79
  of pinworm, 152
  of rat, 339–41
  of *Scypha,* 100
  of shark, 248–49
  of snail, 167–68
  of sponges, 100
  of starfish, 228
  of *Trichinella,* 151
  of *Volvox,* 80, 81
  of water flea, 200–201
Reptilia, 234
Reservoir, of *Peranema,* 79
Resolving power, of compound microscope, 4
Respiratory system
  of Bivalvia, 162, 164
  of cockroach, 210
  of crayfish, 204
  of frog, 279–80
  of *Helix,* 167
  of rat, 333
  of water flea, 200
  of yellow perch, 263
Retractor muscles
  of mussel, 163, 165
  of *Philodina,* 156
Rhabdites, 129
*Rhabditis,* 152
*Rhabditis maupasi,* 155
Rhombencephalon, 254
Ribosomes, 17, 18
Ribs
  of fetal pig, 297, 300
  of rat, 329
Rings, of leeches, 185
*Romalea microptera,* 212, 214–17
Rostellum, 138, 140, 141
Rostrum
  of crayfish, 201
  of lancelets, 236
  of *Philodina,* 156
  of water flea, 199
Rotational cleavage, 52
Rotifera, 146, 147, 155–57
Roundworms, 123
  free-living, 152–55
  parasitic, 147–51

# S

Sacral artery, 336, 337
Sacrum, 271
Safety, xiii–xvii
Sagittal crest, 301
Salamander, 234
Salivary gland
  of cockroach, 210
  of grasshopper, 216, 217
  of leeches, 185
  of *Philodina,* 155, 156
  of rat, 330
  of snail, 168
Sand dollars, 224
Sandworm, 176–80
Sarcodina, 74–77

arcolemma, 27
arcomastigophora, 74–78, 79–81
arcomere, 28
artorius muscle
    of fetal pig, 306
    of frog, 275, 277
Scales, 261
Scallops, 161
Scalpel, 124, 125
Scanning electron microscope, 11
Scaphopoda, 160, 161
Scapula
    of fetal pig, 300
    of rat, 329
Schwann cell, 30
Sciatic artery, 280, 283
Sciatic nerve, 290, 291
Sciatic plexus, 290, 291
Sciatic vein, 282, 284
Sclerites, 194, 214
Sclerospongiae, 97
Scolex, 138, 140, 141, 142
Scopula, 194, 197
Scorpion, 191
Scrotal sac, 312, 313
Scrotum, 328, 339, 340, 341
Scypha, 99–100
Scyphistoma, 115
Scyphozoa, 106, 115, 116
Sea anemones, 106, 116–17
Sea cucumber, 229–30
Sea cucumbers, 224
Sea lilies, 224
Sea spider, 191
Sea squirts, 233, 234–36
Sea stars, 225–28
Sea urchin, 52, 224, 228–29
Sea wasp, 106
Sebaceous gland, 21
Segmental muscles, 261, 262
Segmentation, 123
Segmented worms, 175–87. See also Annelida
Self-fertilization, 153
Semen, 339
Semicircular canals, 254, 255
Semilunar valve
    of fetal pig, 319, 320
    of rat, 335
Semimembranosus muscle, 274, 277
Semimembranosus nerve, 291
Seminal duct, of snail, 168
Seminal receptacle
    of Anguillula aceti, 154
    of crayfish, 201
    of earthworm, 181, 182, 183
    of free-living flatworm, 130, 131
    of liver fluke, 133, 134
    of snail, 168
    of tapeworm, 143
Seminal vesicle
    of Ascaris, 149, 150
    of Ascaris lumbricoides, 150
    of cockroach, 212, 213
    of earthworm, 182, 183
    of fetal pig, 312, 313
    of frog, 286
    of sheep liver fluke, 136, 138
Semitendinosus muscle
    of fetal pig, 304, 306

of frog, 275, 277
Sense organs, of Aurelia, 115
Sensory capsules, 271
Sensory hairs, 207
Sensory neuron, 29, 256
Sensory papillae, 162
Sensory vesicle, 234
Septa
    of corals, 117
    of earthworm, 181
    of Metridium, 116
Septal filaments, 116
Septum, transverse, 243, 261
Serial homology, 202
Setae
    of earthworm, 181
    of sandworm, 177
    of water flea, 199
Setal sac, 179
Sexual dimorphism, 169, 194
    of cockroach, 210
    of spider, 194
Shark, 234, 241–58
Sheep, brain of, 323, 344
Sheep liver fluke, 135–38
Shell gland
    of shark, 248
    of sheep liver fluke, 136, 137
Sinus, of crayfish, 204
Sinus venosus
    embryonic, 62
    of frog, 280, 284
    of shark, 246, 250–51, 253
    of yellow perch, 264
Siphon(s)
    of mussel, 162, 163
    of sea squirts, 234, 235, 236
    of squid, 169
Siphonoglyphs, 116
Siphonophores, 106
Skeletal muscle, 27–28
Skeletal system
    of bony fish, 261
    of cat, 298, 299, 301
    of fetal pig, 297–98, 300
    of frog, 270, 271, 272, 273
    of rat, 328, 329
    of sponges, 97
    of yellow perch, 261
Skin, cross section of, 21
Skull
    of cat, 301
    of frog, 271, 272
Slime glands, 219
Slime papilla, 219
Slugs, 160, 161
Smooth muscle, 27
Snails, 160, 161, 167–68
Snake, 234
Solar plexus, 290, 292
Solenogasters, 161
Somites, 61, 62, 63, 64, 65,
    67, 68
Species, 97
Sperm duct
    of squid, 170
    of water flea, 200
Sperm funnel, 182, 183
Sperm, of Scypha, 100

Spermatheca
    of *Caenorhabditis elegans,* 153
    of cockroach, 212, 213
Spermathecal gland, 212
Spermatic artery
    of fetal pig, 312, 313
    of rat, 336, 337
Spermatic cord, 341
Spermatic vein, 313
Spermatogenesis, 42
Spermatophore
    of cockroach, 212
    of squid, 169, 170
Spermatophore sac, 170
Spermatozoa, 42, 50, 313
Spheroids, of *Volvox,* 79, 80, 81
Spicules, of sponges, 96, 97
Spider, 191, 193–98
Spinal cord
    of fetal pig, 310, 320, 321, 322
    of frog, 285, 290, 291
    of rat, 342, 343
    of shark, 244, 246, 255, 256
Spinal nerves
    of cat, 324
    of fetal pig, 321, 322
    of frog, 290, 291
    of lancelets, 238
    of rat, 344
    of shark, 255, 256
    of yellow perch, 267
Spindle fibers, in mitosis, 38
Spines
    of horseshoe crab, 192
    of sea urchin, 228
Spinnerets, of spider, 193, 196, 197
Spiny-headed worms, 147
Spiracles
    of cockroach, 210
    of grasshopper, 215, 216
    of *Peripatus,* 219
    of shark, 242, 244, 252
    of spider, 196, 197
Spiracular artery, 252
Spiral cleavage, 51, 52
*Spirostomum,* 88, 89
Splanchnic nerves, 290, 292
Spleen
    of fetal pig, 309, 310, 311
    of frog, 279, 283, 284
    of rat, 339
    of shark, 243, 244, 245, 246, 247
    of yellow perch, 262, 263
Splenic artery
    of fetal pig, 315, 316
    of frog, 280, 283
Splenic vein
    of fetal pig, 318, 319
    of frog, 282, 284
Splenius muscle, of fetal pig, 303, 305
Sponges, 95–103
    asconoid, 98, 99
    calcareous, 98
    classification of, 97
    freshwater, 101–2
    glass, 98
    horny, 98
    leuconoid, 98, 101
    morphology of, 96

phylogeny of, 96–97
regeneration of, 103
skeleton of, 97
syconoid, 98, 99–100
Spongin fibers, 96, 101
Spongocoel, 100
Sporocyst
    of liver fluke, 133
    of sheep liver fluke, 136
Spurs, of *Philodina,* 155, 156
*Squalus acanthias,* 241–58
Squamous epithelium
    simple, 18, 19
    stratified, 20–21
Squid, 168–71
Stapedial artery, 252
Starfish, 225–28
    embryo of, 52–54
Statocyst
    of crayfish, 202, 207
    of *Gonionemus,* 111, 112
Statolith
    of crayfish, 207
    of sea squirts, 234
*Stentor,* 88
Stereoscopic microscope, 8–9
Stergites, 210
Sternal artery
    of crayfish, 204, 206
    of fetal pig, 315, 316
Sternal sinus, 204, 206
Sternites, 210
Sternohyoideus muscle
    of fetal pig, 306
    of rat, 330, 331
Sternomastoideus muscle, 330, 331
Sternum
    of fetal pig, 310
    of frog, 278
    of spider, 196, 197
Stigma
    of *Euglena,* 78
    of *Volvox,* 79, 80
Stomach, 205
    of fetal pig, 309, 310, 311, 312, 314
    of frog, 278, 279, 281, 282, 283, 284, 287, 288
    of leeches, 186
    of mussel, 164
    of *Philodina,* 155, 156
    of rat, 331–32, 333
    of shark, 243, 244, 245, 246, 247, 249
    of snail, 167
    of starfish, 226, 227
    of water flea, 200
    of yellow perch, 262, 263
Stomoducal valve, 210
Stone canal, 226, 227, 228
Stony corals, 117
Striatims, 29
Stridulation, 216
Strobila
    of *Aurelia,* 115, 116
    of tapeworm, 138, 140
*Strongylocentrotus drobachiensis,* 229
*Strongylocentrotus purpuratus,* 229
Stylus, of cockroach, 210
Subclavian artery
    of fetal pig, 307, 315, 316, 320
    of frog, 280, 283

of rat, 334, 335, 336, 337
of shark, 252, 253
of yellow perch, 265
Subclavian vein
of fetal pig, 318, 319
of frog, 280, 284
of rat, 337, 338
of shark, 246, 253
Subesophageal ganglion, 212
Subgenital pits, 115
Subgenital plate, 215
Subintestinal vein, 237
Sublingual gland
of fetal pig, 305
of rat, 330
Submaxillary gland, 305, 315
Subneural vessel, 183
Subscapular vein, 280, 284
Suckers
of leeches, 185, 186
of liver fluke, 134
of octopus, 171
of sheep liver fluke, 137, 138
of squid, 169, 170
of tapeworm, 138, 140, 141
of trematodes, 133
Suctorial tentacles, 89
Sulcus, 320
Superficial cleavage, 52
Superficial fascia, 298
Suprabranchial chamber, 162
Supraesophageal ganglion
of cockroach, 212, 214
of crayfish, 207
Supraoccipital bone, 301
Suprapharyngeal ganglion, 183
Sus scrofa, 295–325
skinning of, 298, 302
Sweat gland, 21
Swim bladder, 262, 263, 266
Swimmerets, 201, 202, 203, 204, 206
Syconoid sponge, 98, 99–100
Symbiotic flagellates, 81, 83
Symmetry, body, 122
Sympathetic ganglion, 292
Sympathetic nerves, 292
Symphyla, 191
Synapsis, 44
Synaptic junction, 109
Syncytial theory, of multicellular animal evolution, 92–93
Systemic artery, 292

# T

Tadpole, 56–59, 270
Tadpole larvae, of tunicates, 234
Taenia pisiformis, 141, 142, 143
Tagmata, 198
Tail, 297
Tail bud, of chick embryo, 67
Tail bud stage, of frog, embryo, 59
Tapeworms, 128, 138–41
Tarsals
of fetal pig, 300
of rat, 329
Teeth
of cat, 301
of frog, 276–77
of hookworm, 151

of rat, 328, 329, 330
of sandworm, 177, 178
of sea urchin, 228
of yellow perch, 260
Tegument, of Trematoda, 133
Telencephalon, 62, 68, 69
of fetal pig, 320
of frog, 285, 291
of rat, 342
of shark, 244, 254, 255, 256
of yellow perch, 266, 267
Telolecithal eggs, 51
Telophase
of meiosis, 42–43, 44
of mitosis, 38, 39
Telson
of crayfish, 201, 206
of horseshoe crab, 192
Temporal artery, 253
Temporal bone, 301
Temporal fossa, 301
Temporal muscle, 274
Tensor fasciae latae, 303, 304, 305, 306
Tentacles
of Aurelia, 115
of Cnidaria, 106
of Gonionemus, 111, 112
of Hydra, 107, 108
of lancelets, 236, 237
of Metridium, 116, 117
of Obelia, 113
of Physalia, 114
of Podophyra, 89
of sandworm, 177
of sea cucumber, 229
of snail, 167, 168
of squid, 169, 170
of starfish, 226
Tergites
Terminal nerve, 255, 256, 257
Test
of echinoderms, 223
of sea urchin, 228
Testes, 205
of Anguillula aceti, 154
of Ascaris, 149, 150
of Ascaris lumbricoides, 150
of Caenorhabditis elegans, 153
of cockroach, 212, 213
of earthworm, 182, 183
of fetal pig, 297, 311, 312, 313
of free-living flatworm, 130
of frog, 279, 283, 286, 288
of Hydra, 108, 110, 111
of liver fluke, 133, 134
of rat, 328, 339, 340, 341
of shark, 244, 246, 248
of sheep liver fluke, 136, 138
of tapeworm, 138, 140, 141, 143
of yellow perch, 266, 267
Tetrad, in meiosis, 44
Tetrahymena, 89
Thalamus, 285
Thoracic artery, 206
Thoracic cavity, 305
Thorax
of cockroach, 209
of fetal pig, 297, 307, 309
of water flea, 200

Thymus gland
  of fetal pig, 305, 310
  of rat, 339
*Thyone briareus,* 229
Thyro-cervical artery, 315, 316
Thyroid gland, 307, 308, 310
Tibia
  of fetal pig, 300
  of rat, 329
Tibial artery, 280, 283
Tibial nerve, 291
Tibialis anticus longus muscle, 275
Tibialis anticus muscle, 274, 277
Tibialis posticus muscle, 275, 277
Tick, 191
Tiedemann's bodies, 227, 228
Tissue, 16–31
  connective, 21–26
  epithelial, 17–21
  muscular, 26–28
  nervous, 28–31
Tissue grade, 122
Toad, 234
Toes, of *Philodina,* 155, 156
Tongue
  of fetal pig, 310
  of frog, 276–77
Tooth shells, 160, 161
Torsion, embryonic, 63, 64, 65
Trachea
  of fetal pig, 305, 307, 310
  of rat, 332, 333, 334
Tracheae, of cockroach, 210
Tracheal tubules, 216, 217
Transmission electron microscope, 11
  organelles under, 18
Transverse septum
  of shark, 243
  of yellow perch, 261
Trapezius muscle, 303, 305
Trematoda, 128, 133–38
Triceps brachii muscle
  of fetal pig, 303, 305
  of frog, 274
  of rat, 330, 331
Triceps femoris muscle, 274, 275, 277
*Trichinella spiralis,* 150–51
Trichinosis, 150
Trichocysts, 85–86
Tricuspid valve, 319, 320
Trigeminal nerve
  of cat, 324
  of fetal pig, 322
  of frog, 289, 291
  of rat, 343
  of shark, 255, 256, 257
  of sheep, 323
  of yellow perch, 266
Trilobita, 190
Triploblastic tissue construction, 122
Trivium, 225
Trochal discs, 156
Trochlear nerve
  of cat, 324
  of fetal pig, 322
  of frog, 289, 291
  of rat, 343
  of shark, 255, 256, 257
  of yellow perch, 266

Trochophore larva, 166, 178
Truncus arteriosus
  embryonic, 64, 65
  of frog, 280, 283
*Trypanosoma,* 83, 84
Tube feet
  of sea cucumber, 229
  of sea urchin, 228
  of starfish, 225, 226, 227, 228
Tunic, of sea squirts, 235
Tunicates, 233, 234–36
Turbellaria, 128, 129–33
Turtle, 234
Tusk shells, 160, 161
Tympanic bulla, 301
Tympanic membrane
  of frog, 270
  of grasshopper, 215, 216
Typhlosole
  of earthworm, 182, 184
  of mussel, 164

# U

Ulna
  of fetal pig, 300
  of rat, 329
Ulnaris muscle, 274
Umbilical artery, 297, 308, 309, 310, 311, 312, 313, 315, 316, 317
Umbilical cord, 296, 297, 310, 311, 312, 313, 317
Umbilical vein, 297, 310, 311, 312, 317, 318, 319
Umbilicus, 297, 308, 312
Umbo, 162, 163
Undulating membrane, 83, 84
Uniramia, 191, 208–17
Ureter
  of fetal pig, 309, 310, 311, 312, 313, 317
  of frog, 279, 285, 286
  of rat, 339
Urethra
  of fetal pig, 311, 312, 313
  of rat, 328, 333, 339, 341
Urinary pore, 248, 249
Urochordata, 233, 234–36
Urogenital artery, of frog, 280, 282
Urogenital opening
  of fetal pig, 297
  of yellow perch, 260
Urogenital pore
  of shark, 248
  of yellow perch, 267
Urogenital sinus, 310, 313
Urogenital system
  of fetal pig, 309–13
  of frog, 285, 286–88
  of rat, 339–40
  of shark, 248–49
  of yellow perch, 264, 266, 267
Uropod, 201, 206
Urostyle, 271
Uterine artery, 336, 337
Uterine horns, 340, 341
Uterus
  of *Anguillula aceti,* 154
  of *Ascaris,* 148, 149
  of fetal pig, 310, 311, 312, 313
  of frog, 285
  of rat, 340

of shark, 248
of sheep liver fluke, 136, 137–38
of tapeworm, 138, 140, 143

# V

acuoles
of amoeba, 75, 76
of *Euglena,* 78
of *Paramecium caudatum,* 85, 86
of *Peranema,* 79
of *Volvox,* 79, 80
agina
of *Ascaris,* 148, 149
of cockroach, 212
of fetal pig, 310, 313
of rat, 340, 341
of tapeworm, 138, 143
agus nerve
of cat, 324
of fetal pig, 307, 312, 322
of frog, 289, 291
of rat, 334, 343
of shark, 255, 256, 257
of yellow perch, 266
Vas deferens, 205
of *Anguillula aceti,* 154
of *Ascaris,* 149, 150
of cockroach, 213
of earthworm, 182, 183
of fetal pig, 312, 313
of rat, 339
of sheep liver fluke, 136, 138
of tapeworm, 138, 140, 141, 143
Vas efferens, of tapeworm, 143
Vasa deferentia
of fetal pig, 313
of yellow perch, 267
Vasa efferentia
of frog, 285, 286
of shark, 248
Vastus lateralis muscle, 303, 304, 305
Vastus medialis muscle, 306
Vegetal pole, 51–52, 55, 57
Veins. *See also* Venous system
of cockroach, 210
of grasshopper, wings, 216
of shark, 253
*Velella,* 114
Velum, of *Gonionemus,* 111, 112
Venous system
of fetal pig, 318, 319
of frog, 280, 282, 284, 286
of rat, 337–39
Ventral groove, 198
Ventral nerve cords, 129
Ventral vessel, 183
Ventricle
embryonic, 62
of fetal pig, 307, 309, 310, 312,
   315, 320
of frog, 280
of rat, 333, 334
of shark, 244, 250, 251
of yellow perch, 264
Vertebrae. *See also* Vertebral column
of frog, 271, 272
of yellow perch, 261, 262
Vertebral artery, 280

Vertebral column
of fetal pig, 300
of frog, 271
of rat, 329
of yellow perch, 261
Vertebrata, 234
Vibrissae
of fetal pig, 297
of rat, 328
Vinegar eel, 153–54
Visceral arches, 261
Visceral arteries, 249, 250
Visceral ganglia, 167
Visceral mass, of mussel, 164
Visceral nerve, 255, 256
Visceral skeleton
of frog, 271
of yellow perch, 261
Vitellaria, 155, 156
Vitelline arteries, 62, 63, 64, 66, 67, 69
Vitelline membrane, 55
Vitelline veins, 64, 65, 66, 67, 69
Vocal sacs, 277
*Volvox,* 79–81
Vomer, 301
*Vorticella,* 88–89, 90
Vulva
of *Ascaris,* 149
of *Caenorhabditis elegans,* 153
of fetal pig, 297, 310, 313

# W

Water flea, 195, 198–201
Water scorpion, 191
Water vascular system
of echinoderms, 223–24
of starfish, 226–28
Wet mount, 75
Whelks, 160, 161
Whipworms, 150
Whitefish embryo, 40
Wing bud, 67, 69
Wings
of cockroach, 209
of grasshopper, 215, 216
*Wuchereria bancrofti,* 155

# Y

Yellow perch, 259–67
Yolk, 50, 51
Yolk duct
of liver fluke, 133, 134
of sheep liver fluke, 136, 137
of tapeworm, 143
Yolk gland
of liver fluke, 133, 134
of sheep liver fluke, 136, 137
of tapeworm, 138, 140, 141, 143
Yolk reservoir, 136, 137
Yolk sac, 60

# Z

Z-lines, 28
Zoantharia, 117
Zygomatic arch, 301
Zygomatic bone, 300